Maths
A Student's Survival Guide

This friendly and gentle self-help workbook covers mathematics essential for first-year undergraduate scientists and engineers.

Mathematics underlies all science and engineering degrees. If your mathematics is not strong, it may cause problems for you. This book is the solution. Jenny Olive uses her wide experience of teaching and helping students to give you a clear and confident understanding of the core mathematics you need to start a science or engineering degree.

Each topic is introduced very gently, starting with simple examples that really bring out the basics, then moving on to more challenging problems. The author takes time to give tricks of the trade and short-cuts, but is also careful to show you common errors and how to anticipate and avoid them.

There are over 800 questions for you to do, with full and detailed solutions given, so that if you get stuck you can work through and see exactly where you have gone wrong. Topics covered include trigonometry and hyperbolic functions, sequences and series (with detailed help on binomial series), differentiation and integration, and complex numbers.

This book is a fun and easy way to brush up your maths – either before you start on your university or college course, or while you are getting to grips with it. It's never too late!

Jenny Olive

Maths

A Student's Survival Guide

CAMBRIDGE
UNIVERSITY PRESS

PUBLISHED BY THE PRESS SYNDICATE OF THE UNIVERSITY OF CAMBRIDGE
The Pitt Building, Trumpington Street, Cambridge, United Kingdom

CAMBRIDGE UNIVERSITY PRESS
The Edinburgh Building, Cambridge CB2 2RU, UK   http://www.cup.cam.ac.uk
40 West 20th Street, New York, NY 10011–4211, USA   http://www.cup.org
10 Stamford Road, Oakleigh, Melbourne 3166, Australia
Ruiz de Alarcón 13, 28014 Madrid, Spain

First published 1998
Reprinted 2000

Printed in the United Kingdom at the University Press, Cambridge

A catalogue record for this book is available from the British Library

Library of Congress Cataloguing in Publication data
Olive, Jenny, 1939–
    Maths: a student's survival guide/Jenny Olive
        p.   cm.
    ISBN 0 521 57306 8 (hardback). – ISBN 0 521 57586 9 (paperback)
    1. Mathematics.   I. Title.
QA39.2.0434   1998
511–DC21   97-28653 CIP

ISBN 0 521 57306 8 hardback
ISBN 0 521 57586 9 paperback

Contents

I have split the chapters up in the following way so that you can easily find particular topics. Also, it makes it easy for me to tell you where to go if you need help, and easy for you to find this help.

4   Some trigonometry and geometry of triangles and circles 130

Contents

5 Extending trigonometry to angles of any size 173

8   Differentiation 284

Contents

Acknowledgements

I would particularly like to thank Rodie and Tony Sudbery for their very helpful ideas and comments on large parts of the text. I am also very grateful to Neil Turok, Eleni Haritou-Monioudis, John Szymanski, Jeremy Jones and David Olive for detailed comments on particular sections, and my father, William Tutton, for his helpful advice on my drawings. I would also like to thank the mathematics department of the University of Wales, Swansea, for helpful discussions concerning the needs of incoming students. The referees also all provided detailed and useful input which was very helpful in structuring the book and I thank them for this.

I would also like to thank Rufus Neal, Harriet Millward and Mairi Sutherland for their patient and friendly editorial help and advice, Phil Treble for his great design, and everyone else at Cambridge University Press who has worked on this book.

Finally, I am particularly grateful to my daughter, Rosalind Olive, both for her helpful comments and also for her excellent guinea-pig drawings.

For and because of the people I have taught

Introduction

I have written this book mainly for students who will need to apply maths in science or engineering courses. It is particularly designed to help the foundation or first year of such a course to run smoothly but it could also be useful to specialist maths students whose particular choice of A-level or pre-university course has meant that there are some gaps in the knowledge required as a basis for their University course. Because it starts by laying the basic groundwork of algebra it will also provide a bridge for students who have not studied maths for some time.

The book is written in such a way that students can use it to sort out any individual difficulties for themselves without needing help from their lecturers.

A message to students

I have made this book as much as possible as though I were talking directly to you about the topics which are in it, sorting out possible difficulties and encouraging your thoughts in return. I want to build up your knowledge and your courage at the same time so that you are able to go forward with confidence in your own ability to handle the techniques which you will need. For this reason, I don't just tell you things, but ask you questions as we go along to give you a chance to think for yourself how the next stage should go. These questions are followed by a heavy rule like the one below.

It is very important that you should try to answer these questions yourself, so the rule is there to warn you not to read on too quickly.

I have also given you many worked examples of how each new piece of mathematical information is actually used. In particular, I have included some of the off-beat non-standard examples which I know that students often find difficult.

To make the book work for you, it is vital that you do the questions in the exercises as they come because this is how you will learn and absorb the principles so that they become part of your own thinking. As you become more confident and at ease with the methods, you will find that you enjoy doing the questions, and seeing how the maths slots together to solve more complicated problems.

Always be prepared to think about a problem and have a go at it – don't be afraid of getting it wrong. Students very often underrate what they do themselves, and what they *can* do. If something doesn't work out, they tend to think that their effort was of no worth *but this is not true*. Thinking about questions for yourself is how you learn and understand what you are doing. It is much better than just following a template which will only work for very similar problems and *then* only if you recognise them. If you really understand what you are doing you will be able to apply these ideas in later work, and this is important for you.

Because you may be working from this book on your own, I have given detailed solutions to most of the questions in the exercises so that you can sort out for yourself any problems that you may have had in doing them. (Don't let yourself be tempted just to read through my solutions – you will do infinitely better if you write your own solutions first. This is the most important single piece of advice which I can give you.) Also, if you are stuck and have to look at my solution, don't just read through the whole of it. Stop reading at the point that gets you unstuck and see if you can finish the problem yourself.

I have also included what I have called thinking points. These are usually more open-ended questions designed to lead you forward towards future work.

If possible, talk about problems with other students; you will often find that you can help each other and that you spark each other's ideas. It is also very sensible to scribble down your thoughts as you go along, and to use your own colour to highlight important results or particular parts of drawings. Doing this makes you think about which are the important bits, and gives you a short-cut when you are revising.

There are some pitfalls which many students regularly fall into. These are marked

to warn you to take particular notice of the advice there. You will probably recognise some old enemies!

It often happens in maths that in order to understand a new topic you must be able to use earlier work. I have made sure that these foundation topics are included in the book, and I give references back to them so that you can go there first if you need to. I have linked topics together so that you can see how one affects another and how they are different windows onto the same world. The various approaches, visual, geometrical, using the equations of algebra or the arguments of calculus, all lead to an understanding of how the fundamental ideas interlock. I also show you wherever possible how the mathematical ideas can be used to describe the physical world, because I find that many students particularly like to know this, and indeed it is the main reason why they are learning the maths. (Much of the maths is very nice in itself, however, and I have tried to show you this.)

I have included in some of the thinking points ideas for simple programs which you could write to investigate what is happening there. To do this, you would need to know a programming language and have access to either a computer or programmable calculator. I have also suggested ways in which you can use a graph-sketching calculator as a fast check of what happens when you build up graphs from combinations of simple functions. Although these suggestions are included because I think you would learn from them and enjoy doing them, it is not necessary to have this equipment to use this book.

Much of the book has grown from the various comments and questions of all the students I have taught. It is harder to keep this kind of two-way involvement with a printed book but no longer impossible thanks to the Web. I would be very interested in your comments and questions and grateful for your help in spotting any mistakes which may have slipped through my checking. You can contact me by using my email address of jenolive@netcomuk.co.uk and I look forward to putting little additions on the Web, sparked by your thoughts. My website is at http://www.netcomuk.co.uk/~jenolive

Finally, I hope that you will find that this book will smooth your way forward and help you to enjoy all your courses.

Jenny Olive

1 Basic algebra: some reminders of how it works

In many areas of science and engineering, information can be made clearer and more helpful if it is thought of in a mathematical way. Because this is so, algebra is extremely important since it gives you a powerful and concise way of handling information to solve problems. This means that you need to be confident and comfortable with the various techniques for handling expressions and equations.

The chapter is divided up into the following sections.

1.A Handling unknown quantities
(a) Where do you start? Self-test 1, (b) A mind-reading explained,
(c) Some basic rules, (d) Working out in the right order, (e) Using negative numbers,
(f) Putting into brackets, or factorising

1.B Multiplications and factorising: the next stage
(a) Self-test 2, (b) Multiplying out two brackets,
(c) More factorisation: putting things back into brackets

1.C Using fractions
(a) Equivalent fractions and cancelling down, (b) Tidying up more complicated fractions,
(c) Adding fractions in arithmetic and algebra, (d) Repeated factors in adding fractions,
(e) Subtracting fractions, (f) Multiplying fractions, (g) Dividing fractions

1.D The three rules for working with powers
(a) Handling powers which are whole numbers, (b) Some special cases

1.E The different kinds of numbers
(a) The counting numbers and zero, (b) Including negative numbers: the set of integers,
(c) Including fractions: the set of rational numbers,
(d) Including everything on the number line: the set of real numbers,
(e) Complex numbers: a very brief forwards look

1.F Working with different kinds of number: some examples
(a) Other number bases: the binary system, (b) Prime numbers and factors,
(c) A useful application – simplifying square roots,
(d) Simplifying fractions with $\sqrt{\ }$ signs underneath

1.A   **Handling unknown quantities**

1.A.(a)   **Where do you start? Self-test 1**

All the maths in this book which is directly concerned with your courses depends on a foundation of basic algebra. In case you need some extra help with this, I have included two revision sections at the beginning of this first chapter. Each of these sections starts with a short self-test so that you can find out if you need to work through it.

It's important to try these if you are in any doubt about your algebra. You have to build on a firm base if you are to proceed happily; otherwise it is like climbing a ladder which has some rungs missing, or, more dangerously, rungs which appear to be in place until you tread on them.

Answer each of the following short questions.

(A) Find the value of each of the following expressions if $a = 3$, $b = 1$, $c = 0$ and $d = 2$.
   (1) a^2 (2) b^2 (3) $ab + d$ (4) $a(b + d)$ (5) $2c + 3d$
   (6) $2a^2$ (7) $(2a)^2$ (8) $4ab + 3bd$ (9) $a + bc$ (10) d^3

(B) Find the values of each of the following expressions if $x = 2$, $y = -3$, $u = 1$, $v = -2$, $w = 4$ and $z = -1$.
   (1) $3xy$ (2) $5vy$ (3) $2x + 3y + 2v$ (4) v^2 (5) $3z^2$
   (6) $w + vy$ (7) $2x - 5vw$ (8) $2y - 3v + 2z - w$ (9) $2y^2$ (10) z^3

(C) Simplify (that is, write in the shortest possible form).
   (1) $3p - 2q + p + q$ (2) $3p^2 + 2pq - q^2 - 7pq$ (3) $5p - 7q - 2p - 3q + 3pq$

(D) Multiply out the following expressions.
   (1) $5(2g + 3h)$ (2) $g(3g - 2h)$ (3) $3k^2(2k - 5m + 2n)$ (4) $3k - (2m + 3n - 5k)$

(E) Factorise the following expressions.
   (1) $3x^2 + 2xy$ (2) $3pq + 6q^2$ (3) $5x^2y - 7xy^2$

Here are the answers. (Give yourself one point for each correct answer, which gives a maximum possible score of 30.)

(A)   (1) 9   (2) 1   (3) 5   (4) 9   (5) 6   (6) 18   (7) 36   (8) 18   (9) 3   (10) 8

(B)   (1) -18 (2) 30   (3) -9 (4) 4   (5) 3   (6) 10   (7) 44   (8) -6 (9) 18   (10) -1

(C)   (1) $4p - q$ (2) $3p^2 - 5pq - q^2$ (3) $3p - 10q + 3pq$

(D)   (1) $10g + 15h$ (2) $3g^2 - 2gh$ (3) $6k^3 - 15k^2m + 6k^2n$ (4) $8k - 2m - 3n$

(E)   (1) $x(3x + 2y)$ (2) $3q(p + 2q)$ (3) $xy(5x - 7y)$

If you scored anything less than 25 points then I would advise you to work through Section 1.A. If you made just the odd mistake, and realised what it was when you saw the answer, then go ahead to Section 1.B. If you are in any doubt, it is best to go through Section 1.A. now; these are your tools and you need to feel happy with them.

1.A.(b)   **A mind-reading explained**

Much of what was tested above can be shown in the handling of the following. Try it for yourself. (You may have met this apparently mysterious kind of mind-reading before.)

(1)   Think of a number between 1 and 10. (A small number is easier to use.)
(2)   Add 3 to it.
(3)   Double the number you have now.
(4)   Add the number you first thought of.
(5)   Divide the number you have now by 3.
(6)   Take away the number you first thought of.
(7)   The number you are thinking of now is . . . 2!

How can we lay bare the bones of what is happening here, so that we can see how it is possible for me to know your final answer even though I don't know what number you were thinking of at the start?

It is easier for me to keep track of what is happening, and so be able to arrange for it to go the way I want, if I label this number with a letter. So suppose I call it x. Suppose also that your number was 7 and we can then keep a parallel track of what goes on.

	You	Me
(1)	7	x
(2)	10	$x + 3$ (My unknown number plus 3.)
(3)	20	$2(x + 3) = 2x + 6$ (Each of these show the doubling.)
(4)	27	$2x + 6 + x = 3x + 6$ (I add in the unknown number.)
(5)	9	$\frac{3x + 6}{3} = x + 2$ (The whole of $3x + 6$ is divided by 3.)
(6)	2	2 (The x has been taken away.)

Both your 7 and my x have been got rid of as a result of this list of instructions.

My list uses algebra to make the handling of an unknown quantity easier by tagging it with a letter. It also shows some of the ways in which this handling is done.

1.A.(c)   **Some basic rules**

There are certain rules which need to be followed in handling letters which are standing for numbers. Here I remind you of these.

Adding

$a + b$ means quantity a added to quantity b.

$a + a + b + b + b = 2a + 3b$. Here, we have twice the first quantity and three times the second quantity added together. There is no shorter way of writing $2a + 3b$ unless we know what the letters are standing for.

We could equally have said $b + a$ for $a + b$, and $3b + 2a$ for $2a + 3b$. It doesn't matter what order we do the adding in.

Multiplying

ab means $a \times b$ (that is, the two quantities multiplied together) and the letters are usually, but not always, written in alphabetical order.

> In particular, $a \times 1 = a$, and $a \times 0 = 0$.
> $5ab$ would mean $5 \times a \times b$.

It doesn't matter what order we do the multiplying in, for example $3 \times 5 = 5 \times 3$.

Working out powers

If numbers are multiplied by themselves, we use a special shorthand to show that this is happening.

> a^2 means $a \times a$ and is called a squared.
> a^3 means $a \times a \times a$ and is called a cubed.
> a^n means a multiplied by itself with n lots of a and is called a to the power n.

Little raised numbers, like the 2, 3 and n above, are called **powers** or **indices**. Using these little numbers makes it much easier to keep a track of what is happening when we multiply. (It was a major breakthrough when they were first used.) You can see why this is in the following example.

Suppose we have $a^2 \times a^3$.

Then $a^2 = a \times a$ and $a^3 = a \times a \times a$ so $a^2 \times a^3 = a \times a \times a \times a \times a = a^5$.

The powers are added. (For example, $2^2 \times 2^3 = 4 \times 8 = 32 = 2^5$.)

We can write this as a general rule.

$$a^n \times a^m = a^{n+m}$$

where a stands for any number except 0 and n and m can stand for any numbers.

In this section, n and m will only be standing for positive whole numbers, so we can see that they would work in the same way as the example above.

To make the rule work, we need to think of a as being the same as a^1. Then, for example, $a \times a^2 = a^1 \times a^2 = a^3$ which fits with what we know is true, for example $2 \times 2^2 = 2^3$ or $2 \times 4 = 8$.

Also, this rule for adding the powers when multiplying only works if we have powers of the same number, so $2^2 \times 2^3 = 2^5$ and $7^2 \times 7^3 = 7^5$ but $2^2 \times 7^3$ cannot be combined as a single power.

If we have numbers and different letters, we just deal with each bit separately, so for example $3a^2b \times 2ab^3 = 6a^3b^4$.

Working out mixtures – using brackets

$a + bc$ means quantity a added to the result of multiplying b and c. The multiplication of b and c must be done before a is added.

If $a = 2$ and $b = 3$ and $c = 4$ then $a + bc = 2 + 3 \times 4 = 2 + 12 = 14$.

If we want a and b to be added first, and the result to be multiplied by c, we use a bracket and write $(a + b)c$ or $c(a + b)$, as the order of the multiplication does not matter. This gives a result of $5 \times 4 = 4 \times 5 = 20$.

A bracket collects together a whole lot of terms so that the same thing can be done to all of them, like corralling a lot of sheep, and then dipping them. So $a(b + c)$ means $ab + ac$. The a multiplies every separate item in the bracket.

Similarly, $2x(x + y + 3xy) = 2x^2 + 2xy + 6x^2y$. The brackets show that everything inside them is to be multiplied by the $2x$. It is important to put in brackets if you want the same thing to happen to a whole collection of stuff, both because it tells you that that is what you are doing, and also because it tells anyone else reading your working that that is what you meant. *Many mistakes come from left-out brackets.*

Here is another example of how you need brackets to show that you want different results.

If $a = 2$ then $3a^2 = 3 \times 2 \times 2 = 12$ but $(3a)^2 = 6^2 = 36$. The brackets are necessary to show that it is the whole of $3a$ which is to be squared.

EXERCISE 1.A.1

Try these questions yourself now.

(1) Put the following together as much as possible.

(a) $3a + 2b + 5a + 7c - b - 4c$ (b) $3ab + b + 5a + 2b + 2ba$

(c) $7p + 3pq - 2p + 2pq + 8q$ (d) $5x + 2y - 3x + xy + 3y + 2xy$

(2) If $a = 2$ and $b = 1$, find

(a) a^3 (b) $5a^2$ (c) $(5a)^2$ (d) b^2 (e) $2a^2 + 3b^2$

(3) Multiply the following together.

(a) $(2x)(3y)$ (b) $(3x^2)(5xy)$ (c) $3(2a + 3b)$ (d) $2a(3a + 5b)$

(e) $2p(3p^2 + 2pq + q^2)$ (f) $2x^2(3x + 2xy + y^2)$

1.A.(d)   Working out in the right order

If you are replacing letters by numbers, then you must stick to the following rules to work out the answer from these numbers.

---

(1) In general, we work from left to right.

(2) Any working inside a bracket must be done first.

(3) When doing the working out, first find any powers, then do any multiplying and dividing, and finally do any adding and subtracting.

Here are two examples.

EXAMPLE (1) If $a = 2$, $b = 3$, $c = 4$ and $d = 6$, find $3a(2d + bc) - 4c$.

- Find the inside of the bracket, which is $2 \times 6 + 3 \times 4 = 12 + 12 = 24$.
- Multiply this by $3a$, giving $6 \times 24 = 144$.
- Find $4c$, which is $4 \times 4 = 16$.
- Finally, we have $144 - 16 = 128$.

EXAMPLE (2) If $x = 2$, $y = 3$, $z = 4$ and $w = 6$, work out the value of $x(2y^2 - z) + 3w^2$.

We start by working out the inside of the bracket.
- Find y^2 which is 9.
- The bracket comes to $2 \times 9 - 4 = 14$.
- Multiply this by x, getting 28.
- $w^2 = 6^2 = 36$ so $3w^2 = 108$.
- Finally, we get $28 + 108 = 136$.

EXERCISE 1.A.2

Now try the following yourself.

(1) If $a = 2$, $b = 3$, $c = 4$, $d = 5$ and $e = 0$ find the values of:

(a) $ab + cd$ (b) ab^2e (c) ab^2d (d) $(abd)^2$ (e) $a(b + cd)$

(f) $ab^2d + c^3$ (g) $ab + d - c$ (h) $a(b + d) - c$

(2) Multiply out the following, tidying up the answers by putting together as much as possible.

(a) $3x(2x + 3y) + 4y(x + 7y)$ (b) $5p^2(2p + 3q) + q^2(3p + 5q) + pq(p + 2q)$

Check your answers to these two questions, before going on.

Questions (3) and (4) are very similar to (1) and (2) and will give you some more practice if you need it.

(3) If $a = 3$, $b = 4$, $c = 1$, $d = 5$ and $e = 0$ find the values of:

(a) a^2 (b) $3b^2$ (c) $(3b)^2$ (d) c^2 (e) $ab + c$ (f) $bd - ac$ (g) $b(d - ac)$

(h) $d^2 - b^2$ (i) $(d - b)(d + b)$ (j) $d^2 + b^2$ (k) $(d + b)(d + b)$

(l) $a^2b + c^2d$ (m) $5e(a^2 - 3b^2)$ (n) $a^b + d^a$

(4) Multiply out and collect like terms together if possible:

(a) $3a(2b + 3c) + 2a(b + 5c)$ (b) $2xy(3x^2 + 2xy + y^2)$

(c) $5p(2p + 3q) + 2q(3p + q)$ (d) $2c^2(3c + 2d) + 5d^2(2c + d)$

1.A.(e)   Using negative numbers

We shall need to be able to do more complicated things with minus signs than we have met so far, so here is a reminder about dealing with signed numbers.

Ordinary numbers, such as 6, are written as +6 in order to show that they are different from negative numbers such as −5. If the sign in front of a number is +, then it can sometimes be left out. (We don't speak of having +2 apples, for example.) A negative sign can never be left out, in any working combination of numbers.

One way of understanding how signed numbers work is to think of them in terms of money. Then +2 represents having £2, and −3 represents owing £3, etc.

So using brackets to keep each number and its sign conveniently connected, we have for example:

$(+2) + (+5) = (+7)$	Ordinary addition.
$(-3) + (-7) = (-10)$	Adding two debts.
$(+4) + (-9) = (-5)$	You still have a debt.
$(+3) - (-7) = (+10)$	Taking away a debt means you gain.

The same idea carries through to multiplication (which can be thought of as repeated addition, so 3×2 means 3 lots of 2, or adding 2 to itself three times).

Some examples are:

$(+2) \times (-3) = (-6)$	Doubling a debt!
$(-3) \times (+5) = (-15)$	Taking away 3 lots of 5.
$(-3) \times (-7) = (+21)$	Taking away a debt of 7 three times.

The rule for multiplying signed numbers

Two signs which are the same give plus and two different signs give minus.

Here are two examples of this in action.

(1) $3a - 2(b - 2a) + 7b = 3a - 2b + 4a + 7b = 7a + 5b.$

(2) $2p - (p + 2q - m).$

Here, you can think of the minus sign outside the bracket as meaning −1, so that when the bracket is multiplied by it, all the signs inside it will change.

We get $2p - p - 2q + m = p - 2q + m.$

EXERCISE 1.A.3   Now try the following questions.

Multiply out the following, tidying up the answers as much as possible.

(1) $2x - (x - 2y) + 5y$ (2) $4(3a - 2b) - 6(2a - b)$

(3) $6(2c + d) - 2(3c - d) + 5$ (4) $6a - 2(3a - 5b) - (a + 4b)$

(5) $3x(2x - 3y + 2z) - 4x(2x + 5y - 3z)$ (6) $2xy(3x - 4y) - 5xy(2x - y)$

(7) $2a^2(3a - 2ab) - 5ab(2a^2 - 4ab)$ (8) $-3p - (p + q) + 2q(p - 3)$

1.A.(f) Putting into brackets, or factorising

The process described in the previous section can be done in reverse, so, for example, $xy + xz = x(y + z)$.

This reverse process is called **factorisation** and x is called a **factor** of the expression, that is, something you multiply by to get the whole answer, just as 2, 3, 4, 6 are all **factors** of 12. We can say $12 = 3 \times 4 = 2 \times 6$. Each factor divides into 12 exactly.

Here are three examples showing this process happening.

(1) $3a^2 + 2ab = a(3a + 2b)$. This is as far as we can go.

(2) $3p^2q + 4pq^2 = pq (3p + 4q)$ factorising as much as possible.

(3) $4a^2b^3 - 6a^3b^2 = 2a^2b^2(2b - 3a)$ factorising as far as possible.

$xy + x = x(y + 1)$ *not* $x(y + 0)$ because $x \times 1 = x$ but $x \times 0 = 0$.

HELPFUL HINT

It is useful to remember that factorisation is just the reverse process to multiplying out. If you are at all doubtful that you have factorised correctly, you can check by multiplying out your answer that you do get back to what you started with originally.

Here's an example.

If you factorise $3c^2 + 2cd + c$, which of the following gives the right answer?

(1) $3c(c + 2d + 1)$ (2) $c(3c + 2d)$ (3) $c(3c + 2d + 1)$.

Multiplying out gives (1) $3c^2 + 6cd + 3c$ (2) $3c^2 + 2cd$ and   (3) $3c^2 + 2cd + c$ so (3) is the correct one.

EXERCISE 1.A.4

Factorise the following yourself, taking out as many factors as you can.

(1) $5a + 10b$ (2) $3a^2 + 2ab$ (3) $3a^2 - 6ab$

(4) $5xy + 8xz$ (5) $5xy - 10xz$ (6) $a^2b + 3ab^2$

(7) $4pq^2 - 6p^2q$ (8) $3x^2y^3 + 5x^3y^2$

(9) $4p^2q + 2pq^2 - 6p^2q^2$ (10) $2a^2b^3 + 3a^3b^2 - 6a^2b^2$

1.B Multiplications and factorising: the next stage

1.B.(a) Self-test 2

This section also starts with a self-test. It is sensible to do it even if you think you don't have any problems with these because it won't take you very long to check that you are in this happy state. It's a good idea to cover my answers until you've done yours.

(A) Multiply out the following

(1) $(2x + 3y) (x + 5y)$ (2) $(3a - 5b)(2a - b)$ (3) $(3x + 2)^2$

(4) $(2y - 5)^2$ (5) $(2p^2 + 3pq)(q^2 - 2pq)$

Factorise the following.

(B)   (1) $x^2 + 9x + 14$ (2) $y^2 + 8y + 12$ (3) $x^2 + 8x + 16$ (4) $p^2 + 13p + 22$

(C)   (1) $2x^2 + 7x + 3$ (2) $3a^2 + 16a + 5$ (3) $3b^2 + 10b + 7$ (4) $5x^2 + 8x + 3$

(D)   (1) $x^2 + x - 2$ (2) $2a^2 + a - 15$ (3) $2x^2 + 5x - 12$ (4) $p^2 - q^2$

      (5) $6y^2 - 19y + 10$ (6) $4x^2 - 81y^2$ (7) $6x^2 - 19x + 10$ (8) $4x^2 - 12x + 9$

As in the first test, give yourself one point for each correct answer so that the highest total score is 30. Again, if you got 25 or less, work through this following section.

If you are in any doubt, it is much better to get it sorted out now, because lots of later work will depend on it.

These are the answers that you should have.

(A)   (1) $2x^2 + 13xy + 15y^2$ (2) $6a^2 - 13ab + 5b^2$ (3) $9x^2 + 12x + 4$

      (4) $4y^2 - 20y + 25$ (5) $3pq^3 - 4p^3q - 4p^2q^2$

(B)   (1) $(x+2)(x+7)$ (2) $(y+2)(y+6)$ (3) $(x+4)^2$ (4) $(p+2)(p+11)$

(C)   (1) $(2x+1)(x+3)$ (2) $(3a+1)(a+5)$ (3) $(3b+7)(b+1)$ (4) $(5x+3)(x+1)$

(D)   (1) $(x+2)(x-1)$ (2) $(2a-5)(a+3)$ (3) $(2x-3)(x+4)$ (4) $(p-q)(p+q)$

      (5) $(3y-2)(2y-5)$ (6) $(2x-9y)(2x+9y)$ (7) $(3x-2)(2x-5)$ (8) $(2x-3)^2$

1.B.(b)   Multiplying out two brackets

To multiply out two brackets, each bit of the first bracket must be multiplied by each bit of the second bracket, so

$$(a + b)(c + d) = ac + bd + ad + bc.$$

The $ac + bd + ad + bc$ can be written in any order.

You could also think of this process, if you like, as

$$(a + b)(c + d) = a(c + d) + b(c + d) = ac + ad + bc + bd.$$

You can see this working numerically by putting $a = 1$, $b = 2$, $c = 3$ and $d = 4$.

$$(a + b)(c + d) = (1 + 2)(3 + 4) = 3 \times 7 = 21$$

and

$$ac + ad + bc + bd = 3 + 4 + 6 + 8 = 21.$$

Also, you can see that the order of doing the multiplying doesn't matter, since

$$ac + bd + bc + ad = 3 + 8 + 6 + 4 = 21 \text{ too.}$$

Figure 1.B.1 shows this process happening with areas. $(a + b)(c + d)$ gives the total area of the rectangle.

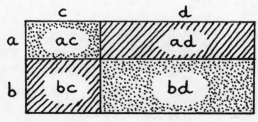

Figure 1.B.1

Exactly the same system is used to work out $(a + b)^2$. We have

$$(a + b)^2 = (a + b)(a + b) = a^2 + ab + ab + b^2 = a^2 + 2ab + b^2$$

We can see this working in Figure 1.B.2.

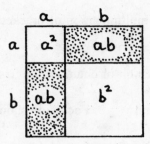

Figure 1.B.2

We can see the two squares and the two same-shaped rectangles.

 Don't forget the middle bit of $2ab$.

The diagram shows that $(a + b)^2$ is not the same thing as $a^2 + b^2$. In a similar way, we have

$$(a - b)^2 = (a - b)(a - b) = a^2 - 2ab + b^2.$$

What happens if the signs are opposite ways round, so we have $(a + b)(a - b)$?

We get

$$(a + b)(a - b) = a^2 - b^2$$

because the middle bits cancel out.

This result is called **the difference of two squares**.

You need to be good at spotting examples of this because it is of very great importance in simplifying and factorising in many different situations.

To help you to get good at this, here are some further examples.

Put back into two brackets (1) $x^2 - 9y^2$,   (2) $49a^2 - 64b^2$.

The answers are (1) $(x + 3y)(x - 3y)$ and (2) $(7a + 8b)(7a - 8b)$.

Check these are true by multiplying them back out, and then try the following ones for yourself.

(1) $x^2 - y^2$ (2) $4a^2 - 9b^2$ (3) $16p^2 - 9q^2$ (4) $16a^2 - 25b^2$ (5) $36p^2 - 100q^2$

These are the answers that you should have.

(1) $(x + y)(x - y)$ (2) $(2a + 3b)(2a - 3b)$ (3) $(4p + 3q)(4p - 3q)$
(4) $(4a + 5b)(4a - 5b)$ (5) $(6p + 10q)(6p - 10q)$

In each case, the brackets can equally well be written the other way round since the letters are standing for numbers.

Here is a more complicated example of multiplication of brackets.

$$(3x + xy)(xy + y^2) = 3x^2y + x^2y^2 + 3xy^2 + xy^3$$

Again, the basic strategy is the same. Each bit or chunk of the first bracket is multiplied by each bit or chunk of the second one.

(This can be checked by putting $x = 2$ and $y = 3$. Each side should come to 180.)

EXERCISE 1.B.1

Multiply out the following pairs of brackets.

(1) $(x + 2)(x + 3)$ (2) $(a + 3)(a - 4)$ (3) $(x - 2)(x - 3)$
(4) $(p + 3)(2p + 1)$ (5) $(3x - 2)(3x + 2)$ (6) $(2x - 3y)(x + 2y)$
(7) $(3a - 2b)(2a - 5b)$ (8) $(3x + 4y)^2$ (9) $(3x - 4y)^2$
(10) $(3x + 4y)(3x - 4y)$ (11) $(2p^2 + 3pq)(5p + 3q)$ (12) $(2ab - b^2)(a^2 - 3ab)$
(13) $(a + b)(a^2 - ab + b^2)$ (14) $(a - b)(a^2 + ab + b^2)$

(15) Try working through the following steps.
    (a) Think of a positive whole number, and write down its square.
    (b) Add 1 to your original whole number, and multiply the result by the original number with 1 taken away from it.
    (c) Repeat this process twice more.
    (d) Describe in words what seems to be happening.
    (e) Must this always happen whatever your starting number is?
Show that it must by taking a starting number of n so that you can see exactly what must happen every time.

1.B.(c)   More factorisation: putting things back into brackets

Again, the reverse process to multiplying out two brackets is called **factorisation**. Very often it is important to be able to replace a more complicated expression by two simpler expressions multiplied together.

We have already done some examples of this, when we were working with the difference of two squares in the previous section.

What happens, though, if there is a middle bit to be sorted out?

For example, suppose we have $x^2 + 7x + 12$.

Can we replace this expression by two multiplied brackets?

We would have $x^2 + 7x + 12 = (something) \, (something)$, and we have to find out what the somethings must be.

We can see that we will need to have x at the beginning of each of the brackets.

Both signs in the brackets are positive since the left-hand side is all positive, so at the ends we need two numbers which when multiplied give $+12$ and which when added give $+7$. What two numbers will do this?

+3 and +4 will do what we want, so we can say $x^2 + 7x + 12 = (x + 3) \, (x + 4)$, giving us an alternative way of writing this expression.

Equally, $x^2 + 7x + 12 = (x + 4)(x + 3)$.

The order of the brackets is not important because multiplication of numbers gives the same answer either way on. For example, $2 \times 3 = 3 \times 2 = 6$.

In all the questions which follow, your answer will be equally correct if you have your brackets in the opposite order from mine.

EXERCISE 1.B.2

Try putting the following into brackets yourself.

(1) $x^2 + 8x + 7$ (2) $p^2 + 6p + 5$ (3) $x^2 + 7x + 6$

(4) $x^2 + 5x + 6$ (5) $y^2 + 6y + 9$ (6) $x^2 + 6x + 8$

(7) $a^2 + 7a + 10$ (8) $x^2 + 9x + 20$ (9) $x^2 + 13x + 36$

Now, a step further! Suppose we have $2x^2 + 7x + 3 = (something)\,(something)$. This time we need $2x$ and x at the fronts of the brackets to give the $2x^2$. If it is possible to factorise this with whole numbers then the ends will need 1 and 3 to give $1 \times 3 = 3$.

Do we need $(2x + 3)(x + 1)$ or $(2x + 1)(x + 3)$?

Multiplying out, we see that

$(2x + 3)(x + 1) = 2x^2 + 5x + 3$ which is wrong,

$(2x + 1)(x + 3) = 2x^2 + 7x + 3$ so this is the one we need.

EXERCISE 1.B.3

Try factorising these for yourself now.

(1) $3x^2 + 8x + 5$ (2) $2y^2 + 15y + 7$ (3) $3a^2 + 11a + 6$

(4) $3x^2 + 19x + 6$ (5) $5p^2 + 23p + 12$ (6) $5x^2 + 16x + 12$

The system is exactly the same if the expression involves minus signs. Here are two examples showing what can happen.

EXAMPLE (1) Factorise $x^2 - 10x + 16$.

Here we require two numbers which when multiplied give $+16$, and which when put together give -10. Can you see what they will be?

Both the numbers must be negative, and we see that -2 and -8 will fit the requirements. This gives us $x^2 - 10x + 16 = (x - 2)(x - 8) = (x - 8)(x - 2)$.

EXAMPLE (2) Factorise $x^2 - 3x - 10$.

Now we require two numbers which when multiplied give -10 and which when put together give -3. Can you see what we will need?

This time, to give the -10, they need to be of different signs. We see that -5 and $+2$ will do what we want, so we have

$$x^2 - 3x - 10 = (x - 5)(x + 2) = (x + 2)(x - 5).$$

Remember that it makes no difference which way round you write the brackets.

Now try factorising the following yourself.

(1) $x^2 - 11x + 24$ (2) $y^2 - 9y + 18$ (3) $x^2 - 11x + 18$

(4) $p^2 + 5p - 24$ (5) $x^2 + 4x - 12$ (6) $2q^2 - 5q - 3$

(7) $3x^2 - 10x - 8$ (8) $2a^2 - 3a - 5$ (9) $2x^2 - 5x - 12$

(10) $3b^2 - 20b + 12$ (11) $9x^2 - 25y^2$ (12) $16x^4 - 81y^4$, a sneaky one!

1.C   Using fractions

Very many students find handling fractions in algebra quite difficult, but it is important to be able to simplify these fractions as far as possible. This is because they often come into longer pieces of working and, if you do not simplify as you go along, the whole thing will become hideously complicated. It is only too likely then that you will make mistakes.

This section is designed to save you from this. You will find that if you understand how arithmetical fractions work then using fractions in algebra will be easy. If you have been using a calculator to do fractions, it's likely that you will have forgotten how they actually work, so I've drawn some little pictures of what is happening to help you.

If you think that you can already work well with fractions, try some of each exercise to be sure that there are no problems before you move on to the next section.

Because we are looking here at what we can and can't do with fractions, we shall need to use the sign ≠.

> The sign ≠ means 'is not equal to'.

1.C.(a)   Equivalent fractions and cancelling down

> $\dfrac{a}{b}$ means a divided by b.
>
> a is called the **numerator** and b is called the **denominator**.

In dividing, the order that the letters are written in matters, unlike $a \times b$, which is the same as $b \times a$.

The order also matters with subtraction; $a - b$ is not the same as $b - a$ unless both a and b are zero. But $a + b = b + a$ always.

For example, $2 \times 3 = 3 \times 2$ and $2 + 3 = 3 + 2$, but $\frac{2}{3} \neq \frac{3}{2}$ and $2 - 3 \neq 3 - 2$.

Also, $\dfrac{a + b}{c} = \dfrac{a}{c} + \dfrac{b}{c}$. For example, $\dfrac{2 + 3}{7} = \dfrac{2}{7} + \dfrac{3}{7} = \dfrac{5}{7}$.

The whole of $a + b$ is divided by c, and so we can get the same result by splitting this up into two separate divisions. The line in the fraction is effectively working as a bracket.

In fact, it is safer to write $\dfrac{a + b}{c}$ as $\dfrac{(a + b)}{c}$ if it is part of some working.

In $\dfrac{a}{b + c}$, the number a is divided by the whole of the number $(b + c)$.

From this, we see that

$$\frac{a}{b + c} \neq \frac{a}{b} + \frac{a}{c}.$$

You can check this by putting $a = 4$, $b = 2$, $c = 3$, say.

Dividing by c is the same as multiplying by $1/c$, so

$$\frac{a + b}{c} = \frac{1}{c}(a + b).$$

For example, if $a = 6$, $b = 4$, and $c = 2$ then

$$\frac{6 + 4}{2} = \tfrac{1}{2}(6 + 4) = 5.$$

If you find half of 10, it is the same as dividing 10 by 2.

Fractions always keep the same value if they are multiplied or divided top and bottom by the same number, so

$$\frac{4}{6} = \frac{8}{12} = \frac{6}{9} = \frac{2}{3}, \text{ etc.}$$

These are shown in the drawings in Figure 1.C.1.

These four equal fractions are said to be **equivalent** to each other. The process of dividing the top and bottom of a fraction by the same number is called **cancellation** or cancelling down.

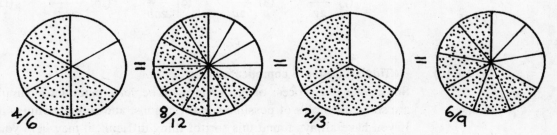

Figure 1.C.1

1.C Using fractions

$$a\left(\frac{b}{c}\right) = \frac{ab}{c} \quad \text{not} \quad \frac{ab}{ac}.$$

For example, $4\left(\frac{2}{3}\right) = \frac{4 \times 2}{3}$ not $\frac{4 \times 2}{4 \times 3}$ which is still $\frac{2}{3}$.

In words, four lots of two thirds is eight thirds.

This works in exactly the same way with fractions in algebra.

So, for example:

$$\frac{2a}{5a} = \frac{2}{5} \quad \text{(dividing top and bottom by } a\text{)}$$

$$\frac{xw}{yw} = \frac{x}{y} \quad \text{(dividing top and bottom by } w\text{)}$$

and $\quad \dfrac{2a^3b}{a^2b^2} = \dfrac{2a}{b} \quad$ (dividing top and bottom by a^2b).

Check these three results by giving your own values to the letters.

When doing this, it is important to avoid values which would involve you in trying to divide by zero, because this cannot be done.

You can use a calculator to investigate this by dividing 4, say, by a very small number, say 0.00001.

Now repeat the process, dividing 4 by an even smaller number.

The closer the number you divide by gets to zero, the larger the answer becomes. In fact, by choosing a sufficiently small number, you can make the answer as large as you please.

If you try to divide by zero itself, you get an ERROR message.

EXERCISE 1.C.1

Cancel down the following fractions yourself as far as possible.

(1) $\dfrac{9}{12}$ (2) $\dfrac{6}{30}$ (3) $\dfrac{25}{95}$ (4) $\dfrac{24}{64}$ (5) $\dfrac{5x}{8x}$ (6) $\dfrac{ab}{ac}$

(7) $\dfrac{3y^2}{2y}$ (8) $\dfrac{8pq}{2q}$ (9) $\dfrac{4a^2}{2ab}$ (10) $\dfrac{3x^2y^3}{2xy^4}$ (11) $\dfrac{6p^2q}{5pq^2}$ (12) $\dfrac{5ab}{b^3}$

1.C.(b)   **Tidying up more complicated fractions**

Sometimes, the process of factorising will be very important in simplifying fractions. Here are some examples of possible simplifications, and some warnings of what *can't* be done. If you have always found this sort of thing difficult, it may help you here to highlight the matching parts which are cancelling with each other in the same colour.

(1) $\dfrac{xy + xz}{xw} = \dfrac{x(y + z)}{xw} = \dfrac{y + z}{w}$

dividing top and bottom by x.

(2) $\dfrac{ab + ac}{b + c} = \dfrac{a(b + c)}{b + c} = a$

dividing top and bottom by the whole chunk of $(b + c)$.

(3) $\dfrac{ab + c}{b + c}$ can't be simplified.

We can't cancel the $(b + c)$ here because a only multiplies b.

(4) $\dfrac{x + xy}{x^2} = \dfrac{x(1 + y)}{x^2} = \dfrac{1 + y}{x}$

dividing top and bottom by x.

(5) $\dfrac{x^2 + 5x + 6}{x^2 - 3x - 10} = \dfrac{(x + 3)(x + 2)}{(x - 5)(x + 2)} = \dfrac{x + 3}{x - 5}$

dividing top and bottom by $(x + 2)$.

(6) $\dfrac{x^2(x^2 + xy)}{x} = x(x^2 + xy)$

dividing top and bottom by x.

 It is not true that $\dfrac{x(x^2 + xy)}{x} = x + y.$

This wrong answer comes from cancelling the x twice on the top of the fraction, but only once underneath.

It is like saying $\frac{1}{2}(4)(6) = (2)(3) = 6$ but really $\frac{1}{2}(4)(6) = \frac{1}{2}(24) = 12$.

You can halve either the 4 or the 6 but not both!

 (7) $\dfrac{xy + z}{xw}$ is *not* the same as $\dfrac{y + z}{w}.$

We cannot cancel the x here because x is only a factor of *part* of the top. You can check this by putting $x = 2$, $y = 3$, $z = 4$, and $w = 5$. Then

$$\dfrac{xy + z}{xw} = \dfrac{10}{10} = 1 \quad \text{and} \quad \dfrac{y + z}{w} = \dfrac{7}{5}$$

DELICATE POINT

If we had put $x = 1$, the difference would not have shown up, since both answers would have been $\frac{7}{5}$.

This is because multiplying by 1 actually leaves numbers unchanged.

This example shows that checking with numbers is only a check, and never a proof that something is true.

EXERCISE 1.C.2

Try these questions yourself now.

(1) Which of the following fractions are the same as each other (equivalent)?

(a) $\dfrac{2}{3}, \dfrac{4}{9}, \dfrac{12}{18}, \dfrac{10}{15}, \dfrac{2}{6}, \dfrac{6}{9}$

(b) $\dfrac{ax}{bx}, \dfrac{a}{b}, \dfrac{a(c+d)}{b(c+d)}, \dfrac{a^2x}{abx}$

(c) $\dfrac{ab+ac}{ad}, \dfrac{ab+c}{ad}, \dfrac{b+c}{d}$

(d) $\dfrac{x}{x+y}, \dfrac{xz}{xz+yz}, \dfrac{xp}{x+yp}$

(2) Factorise and cancel down the following fractions if possible.

(a) $\dfrac{2x+6y}{6x-8y}$

(b) $\dfrac{6a-9b}{4a-6b}$

(c) $\dfrac{px-pq}{p^2-px}$

(d) $\dfrac{3x+2y}{6x}$

(e) $\dfrac{2xy+5xz}{6x}$

(f) $\dfrac{4xz+6yz}{2x+3y}$

(g) $\dfrac{2p-3q}{2p+3q}$

(h) $\dfrac{x^2-y^2}{(x+y)^2}$

(i) $\dfrac{x^2+5x+6}{x^2+x-2}$

1.C.(c)   Adding fractions in arithmetic and algebra

It is particularly easy to add fractions which have the same number underneath.

For example, $\frac{2}{7} + \frac{3}{7} = \frac{5}{7}$. I've drawn this one in Figure 1.C.2 below.

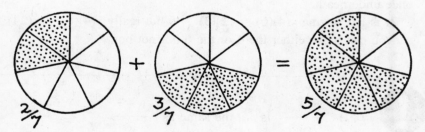

Figure 1.C.2

If the fractions which we want to add don't have the same denominator then we have to first rewrite them as equivalent fractions which do share the same denominator.

For example, to find $\dfrac{2}{3} + \dfrac{3}{4}$ we use $\dfrac{2}{3} = \dfrac{8}{12}$ and $\dfrac{3}{4} = \dfrac{9}{12}$.

Basic algebra: some reminders of how it works

The two fractions have both been written as parts of 12. The number 12 is called the **common denominator**. It's now very easy to add them, and we have

$$\frac{2}{3} + \frac{3}{4} = \frac{8}{12} + \frac{9}{12} = \frac{17}{12}.$$

The answer of $\frac{17}{12}$ can also be written as $1\frac{5}{12}$, but in general, for scientific and engineering purposes, it is better to leave such arithmetical fractions in their top-heavy state.

You should be safe now from the most usual mistake made when adding fractions, which is to add the tops and add the bottoms.

 $\frac{1}{6} + \frac{3}{4}$ (for example) is *not* $\frac{1+3}{6+4} = \frac{4}{10}.$

We can see that this must be wrong from Figure 1.C.3.

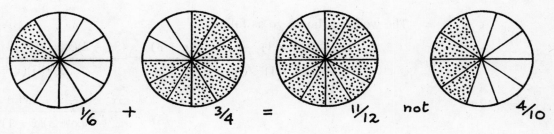

Figure 1.C.3

EXERCISE 1.C.3 Since the process in arithmetic is exactly the same as the process we use to add fractions in algebra, it is worth practising adding some numerical fractions yourself without using a calculator, before we move on to this.

Try adding these three.

(1) $\dfrac{3}{4} + \dfrac{2}{7}$ (2) $\dfrac{2}{3} + \dfrac{4}{5}$ (3) $\dfrac{1}{2} + \dfrac{2}{3} + \dfrac{4}{5}$

The letters work in exactly the same way as the numbers. We can say

$$\frac{a}{b} + \frac{c}{d} = \frac{ad}{bd} + \frac{bc}{bd} = \frac{ad + bc}{bd}$$

where a, b, c and d are standing for unknown numbers, and neither b nor d are zero. We have written both fractions as parts of bd to make it easy to add them.

Indeed, we can say

$$\frac{A}{B} + \frac{C}{D} = \frac{AD}{BD} + \frac{BC}{BD} = \frac{AD + BC}{BD}$$

where A, B, C and D are standing for whole lumps or chunks of letters and numbers.

As an example of this, we will find

$$\frac{x + 2y}{x - y} + \frac{3x + 2y}{x + 3y}.$$

Here, $A = x + 2y$, $B = x - y$, $C = 3x + 2y$ and $D = x + 3y$. So we have:

$$\frac{(x + 2y)(x + 3y)}{(x - y)(x + 3y)} + \frac{(3x + 2y)(x - y)}{(x + 3y)(x - y)} = \frac{(x + 2y)(x + 3y) + (3x + 2y)(x - y)}{(x - y)(x + 3y)}$$

$$= \frac{x^2 + 5xy + 6y^2 + 3x^2 - xy - 2y^2}{(x - y)(x + 3y)}$$

$$= \frac{4x^2 + 4xy + 4y^2}{(x - y)(x + 3y)} = \frac{4(x^2 + xy + y^2)}{(x - y)(x + 3y)}.$$

We don't usually multiply out the brackets on the bottom, because then we might miss a possible cancellation. (This saves you some work.)

Try combining $\dfrac{3x - 2}{x + 3} + \dfrac{2x - 3}{x + 1}$ into a single fraction, yourself.

The working should go as follows:

$$\frac{(3x - 2)(x + 1)}{(x + 3)(x + 1)} + \frac{(2x - 3)(x + 3)}{(x + 1)(x + 3)} = \frac{(3x - 2)(x + 1) + (2x - 3)(x + 3)}{(x + 3)(x + 1)}$$

$$= \frac{3x^2 + x - 2 + 2x^2 + 3x - 9}{(x + 3)(x + 1)}$$

$$= \frac{5x^2 + 4x - 11}{(x + 3)(x + 1)}.$$

(Remember that the order in which we multiply the brackets doesn't matter.)

1.C.(d)   Repeated factors in adding fractions

Sometimes, the addition is a little easier because there is a repeated factor. Here's a numerical example of this.

$\dfrac{3}{4} + \dfrac{5}{6}$ has a repeated factor of 2 underneath.

So, instead of saying

$$\frac{3}{4} + \frac{5}{6} = \frac{18}{24} + \frac{20}{24} = \frac{38}{24} = \frac{19}{12}$$

we can say more directly

$$\frac{3}{4} + \frac{5}{6} = \frac{9}{12} + \frac{10}{12} = \frac{19}{12}.$$

The number 12, which is the smallest number which both 4 and 6 will divide into, is called the **lowest common denominator** or **l.c.d.** for short.

This same simplification applies to fractions in algebra.

EXAMPLE (1) $\dfrac{2}{x(x+3)} + \dfrac{3}{x(2x-1)}$

There is a repeated factor of x underneath, so we say

$$\frac{2}{x(x+3)} = \frac{2(2x-1)}{x(x+3)(2x-1)}$$

and

$$\frac{3}{x(2x-1)} = \frac{3(x+3)}{x(2x-1)(x+3)}.$$

So

$$\frac{2}{x(x+3)} + \frac{3}{x(2x-1)} = \frac{2(2x-1) + 3(x+3)}{x(x+3)(2x-1)}$$

$$= \frac{7x+7}{x(2x-1)(x+3)} = \frac{7(x+1)}{x(2x-1)(x+3)}.$$

You can follow through this example experimentally, converting it into arithmetical fractions by putting in some value of your choice for x.

Be careful though! There are three values which you mustn't choose. Can you see what they are?

You can't have $x = 0$ or $x = -3$ or $x = \frac{1}{2}$, because each of these values would involve trying to divide by zero, which is impossible as we saw at the end of Section 1.C.(a).

In this example, it would not have been wrong to put everything over the common denominator of $x(x + 3)x(2x + 1)$ or $x^2(x+3)(2x+1)$. It would just have taken longer to work out.

EXAMPLE (2) $\dfrac{2x}{y(3x-2y)} + \dfrac{3y}{4x(3x-2y)}$

Here, $(3x - 2y)$ is a repeated factor underneath, so the expression is equal to

$$\frac{(2x)(4x)}{y(3x-2y)(4x)} + \frac{3y(y)}{4x(3x-2y)(y)} = \frac{8x^2 + 3y^2}{4xy(3x-2y)}.$$

Check this example by putting $x = 4$, $y = 2$ and $z = 5$.

You should get

$$\frac{8}{2(8)} + \frac{6}{16(8)} = \frac{8(16) + 6(2)}{32(8)} = \frac{128 + 12}{256} = \frac{140}{256} = \frac{35}{64},$$

$$\frac{8(16) + 3(4)}{32(8)} = \frac{128 + 12}{256} = \frac{140}{256} = \frac{35}{64}.$$

Try these for yourself.

(1) $\dfrac{2}{9} + \dfrac{7}{15}$

(2) $\dfrac{5}{6} + \dfrac{3}{8}$

(3) $\dfrac{1}{3} + \dfrac{3}{4} + \dfrac{5}{6}$

(4) $\dfrac{3x}{y(2x-y)} + \dfrac{5y}{x(2x-y)}$

(5) $\dfrac{2}{x(3x+1)} + \dfrac{5}{x(2x-1)}$

(6) $\dfrac{4}{x^2 - y^2} + \dfrac{3}{(x+y)^2}$

1.C.(e)   Subtracting fractions

Subtraction works in exactly the same kind of way as addition, so, for example

$$\frac{2}{3} - \frac{5}{8} = \frac{2 \times 8}{3 \times 8} - \frac{5 \times 3}{8 \times 3} = \frac{16}{24} - \frac{15}{24} = \frac{1}{24}.$$

In just the same way,

$$\frac{a}{b} - \frac{c}{d} = \frac{ad}{bd} - \frac{cb}{db} = \frac{ad - bc}{bd},$$

where a, b, c and d are standing for numbers such as the 2,3,5 and 8 we had in the first example.

Equally, just as in adding fractions, we can say that

$$\frac{A}{B} - \frac{C}{D} = \frac{AD - BC}{BD}$$

where A, B, C and D stand for any chunks of letters and numbers.

The line in a fraction works in the same way as a bracket. If we are adding fractions this won't affect what happens, but if we are subtracting them we have to be careful. For example, suppose we have

$$\frac{4x - 3}{2} - \frac{2x + 1}{3}.$$

The minus sign in the middle is affecting the whole right-hand chunk. We can show this most safely by rewriting using brackets. Then we have:

$$\frac{(4x - 3)}{2} - \frac{(2x + 1)}{3} = \frac{3(4x - 3)}{3 \times 2} - \frac{2(2x + 1)}{2 \times 3}$$

$$= \frac{3(4x - 3) - 2(2x + 1)}{6}$$

$$= \frac{12x - 9 - 4x - 2}{6}$$

$$= \frac{8x - 11}{6}$$

The safest strategy is always to put the brackets in, because then they will be there on the occasions when their presence is vital.

Try these mixed additions and subtractions yourself.

(1) $\dfrac{3x - 5}{10} + \dfrac{2x - 3}{15}$

(2) $\dfrac{3a + 5b}{4} - \dfrac{a - 3b}{2}$

(3) $\dfrac{3m - 5n}{6} - \dfrac{3m - 7n}{2}$

(4) $\dfrac{2b}{a(2a + b)} + \dfrac{3a}{b(2a + b)}$

(5) $\dfrac{2a}{(a + b)(3a + b)} + \dfrac{3b}{(a - b)(3a + b)}$

(6) $\dfrac{5}{x^2 - y^2} - \dfrac{2}{x(x + y)}$

1.C.(f) Multiplying fractions

This is very straightforward. (It is much easier than adding!) We simply say

$$\frac{a}{b} \times \frac{c}{d} = \frac{ac}{bd}.$$

That is, we multiply the tops, and multiply the bottoms.

We can take $\frac{2}{3} \times \frac{3}{4} = \frac{6}{12} = \frac{1}{2}$ as a numerical example of what's happening. If you take two thirds of three quarters, you get one half. I show this happening in Figure 1.C.4.

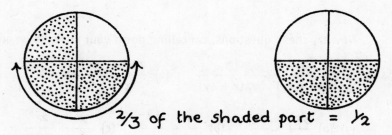

⅔ of the shaded part = ½

Figure 1.C.4

If A, B, C and D are standing for any chunks of letters and numbers,

then we can say $\dfrac{A}{B} \times \dfrac{C}{D} = \dfrac{AC}{BD}.$

It may then be possible to cancel down, for example

$$\frac{x(b + c)}{y^2} \times \frac{y}{x^2(b + c)} = \frac{xy\,(b + c)}{x^2 y^2 (b + c)} = \frac{1}{xy}$$

dividing top and bottom by $xy(b + c)$. You should always cancel down the answer like this if it is possible. The reason for this is that often fractions like this come in as part of the working out of a larger problem, and it pays to simplify them as much as possible before going on to the next step, to make that next step as easy as possible for yourself.

You can also do the cancelling *before* you do the multiplying if you want; I show the working done this way in Figure 1.C.5. Cancellations are usually shown by diagonal lines. Notice that, when everything on the top cancels, we finish up with 1 not 0.

$$\frac{\cancel{x}(\cancel{b+c})}{y^{\cancel{2}}} \times \frac{\cancel{y}}{\cancel{x^2}(\cancel{b+c})} = \frac{1}{xy}$$

Figure 1.C.5

1.C.(g) Dividing fractions

The rule for dividing fractions is to turn the second fraction upside down and then multiply.

$$\frac{a}{b} \div \frac{c}{d} = \frac{a}{b} \times \frac{d}{c} = \frac{ad}{bc}.$$

We can see that this works by taking the numerical example of one and one half divided by one half. We get

$$\frac{3}{2} \div \frac{1}{2} = \frac{3}{2} \times \frac{2}{1} = 3 \quad \text{(that is, there are three halves in } 1\tfrac{1}{2}.)$$

EXERCISE 1.C.6

Now try these questions, cancelling down your answers where possible.

(1) $\dfrac{2}{x(2x - 3y)} - \dfrac{3}{2x(x + 4y)}$ (2) $\dfrac{2x - 1}{3} - \dfrac{x - 7}{5}$

(3) (a) $\dfrac{3a^2}{2b} \times \dfrac{ab}{6c}$ (b) $\dfrac{2a}{3b} \div \dfrac{b^2}{9a^2}$ (c) $\dfrac{3x}{y^2z} \div \dfrac{2x^2}{5yz^2}$

(4) (a) $\dfrac{3x^2(2x + 3y)}{2y(x - y)} \times \dfrac{y^2(x - y)}{(x + 3y)}$ (b) $\dfrac{5pq(p + q)}{(3p + 2q)} \times \dfrac{(3p + 2q)}{q^2(5p - q)}$

(c) $\dfrac{(a^2 - b^2)^4}{(a^2 + b^2)} \times \dfrac{(a^4 - b^4)}{(a + b)^4}$ **Be cunning!**

1.D The three rules for working with powers

1.D.(a) Handling powers which are whole numbers

It will be useful for us now to spend some time looking in more detail at how numbers written as powers of other numbers can be combined with each other. (We have already looked briefly at the rules for multiplying such numbers in Section 1.A.(c).)

We'll use the four numbers $8 = 2^3$, $32 = 2^5$, $9 = 3^2$ and $81 = 3^4$ as examples.

We could combine these numbers in many ways, some of which I have written down here.

(1) 32×8 (2) 9×81 (3) $32 \div 8$

(4) $81 \div 9$ (5) 8×9 (6) $81 \div 32$ (7) 8^2

If we rewrite the numbers as powers, we get the following results.

(1) $32 \times 8 = 2^5 \times 2^3 = (2 \times 2 \times 2 \times 2 \times 2) \times (2 \times 2 \times 2) = 2^{5+3} = 2^8 = 256$.
The answer to the multiplication can be obtained by *adding* the powers.

(2)   Similarly, $9 \times 81 = 3^2 \times 3^4 = (3 \times 3) \times (3 \times 3 \times 3 \times 3) = 3^{2+4} = 3^6 = 729$.
Again, the result can be obtained by adding the powers.

(3) $32 \div 8 = 2^5 \div 2^3 = \dfrac{2 \times 2 \times 2 \times 2 \times 2}{2 \times 2 \times 2} = 2 \times 2 = 2^{5-3} = 2^2 = 4$

This time, the answer has been obtained by subtracting the powers.

(4)   Similarly, $81 \div 9 = 3^4 \div 3^2 = \dfrac{3 \times 3 \times 3 \times 3}{3 \times 3} = 3 \times 3 = 3^2 = 9$

and again the result is obtained by subtracting the powers.

(5) $8 \times 9 = 2^3 \times 3^2$. This time, the calculation is made no easier by writing the numbers in this form. As they are powers of different numbers, we cannot use the same system as we did in (1) and (2).
Returning to the original form, $8 \times 9 = 72$.

(6)   Similarly, there is no advantage to be gained by writing $81 \div 32$ as $3^4 \div 2^5$.
$\dfrac{81}{32}$ can be left like this, or written in decimal form as 2.53125.

(7) $8^2 = (2 \times 2 \times 2)^2 = (2 \times 2 \times 2) \times (2 \times 2 \times 2) = 2^6$ and $8^2 = (2^3)^2 = 2^6$.
The answer comes from multiplying the two powers.

Any powers which are whole numbers will work in the same kind of way, so we will now write down the three rules or laws for working with powers.

The three rules for powers

Rule (1) $a^m \times a^n = a^{m+n}$

           Example: $a^2 \times a^3 = (a \times a) \times (a \times a \times a) = a^5$.

Rule (2) $a^m \div a^n = a^{m-n}$

           Example: $a^5 \div a^2 = \dfrac{a \times a \times a \times a \times a}{a \times a} = a^3$.

Rule (3) $(a^m)^n = a^{mn}$

           Example: $(a^2)^3 = (a \times a) \times (a \times a) \times (a \times a) = a^6$.

We saw from the numerical examples that we must have powers of the same number for these rules to work. There, we used either 2 or 3, and for the rules above I have used a. The number a is called the **base** that we are working with.

1.D.(b)   **Some special cases**

It can be shown that the three rules above are true for *any* values of m and n, provided that $a \neq 0$, but it is not possible for us to prove this yet. However, by using powers which are whole numbers we can see how some particular cases will have to go.

(1) $a^3 \div a^2 = \dfrac{a \times a \times a}{a \times a} = a$ and, by Rule (2), $a^3 \div a^2 = a^{3-2} = a^1$.

So we must have

$$a^1 = a.$$

(2) $a^3 \div a^3 = \dfrac{a \times a \times a}{a \times a \times a} = 1$ and, by Rule (2), $a^3 \div a^3 = a^{3-3} = a^0$.

So we must have

$$a^0 = 1.$$

(3) $a^2 \div a^3 = \dfrac{a \times a}{a \times a \times a} = \dfrac{1}{a}$ and, by Rule (2), $a^2 \div a^3 = a^{2-3} = a^{-1}$.

So we must have

$$a^{-1} = \frac{1}{a}.$$
In fact, more generally, $a^{-n} = \dfrac{1}{a^n}.$

(4) $a^{1/2} \times a^{1/2} = a^1$ by Rule (1), and $a^1 = a$.

So $a^{1/2}$ is the number which multiplied by itself gives a.

$a^{1/2}$ means the square root of a.

Similarly, $a^{1/3} \times a^{1/3} \times a^{1/3} = a^1$ by Rule (1).

So $a^{1/3}$ means the cube root of a, or $\sqrt[3]{a}$.

and $a^{1/n}$ means the nth root of a or $a^{1/n} = \sqrt[n]{a}$.

Here are four examples.

What are   (1) $4^{1/2}$ (2) $8^{1/3}$ (3) $27^{2/3}$ (4) $16^{1/4}$?

(1) $4^{1/2}$ means the square root of 4, so it means the number which multiplied by itself gives 4. There are *two* numbers which do this. What are they?

They are $+2$ and -2. So $4^{1/2} = +2$ or -2.

We can write this as $4^{1/2} = \pm 2$. (The symbol $\pm$ means $+$ or $-$.)

(2) $8^{1/3}$ means the cube root of 8 so it means finding a number a so that $a \times a \times a = 8$. What can a be?

There is only one possible value for a in ordinary numbers, which is $+2$.

(I say 'ordinary numbers' here because it is possible to extend the number system so that other possibilities open up. In fact, as we shall see in Chapter 10, we then rather pleasingly get three cube roots. But for the present, we are only interested in solutions in ordinary numbers.)

(3) $27^{2/3} = (27^{1/3})^2$ by Rule (3). But $27^{1/3} = 3$ so $(27^{1/3})^2 = 3^2 = 9$.

(4) $16^{1/4}$ means the fourth root of 16. What are the possibilities here?

There are two possibilities using ordinary numbers.

We have $2 \times 2 \times 2 \times 2 = 16$ and $-2 \times -2 \times -2 \times -2 = 16$ so $16^{1/4} = \pm 2$.

In general we can say that each even root of a positive number has two possible solutions, and each odd root of either a positive or a negative number has just one solution.

At present, we cannot find any even roots of negative numbers, although in Chapter 10 we will find out how it is possible to extend the number system so that we can have roots for these numbers too. Have a guess at how many fourth roots of 16 we shall then have.

Yes, it is most satisfyingly four.

EXERCISE 1.D.1   It is very useful to get a feeling for what these powers do, so that you can quickly recognise alternative ways of writing them, or possible simplifications.

Try these numerical examples without a calculator to help you develop this feel.

Then go through, checking all your answers on your calculator. If you have a mismatch, try to spot which one has gone wrong. Maybe the answers are the same but just in a different form? (Your calculator will only give you positive values for roots; you have to add possible alternative negative answers yourself.) Make sure that you know how powers work on your calculator; read its little instruction book if necessary!

(1) 3^{-1}	(2) $16^{1/2}$	(3) $9^{3/2}$	(4) $27^{-1/3}$	(5) 4^0	(6) 7^1
(7) 7^{-2}	(8) $4^{-1/2}$	(9) $32^{1/5}$	(10) $16^{-3/4}$	(11) $25^{3/2}$	(12) $49^{-1/2}$

1.D The three rules for working with powers

The different kinds of numbers

The number system has been invented and extended as people needed ways to describe ever more complicated situations and transactions. This procedure took thousands of years, so I have to compress it somewhat in this brief description.

1.E.(a)   The counting numbers and zero

By inventing names, with symbols for those names, it became possible to count how many distinct objects there were when they were collected together. It was also then possible to count the totals when collections were combined together, provided enough names or symbols had been invented.

Having a symbol for zero was a great advance. The oldest written record with a symbol for zero dates from the ninth century in a Hindu manuscript.

We don't very often have to say that we have none of something. So why is having a symbol for zero so important?

It makes it possible to put in all the necessary place values in our system for writing numbers, for example 301. Having a place value system means that once the symbols for 1 to 9 are learnt, a number of any size can be written. This use of the symbol for zero was ridiculed by some people when it was first adopted. How could it be possible to write a large number, they said, by using lots of symbols which each individually stand for nothing?

The fact that it took two centuries before this symbol for zero was invented shows what a subtle development it was.

1.E.(b)   Including negative numbers: the set of integers

The first important extension to the system of counting numbers for a collection of objects is having some arrangement to represent what happens if we want to take away more than we have, so that we owe.

If we include the negative numbers we can do this.

We now have the number system of **integers** given by

$$\ldots \quad -4, \quad -3, \quad -2, \quad -1, \quad 0, \quad 1, \quad 2, \quad 3, \quad 4, \quad \ldots$$

The German mathematician Kronecker said of these numbers: 'God made the whole numbers; everything else is the work of man.'

Also now we have a nice symmetry.

For every number there is another number so that put together they make zero, so each number has its matching pair. These pairs of numbers are reflections of each other around zero. What are the pairs of (a) +7, (b) −9, and (c) 0?

(a)   +7 has the pair −7.   (b) −9 has the pair +9.   (c) 0 is its own pair.

Putting together *any* two numbers in this system gives us another number in the system. It has a nice completeness about it.

1.E.(c)   Including fractions: the set of rational numbers

The next major extension to the number system results from the requirement of being able to divide quantities up. To do this, we have to include fractions, that is, numbers which can be written in the form a/b where a and b are integers or whole numbers, excluding the case when $b = 0$. These numbers are called the **rational numbers**. Then the integers themselves come from the special case in which $b = 1$, so they are included in this description.

We can now divide quantities into smaller amounts, even if the numbers involved mean that the result of the division is not a whole number (provided of course that the quantity concerned *is* physically divisible into non-integer amounts).

We have a second nice symmetry here, this time about 1.

For every number except zero, there is now another number so that multiplied together we get 1. For example, $\frac{2}{3}$ has the pair $\frac{3}{2}$.

What are the pairs of (a) $\frac{3}{7}$, (b) $-\frac{3}{5}$ and (c) 1?

(a) $\frac{3}{7}$ has the pair $\frac{7}{3}$. (b) $-\frac{3}{5}$ has the pair $-\frac{5}{3}$. (c) 1 is its own pair.

Putting together any two numbers in this system by multiplying them together gives us another number in the system, so we have exactly the same sort of completeness that we had above with adding. The two systems have the same underlying structure of each number having its own individual partner so that each pair together gives a special number, zero in the case of adding and 1 in the case of multiplying.

If we put little tiny points for the value of each possible fraction on a number line how close will these points be together? Will there be any gaps?

Suppose we have two fractions F_1 and F_2 which are very close together, say $F_1 = \frac{1}{100}$ and $F_2 = \frac{1}{101}$.

Then, there must be at least one fraction which lies between these two. Can you think of one?

There are lots of possibilities for this. In particular, we could take $(F_1 + F_2)/2$.

This is exactly midway between F_1 and F_2. Here, it would be $\frac{201}{10100}$.

This system of insertion can be infinitely repeated, so we see that there can't be any spaces between these fractions.

1.E.(d)   **Including everything on the number line: the set of real numbers**
If the fractions are packed infinitely closely together, where is $\sqrt{2}$?

Is it a fraction? Trying a few possibilities doesn't look very promising, but maybe we just haven't got the right numbers.

Suppose that it *is* possible, and we have found a and b so that

$$\frac{a}{b} = \sqrt{2} \quad \text{so} \quad \frac{a^2}{b^2} = 2$$

and therefore

$$a^2 = 2b^2.$$

We'll also suppose that any possible cancelling down of the fraction a/b has already been done, so it is tidied up as much as possible.

What kind of number must $2b^2$ be?

It must be even, so a^2 must be even as well.

What happens if you square (a) even numbers (b) odd numbers?

An even number squared gives another even number and an odd number squared gives an odd number. We can show this by writing even numbers as $2n$ (with n standing for any whole number) and odd numbers as $2n + 1$.

Then $(2n)^2 = 4n^2$ and $(2n + 1)^2 = 4n^2 + 4n + 1$.

Because of this, we see that the number a must be even. We could call it $2a_1$ to show this. Then

$$a^2 = (2a_1)(2a_1) = 4a_1^2 = 2b^2 \quad \text{which means that} \quad b^2 = 2a_1^2.$$

Now, by the same argument as before, b must also be even, so a and b *could* have been cancelled down.

But if we cancel them, we can use exactly the same argument to show that they would cancel down again, and so on for ever.

So there is no fraction which is exactly equal to $\sqrt{2}$.

This argument is due to the Pythagoreans of Ancient Greece. They were disconcerted and alarmed by such numbers, which they called 'incommensurable'. There is a story that the first Pythagorean to show their existence was thrown into the sea for his pains.

In fact, $\sqrt{2}$ is somewhere between $\frac{1414}{1000}$ and $\frac{1415}{1000}$. So although the fractions are packed infinitely closely, there are still gaps where the numbers like $\sqrt{2}$, $\sqrt{7}$, etc. are.

(This is one of the mysteries of maths and is because infinite numbers of things behave in very peculiar ways.)

These numbers, together with π and similar numbers, are called **irrational numbers**. The rational and irrational numbers together are called the set of **real numbers**.

Here's another example of how infinite quantities of things behave in unexpected ways.

If we have two collections or sets of objects and we can tally off each object in the first set with a corresponding object in the second set and vice versa, like knives and forks in place settings, then the two sets must have an equal number of objects in them.

Or must they?

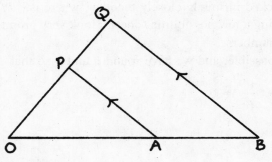

Figure 1.E.1

Suppose we start with the two lines meeting at O which I have drawn above in Figure 1.E.1, and we then draw parallel lines like AP and BQ so that point A is matched with point P and point B is matched with point Q. All the points on the two lines can be paired off in this way, so the two lines must be equal in length. But clearly they are not! We can no longer say that the sets are equal because now there are an infinite number of objects involved and the usual rules no longer apply.

1.E.(e)   **Complex numbers: a very brief forwards look**

Finally, to make the list complete, we will jump ahead of ourselves briefly. We know that $2 \times 2 = 4$ and $-2 \times -2 = 4$. So the square root of 4 is $+2$ or -2. But we have no number for the square root of -4.

In Chapter 10, we shall find out how it is possible to extend the number system even further so that we can have an answer for $\sqrt{-4}$. In fact, even better, we get *two* answers, just like $\sqrt{4}$ has two answers.

We get this extension by including the so-called **imaginary numbers**. The real and imaginary numbers together form the set of **complex numbers**.

1.F   **Working with different kinds of number: some examples**

1.F.(a)   **Other number bases: the binary system**

We have to use ten symbols for writing numbers because our counting system is based on 10. Our whole system is therefore called the decimal system, although in ordinary speech we use 'decimals' for just the fractions written in this system.

However, other bases can be used. One of the most important of these is the system based on 2, the **binary system**. This involves counting in place values given by powers of 2 instead of powers of 10.

So, for example,

$$324 \text{ in the decimal system} = 4(10^0) + 2(10^1) + 3(10)^2 = 4 + 2(10) + 3(100).$$

$$11011 \text{ in the binary system} = 1(2^0) + 1(2^1) + 0(2^2) + 1(2^3) + 1(2^4)$$

$$= 1 + 1(2) + 0(4) + 1(8) + 1(16)$$

$$= 1 + 2 + 8 + 16 = 27 \text{ in the decimal system.}$$

Notice that, in each case, we have processed the number from right to left, instead of from left to right.

In each case, we wrote down the number of units, the number of 'tens', the number of 'hundreds', etc., where the 'ten' or 10 of the binary system is 2, the 'hundred' or 100 of the binary system is 2^2 or 4, and so on.

Counting in binary goes 1, 10, 11, 100, 101, 110, 111, 1000, etc. instead of the decimal 1,2,3,4,5,6,7,8, etc.

The binary system only requires two symbols to write, those for one and zero, which is why it is so important. The separate digits of numbers written in this system can be represented by electric current either flowing or not flowing in a circuit, and therefore numbers can be handled in this form by computers.

EXERCISE 1.F.1   Try converting these three binary numbers into decimal numbers for yourself.
(1) 10111   (2) 1111   (3) 111011

How can we go the other way, and convert decimal numbers into binary numbers?

If we have the number 109, say, we *could* do it just by splitting it up into powers of two.

$$109 = 64 + 45 = 64 + 32 + 13 = 64 + 32 + 8 + 5$$

$$= 64 + 32 + 8 + 4 + 1 = 2^6 + 2^5 + 2^3 + 2^2 + 1$$

$$= 1101101 \text{ in binary or base 2.}$$

(A useful way of showing that this number is in base 2 is to write the 2 as a little subscript, so we write the number as 1101101_2.)

This is good for seeing what is happening, but not so good as a standard method of conversion.

What we have actually done here is to split the number up into progressively higher powers of 2, which we can do equally well by repeatedly dividing it by 2, recording the remainder at each stage so we get the smaller powers as they shed off.

I show the working for this below.

		Remainder
2	109	
2	54	1
2	27	0
2	13	1
2	6	1
2	3	0
2	1	1
	0	1

The answer is:

109_{10} is the same as 1101101_2.

EXERCISE 1.F.2

Try converting these three decimal numbers to binary numbers for yourself.
(1) 72   (2) 2431   (3) 3251

THINKING POINT

If you have the use of a computer and know a programming language, you could write a program to do this, since the process of dividing by 2 is a repeated loop until the number being divided is itself less than 2. You just have to record the remainders so that you can display or print out your binary conversion at the end.

This system works equally well in other number bases.

For example, in base 8, we have a 'ten' of 8 and a 'hundred' of 8^2, etc.

So $237_8 = 7 + 3(8^1) + 2(8^2) = 7 + 24 + 128 = 159_{10}$.

Working the other way round is done by repeated division by 8.

So, for example, to convert 397_{10} into base 8, you would do the working shown below.

		Remainder
8	397	
8	49	5
8	6	1
	0	6

$397_{10} = 615_8$.

Check: $615_8 = 5 + 1 \times 8 + 6 \times 8^2$
$= 397_{10}$.

1.F.(b)   Prime numbers and factors

In this section, we look briefly at how the different numbers are built up.

Many numbers can be written as products (i.e. multiplications) of smaller numbers or factors in quite a few different ways, for example

$$12 = 2 \times 6 = 2 \times 3 \times 2 = 3 \times 4 = 12 \times 1.$$

Numbers which have no factors other than themselves and one are called **prime numbers**. No smaller number (except for 1) will divide into them exactly. 7, 11 and 19 are all examples of prime numbers.

Are there any even prime numbers?

Every even number can be divided exactly by 2, so there is just one even prime number, which is 2 itself.

Every number can be written as a product of its prime factors, so for example

$$15 = 3 \times 5 \quad \text{and} \quad 12 = 2^2 \times 3.$$

Mathematicians have shown that every number can only be broken down into a product of prime factors in one way, so, if we split 126 as $2 \times 3^2 \times 7$, we don't have to worry that maybe it could also be split so that it has some completely different prime factors.

Is there a pattern for how prime numbers slot into the other numbers? Figure 1.F.1 shows all the prime numbers between 1 and 50, as shaded squares.

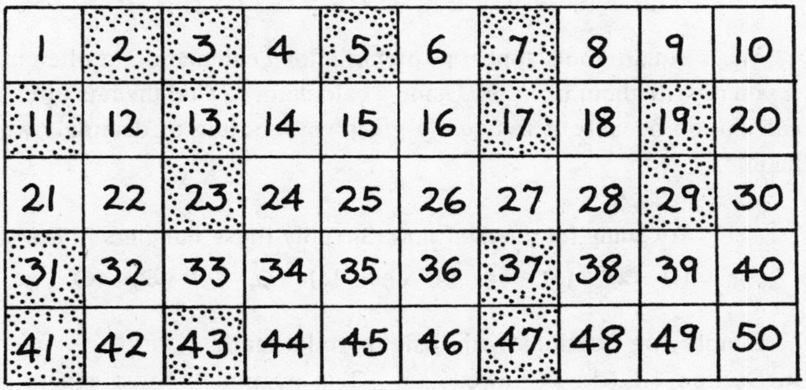

1	2	3	4	5	6	7	8	9	10
11	12	13	14	15	16	17	18	19	20
21	22	23	24	25	26	27	28	29	30
31	32	33	34	35	36	37	38	39	40
41	42	43	44	45	46	47	48	49	50

Figure 1.F.1

It doesn't look as though there *is* a pattern, although we do notice that many of them seem to come in pairs with just one number in between. We also see that, as we go down through the numbers, we are getting more and more possible prime factors for the numbers which we haven't yet reached. Does this mean that after a while we will have collected all the building blocks that we need to make future numbers, so that there will no longer be any new prime numbers?

The answer to this question is that we will never have enough building blocks to make all the possible future numbers. Given any prime number, however large, it is always possible to find at least one larger one.

We can show that this is true in the following neat way.

We start by taking a numerical example, because it is easier then to explain how the argument goes.

Suppose we think that 23 might be the largest prime number. (I have deliberately chosen quite a small number here. It is, in fact, easy to find larger prime numbers than 23, but it will do very nicely to show how the general argument goes.) First, we list all the prime numbers up to 23. (We don't normally include the number 1 in these – 1 is its own special unique case of a number.) Doing this gives us 2, 3, 5, 7, 11, 13, 17, 19 and 23 itself.

Next, we use all these prime numbers to write down a new number. This new number is

$$(2 \times 3 \times 5 \times 7 \times 11 \times 13 \times 17 \times 19 \times 23) + 1.$$

What kind of number is this? None of the prime numbers up to 23 will divide into it exactly, because each of these divisions would leave a remainder of 1. So *either* it is itself a prime number, *or* it has prime factors which are larger than 23.

Either way round, we have shown that there must be at least one prime number which is larger than 23, and we could use this argument in exactly the same way to show that if we start with any prime number N, then there must be at least one prime number larger than N.

This very nice ingenious method is due to Euclid, a mathematician from Ancient Greece.

1.F.(c) A useful application – simplifying square roots

We can often use a number's prime factors to simplify its square root.

For example,

$$\sqrt{12} = \sqrt{2 \times 2 \times 3}, \quad \text{but} \quad \sqrt{2 \times 2} = 2, \quad \text{so we can say} \quad \sqrt{12} = 2\sqrt{3}.$$

Here is another example.

$$\sqrt{72} = \sqrt{2 \times 2 \times 2 \times 3 \times 3} = \sqrt{2 \times 2^2 \times 3^2} = 2 \times 3 \times \sqrt{2} = 6\sqrt{2}.$$

When square roots appear as part of a long calculation, it often makes things much easier if you rewrite them like this. Using a calculator to find them is often not very helpful in mid-calculation because it frequently gives you a string of decimals which is very awkward to handle.

EXERCISE 1.F.3

Try some for yourself now. Simplify these numbers in the same way.

(1) $\sqrt{28}$ (2) $\sqrt{45}$ (3) $\sqrt{50}$ (4) $\sqrt{44}$ (5) $\sqrt{63}$ (6) $\sqrt{40}$

1.F.(d) Simplifying fractions with $\sqrt{}$ signs underneath

In Section 1.E.(d), I showed that $\sqrt{2}$ is irrational. Most square roots are irrational, the exceptions being numbers such as $2 = \sqrt{4}$, $6 = \sqrt{36}$, etc. Numbers such as 4 and 36 are called **perfect squares**.

If we have a number made up of two separate bits, one of which is rational and one of which is irrational, like $3 + \sqrt{5}$, then the combined number will be irrational.

But the matching pair of numbers of $3 + \sqrt{5}$ and $3 - \sqrt{5}$ have two rather nice properties.

We can see the first of these by adding them.

This gives us $(3 + \sqrt{5}) + (3 - \sqrt{5}) = 6$. (We have lost the irrational part.)

Can you see what other good possibility we have?

Multiplying them together also works very nicely.

We get $(3 + \sqrt{5})(3 - \sqrt{5}) = 9 + 3\sqrt{5} - 3\sqrt{5} - 5 = 4$.

Basic algebra: some reminders of how it works

This is another application of the ubiquitous difference of two squares. (We have also used $(\sqrt{5})^2 = 5$.)

Fractions such as $5/(2 - \sqrt{3})$ are particularly unwelcome because they involve dividing by a number which is partly rational and partly irrational. We can get round this problem in the following way.

$$\frac{5}{2 - \sqrt{3}} = \frac{5(2 + \sqrt{3})}{(2 - \sqrt{3})(2 + \sqrt{3})}$$

multiplying top and bottom of the fraction by $2 + \sqrt{3}$. This gives

$$\frac{10 + 5\sqrt{3}}{(2)^2 - (\sqrt{3})^2} = \frac{10 + 5\sqrt{3}}{4 - 3} = 10 + 5\sqrt{3}.$$

We have cleverly got the $\sqrt{}$ signs on the bottom to cancel out, by multiplying the fraction top and bottom by $(2 + \sqrt{3})$. Then we use the fact that $(\sqrt{3})^2 = 3$.

As another example, we will simplify $\dfrac{3 - \sqrt{2}}{\sqrt{5} - \sqrt{2}}$.

The denominator (or underneath number) is particularly unpleasant this time.

Can you see what we could multiply by to get rid of the $\sqrt{}$ signs on the bottom? Look again at the previous example if necessary.

We multiply the top and the bottom by $(\sqrt{5} + \sqrt{2})$ and get:

$$\frac{(3 - \sqrt{2})(\sqrt{5} + \sqrt{2})}{(\sqrt{5} - \sqrt{2})(\sqrt{5} + \sqrt{2})} = \frac{3\sqrt{5} - 2 - \sqrt{10} + 3\sqrt{2}}{5 - 2} = \frac{3\sqrt{5} - 2 - \sqrt{10} + 3\sqrt{2}}{3}.$$

It may help you to recognise references to this process if you know that this process of removing the $\sqrt{}$s on the bottom is called **rationalising the denominator**. Numbers like $\sqrt{2}$ are called **surds**.

We shall use exactly this process in Chapter 10 to simplify complex numbers.

EXERCISE 1.F.4

Try simplifying these three for yourself.

(1) $\dfrac{5}{3 + \sqrt{2}}$ (2) $\dfrac{3 - \sqrt{5}}{3 + \sqrt{5}}$ (3) $\dfrac{3 - 2\sqrt{3}}{5 + 3\sqrt{2}}$

2 Graphs and equations

In this chapter we look at different ways of solving equations. We shall do this both by using the algebra from the first chapter and also by seeing what the solutions we find mean when we look at them graphically.

The chapter is split up into the following sections.

2.A Solving simple equations
(a) Do you need help with this? Self-test 3, (b) Rules for solving simple equations, (c) Solving equations involving fractions, (d) A practical application – rearranging formulas to fit different situations

2.B Introducing graphs
(a) Self-test 4, (b) A reminder on plotting graphs, (c) The midpoint of the straight line joining two points, (d) Steepness or gradient, (e) Sketching straight lines, (f) Finding equations of straight lines, (g) The distance between two points, (h) The relation between the gradients of two perpendicular lines, (i) Dividing a straight line in a given ratio

2.C Relating equations to graphs: simultaneous equations
(a) What do simultaneous equations mean? (b) Methods of solving simultaneous equations

2.D Quadratic equations and the graphs which show them
(a) What do the graphs which show quadratic equations look like? (b) The method of completing the square, (c) Sketching the curves which give quadratic equations, (d) The 'formula' for quadratic equations, (e) Special properties of the roots of quadratic equations, (f) Getting useful information from '$b^2 - 4ac$', (g) A practical example of using quadratic equations, (h) All equations are equal – but are some more equal than others?

2.E Further equations – the Remainder and Factor Theorems
(a) Cubic expressions and equations, (b) Doing long division in algebra, (c) Avoiding long division – the Remainder and Factor Theorems, (d) Three examples of using these theorems, and a red herring

---

2.A   Solving simple equations

2.A.(a)   Do you need help with this? Self-test 3

In the first chapter, we revised the various methods for using the rules of algebra to handle and simplify unknown quantities. We now see how we can use these rules to find information from different kinds of equation. In case you need to be reminded how to solve simple equations, I have put in another self-test here. As before, if you are in any doubt about how much you remember, you should try the test now because it is much easier to go forward happily if any problems are sorted out at the beginning.

Self-test 3

Answer each of the following short questions by finding the value which the letter is standing for in each case.

(1) $x + 7 = 4$

(2) $3y = 27$

(3) $5y = 12$

(4) $2p + 3 = 8$

(5) $2a + 3 = 5a - 2$

(6) $10 - 2b = b + 7$

(7) $3(2x - 1) = 2(2x + 3)$

(8) $\dfrac{x}{4} = \dfrac{3}{5}$

(9) $\dfrac{3x}{8} = \dfrac{5}{9}$

(10) $\dfrac{8}{x} = 2$

(11) $2x + \dfrac{1}{2} = \dfrac{3}{5}$

(12) $\dfrac{5}{y} = \dfrac{3}{7}$

(13) $\dfrac{x + 1}{2} = 5$

(14) $\dfrac{2y + 3}{4} = 5$

(15) $\dfrac{2y + 1}{3} = \dfrac{y + 3}{2}$

(16) $\dfrac{3x}{5} + 3 = x - 5$

(17) $\dfrac{2x}{3} - 3 = \dfrac{x}{2}$

(18) $\dfrac{5}{3a - 2} = 3$

(19) $\dfrac{3}{p + 3} = \dfrac{2}{p + 4}$

(20) $\dfrac{2}{2a + 1} = \dfrac{5}{3a - 2}$.

Save your working on this test because I shall do most of these questions as examples, and you will be able to compare what you did with my solutions. Indeed, you might find as we go through that you can change some to make them right before you look at my version.

If your present answers are right, give yourself one mark each for questions (1) to (10), and two marks each for questions (11) to (20), so the test has a possible total of 30 marks. If you have less than 25 marks, you should work through the next section. Remember that if you are in any doubt about your handling of these equations, it is best to get the difficulties sorted out straight away.

The answers to the test are as follows:

(1) $x = -3$ (2) $y = 9$ (3) $y = \frac{12}{5}$ (4) $p = \frac{5}{2}$ (5) $a = \frac{5}{3}$

(6) $b = 1$ (7) $x = \frac{9}{2}$ (8) $x = \frac{12}{5}$ (9) $x = \frac{40}{27}$ (10) $x = 4$

(11) $x = \frac{1}{20}$ (12) $y = \frac{35}{3}$ (13) $x = 9$ (14) $y = \frac{17}{2}$ (15) $y = 7$

(16) $x = 20$ (17) $x = 18$ (18) $a = \frac{11}{9}$ (19) $p = -6$ (20) $a = -\frac{9}{4}$.

2.A.(b) **Rules for solving simple equations**

Since the two sides of an equation are equal, in general you are safe if you do the same thing to each side. For example, the equation is still true if:

- we add the same amount to each side;
- we subtract the same amount from each side;
- we multiply both sides by the same amount;
- we divide both sides by the same amount, remembering that we must not try to divide by zero. (See the end of Section 1.C.(a) for what happens then.)

We can use these rules to simplify equations to the point where it is easy to see the solution.

Here is an example:

$$3x + 17 = x + 7.$$

Taking 17 from both sides gives

$$3x = x + 7 - 17,$$

so $\qquad 3x = x - 10.$

Taking x from both sides gives

$$2x = -10.$$

Dividing both sides by 2 gives

$$x = -5.$$

We see from this example that adding or subtracting the same amount from each side has the same effect as shifting bits from one side of the equation to the other provided that we change the signs from + to − or − to + as we do so.

We can now check the solution we have found by putting it back into the original equation. If it is correct then the two sides should indeed be equal, so we look at each side in turn. It is helpful to have a shorthand for this, and I shall use LHS to stand for the left-hand side and RHS to stand for the right-hand side.

Here, putting $x = -5$, the LHS $= 3 \times -5 + 17 = 2$, and the RHS $= -5 + 7 = 2$ also.

As further examples, here are the solutions of the first seven questions of Self-test 3.

(1) $x + 7 = 4$ so $x = 4 - 7 = -3$.
(2) $3y = 27$ so $y = \frac{27}{3} = 9$ (dividing both sides by 3).
(3) $5y = 12$ so $y = \frac{12}{5}$ (dividing both sides by 5).
(4) $2p + 3 = 8$ so $2p = 8 - 3 = 5$ and $p = \frac{5}{2}$.
(5) $2a + 3 = 5a - 2$ so $3 + 2 = 5a - 2a = 3a$ and $a = \frac{5}{3}$.
       (Notice, it was easier to rearrange here so that we had a positive number of the unknown amount.)
(6) $10 - 2b = b + 7$ so $10 - 7 = b + 2b = 3b$ and $b = 1$.
(7) $3(2x - 1) = 2(2x + 3)$ so $6x - 3 = 4x + 6$ so $2x = 9$ and $x = \frac{9}{2}$.

EXERCISE 2.A.1
Try these for yourself now. The best method is to do what you comfortably can in your head, without chopping out so many steps that mistakes begin to creep in. Check that all your answers fit their equations.

(1) $x + 8 = 5$ $\qquad$ (2) $5y = 40$ $\qquad$ (3) $2y = 7$
(4) $7 + 2x = 5 - x$ $\qquad$ (5) $4 + 2b = 5b + 9$ $\qquad$ (6) $3(x - 3) = 6$
(7) $3(y - 2) = 2(y - 1)$ $\qquad$ (8) $2(3a - 1) = 3(4a + 3)$ $\qquad$ (9) $3x - 1 = 2(2x - 1) + 3$
(10) $2(p + 2) = 6p - 3(p - 4)$.

2.A.(c) Solving equations involving fractions

I think that the easiest way to solve this kind of equation is to start by getting rid of the fractions. We can do this by multiplying both sides of the equation by a number chosen so that, after cancelling, we have only whole numbers to deal with.

I shall now use some further questions from Self-test 3 as examples of this.

(8) $\dfrac{x}{4} = \dfrac{3}{5}$

Multiplying both sides of the equation by $4 \times 5 = 20$, and cancelling, gives

$$5x = 4 \times 3 = 12 \quad \text{so} \quad x = \dfrac{12}{5}.$$

(11) $2x + \dfrac{1}{2} = \dfrac{3}{5}$

Multiplying both sides by $2 \times 5 = 10$ gives

$$10\left(2x + \tfrac{1}{2}\right) = 10 \times \tfrac{3}{5} \quad \text{so} \quad 20x + 5 = 6 \quad \text{so} \quad x = \tfrac{1}{20}.$$

Notice that I used a bracket to make sure that every separate piece of the original equation got multiplied by 10.

(12) $\dfrac{5}{y} = \dfrac{3}{7}$

Multiplying both sides by $7y$ gives

$$7 \times 5 = 3y \quad \text{so} \quad y = \dfrac{35}{3}.$$

This has the same effect as doing a sort of cross-multiplying of bottoms to tops. It is fine to use this method so long as you only do it for equations with single fractions each side. It wouldn't work for (11), for example.

(14) $\dfrac{2y + 3}{4} = 5$

Multiplying both sides by 4 gives

$$2y + 3 = 20 \quad \text{so} \quad 2y = 17 \quad \text{and} \quad y = \dfrac{17}{2}.$$

(15) $\dfrac{2y + 1}{3} = \dfrac{y + 3}{2}$

Multiplying both sides by $3 \times 2 = 6$ gives

$$2(2y + 1) = 3(y + 3) \quad \text{so} \quad 4y + 2 = 3y + 9 \quad \text{and} \quad y = 7.$$

(17) $\dfrac{2x}{3} - 3 = \dfrac{x}{2}$

Multiplying both sides by $3 \times 2 = 6$ gives

$$6\left(\dfrac{2x}{3} - 3\right) = 6 \times \dfrac{x}{2} \quad \text{so} \quad 4x - 18 = 3x \quad \text{and} \quad x = 18.$$

 It is important to remember that the -3 also gets multiplied by the 6. Again, I've used a bracket to make clear that this is what I must do.

(18) $\dfrac{5}{3a - 2} = 3$

Multiplying both sides by $(3a - 2)$ and cancelling on the left-hand side gives

$5 = 3(3a - 2)$ so $5 = 9a - 6$ so $11 = 9a$ and $a = \dfrac{11}{9}$.

(20) $\dfrac{2}{2a + 1} = \dfrac{5}{3a - 2}$

Multiplying both sides by $(2a + 1)(3a - 2)$, and cancelling, gives

$2(3a - 2) = 5(2a + 1)$ so $6a - 4 = 10a + 5$ so $-9 = 4a$ and $a = -\dfrac{9}{4}$.

My last example involves three fractions. Solve

$$\dfrac{2x + 1}{3} - \dfrac{3x - 2}{4} = \dfrac{x - 1}{6}.$$

What should we multiply by to get rid of the fractions this time?

Did you think of $3 \times 4 \times 6 = 72$? This will do, but we could use the more delicate instrument of 12 since 3, 4 and 6 are all factors of 12.

 This equation has a tricky bit which often leads to mistakes. Can you see what it is? It was mentioned as a warning in Section 1.C.(e). Try the next step yourself before looking at what I've done to see if you can avoid this pitfall.

The whole of $\dfrac{3x - 2}{4}$ is being subtracted from $\dfrac{2x + 1}{3}$.

The line of the fraction is acting in the same way as a bracket, and it is safest to put brackets round each fraction chunk to keep the working clear and the signs correct.

Then, multiplying through by 12, we have

$$12\left(\dfrac{2x + 1}{3}\right) - 12\left(\dfrac{3x - 2}{4}\right) = 12\left(\dfrac{x - 1}{6}\right).$$

Cancelling each fraction in turn, we get

$$4(2x + 1) - 3(3x - 2) = 2(x - 1) \quad \text{so} \quad 8x + 4 - 9x + 6 = 2x - 2$$

(Leaving out the brackets could mean that you would wrongly have a -6 in this last equation.)

So $4 + 6 + 2 = -8x + 9x + 2x$ therefore $12 = 3x$ and $x = 4$.

Checking back, the LHS $= \dfrac{9}{3} - \dfrac{10}{4} = \dfrac{1}{2}$ and the RHS $= \dfrac{3}{6} = \dfrac{1}{2}$.

It is important that we can only get rid of fractions by multiplying if we are dealing with an *equation*. It will not work if we just have an expression such as

$$\frac{x+4}{2} + \frac{x+3}{5}.$$

Here we would have no justification for making this 10 times larger.

The best we can do is to simplify as we did in Section 1.C.(c). Then

$$\frac{x+4}{2} + \frac{x+3}{5} = \frac{5(x+4)}{10} + \frac{2(x+3)}{10} = \frac{5(x+4) + 2(x+3)}{10} = \frac{7x+26}{10}.$$

I've put in quite a lot of detail in these examples so that you can see exactly what's happening. As you get more confident, you'll find you probably don't need to write down all the steps. This is fine, but it's a good idea to check your answers to make sure that they do fit the given equations.

EXERCISE 2.A.2

Try these questions for yourself now.
Solve each of the following equations.

(1) $\dfrac{5x}{3} = 2$

(2) $5 + x = \dfrac{2x}{3}$

(3) $\dfrac{x}{3} - \dfrac{x}{4} = 1$

(4) $\dfrac{y}{3} - \dfrac{3y-7}{5} = \dfrac{y-2}{6}$

(5) $\dfrac{3m-5}{4} - \dfrac{9-2m}{3} = 0$

(6) $\dfrac{x-1}{2} - \dfrac{x-2}{3} = 1$

(7) $\dfrac{p+1}{p-1} = \dfrac{3}{4}$

(8) $\dfrac{2}{y} = \dfrac{3}{y+1}$

(9) $\dfrac{4}{2x+3} = \dfrac{3}{x-2}$

(10) $\dfrac{2x}{x+2} = \dfrac{3x}{x+5} - 1$

(11) $\dfrac{2x+1}{3} + \dfrac{x+5}{2} = \dfrac{3x-1}{7}$

(12) $\dfrac{x+3}{4} - \dfrac{x-1}{5} = \dfrac{2x-1}{10}$

2.A.(d) **A practical application – rearranging formulas to fit different situations**
We can also use the rules for solving equations to rearrange formulas so that they are in a more convenient form to use in changed situations.

EXAMPLE (1) The formula

$$T = 2\pi \sqrt{\frac{l}{g}}$$

gives the period T of a pendulum of length l. The period is the length of time for a complete to-and-fro swing. π is the π of circles, and g stands for the acceleration due to gravity.

If we want to find the length of a pendulum which has a given period, it would be more convenient to have the formula rearranged so that the length l is given in terms of the other quantities. This is sometimes called **changing the subject of the formula** to l. We have

$$T = 2\pi \sqrt{\frac{l}{g}}.$$

Since the two sides of an equation are equal, they must still be equal if we square both of them. Therefore

$$T^2 = 4\pi^2 \left(\frac{l}{g}\right).$$

(Notice that everything must be squared, including the 2π.) So now we have

$$l = \frac{gT^2}{4\pi^2} \quad \text{(multiplying both sides by } g \text{ and dividing by } 4\pi^2\text{)}$$

and this gives us the new formula we wanted.

EXAMPLE (2) For this, I'll take the formula relating the distance u of an object from a lens of focal length f to the distance v of its image from the lens. This is

$$\frac{1}{u} + \frac{1}{v} = \frac{1}{f}.$$

Suppose you want to find the distance of the image from the lens for certain given distances of the object from the lens; you need a formula for v in terms of u and f.

 Students sometimes think that they can go through the equation above turning everything upside down and it will still be true. This is not so!

It is true that $\dfrac{1}{3} + \dfrac{1}{6} = \dfrac{1}{2}$ but $3 + 6 \neq 2$.

Remember, it is only possible to turn both sides of an equation upside down if there is just one fraction on each side. For example we *can* say that

$$\frac{2}{3} = \frac{4}{6} \quad \text{so} \quad \frac{3}{2} = \frac{6}{4}.$$

What do you think we should do to help us rearrange

$$\frac{1}{u} + \frac{1}{v} = \frac{1}{f}$$

if we can't turn it all upside down?

We can get rid of all the fractions by multiplying both sides of the equation by uvf. Then we have

$$uvf\left(\frac{1}{u} + \frac{1}{v}\right) = uvf\left(\frac{1}{f}\right)$$

so, cancelling down,

$$vf + uf = uv.$$

We want a formula for v, so we put everything with a v in it on the same side of the equation. This gives $uf = uv - vf$ so, factorising, $uf = v(u - f)$. Now, dividing both sides by $(u - f)$, we have

$$v = \frac{uf}{u - f}$$

which gives us the new formula for v that we wanted.

We shall use exactly these same techniques for shifting stuff around when we find inverse functions in Section 3.B.(h).

EXERCISE 2.A.3

Try some rearranging of actual formulas for yourself now.

(1) The surface area, S, of a sphere of radius r is given by the formula $V = 4\pi r^2$. Its volume, V, is given by $V = \frac{4}{3}\pi r^3$. Rearrange these two formulas to give (a) the radius in terms of the surface area, and (b) the radius in terms of the volume.

(2) The volume, V, of a closed cylinder of radius r and height h is given by the formula $V = \pi r^2 h$. Its surface area S is given by $S = 2\pi r^2 + 2\pi rh$. Rearrange these two formulas to give (a) the height in terms of the radius and the volume, and (b) the radius in terms of the height and the volume, and (c) the height in terms of the radius and the surface area.

(3) $v^2 = u^2 + 2as$ is a formula which relates the final velocity v to the initial velocity u of a body which travels a distance s with constant acceleration a.
   Find (a) a formula for a in terms of u, v and s, and (b) a formula for u in terms of v, a and s.

(4) If two resistances, R_1 and R_2, in an electric circuit are arranged in parallel then they are equivalent to a single resistance R, with the relation between them being given by the formula

$$\frac{1}{R} = \frac{1}{R_1} + \frac{1}{R_2}.$$

Find a formula which will give the value of R_2 in terms of R and R_1, in the form $R_2 = \ldots$ Use this formula to find out what resistance should be put in parallel with a resistance of $3\,\Omega$ to give an effective resistance of $2\,\Omega$. (Ω is the symbol used for ohms, the unit in which resistance is measured.)

2.B Introducing graphs

It can be very helpful when thinking about how equations work if we can show them graphically, so that we can see what is happening in another way. I shall start by considering equations which can be shown by straight lines. This section is here in case you need any reminders on how to handle straight line graphs. I have put in another self-test here, so that you can see if you need to work through this.

2.B.(a)   Self-test 4

Try answering each of the following questions.

(1)   What are the coordinates of the midpoints of the straight lines joining
      (a) (2, −1) and (8, 5) (b) (−3, 1) and (2, −8)?
(2)   What is the steepness or gradient of the straight lines joining
      (a) (2, 5) to (8, 17) (b) (−1, 3) to (8, −6)?
(3)   What are the gradients of the following straight lines?
      (a) $y = 3x + 4$ (b) $y + 4x = 2$ (c) $2y = x - 4$ (d) $3y + 4x = 0$.
(4)   Find the equations of the following straight lines:
      (a) with gradient 2 and passing through (1, 3)
      (b) with gradient −1 and passing through (2, −1)
      (c) with gradient $\frac{2}{3}$ and passing through (2,4)
      (d) passing through (2, 5) and (8, 10)
      (e) passing through (−4, −2) and (−1, 5).
(5)   What is the distance between each of the two pairs of points given in the first
      question? (Give your answers to two decimal places or d.p.)
(6)   Find the equations of the straight lines which pass through (1, 4) and are
      perpendicular to (a) $y = 2x + 5$ (b) $3y + 2x = 1$ (c) $4y + x = 0$.
(7)   What are the coordinates of the point which divides the straight line joining the
      points (1, 3) and (6, 18) in the ratio $2 : 3$?

Here are the answers which you should have.

Give yourself one mark for each correct part of (1), (2) and (3), and two marks for each
correct part of (4), (5), (6) and (7).

(1)   (a) (5, 2)   (b) $(-\frac{1}{2}, -\frac{7}{2})$
(2)   (a) 2   (b) −1
(3)   (a) 3   (b) −4   (c) $\frac{1}{2}$ (d) $-\frac{4}{3}$
(4)   (a) $y = 2x + 1$ (b) $y + x = 1$ (c) $3y = 2x + 8$ (d) $6y = 5x + 20$ (e) $3y = 7x + 22$
(5)   (a) $\sqrt{72} = 8.49$ to 2 d.p.   (b) $\sqrt{106} = 10.30$ to 2 d.p.
(6)   (a) $2y + x = 9$ (b) $2y = 3x + 5$ (c) $y = 4x$ (7) (3, 9)

As with the other self-tests, if you have less than 25 marks you should certainly work
through this next section. Each particular point is dealt with here in the same order as the test
questions, so it is also possible to go directly to any particular area where you need help.

2.B.(b)   A reminder on plotting graphs

Here is a brief reminder of how graph plotting works. Suppose we have the equation $y = 2x + 3$.
Then, for each value of x that we might choose, there will be a corresponding value of y. The
values of y *depend* on the values of x, and we call y the **dependent variable** and x the
independent variable. We could show some of these pairs of values in a table, as below.

x	−2	−1	0	1	2	3
y	−1			5		9

Fill in the three missing y values yourself.

You should have 1, 3 and 7.

We can write these pairs of values grouped together as (−2, −1), (−1, 1), (0, 3), (1, 5), (2, 7) and (3, 9). The independent value always comes first, and belongs to the variable which is plotted from side to side on a piece of graph paper, using the horizontal axis. The dependent variable is plotted from top to bottom, using the vertical axis. Because it matters what order we write these pairs of numbers in, they are often called **ordered pairs**.

To plot them, we mark out a piece of graph paper with suitable scales to include all of the points which we are interested in.

The point (0, 0) where the axes cross is called the **origin**.

If the point P is (2, 7) then the numbers 2 and 7 are called the **coordinates** of P. 2 is its x-coordinate and 7 is its y-coordinate.

The scales do not have to be equal. Here, it was more convenient to make the scale on the y-axis smaller, and we get a graph which looks like the one in Figure 2.B.1.

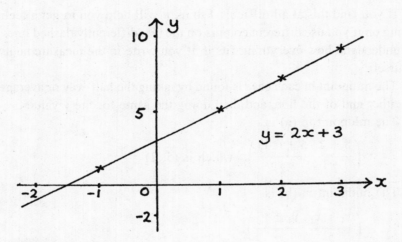

Figure 2.B.1

It is important always to label the axes of your graphs with the letters of the variables you are using, so here I have labelled them x and y.

I have joined the points with a straight line. I've done this because I am thinking that for *every* value of x there is a corresponding value of y, and all these points together make the line. (For example, if $x = \frac{3}{2}$, then $y = 6$ and $(\frac{3}{2}, 6)$ is also a point on the line.)

When you plot a graph accurately on graph paper, you should use a well-sharpened pencil to mark each point with a small cross as accurately as you can. Then, if it is a straight line, draw this through the points in pencil. Of course, for any particular straight line, you only need to find two points, but it is always safer to work out three because this allows you to check your arithmetic if they turn out not to be in line.

2.B.(c)   The midpoint of the straight line joining two points

To show this, I shall draw two diagrams for you. Figure 2.B.2(a) shows the special case of (1)(a) from the self-test, and Figure 2.B.2(b) shows two general points which I shall call (x_1, y_1) and (x_2, y_2).

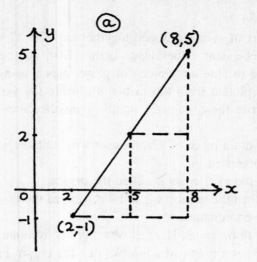
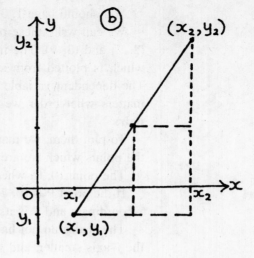

Figure 2.B.2

If you find this at all difficult, I think it will help you to get a feeling of exactly what is going on if you use different colours on the two differently dashed lines. It may also help you to understand how everything fits in if you write in the measurements for the separate bits yourself.

The midpoint in each case is found by taking the half-way or average value of the x values at either end of the line, and then doing the same for the y values.

The midpoint in (a) is

$$\left(\frac{8 + 2}{2}, \frac{5 + (-1)}{2}\right) \quad \text{which is (5, 2).}$$

The midpoint in (b) is

$$\left(\frac{x_1 + x_2}{2}, \frac{y_1 + y_2}{2}\right).$$

We can now use this to find the midpoint of the line joining $(-3, 1)$ and $(2, -8)$. (This was question (1)(b).) We let $(-3, 1)$ be (x_1, y_1) and $(2, -8)$ be (x_2, y_2), which gives us the midpoint as

$$\left(\frac{-3 + 2}{2}, \frac{1 + (-8)}{2}\right) \quad \text{or} \quad \left(\frac{-1}{2}, \frac{-7}{2}\right).$$

It would have worked equally well if we had taken (x_1, y_1) as $(2, -8)$ and (x_2, y_2) as $(-3, 1)$. (Try it and see.)

(If you have any problems with putting together the positive and negative numbers, you should go back to Section 1.A.(e) in the first chapter. It will also help you if you make your own drawings of the pairs of points and their midpoints. Then you can actually see how the numbers are combining together to work.)

The midpoint of the line joining (x_1, y_1) and (x_2, y_2) is given by

$$\left(\frac{x_1 + x_2}{2}, \frac{y_1 + y_2}{2}\right).$$

Find the coordinates of the midpoints of the straight lines joining these pairs of points.

> (1) (−3, 2) and (1, −6)
> (2) (−2, −1) and (3, 4)
> (3) (−1, −5) and (−4, −6)

2.B.(d)   Steepness or gradient

Straight lines have the same steepness or gradient all the way along. This gradient can be measured by the distance moved vertically in the y direction for a unit distance moved from left to right in the x direction. If the line goes uphill from left to right so that this vertical distance is being measured in the positive direction up the y-axis, then the gradient is positive. If the line goes downhill from left to right then the vertical distance and the gradient are negative. We could think of the gradient as telling us the rate of change of y as x changes.

Figure 2.B.3(a) shows the line joining (2, 5) and (8, 17) (question 2(a) from Self-test(4)), and Figure 2.B.3(b) shows the line joining the two points (x_1, y_1) and (x_2, y_2).

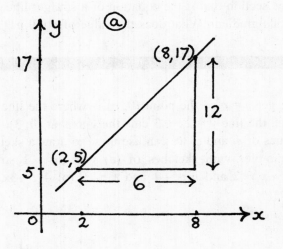 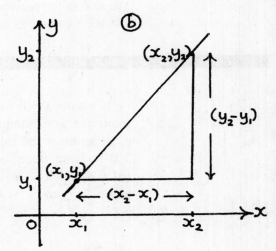

Figure 2.B.3

The gradient in (a) is given by

$$\frac{\text{distance up}}{\text{distance along}} = \frac{12}{6} = 2$$

The gradient in (b) is given by

$$\frac{\text{distance up}}{\text{distance along}} = \frac{y_2 - y_1}{x_2 - x_1}$$

The gradient of a straight line is often written as the single letter m. Using this, we can now write down the following formula:

The gradient, m, of the straight line joining (x_1, y_1) to (x_2, y_2) is given by

$$m = \frac{y_2 - y_1}{x_2 - x_1}.$$

The m gives us the measure of how y is changing relative to x. We have already seen that the line $y = 2x + 3$ has a gradient of 2, with y increasing twice as fast as x. Similarly, the line $y = mx + c$ has a gradient of m. Rewriting the equation of any straight line in this form enables us to read off its gradient. For example, in question (3) of Self-test 4, the line (a), $y = 3x + 4$, has a gradient of 3.

Line (b), $y + 4x = 2$, can be rewritten as $y = -4x + 2$ so m, the gradient, is -4.

Line (c), $2y = x - 4$, can be rewritten as $y = \frac{1}{2}x - 2$ so $m = \frac{1}{2}$.

Line (d), $3y + 4x = 0$, can be rewritten as $y = -\frac{4}{3}x$ so $m = \frac{-4}{3}$.

EXERCISE 2.B.2

Find the gradients of the following straight lines.
(1) $y = 3 - 5x$
(2) $2y = 3x + 7$
(3) $3y + x = 1$
(4) $4y - 5x = 2$

2.B.(e)   Sketching straight lines

We said in the previous section that if the equation of a straight line is written in the form $y = mx + c$ them m is its gradient. What does the value of c tell us?

If we put $x = 0$ we get $y = c$ so the point $(0, c)$ is where the line cuts the y-axis (its y intercept). For example, the line $y = 2x + 3$ cuts the y-axis at $(0, 3)$.

If we know the values of m and c, we can use these to draw a sketch of the line. Figure 2.B.4 shows three examples with sketches of (a) $y = 3x + 1$ so $m = 3$ and $c = 1$, (b) $y + x = 2$ so $y = -x + 2$ and $m = -1$ and $c = 2$, (c) $4y = 3x + 4$ so $y = \frac{3}{4}x + 1$ and $m = \frac{3}{4}$ and $c = 1$.

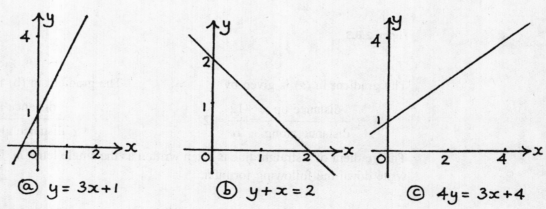

Figure 2.B.4

EXERCISE 2.B.3

Each of the following sketches in Figure 2.B.5 fits one of the lines whose equations are given below. Pair each equation up with its correct sketch.

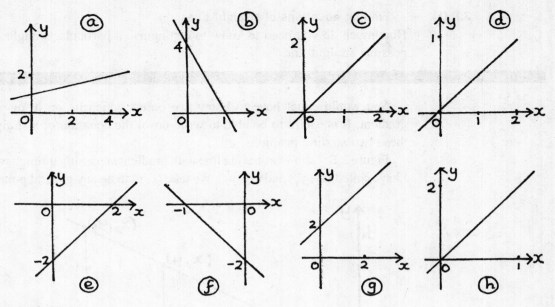

Figure 2.B.5

(1) $y = x$ (2) $y + 4x = 4$ (3) $4y = x + 4$ (4) $y = x - 2$

(5) $y = 2x$ (6) $y = x + 2$ (7) $y = \frac{1}{2}x$ (8) $y + 2x = -2$

SPECIAL CASES

How can we write the equations of the lines shown in the four sketches in Figure 2.B.6?

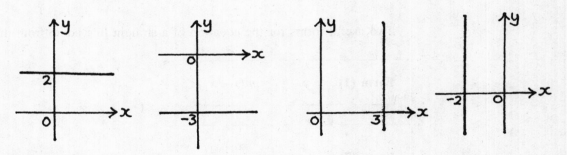

Figure 2.B.6

The first sketch shows a line every point of which has a y-coordinate of 2, so it can be written as $y = 2$. (The value of x can be anything you like, since you can choose any point on this line.) Similarly, the second sketch shows $y = -3$. What do the third and fourth sketches show?

The third sketch shows $x = 3$ and the fourth sketch shows $x = -2$.

The lines in the first two sketches are flat, so their gradient, m, is zero.

We can't write down the gradient for the last two lines because they are infinitely steep and we can't divide by zero.

Finding equations of straight lines

How much do you need to know to distinguish a particular straight line from all the other possible straight lines?

You would either have to know two points which lie on it, or one point on it and its gradient. It is useful to be able to write down the equation of a straight line from either of these two starting positions.

Figure 2.B.7 shows a straight line with gradient m passing through two known points which I have called (x_1, y_1) and (x_2, y_2). We take (x, y) to be any general point on this line.

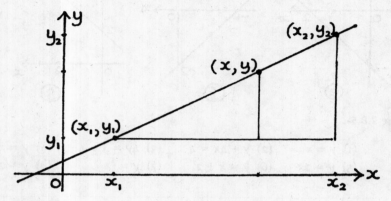

Figure 2.B.7

We have

$$\frac{y_2 - y_1}{x_2 - x_1} = \frac{y - y_1}{x - x_1} = m.$$

Two useful forms for the equation of a straight line come from this.

Form (1)	$y - y_1 = m(x - x_1)$

Form (2)	$\dfrac{y - y_1}{y_2 - y_1} = \dfrac{x - x_1}{x_2 - x_1}$

Form (2) comes from rearranging

$$\frac{y_2 - y_1}{x_2 - x_1} = \frac{y - y_1}{x - x_1}$$

in the same way that we can rearrange

$$\frac{8}{12} = \frac{6}{9} \quad \text{as} \quad \frac{6}{8} = \frac{9}{12}.$$

EXAMPLE (1)   Find the equation of the line with gradient $\frac{1}{2}$ which passes through $(3, 2)$.

Substituting in form (1) gives $y - 2 = \frac{1}{2}(x - 3)$ so $2y = x + 1$.

EXAMPLE (2) Find the equation of the line passing through (3, 2) and (9, 5).

Substituting in form (2) gives

$$\frac{y - 2}{5 - 2} = \frac{x - 3}{9 - 3} \quad \text{so} \quad 6(y - 2) = 3(x - 3) \quad \text{and} \quad 2y = x + 1.$$

Notice that this is the *same* line as we got from the first example. The reason for this is that I have chosen the points (3, 2) and (9, 5) because they fit nicely on Figure 2.B.7 above. If you have found any difficulty with the general rules in the two boxes above, you can feed these numbers in and mark the different numerical distances on the diagram to help you.

For completeness, I also include the equation of a straight line written in the form $y = mx + c$ which we have already used in Section 2.B.(d). This gives us

---

Form (3) $y = mx + c$.

Writing the numerical example of $2y = x + 1$ in the form $y = \frac{1}{2}x + \frac{1}{2}$, we have $m = \frac{1}{2}$ and $c = \frac{1}{2}$.

EXERCISE 2.B.4

Have another go at question (4) from Self-test 4 if you couldn't do it earlier. You should be able to do it now.

2.B.(g) **The distance between two points**

Suppose we need to find the distance D between the two points (x_1, y_1) and (x_2, y_2) as I have shown in Figure 2.B.8(a).

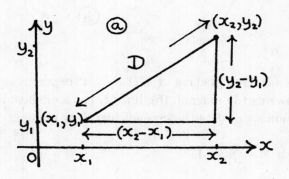
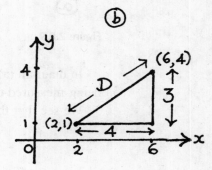

Figure 2.B.8

We use Pythagoras' Theorem which says that:

The distance between the two points (x_1, y_1) and (x_2, y_2) is given by

$$D^2 = (y_2 - y_1)^2 + (x_2 - x_1)^2$$

so $\quad D = \sqrt{(y_2 - y_1)^2 + (x_2 - x_1)^2}.$

In the numerical example of Figure 2.B.8(b), this will give us

$$D = \sqrt{(4-1)^2 + (6-2)^2} = \sqrt{3^2 + 4^2} = \sqrt{25} = 5.$$

(Pythagoras' Theorem is shown to be true in Section 4.A.(b).)

EXERCISE 2.B.5   Try question (5) of Self-test 4 again if you couldn't do it earlier.

2.B.(h)   **The relation between the gradients of two perpendicular lines**
If we know the gradient of a line, surely it must be possible to write down the gradient of a line perpendicular to it. Suppose we start with the line $y = \frac{1}{2}x$. What is the gradient of any line perpendicular to this?

We can see the way in which we can find the answer to this question by looking at Figure 2.B.9 below. Figure 2.B.9(a) shows the special case of line (1) being $y = \frac{1}{2}x$ and Figure 2.B.9(b) shows the general case of line (1) having a gradient of $p/q = m_1$, say. (I have only shown where the two lines cross each other in the two diagrams.)

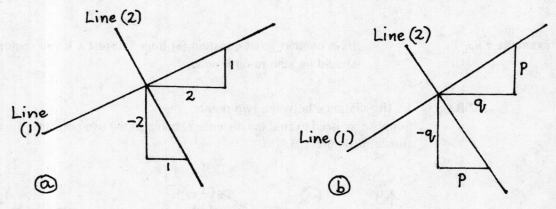

Figure 2.B.9

In diagram (a), line (2) has a gradient of $-2/1 = -2$. (The minus sign is because the 2 is being measured downwards.) In diagram (b), line (2) has a gradient m_2 of $-q/p$.
We see that the gradients of the two perpendicular lines multiplied together give

$$\frac{1}{2} \times -2 = p/q \times -q/p = -1.$$

If two lines with gradients m_1 and m_2 are perpendicular, then $m_1 m_2 = -1$.

EXERCISE 2.B.6   Do question (6) from Self-test 4 again if you couldn't do it earlier.

2.B.(i)   **Dividing a straight line in a given ratio**
In Section 2.B.(c) we found that the midpoint of the line joining (x_1, y_1) and (x_2, y_2) is given by

$$\left(\frac{x_1 + x_2}{2}, \frac{y_1 + y_2}{2} \right).$$

We now look at how to find the coordinates of a point which divides a line in any proportion or ratio.

Figure 2.B.10(a) shows the special case of question (7) of Self-test 4, where we are looking for the point which divides the straight line joining the points (1, 3) and (6, 18) in the ratio $2:3$.

Figure 2.B.10(b) shows the point (x, y) which divides the straight line joining (x_1, y_1) to (x_2, y_2) in the ratio $p:q$. We shall use this to find a general formula.

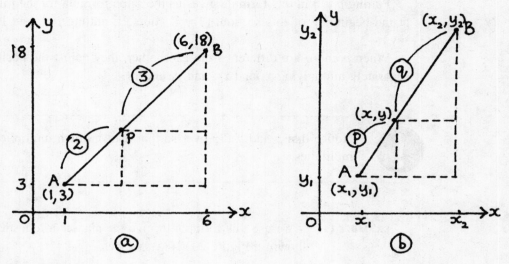

Figure 2.B.10

In (a), the point P is $\frac{2}{5}$ of the way along line AB so each of its x- and y-coordinates is given by moving on from A by $\frac{2}{5}$ of the total change from A to B.

So we could say that P is given by $(1 + \frac{2}{5}(6 - 1), 3 + \frac{2}{5}(18 - 3))$ which is (3, 9).

Similarly, we can see in (b) that P is given by

$$\left(x_1 + \frac{p}{p + q}(x_2 - x_1), y_1 + \frac{p}{p + q}(y_2 - y_1)\right).$$

This looks rather clumsy. Perhaps we can make it nicer if we put the whole of each coordinate over $(p + q)$. Then we get

$$x_1 + \frac{p}{p + q}(x_2 - x_1) = \frac{x_1(p + q) + p(x_2 - x_1)}{p + q}$$

$$= \frac{x_1 q + x_2 p}{p + q}$$

and, similarly, the y coordinate of P is

$$\frac{y_1 q + y_2 p}{p + q}.$$

This gives us a much neater form for the coordinates of P.

The point P which divides (x_1, y_1) and (x_2, y_2) in the ratio $p : q$ is given by

$$\left(\frac{x_1 q + x_2 p}{p + q}, \frac{y_1 q + y_2 p}{p + q} \right).$$

Putting $p = q$ in this formula gives us the same formula for the midpoint that we quoted at the beginning of this section. (Try it yourself, putting $p = q = 1$, and also $p = q = 3$, say.)

When p and q are different from each other, they adjust the position of the point P by separately multiplying x_1 and x_2, and y_1 and y_2.

 Notice that p and q flip over so that it is q which multiplies x_1 and p which multiplies x_2.

EXAMPLE (1) If we use this formula to give the answer to question (7) of Self-test 4, shown in Figure 2.B.10(a), we get

$$P \text{ is given by } \left(\frac{1 \times 3 + 6 \times 2}{2 + 3}, \frac{3 \times 3 + 18 \times 2}{2 + 3} \right) = (3, 9).$$

EXERCISE 2.B.7

Find the coordinates of the points which divide
(1) the line joining $(-1, 2)$ and $(5, 14)$ in the ratio $2 : 1$,
(2) the line joining $(-2, -3)$ and $(6, 9)$ in the ratio $1 : 3$.

2.C   Relating equations to graphs: simultaneous equations

2.C.(a)   What do simultaneous equations mean?

We now have two ways in which we can look at equations. We can find ways of solving them using algebra and we can also see what the meaning of these solutions is graphically.

We will use this double approach first on pairs of equations like the following:

$$\begin{cases} 2x + 3y = 5 & (1) \\ x - 2y = 6 & (2) \end{cases}$$

These are two equations which are true together, so that we have two pieces of information about the two unknowns, x and y.

Such pairs of equations are called **simultaneous equations**.

We could show these as two straight lines on a graph sketch. (See Figure 2.C.1.). To draw this sketch, I have rearranged $2x + 3y = 5$ as $y = -\frac{2}{3}x + \frac{5}{3}$ and $x - 2y = 6$ as $y = \frac{1}{2}x - 3$. Then we can see that there is just one possible pair of values for x and y which fit both equations. These are the coordinates of the point where the two lines cross each other (here this is at about $(4, -1)$).

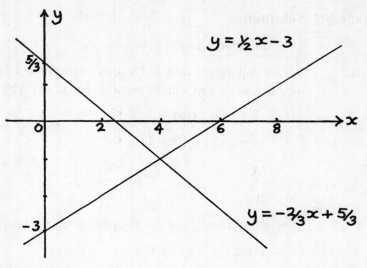

$$y = \tfrac{1}{2}x - 3$$

$$y = -\tfrac{2}{3}x + \tfrac{5}{3}$$

Figure 2.C.1

Does this mean that *any* two equations which give straight lines on a graph will have a solution which can be shown in this way? What might happen which would make this impossible?

If the two lines have the same gradient so that they are parallel there will be no solutions which will fit both. (For example, there is no solution which fits $2x + 3y = 1$ and $2x + 3y = 5$.)

What happens if we have the two equations $x - 2y = 6$ and $2x - 4y = 12$?

We only actually have one piece of information here since the second equation is just the first one multiplied by 2, and so we have the same line drawn on top of itself. Every point on this line fits both equations and we therefore have an infinite number of possible answers.

What happens if we have a third equation which we want to be true at the same time as the original pair?

Geometrically, it is easy to see what happens. Either its line passes through the same crossing point as the other two, in which case it agrees with them or is *consistent* with them, but doesn't add any new information. Or its line does not pass through this crossing point at all. In this case, it is inconsistent with the other two equations, and the three equations cannot be simultaneously true.

2.C.(b) Methods of solving simultaneous equations

Although the graph method makes it easy to *see* what is happening, it can be very difficult to read off an accurate answer. A far simpler way to find this answer is to use algebra. There are various methods which can be used, and the best choice depends on the actual equations and comes with practice. I will show you two different ways of solving the pair of equations which were shown in Figure 2.C.1 above.

METHOD (A) **Substitution.**

From equation (2), we have $x = 2y + 6$.

We are looking for values of x and y so that both the equations are true together, so we can replace the 'x' in equation (1) by $2y + 6$. We then have

$$2(2y + 6) + 3y = 5$$

so $4y + 12 + 3y = 5$

so $7y = -7$

and $y = -1$.

Now, substituting -1 for y in equation (2) we have

$$x + 2 = 6 \quad \text{so} \quad x = 4.$$

Checking in equation (1), LHS = $8 - 3 = 5$ = RHS.

I am again using the shorthand LHS for the left-hand side of an equation, and RHS for its right-hand side.

METHOD (B) **Elimination.**   Returning to the beginning, multiply equation (1) by 2 and equation (2) by 3. Then we have

$$\begin{cases} 4x + 6y = 10 & (3) \\ 3x - 6y = 18 & (4) \end{cases}$$

Adding equations (3) and (4) gives $7x = 28$ so $x = 4$ and, by substitution, $y = -1$ as before.

Method (B) could also have been done by multiplying equation (2) by -2. Then

$$\begin{cases} 2x + 3y = 5 & (3a) \\ -2x + 4y = -12 & (4a) \end{cases}$$

and adding equations (3a) and (4a) gives $7y = -7$ and $y = -1$ as before.

Alternatively, you could multiply equation (2) by $+2$ and subtract. This gives

$$\begin{cases} 2x + 3y = 5 & (3b) \\ 2x - 4y = 12 & (4b) \end{cases}$$

Subtracting equation (4b) from (3b) gives $7y = -7$ and $y = -1$.

HELPFUL HINT   It is easier to make mistakes when subtracting negative quantities, so it is usually better to choose your numbers so that you can get rid of one of the letters by adding.

It is likely, if a real-life situation is being modelled, that we would have to solve more equations in more variables. If there is the same number of equations as there are variables, and provided we don't have a situation similar to the two equations being either parallel or just the same equation, as described above, then we can usually solve them by successive elimination until just one variable is left. Once this is known, the other variables can be found in turn by substituting back into the equations. Such sets of equations, and their more complicated cousins in which the number of variables does not tally with the number of equations, can be dealt with more systematically by using matrix methods which I plan to show you in a later book.

Try solving these two pairs of simultaneous equations yourself before continuing.

$$\text{Qu(1)} \begin{cases} 3x - 2y = 21 & (1) \\ 2x + 5y = -5 & (2) \end{cases} \qquad \text{Qu(2)} \begin{cases} \dfrac{x}{3} - \dfrac{y}{2} + 1 = 0 & (1) \\ 6x + y + 8 = 0 & (2) \end{cases}$$

These are possible routes to solutions.

For Qu(1), multiply equation (1) by 2 and equation (2) by −3. This gives

$$\begin{cases} 6x - 4y = 42 & (3) \\ -6x - 15y = 15 & (4) \end{cases}$$

Equation (3) added to (4) gives $-19y = 57$ so $y = -3$.
Putting $y = -3$ in equation (1) gives $3x + 6 = 21$ so $x = 5$.
Now check in equation (2). LHS $= 10 - 15 = -5 =$ RHS.

In Qu(2), we start by getting rid of the fractions in equation (1) by multiplying by 6. Then we multiply equation (2) by 3. This gives us

$$\begin{cases} 2x - 3y + 6 = 0 & (3) \\ 18x + 3y + 24 = 0 & (4) \end{cases}$$

Adding equations (3) and (4) gives $20x + 30 = 0$ so $20x = -30$ and $x = -\frac{3}{2}$.
Putting this value in (2) gives $-9 + y + 8 = 0$ so $y = 1$.
Checking in (1) gives LHS $= -\frac{1}{2} - \frac{1}{2} + 1 = 0 =$ RHS.

Sometimes we can use these techniques in situations which at first sight don't look very promising. Here is an example.

$$\begin{cases} \dfrac{6}{x} - \dfrac{2}{y} = \dfrac{1}{2} & (1) \\ \dfrac{4}{x} - \dfrac{3}{y} = 0 & (2) \end{cases}$$

Our usual method is to get rid of fractions first. To do this, we would have to multiply equation (1) by $2xy$ and equation (2) by xy. Then we would have:

$$\begin{cases} 12y - 4x = xy & (3) \\ 4y - 3x = 0 & (4) \end{cases}$$

which looks rather unpleasant.

2.C Relating equations to graphs

But if we put

$$X = \frac{1}{x} \quad \text{and} \quad Y = \frac{1}{y}$$

the original equations become

$$\begin{cases} 6X - 2Y = \frac{1}{2} & (3) \\ 4X - 3Y = 0 & (4) \end{cases}$$

Then multiplying equation (3) by 2 and equation (4) by -3 gives

$$\begin{cases} 12X - 4Y = 1 & (5) \\ -12X + 9Y = 0 & (6) \end{cases}$$

Adding these two equations gives $5Y = 1$ so $Y = \frac{1}{5}$ and $y = 5$. Now (2) becomes

$$\frac{4}{x} - \frac{3}{5} = 0 \quad \text{so} \quad \frac{4}{x} = \frac{3}{5} \quad \text{so} \quad 20 = 3x \quad \text{and} \quad x = \frac{20}{3}.$$

Checking in (1) gives LHS $= \frac{18}{20} - \frac{2}{5} = \frac{1}{2} =$ RHS.

EXERCISE 2.C.1

Solve the following pairs of simultaneous equations.

(1) $\begin{cases} 5a - 2b = 68 & (1) \\ 3a + b = 10 & (2) \end{cases}$ (2) $\begin{cases} 5p - 2q = 9 & (1) \\ 2p + 5q = -8 & (2) \end{cases}$

(3) $\begin{cases} \dfrac{x}{8} - y = -\dfrac{5}{2} & (1) \\[2mm] 3x + \dfrac{y}{3} = 13 & (2) \end{cases}$ (4) $\begin{cases} \dfrac{3}{x} + \dfrac{4}{y} = 0 & (1) \\[2mm] \dfrac{2}{x} - \dfrac{2}{y} = 7 & (2) \end{cases}$

2.D Quadratic equations and the graphs which show them

Because quadratic equations have many applications, I have emphasised the particular aspects of them here which will help you later on. For this reason, I haven't started this section with a self-test. You will be able to check through quite quickly to see what is here, doing some of the exercises to be sure you understand. As usual, I am starting from scratch just in case some of you do need this basic help.

2.D.(a) What do the graphs which show quadratic equations look like?

So far, we have only looked at graphs of straight lines. These all have equations of the form $y = mx + c$ where, as we have seen, m tells us the relative change in the y values for a given change in the x values, and c tells us where the line cuts the y-axis.

What effect will it have if we include an x^2 term as well?

We will look at $y = x^2 - x - 6$ as a first example and we start by making a table of some values below.

$y = x^2 - x - 6$

x	-3	-2	-1	0	1	2	3	4
y	6	0	-4		-6		0	

(Fill in the three missing ones yourself.)

You should have -6, -4 and 6.

If we plot these pairs of values we will get the graph I show in Figure 2.D.1.

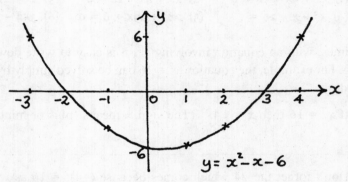

Figure 2.D.1

Clearly, this is not a straight line. Because of the x^2, the y values no longer change evenly in proportion to the x values.

If we join the points smoothly, we get a curve. (We can justify doing this because working out intermediate values gives us more points which lie on the same curve.) This curve that we get is called a **parabola**.

Factorising as we did in Section 1.B.(c), we can also say that

$$x^2 - x - 6 = (x - 3)(x + 2).$$

Now, if $y = 0$ then $x^2 - x - 6 = (x - 3)(x + 2) = 0$.

$x^2 - x - 6 = 0$ is an example of a **quadratic equation**.

We can see from the graph that $y = 0$ when $x = 3$ or $x = -2$. We also see that each of these values for x makes one of the brackets $(x - 3)$ and $(x + 2)$ equal to zero.

If two numbers multiplied together give zero, then one of them must itself be zero. (There is no other number which behaves like this; we saw in Section 1.E.(c) that there are infinitely many pairs of numbers which multiply together to give the number 1, and the same is true for any other number but zero. Zero drops any number it multiplies into a black hole of zero.)

We now use this special property of zero to find solutions for quadratic equations like $x^2 - x - 6 = 0$ directly by algebra, without having to draw a graph.

For example, suppose we have the equation $x^2 - x + 12 = 0$.

Factorising, we get

$$x^2 - x + 12 = (x - 4)(x + 3) = 0.$$

Therefore, either $x - 4 = 0$ giving $x = 4$, or $x + 3 = 0$ giving $x = -3$.

Notice that the signs of the solutions for x are the opposite of the signs in the corresponding brackets.

(If you need help with factorising, go back to Section 1.B.(c) in Chapter 1.)

EXERCISE 2.D.1

Try solving these for yourself.
(1) $x^2 + 9x + 14 = 0$ (2) $x^2 + 4x - 12 = 0$ (3) $x^2 - 11x + 18 = 0$
(4) $x^2 - x - 20 = 0$ (5) $2x^2 + 13x + 6 = 0$ (6) $3x^2 - 7x - 6 = 0$

Sometimes, with an equation involving x^2, it is easy to write down the answers without factorising. For example, the equation $x^2 = 16$ can be solved simply by taking the square root of both sides.

If $x^2 = 16$ then $x = \pm 4$.   (The sign $\pm$ means 'plus or minus'.)

Don't forget the -4 which comes because $(-4)^2 = 16$ as well as $(+4)^2$. Notice, too, that you only need the $\pm$ one side; putting it both sides will just give you the same pair of answers twice over.

So that you can see that we get the same answers, I will also show you how to solve this equation by factorising.

We would say $x^2 - 16 = 0$ so $(x - 4)(x + 4) = 0$ so $x = \pm 4$. This factorising is another example of the difference of two squares.

Now I shall take the slightly more complicated equation of $(x + 2)^2 = 16$ as a second example.

Again, we square-root both sides. This gives us the following working:

$$(x + 2)^2 = 16 \quad \text{so} \quad x + 2 = \pm 4 \quad \text{so} \quad x = 2 \quad \text{or} \quad x = -6.$$

This is quicker than the working needed for factorising which goes

$$(x + 2)^2 = 16 \quad \text{so} \quad x^2 + 4x + 4 = 16 \quad \text{so} \quad x^2 + 4x - 12 = 0$$

$$\text{so} \quad (x - 2)(x + 6) = 0 \quad \text{so} \quad x = 2 \quad \text{or} \quad x = -6.$$

EXERCISE 2.D.2

Solve the following equations yourself.
(1) $x^2 = 9$ (2) $x^2 = \frac{16}{25}$ (3) $(x - 3)^2 = 4$
(4) $(2x - 3)^2 = 25$ (5) $(3x - 2)^2 = 36$

2.D.(b)   The method of completing the square

There is another way of finding the solutions for quadratic equations which is called **completing the square**. This method may seem clumsy at first, but it is worth persevering with it because it has other very useful applications. In particular, we shall use it to handle the equations of circles in Section 4.C.(d), Section 8.F.(a) and Section 10.E.(c). We shall also use it in Section 9.B.(d) to help us with integration, and in Section 2.D.(d) to show how we get the 'formula' for quadratic equations. Finally, we shall need it in the next section to help us to sketch graphs, so altogether we see that it will be worth the effort we put into it.

The following example shows how it works.

Suppose we have the equation $x^2 + 6x - 16 = 0$. Then either we can say

$$x^2 + 6x - 16 = (x + 8)(x - 2) = 0 \quad \text{so} \quad x = -8 \quad \text{or} \quad x = 2,$$

which is the method that we have been using so far, or we can rearrange the equation so that it looks like the equation $(x + 2)^2 = 16$ which we solved in the previous section.

We do this as follows. We have

$$x^2 + 6x - 16 = 0 \quad \text{so} \quad x^2 + 6x = 16.$$

Now we say that $x^2 + 6x$ could have come from $(x + 3)^2$ except that $(x + 3)^2$ gives us an extra term of 9 since $(x + 3)^2 = x^2 + 6x + 9$.

So, taking account of this, we can replace $x^2 + 6x$ by $(x + 3)^2 - 9$. We have written $x^2 + 6x$ by completing a square and then taking off the extra $+9$ which this has given us.

The equation now becomes

$$(x + 3)^2 - 9 = 16 \quad \text{so} \quad (x + 3)^2 = 25.$$

Square-rooting both sides, as we did in the last section, we have $x + 3 = \pm 5$ so $x = 2$ or $x = -8$.

Here is a second example in which I have shown the working more briefly.

I will solve the equation $x^2 - 2x - 3 = 0$ by completing the square.

$$x^2 - 2x - 3 = 0 \quad \text{so} \quad x^2 - 2x = 3 \quad \text{but} \quad x^2 - 2x = (x - 1)^2 - 1$$

Therefore we have

$$(x - 1)^2 - 1 = 3 \quad \text{so} \quad (x - 1)^2 = 4.$$

Square-rooting both sides gives us

$$x - 1 = \pm 2 \quad \text{so} \quad x = 3 \quad \text{or} \quad x = -1.$$

We see from this and the previous example that all we have to do to get the correct bracket for completing the square is to halve the coefficient of x. In the first example, we halved 6 to get 3, and in the second we halved -2 to get -1.

We must also remember to take off the extra bit which we have added on by squaring the bracket. These were $3^2 = 9$ in the first example, and $1^2 = 1$ in the second.

EXERCISE 2.D.3   Now try solving these three quadratic equations yourself by completing the square.

(1) $x^2 + 4x = 21$ (2) $x^2 - 6x + 8 = 0$ (3) $x^2 - 3x - 10 = 0$

2.D Quadratic equations and their graphs

2.D.(c)   **Sketching the curves which give quadratic equations**

The method of completing the square gives us a neat way of sketching the curves connected with quadratic equations. We shall now look at how this is done by taking $y = x^2 - 2x - 3$ as an example.

We can rewrite $x^2 - 2x - 3$ as

$$(x - 1)^2 - 1 - 3 \quad \text{or} \quad (x - 1)^2 - 4.$$

Using this rewritten form of $y = (x - 1)^2 - 4$, what is the smallest possible value which y can take, and what value of x makes this happen?

Since we can't get a negative result when we square a number, the smallest possible value of $(x - 1)^2$ is zero, and this happens when $x = 1$. So the smallest possible value of y is -4 and the lowest point on the curve of $y = x^2 - 2x - 3$ has the coordinates $(1, -4)$.

As the values taken by x move further and further away either side from $x = 1$, the value of y becomes increasingly large since the value of x^2 becomes increasingly large. (It very soon swamps out the effect of the $-2x - 3$.) If you are unsure about this behaviour of y, test it for yourself using your calculator by choosing pairs of values of x symmetrically placed either side of $x = 1$. The further away you go, the larger the value of y becomes.

We can also use two other pieces of information to help us to draw the sketch of $y = x^2 - 2x - 3$.

The first is the value of the y-intercept, that is, the place where the curve crosses the y-axis. For this curve, this is $(0, -3)$, since $y = -3$ when $x = 0$.

The second is the values of x for which $y = 0$. These are called the **roots** of the equation $y = 0$. Here, putting

$$y = (x - 1)^2 - 4 = 0$$

gives

$$(x - 1)^2 = 4$$

so $x - 1 = \pm 2 \quad \text{giving} \quad x = 3 \quad \text{or} \quad x = -1.$

We can now draw a sketch of the parabola $y = x^2 - 2x - 3$ using all the information which we have found above. I show this in Figure 2.D.2.

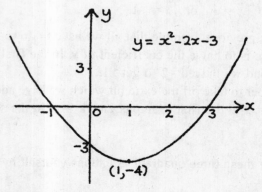

Figure 2.D.2

The roots are the values of x which are the solutions of the equation $x^2 - 2x - 3 = 0$. It is very important to remember to write this *as* an equation by including the '$= 0$'. The expression $x^2 - 2x - 3$ on its own can have infinitely many values, some of which are shown by the y values in the graph sketch of $y = x^2 - 2x - 3$ shown above.

Notice that all the important information is clearly labelled on the graph.

What will happen if we have to sketch a graph which starts off with $-x^2$?

For instance, what happens if we sketch $y = -x^2 + 2x + 3$ (the same as the one which we have just done, but with all the signs changed? Try doing this for yourself before reading on.)

The whole curve is simply turned upside down, because each positive value for y is changed to the corresponding negative value, and vice versa.

The roots of $x = -1$ and $x = 3$ are still the same, but now the *highest* point is given by $(1, +4)$, and the y-intercept is $(0, 3)$.

If you weren't able to sketch it before reading this, sketch it on top of my graph of $y = x^2 - 2x - 3$ now.

Whenever we have an equation for y which starts with a negative quantity of x^2, we will get an upside-down or inverted U-shaped curve like this one. (The negative changes the smiley parabola into a sad parabola.)

EXERCISE 2.D.4

Try using the same techniques to sketch the following two pairs of graphs.
(1) (a) $y = x^2 - 4x + 3$ (b) $y = -x^2 + 4x - 3$
(2) (a) $y = x^2 + 2x - 8$ (b) $y = -x^2 - 2x + 8$

(The general rules for sketching curves like this are given at the end of Section 2.D.(f) as they also involve results which come from the formula for solving quadratic equations.)

2.D.(d)   The 'formula' for quadratic equations

So far, all the quadratic equations we have looked at have turned out to have roots which are either whole numbers or fractions. Surely this will not always be true? The square roots of most numbers cannot be written as exact fractions or whole numbers. (In Section 1.E.(d) we showed that $\sqrt{2}$ can't be written in this way.)

Also, how can we tell if the curve of a particular equation never actually crosses the x-axis without drawing it?

It will be much easier for us to answer these questions if we can find a general rule for solving quadratic equations. Then we shall be able to see exactly what makes particular problems arise.

We start with $ax^2 + bx + c = 0$ with a, b and c standing for numbers and $a \neq 0$.

We want to find a formula from this which will give us a rule for finding the possible values of x if we know the values of the numbers a, b and c.

First, we divide through by a as it is easier then to complete the square. Then we have

$$x^2 + \frac{b}{a}x + \frac{c}{a} = 0 \quad \text{so} \quad x^2 + \frac{b}{a}x = -\frac{c}{a}.$$

Now we complete the square, halving the coefficient of x, and taking off the square of this amount just like we did in the numerical examples in Section 2.D.(b). This gives us

$$\left(x + \frac{b}{2a}\right)^2 - \left(\frac{b}{2a}\right)^2 = -\frac{c}{a} \quad \text{so} \quad \left(x + \frac{b}{2a}\right)^2 = \left(\frac{b}{2a}\right)^2 - \frac{c}{a}$$

so
$$\left(x + \frac{b}{2a}\right)^2 = \frac{b^2}{4a^2} - \frac{c}{a} \quad \text{so} \quad \left(x + \frac{b}{2a}\right)^2 = \frac{b^2 - 4ac}{4a^2}.$$

Now, taking the square root of both sides, we get

$$x + \frac{b}{2a} = \pm \sqrt{\frac{b^2 - 4ac}{4a^2}} = \frac{\pm\sqrt{b^2 - 4ac}}{2a}$$

so
$$x = -\frac{b}{2a} \pm \frac{\sqrt{b^2 - 4ac}}{2a}.$$

Finally, we get

$$x = \frac{-b \pm \sqrt{b^2 - 4ac}}{2a}.$$

This is the so-called 'formula' for solving quadratic equations.

If you have seen this before, you may have realised that the right-hand side of the above working was growing more and more familiar.

All we have to do to make use of it is to substitute the values of a, b and c from the particular equation that we want to solve.

For example, to solve $2x^2 - 5x + 1 = 0$ we put $a = 2$, $b = -5$ and $c = 1$. Then

$$x = \frac{+5 \pm \sqrt{25 - 4(2)(1)}}{4} = \frac{5 \pm \sqrt{17}}{4} = 2.28 \text{ or } 0.22 \text{ to 2 d.p.}$$

Because $\sqrt{17}$ is irrational, that is, it has no exact square root, it would not have been possible to factorise this equation in any simple way.

Even equations which can be solved by factorising are often more easily dealt with by using the formula, if the factorisation is at all difficult.

For example, the equation $12x^2 + 19x - 18 = 0$ *will* factorise into brackets with whole number coefficients. We know that this is possible from working out the value of '$b^2 - 4ac$'. Here $b = 19$, $a = 12$ and $c = -18$, so $b^2 - 4ac = 1225 = (35)^2$. (The number 1225 is called a **perfect square** because it has an exact square root.)

In fact, $12x^2 + 19x - 18 = (4x + 9)(3x - 2)$ but these brackets may not spring immediately into your head. Substitution into the formula gives

$$x = \frac{-19 \pm 35}{24} = -\frac{9}{4} \text{ or } \frac{2}{3}$$

just as we would obtain from the factorised form. So the equation $12x^2 + 19x - 18 = 0$ has the two roots or solutions of $-\frac{9}{4}$ and $\frac{2}{3}$.

Use the formula to solve the following quadratic equations. (If the answers are not exact fractions, give them correct to 2 d.p.)

(1) $x^2 + 10x + 16 = 0$ (2) $x^2 - 2x - 8 = 0$ (3) $2x^2 + 5x - 3 = 0$

(4) $x^2 + 4x + 2 = 0$ (5) $3x^2 - x - 2 = 0$ (6) $2x^2 - x - 7 = 0$

You should try this now as you will need your answers for the next section.

(a) For each equation which you have just solved, find what you get if you add the two solutions or roots together. Can you connect this answer with the a, b and c of the particular equation in any way?

(b) Now find what you get if you multiply each of the pairs of roots together. Then again see if you can connect the results with the a, b and c of the particular equation. If your answers aren't exact fractions or whole numbers, you will find that the more decimal places you take, the closer you will get to a nice result, because you will be lessening the rounding errors.

(c) Now for the tricky bit. Can you see why you are getting these neat results from adding and multiplying the pairs of roots even when the roots themselves are *not* simple numbers? Try looking at how your working went when you used the formula to get your two answers.

2.D.(e)   Special properties of the roots of quadratic equations

This section is based on your answers to the thinking point at the end of the previous section.

When you add the pairs of roots for each of the equations in Exercise 2.D.5, you should find each time that you get the answer of $-b/a$ for that equation.

For example, in question (3), the two roots are $\frac{1}{2}$ and -3, and $a = 2$, $b = 5$ and $c = -3$. Adding the roots gives $\frac{1}{2} - 3 = -2\frac{1}{2} = -\frac{5}{2}$.

We can see exactly why this should be so by looking at the roots of the equation $ax^2 + bx + c = 0$. These are

$$\frac{-b + \sqrt{b^2 - 4ac}}{2a} \quad \text{and} \quad \frac{-b - \sqrt{b^2 - 4ac}}{2a}.$$

Splitting each of them into two parts and adding them gives

$$\left(\frac{-b}{2a} + \frac{\sqrt{b^2 - 4ac}}{2a}\right) + \left(\frac{-b}{2a} - \frac{\sqrt{b^2 - 4ac}}{2a}\right) = \frac{-b}{2a} + \frac{-b}{2a} = -\frac{b}{a}.$$

The two complicated bits have cancelled out.

When you multiply the pairs of roots for each of the equations in Exercise 2.D.5, you should find that you get the answer of $+c/a$ for that equation. (For example, in question (3) you get $\frac{1}{2} \times -3 = -\frac{3}{2}$. The minus agrees with c being negative here.)

We can see why this happens if we multiply the two roots of $ax^2 + bx + c = 0$ together, though it's a bit more complicated this time. We have

$$\left(\frac{-b}{2a} + \frac{\sqrt{b^2 - 4ac}}{2a}\right)\left(\frac{-b}{2a} - \frac{\sqrt{b^2 - 4ac}}{2a}\right) = \left(\frac{-b}{2a}\right)^2 - \left(\frac{\sqrt{b^2 - 4ac}}{2a}\right)^2.$$

The two middle bits have cancelled out, because of the + and − signs. This is the difference of two squares of Section 1.B.(b) again. Tidying up gives us

$$\frac{b^2}{4a^2} - \frac{(b^2 - 4ac)}{4a^2} = \frac{4ac}{4a^2} = \frac{c}{a}.$$

When we either add or multiply any pair of roots, we get rid of the square root of the number $b^2 - 4ac$. We therefore also get rid of any complications which might arise from trying to find this square root.

Two special properties of the quadratic equation $ax^2 + bx + c = 0$

- Adding its two roots together gives $-b/a$.
  This is called **the sum of the roots**.
- Multiplying its two roots together gives c/a.
  This is called **the product of the roots**.

We shall also get this same pair of results by following a different route in Section 2.D.(h).

EXERCISE 2.D.6

This is an exercise of mixed questions on solving quadratic equations. If the answers to any question are not exact, give them correct to three decimal places.

(1) Solve these in whatever way seems suitable.

(a) $2x^2 + 7x + 3 = 0$ (b) $3x^2 + 4x + 1 = 0$ (c) $2x^2 + x - 4 = 0$
(d) $6x^2 - 7x + 2 = 0$ (e) $x^2 - 5x + 3 = 0$ (f) $6x^2 + 5x - 6 = 0$
(g) $x^2 - 81 = 0$ (h) $6x^2 - x + 12 = 0$ (i) $x^2 - 2 = 0$ (j) $x^2 - 5x = 0$

Check that the sum and product of the roots of each equation *do* fit the results given in the box above.

(2) Solve the following equations.

(a) $\dfrac{2x-3}{2x+3} = \dfrac{x-1}{x+1}$ (b) $\dfrac{2}{y+1} + \dfrac{1}{y-1} = \dfrac{3}{y}$ (c) $\dfrac{2x+4}{x+1} = \dfrac{x-8}{2x-1}$

2.D.(f)   Getting useful information from '$b^2 - 4ac$'

From the quadratic equations which we have solved and the work of the last section, we have seen that it is having to find the square root of $b^2 - 4ac$ which can make us sometimes get complicated answers.

The $b^2 - 4ac$ in the quadratic equation formula works as a kind of litmus paper or probe to tell us what kind of roots any particular equation will have.

We look now at the different possibilities.

(1)   If $b^2 - 4ac$ is positive then the equation will have two distinct roots. Geometrically, the curve of $y = ax^2 + bx + c$ cuts the x-axis in two separate places.

     If $b^2 - 4ac$ has an exact square root, then the two roots will be either whole numbers or fractions. This means that it must be possible to solve the equation by factorising and so gives a good quick test for this.

(2) If $b^2 - 4ac = 0$ then the two roots will come together as one root. For example,

if we have $x^2 - 6x + 9 = 0$ then $x = \dfrac{6 \pm \sqrt{36 - 36}}{2} = 3$.

Also

$$x^2 - 6x + 9 = (x - 3)(x - 3) = (x - 3)^2.$$

It is as though we have the root of 3 repeated twice. Geometrically, this is because $y = (x - 3)^2$ just touches the x-axis when $x = 3$. (See Figure 2.D.3.) The usual two roots have met up together to make just one root.

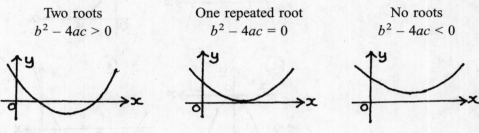

Figure 2.D.3

We shall use this property geometrically in Section 4.C.(e).

(3) If $b^2 - 4ac$ is negative, we cannot find a square root for it. The curve of the equation does not cut the x-axis at all. It is either completely above or completely below it so there are no values of x on the x-axis which fit the equation $y = ax^2 + bx + c = 0$.

For some purposes, this lack of roots is not very satisfactory, and we cleverly get round it in Chapter 10 by inventing a new sort of number.

A summary of everything that we now know which will help us to sketch curves of the form $y = ax^2 + bx + c$

- If a is positive, the curve is U-shaped.
  If a is negative, the curve is an upside-down U.

- The value of c tells us the y-intercept.
  The curve crosses the y-axis at $(0, c)$.

- We can factorise (or use the formula) to find whether and where the curve cuts the x-axis.
  If $b^2 - 4ac$ is negative, the curve does not cut the x-axis at all.

- We can complete the square to find where the least value of the curve is (or the greatest value, if it is an inverted U-shape). We shall see in Section 8.E.(b) that this can also be found by using calculus.
  If the curve *does* cut the x-axis, substituting the midway value of x between the cuts gives the least value of y (or the greatest value of y if the curve has an inverted U-shape).

Each of the six sketches shown below in Figure 2.D.4 comes from one of the ten curves whose equations are given. Fit each sketch to its correct equation, and then draw your own sketches for the four equations which are left over.

(1) $y = x^2 + 6x + 5$ (2) $y = x^2 - 6x + 5$ (3) $y = x^2$ (4) $y = -x^2$

(5) $y = x^2 - 4x + 4$ (6) $y = 4x - x^2 - 4$ (7) $y = x^2 - 8x + 16$

(8) $y = x^2 + 1$ (9) $y = x^2 - 3x - 4$ (10) $y = 3x + 4 - x^2$

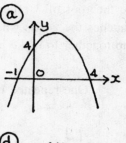
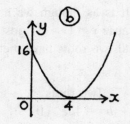
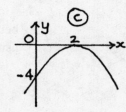

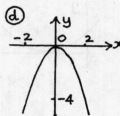
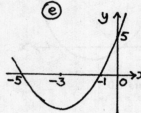
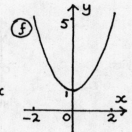

Figure 2.D.4

2.D.(g) A practical example of using quadratic equations

$s = ut - \frac{1}{2}gt^2$ is a formula which gives the distance s in metres travelled by a ball from the thrower's hands if it is thrown upwards with an initial velocity of u m s^{-1} (metres per second), after a time of t seconds. g is the acceleration due to gravity and is 9.8 m s^{-2} (metres per second per second) to 1 d.p.

We shall now use this formula to answer the following questions.

(1) If a rubber ball is thrown upwards at 14 m s^{-1}, how high has it gone after 1 second?

(2) How long does it take for the ball to reach a height of (a) 5 m, (b) 10 m, (c) 15 m from the thrower's hands?

(3) Using the information you have now found, draw a sketch showing the relation between s and t.

(4) How long does the ball take to fall back into the thrower's hands, which we will assume are ready and waiting?

(5) Where is the ball after 2.9 seconds?

(1) Using $s = ut - \frac{1}{2}gt^2$, we have $u = 14$, $t = 1$ and $g = 9.8$ so $s = 14 - (9.8/2) = 9.1$; the ball has reached a height of 9.1 metres after 1 second.

(2) (a) Putting $s = 5$, we have $5 = 14t - (9.8/2)t^2$ so $4.9t^2 - 14t + 5 = 0$. Solving this using the formula for quadratic equations gives

$$t = \frac{14 \pm \sqrt{196 - 98}}{9.8} = \frac{14 \pm \sqrt{98}}{9.8}$$

which gives $t =$ either 2.4 or 0.4 to 1 d.p.

(b) Putting $s = 10$ gives $10 = 14t - 4.9t^2$ so $4.9t^2 - 14t + 10 = 0$.

Again using the formula, we get

$$t = \frac{14 \pm \sqrt{196 - 196}}{9.8} = 1.4 \text{ to 1 d.p. or } 1.43 \text{ to 2 d.p.}$$

(c) Putting $s = 15$ gives $15 = 14t - 4.9t^2$ so $4.9t^2 - 14t + 15 = 0$.

Using the formula gives

$$t = \frac{14 \pm \sqrt{196 - 294}}{9.8} = \frac{14 \pm \sqrt{-98}}{9.8}.$$

Because we have a negative square root here, it is impossible to find any value of t on the horizontal t axis which fits this equation.

What is the physical meaning of the three answers we have found for question (2)?

• Why are there two possible times to reach a height of 5 metres?
• Why is there just one time to reach a height of 10 metres?
• Why couldn't we find a time to reach a height of 15 metres?

Try answering each of these questions yourself.

The ball reaches a height of 5 metres from the thrower's hands both on the way up and on the way down, so there are two possible answers for the time.

The single answer for the time taken to reach 10 metres means that this was the highest point the ball reached. So it never reached a height of 15 metres and it was impossible to find a time for this.

The mathematics of the quadratic equations has exactly corresponded back to the physical situation.

(3)   With this information we can now draw a sketch of the relation between s and t. I show this below in Figure 2.D.5.

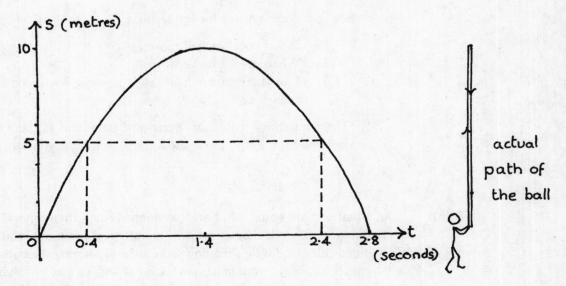

Figure 2.D.5

Notice that the graph sketch shows the height of the ball after time t. The little sketch at the side shows the actual path of the ball which is straight up and then straight back down.

(4)   Because the curves giving quadratic equations are symmetrical, if we know that the time taken for the ball to reach its highest point is 1.4 seconds, then the time taken for it to fall back into the thrower's hands will be 2.8 seconds.

(5)   Clearly, from the sketch, after 2.9 seconds the ball should have been safely caught. If we put $t = 2.9$ in $s = ut - \frac{1}{2}gt^2$, we get $s = -0.6$ to 1 d.p. This describes what has happened to the ball if the thrower completely misses it and it just carries on downwards. It will be 0.6 metres below the thrower's hands after 2.9 seconds.

Now see if you can answer this question.

What is the meaning of the quadratic equation $0 = ut - \frac{1}{2}gt^2$?

Solving this equation tells us when the ball is in the thrower's hands, that is, when $s = 0$. Factorising, we have

$$0 = ut - \tfrac{1}{2}gt^2 = t(u - \tfrac{1}{2}gt)$$

so either $t = 0$ (the ball is just about to be thrown up) or $u - \frac{1}{2}gt = 0$ so $t = 2u/g$ which is the time taken for the ball to return to the thrower's hands. When $u = 14$, $t = 2.86 = 2 \times 1.43$ seconds. Strictly speaking, the time of 2.8 seconds is an underestimate.

The above working has ignored air resistance. It describes the motion of a rubber ball quite well but would be of no use to describe the motion of a feather. We are using the formula $s = ut - \frac{1}{2}gt^2$ as a mathematical model we can work with and which approximates quite well to the actual physical situation.

THINKING POINT

If the ball is thrown up at $14\,\text{m s}^{-1}$ we know that $s = 14t - \frac{1}{2}gt^2$. Therefore we know the ball's height at any time during the throw. Surely, if we know this, we ought to be able to find out how fast it is moving at any particular time?

See if you can answer these questions.

(1)   When does the ball move fastest?
(2)   When does it move slowest?
(3)   Can you estimate how fast it is going one second after it has been thrown up?

(These questions will be answered in Section 8.A.(a) later on but it would be very good for you to think about the possibilities yourself here.)

2.D.(h)   **All equations are equal – but are some more equal than others?**
In the last section, we looked at some of the physical meanings which equations can hold. We will end this chapter by spending some time examining the equations themselves.

Do equations always work in the same kind of way, so that by solving them we find some specific answers which fit these particular circumstances?

Or, if not, what else can happen?

The following examples all look straightforward at first sight, but try solving each of them yourself. Things are not always quite as they seem.

(1) $x^2 + 5x + 6 = x^2 + x - 2$ (2) $x^2 - x - 6 = x^2 + 3x - 4$

(3) $2x^2 - 8x + 8 = x^2 - 4x + 5$ (4) $x^2 - 6x + 8 = (x - 2)(x - 4)$.

It will help you to see what is happening if you also sketch the graph of each side of each equation. Then you can see whether, and if so where, these graphs cross.

You should try doing this for yourself before looking at my solutions.

(1) $x^2 + 5x + 6 = x^2 + x - 2$ so $4x = -8$ and $x = -2$.

To show this single solution graphically, we sketch, using the same axes,

(a) $y = x^2 + 5x + 6 = (x + 3)(x + 2)$ and (b) $y = x^2 + x - 2 = (x + 2)(x - 1)$.

The sketch in Figure 2.D.6 shows that $y = 0$ for both (a) and (b) when $x = -2$.

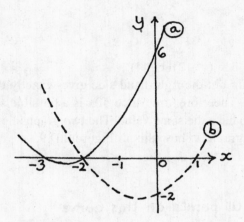

Figure 2.D.6

(2) $x^2 - x - 6 = x^2 + 3x - 4$ so $-2 = 4x$ and $x = -\frac{1}{2}$.

The sketch in Figure 2.D.7 of (a) $y = x^2 - x - 6 = (x - 3)(x + 2)$ and

(b) $y = x^2 + 3x - 4 = (x - 1)(x + 4)$ shows that there is the single solution of

$x = -\frac{1}{2}$ which gives equal y values for both (a) and (b).

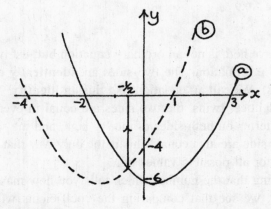

Figure 2.D.7

2.D Quadratic equations and their graphs

(3) $2x^2 - 8x + 8 = x^2 - 4x + 5$ so $x^2 - 4x + 3 = 0$ so $(x-3)(x-1) = 0$ and $x = 3$ or $x = 1$.
The sketch in Figure 2.D.8 of (a) $y = 2x^2 - 8x + 8 = 2(x^2 - 4x + 4) = 2(x-2)^2$ and
(b) $y = x^2 - 4x + 5 = (x-2)^2 - 4 + 5 = (x-2)^2 + 1$ shows the two possible values of x
which make the y values of (a) and (b) the same. These are $x = 1$ and $x = 3$.

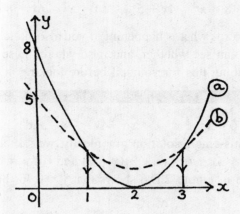

Figure 2.D.8

(4) $x^2 - 6x + 8 = (x - 2)(x - 4)$
Multiplying out the right-hand side gives exactly the same expression as the left-
hand side. Therefore, *any* value of x is a possible solution since it will make each
side of (4) have the same value. The two graphs lie on top of each other – they are
the *same* graph. I show this in Figure 2.D.9.

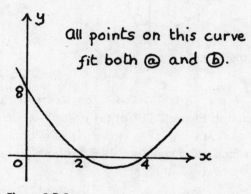

All points on this curve
fit both ⓐ and ⓑ.

Figure 2.D.9

What we have here is not an ordinary equation but just two different ways of writing the
same piece of information. The two sides are **identically equal** to each other (rather like
identical twins). We call an equation like this an **identity**.

Just like identical twins, the two sides are equal in every detail, so there are the same
number of x^2 terms on both sides of the '=' sign, and the same number of xs. The number
terms on each side are also equal. This is the only way that the two sides can remain equal
to each other for all possible values of x.

Remembering that the number which tells you how many you have of x^2, say, is called
its coefficient, we see that comparing the coefficients will give us three equal pairs of
values.

> If two expressions are identically equal to each other, the coefficients of each separate power of x on each side of the '=' sign must be the same as each other.

This rule gives us a very neat method of finding out how to write expressions in different ways. We'll use it in the next section to factorise expressions which involve terms with x^3, and then later on in Section 10.D.(c) to find complex roots of equations. Also, we'll see in Section 6.E.(d) that it will make finding some kinds of partial fraction much easier.

I'll now finish this section by showing you how to use this rule to find the special properties of the sum and product of the roots of quadratic equations. We have already found these properties in Section 2.D.(e) by working directly from the roots themselves, but this new method will avoid the tricky algebra which we had to use there.

Suppose that the equation $ax^2 + bx + c = 0$ has the two solutions $x = \alpha$ and $x = \beta$ so that its two roots are α and β. (α and β are the Greek letters for a and b and are called alpha and beta. They are very often used to stand for the roots of quadratic equations.)

We start by dividing both sides of the equation $ax^2 + bx + c = 0$ by a. This gives us

$$x^2 + \frac{b}{a}x + \frac{c}{a} = 0.$$

(We do this division because it will simplify the working which follows.)

Now, $(x - \alpha)(x - \beta) = 0$ is just another way of writing

$$x^2 + (b/a)x + c/a = 0.$$

Also,

$$(x - \alpha)(x - \beta) = x^2 - \alpha x - \beta x + \alpha\beta = x^2 - (\alpha + \beta)x + \alpha\beta$$

so $y = x^2 - (\alpha + \beta)x + \alpha\beta$ gives exactly the same curve as $y = x^2 + (b/a)x + c/a$.

(The earlier division by a means that we now have two curves which are identical for every value of x. You can see exactly how this works if you take the numerical example of $2x^2 - 6x + 4 = 0$ which has the two roots $x = 1$ and $x = 2$.)

We already have matching terms of x^2 on both sides.

Comparing the coefficients of x (which must also be equal), we have

$$-(\alpha + \beta) = \frac{b}{a} \quad \text{so} \quad \alpha + \beta = -\frac{b}{a}.$$

Also, comparing the two number terms, we have $\alpha\beta = c/a$.

This gives us the following two rules.

> If we have the quadratic equation $ax^2 + bx + c = 0$,
> then the sum of its roots $= -b/a$ and the product of its roots $= c/a$.

A note on writing identities

The special form of equality called an identity in maths, where the two sides of the expression remain equal for all possible values of x, is sometimes written using the triple equals sign '$\equiv$'. You can think of the sign '$\equiv$' as meaning 'is the same as' or 'is equivalent to'. Mathematicians often speak of the two sides as being identically equal to each other.

2.E.(a)   Cubic expressions and equations

How could we set about solving an equation like $2x^3 - 5x^2 - 6x + 9 = 0$?

This is called a cubic equation since the highest power of x is x^3. There isn't a very simple formula for solving cubic equations, so we see if we can successfully guess one answer to start us off. (The following method will only work for equations which have exact solutions which are also not too hard to guess; if this is not the case, other methods involving closer and closer approximations to the true solutions would have to be used.)

Here, if we try putting $x = 1$, we get $2x^3 - 5x^2 - 6x + 9 = 2 - 5 - 6 + 9 = 0$, so we immediately have one solution of our equation.

It will make the working much shorter and easier to follow if we now introduce a shorthand way of describing $2x^3 - 5x^2 - 6x + 9$. We will call it $f(x)$, with the name $f(x)$ meaning this particular collection of terms whose value changes as x changes.

This gives us a neat way of showing particular values of $f(x)$ associated with their corresponding values of x.

For example, if $x = 2$ we have $f(2) = 2(2^3) - 5(2^2) - 6(2) + 9 = -7$ so $f(2) = -7$.

(In fact, $f(x)$ is what is called a function of x. In Section 3.B, we shall look at what functions are in more detail.)

We can now say that $f(x) = 2x^3 - 5x^2 - 6x + 9$ and we know that $f(1) = 0$.

Since $x = 1$ is a solution or root of this equation, it seems reasonable to think that $(x - 1)$ must be a factor of $f(x)$, just as we found with quadratic equations.

(We will show that it *is* all right to say this in Section 2.E.(c).)

If $(x - 1)$ is a factor, we can say that

$$f(x) = 2x^3 - 5x^2 - 6x + 9 = (x - 1) \text{ (something)}.$$

Since the right-hand side is just another way of writing the left-hand side, the two sides must be exactly the same as each other. Therefore we must have the same matching quantities of x^3, x^2, x and numbers on both sides. This means that it is easy to match up the two end terms in the right-hand bracket. It is just the middle one which will take a bit more thought. We can say

$$2x^3 - 5x^2 - 6x + 9 = (x - 1)(2x^2 + px - 9)$$

where p is standing for the number which we haven't found yet.

Now, matching the terms in x^2, we have $-5x^2 = -2x^2 + px^2$, picking out the ways in which we can get x^2 on the right-hand side.

Therefore, $-5 = -2 + p$ so $p = -3$.

We can check that this is correct by matching the terms in x.

Doing this gives us $-6x = -px - 9x$ which does indeed work for $p = -3$.

So now we can say $2x^3 - 5x^2 - 6x + 9 = (x - 1)(2x^2 - 3x - 9)$.

What we have here is an example of an **identity**, like the ones which we described in Section 2.D.(h) where we also matched up terms in this way.

We can find the other two solutions or roots of the equation $f(x) = 0$ by solving $2x^2 - 3x - 9 = 0$. Factorising,

$$2x^2 - 3x - 9 = (2x + 3)(x - 3) = 0$$

so $x = 3$ or $-\frac{3}{2}$.

The three solutions or roots of $f(x) = 2x^3 - 5x^2 - 6x + 9 = 0$ are therefore given by $x = 1$, $x = 3$ and $x = -\frac{3}{2}$.

What will the graph of $y = f(x) = 2x^3 - 5x^2 - 6x + 9$ look like?

We know that it must cut the x-axis three times, at $x = 1$, $x = 3$ and $x = -\frac{3}{2}$.

It also seems reasonable to say that, if we find enough values of y from feeding in values for x into $f(x)$, the graph would be able to be drawn in one continuous line.

If we put $x = 0$, we get $f(x) = 9$, so we know that the curve cuts the y-axis at the point $(0,9)$.

If x is large and positive, which has the most powerful effect: the $2x^3$ or the $-5x^2$? Try putting $x = 2$, $x = 10$ and $x = 100$. You will see that, as x gets larger, the $2x^3$ term swamps out the $-5x^2$ term. So y will also become large and positive.

In just the same way, if x is large and negative, $2x^3$ will also be large and negative, and so y also is large and negative.

We now know enough to make a sketch of the graph of $f(x)$ and I show this below in Figure. 2.E.1.

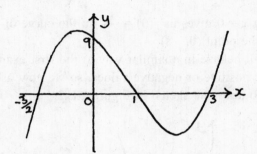

Figure 2.E.1 $f(x) = 2x^3 - 5x^2 - 6x + 9$

This is the best that we can do at the moment. With straight lines, we could also use the steepness or gradient to help us with the graph sketch. With quadratic graphs, we were able to complete the square to find the least (or greatest) value of the graph. You might perhaps feel that, since we can find the value of y for any value of x here, surely we ought to be able to find out a bit more about the size of the greatest value coming between $x = -\frac{3}{2}$ and $x = 1$, and the least value coming between $x = 1$ and $x = 3$. We can't discover these values yet, except approximately by trying lots of values of x, but we shall find out how it *is* possible to do it in Section 8.E.(b).

EXAMPLE (1) We will now solve the equation $f(x) = 3x^3 + 2x^2 - 12x - 8$ and use the roots to sketch the graph of $y = f(x)$.

($f(x)$ is now referring to the new collection of terms of $3x^3 + 2x^2 - 12x - 8$. We could also have used some other letter, calling it, say, $g(x)$ or $h(x)$ if we had wished.)

First, we hope to find a root of $f(x) = 0$. Can you find one?

This time, if we try $x = 1$, we get $f(1) = 3 + 2 - 12 - 8 \neq 0$ so $x = 1$ is not a solution of $f(x) = 0$.

Putting $x = 2$ gives $\quad f(2) = 3 \times 8 + 2 \times 4 - 12 \times 2 - 8 = 0 \quad$ so $\quad x = 2$ is a root. This means that $(x - 2)$ is a factor of $f(x)$. We can now say

$$f(x) = 3x^3 + 2x^2 - 12x - 8 = (x - 2)(3x^2 + px + 4)$$

matching up the two end terms in the right-hand bracket and letting p stand for the number which we still have to find.

Matching up the terms in x^2, we have $2x^2 = -6x^2 + px^2$ so $p = 8$.

Checking with the terms in x, we have $-12x = -2px + 4x$ so $p = 8$ is correct.

(It is also possible to find the second bracket here of $(3x^2 + 8x + 4)$ by long division of $(x - 2)$ into $3x^3 + 2x^2 - 12x - 8$, but I think the method above is easier. I shall show you how to do long division in algebra in the next section.)

We now have

$$f(x) = (x - 2)(3x^2 + 8x + 4) = (x - 2)(3x + 2)(x + 2)$$

factorising the second bracket, and the equation $f(x) = 0$ has the three solutions or roots: $x = 2$ or $x = -\frac{2}{3}$ or $x = -2$.

We will now use these three roots to help us to sketch the graph of $y = f(x)$.

Putting $x = 0$ gives us $f(0) = -8$, so the curve of $y = f(x)$ cuts the y-axis at the point $(0, -8)$.

$f(x)$ will behave in a similar way to the first example when x takes very large positive or negative values, so we now use all the information we have to draw the sketch in Figure 2.E.2.

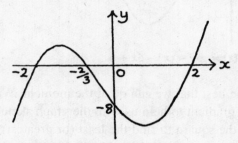

Figure 2.E.2 $f(x) = 3x^3 + 2x^2 - 12x - 8$

EXERCISE 2.E.1

For each of the following, first find the roots of $f(x) = 0$ and then use these to help you to sketch the graphs of $y = f(x)$ in each case. For each graph, you will also need to find out where it cuts the y-axis, and how $f(x)$ behaves when x takes either very large positive values or very large negative values.

(1) $y = f(x) = 3x^3 + 2x^2 - 3x - 2$ (2) $y = f(x) = 2 + 3x - 3x^2 - 2x^3$
(3) $y = f(x) = 4x^3 - 15x^2 + 12x + 4$ (4) $y = f(x) = x^3 - 3x^2 + 3x - 1$

We could use exactly the same method to solve equations which start with a term in x^4. The only problem is that it depends upon being able to guess some roots correctly to start with. Often, none of the roots of $f(x) = 0$ will be simple whole numbers, and indeed they may not even be real numbers, as we have already found with some quadratic equations. If this happens, the graph sketches will no longer look like the ones we have drawn, though in the case of a cubic graph it will have to cross the x-axis at least once, because the y values must go from large negative to large positive or vice versa, and the graph itself is a continuous line. So a cubic equation will always have at least one real root (that is, a root which can be found on the x-axis).

Also, once we have got beyond quadratic equations, general formulas for finding the roots are either far more complicated or do not exist at all. It is, however, possible to use numerical methods for solving such equations by approximating to the roots with any desired degree of accuracy.

2.E.(b)   Doing long division in algebra

Usually long division in algebra can be avoided (as we did in the last section when we used the method of matching up the terms on the two sides for factorising), but sometimes this isn't possible, so we will now look at how this process works.

We will take as a first example $\dfrac{2x^3 + 9x^2 - 3x - 20}{x + 3}$

We will have:

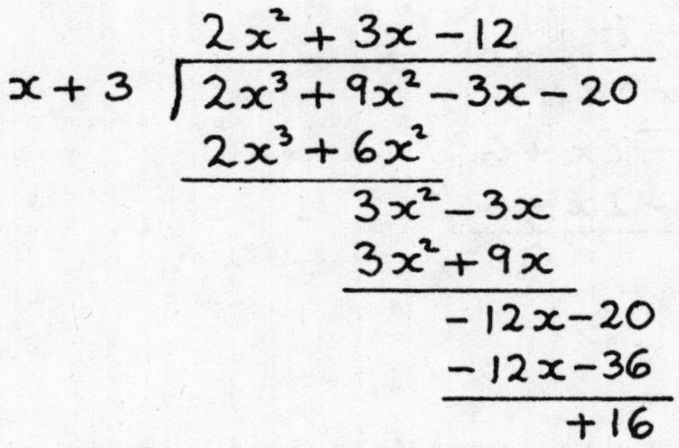

$$
\begin{array}{r}
2x^2 + 3x - 12 \\
x + 3 \overline{\smash{\big)}\, 2x^3 + 9x^2 - 3x - 20} \\
\underline{2x^3 + 6x^2} \\
3x^2 - 3x \\
\underline{3x^2 + 9x} \\
-12x - 20 \\
\underline{-12x - 36} \\
+16
\end{array}
$$

Figure 2.E.3

The working for the division is set out as I have shown in Figure 2.E.3.
$x + 3$ is called the **divisor** and $2x^3 + 9x^2 - 3x - 20$ is called the **dividend**.
The process consists of the following.

- **Divide the highest power by the highest power in the divisor.**
  Here, divide $2x^3$ by x, which gives us $2x^2$.
- **Multiply the divisor by this quantity.**
  Here, we multiply $x + 3$ by $2x^2$ to get $2x^3 + 6x^2$.
- **Subtract. This gives us the mismatch at each stage.**
  Here, we get $3x^2$.
- **Bring down the next term in the quantity being divided to the working level.**
  Here, we now get $3x^2 - 3x$.
- **Repeat the process until the highest power of x in the divisor is greater than the highest power of x it would be divided into.**

What is then left is called the **remainder**, and the result of the division is called the **quotient**.
Here we have the result

$$
\frac{2x^3 + 9x^2 - 3x - 20}{x + 3} = 2x^2 + 3x - 12 + \frac{16}{x + 3}.
$$

The **quotient** is $2x^2 + 3x - 12$ and the **remainder** is $+16$.

Compare this with the numerical example

$$
\frac{187}{15} = 12 + \frac{7}{15}.
$$

We see that 15 goes 12 times into 187 with a remainder of 7.

Here is another example of long division, this time with no remainder.

If $(x - 3)$ is a factor of $2x^3 - 9x^2 + 7x + 6$ then it must divide into it exactly, (just as 3 is a factor of 12 and divides into it exactly four times).

We will now prove that $(x - 3)$ *is* a factor of $2x^3 - 9x^2 + 7x + 6$ by using long division. The working is shown in Figure 2.E.4.

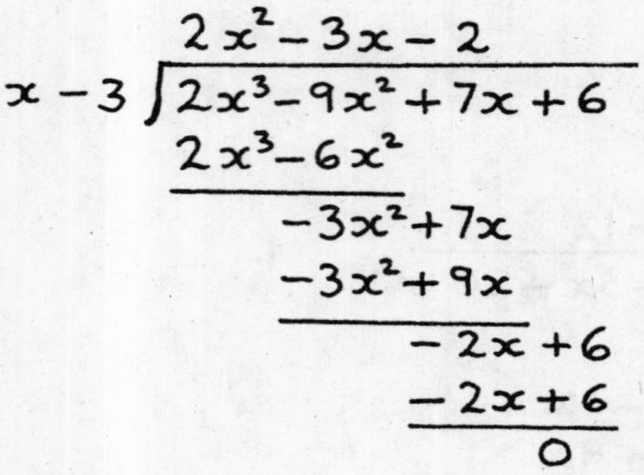

$$
\begin{array}{r}
2x^2 - 3x - 2 \\
x - 3 \overline{\smash{)}2x^3 - 9x^2 + 7x + 6} \\
\underline{2x^3 - 6x^2 } \\
-3x^2 + 7x \\
\underline{-3x^2 + 9x } \\
-2x + 6 \\
\underline{-2x + 6} \\
0
\end{array}
$$

Figure 2.E.4

In practice, it is almost always possible to avoid long division if you do not take kindly to it; we managed to do this when we were doing the factorising earlier, and there are other ingenious methods which can be used, which I will show you as you need them.

2.E.(c)   **Avoiding long division – the Remainder and Factor Theorems**
In Section 2.E.(a), we found that if $f(x) = 2x^3 - 5x^2 - 6x + 9$ then $f(1) = 0$.

It is certainly true that if $(x - 1)$ is a factor of $f(x)$ then putting $x = 1$ will make $f(x) = 0$. We assumed in that section that this would work the other way round too, so that if $f(1) = 0$ then $(x - 1)$ must be a factor of $f(x)$. We shall now prove that this assumption was justified, and we shall also find a very neat way of finding the remainder from doing an algebra long division without actually having to *do* this rather tedious process.

We prove these useful results as follows:

Suppose we have some general cubic equation $f(x) = ax^3 + bx^2 + cx + d$, and we divide it by $(x - k)$. (Here, a, b, c, d and k are all standing for whatever particular numbers we might have.) We will get

$$
\frac{ax^3 + bx^2 + cx + d}{(x - k)} = q(x) + \frac{R}{(x - k)} \tag{1}
$$

where $q(x)$ corresponds to the $2x^2 + 3x - 12$ of the first example in the last section, and R corresponds to the remainder of +16.

Now we multiply all through by $(x - k)$. This gives us

$$
ax^3 + bx^2 + cx + d = (x - k)q(x) + R.
$$

We can compare this with an arithmetical example.

$$
\frac{79}{15} = 5 + \frac{4}{15} \quad \text{so} \quad 79 = 5 \times 15 + 4.
$$

15 goes 5 times into 79 with a remainder of 4. In other words, 79 is made up of 5 lots of 15 with an extra 4 added on.

Here, $ax^3 + bx^2 + cx + d$ is made up of $(x - k)$ lots of $q(x)$ with an extra R added on.

Since we have $f(x) = ax^3 + bx^2 + cx + d = (x - k) q(x) + R$, putting $x = k$ gives us $f(k) = ak^3 + bk^2 + ck + d = (k - k) q(k) + R$, that is, $f(k) = R$.

From this, if $f(k) = 0$ then $R = 0$ also, which means that $(x - k)$ divides into $f(x)$ exactly. It is a **factor** of $f(x)$.

We now have the following pair of results.

If we have $f(x) = ax^3 + bx^2 + cx + d$

then dividing $f(x)$ by $(x - k)$ gives a remainder of $f(k)$.

This is the **Remainder Theorem for cubics**.

If $f(k) = 0$, then $(x - k)$ is a factor of $f(x)$.

This is the **Factor Theorem for cubics**.

We now see how we can use these results by looking at the two long division examples from the previous section.

In the first example, we divided $f(x) = 2x^3 + 9x^2 - 3x - 20$ by $(x + 3)$. To find the remainder, we no longer need to do this division. All we have to do is to work out $f(-3) = 2(-3)^3 + 9(-3)^2 - 3(-3) - 20 = -54 + 81 + 9 - 20 = 16$ which agrees with the answer that we found there.

 Notice the switch in sign from $x + 3$ to $f(-3)$.

This is because $x + 3 = x - (-3)$ which corresponds to the $x - k$.

If we only need to know the remainder from a long division, we can now find this just by working out $f(k)$.

In the second example, putting $x = 3$ in $f(x) = 2x^3 - 9x^2 + 7x + 6$ gives us $f(3) = 54 - 81 + 21 + 6 = 0$ so therefore $(x - 3)$ must be a factor of $f(x)$.

Again, we don't need to do the long division to prove this.

Although we have taken the special case of $f(x)$ being a cubic expression, the argument would have worked in exactly the same way for higher whole number powers of x, so these two theorems are true for any such expression.

2.E.(d)   **Three examples of using these theorems, and a red herring**

EXAMPLE (1)   Find the remainder when $f(x) = 3x^3 - 4x^2 + 5x - 2$ is divided by $(x - 2)$.

We simply find $f(2)$. This is $3(8) - 4(4) + 5(2) - 2 = 16$ so the remainder is 16 and we have not had to do the actual division to find this out.

EXAMPLE (2)   Given that $(x - 4)$ is a factor of $f(x) = 6x^3 + ax^2 + bx + 8$ and that the remainder when $f(x)$ is divided by $(x + 1)$ is -15, find a and b and the other two factors.

We have

$$f(x) = 6x^3 + ax^2 + bx + 8.$$

We are told that $(x - 4)$ is a factor, therefore $f(4) = 0$. So

$$f(4) = 384 + 16a + 4b + 8 = 0 \quad \text{and} \quad 4a + b = -98. \tag{1}$$

The remainder when $f(x)$ is divided by $(x + 1)$ is -15. So $f(-1) = -15$. We have

$$f(-1) = -6 + a - b + 8 = -15 \quad \text{so} \quad a - b = -17. \tag{2}$$

Adding equations (1) and (2) gives $5a = -115$ so $a = -23$.
Substituting in (1) gives $-92 + b = -98$ so $b = -6$.
Check in (2): LHS $= -23 - (-6) = -17 =$ RHS.

Now we have

$$f(x) = 6x^3 - 23x^2 - 6x + 8 = (x - 4)(\text{something}).$$

Comparing the two sides, the first term in the second bracket must be $6x^2$. The last term of the second bracket must be -2. Let the middle term be px. Then we have

$$6x^3 - 23x^2 - 6x + 8 = (x - 4)(6x^2 + px - 2).$$

Matching the terms in x^2 gives

$$-23x^2 = -24x^2 + px^2 \quad \text{so} \quad p = 1.$$

Checking with the term in x we have $-6x = -4px - 2x$ so again we have $p = 1$. So we have

$$f(x) = (x + 1)(6x^2 + x - 2) = (x + 1)(2x - 1)(3x + 2)$$

factorising the second bracket.

The other two factors are $(2x - 1)$ and $(3x + 2)$.

EXAMPLE (3) This example is just sufficiently different that you might find it a little difficult.

Suppose you have been asked to show that $x^2 - 4$ is a factor of $3x^3 + 4x^2 - 12x - 16$. Can you see that you have actually been asked about *two* factors? What are they?

We can use the difference of two squares to say $x^2 - 4 = (x - 2)(x + 2)$.
Now, $f(2) = 24 + 16 - 24 - 16 = 0$ so $(x - 2)$ is a factor.
$f(-2) = -24 + 16 + 24 - 16 = 0$ so $(x + 2)$ is a factor also.

If two factors are multiplied together, then the resulting expression is also a factor.

EXAMPLE (4) (This is the red herring.) Solve the equation $4x^4 - 37x^2 + 9 = 0$.

It is possible to solve this equation by finding two solutions by guessing, but they are quite hard to find, and there is a much neater and quicker way of finding the answers.

This is because what we have been asked to solve is really a heavily disguised quadratic equation.

If we put $y = x^2$, the equation becomes $4y^2 - 37y + 9 = 0$.

Factorising, we get $(y - 9)(4y - 1) = 0$ so $y = 9$ or $y = \frac{1}{4}$. (If you couldn't spot these factors, you could have used the quadratic equation formula to find y.)

Replacing y by x^2, we get $x^2 = 9$ or $x^2 = \frac{1}{4}$.

This gives us the four solutions of $x = \pm 3$ or $x = \pm \frac{1}{2}$.

Try these questions for yourself now.

(1) Show that $(x - 2)$ is a factor of $x^3 + 2x^2 - 5x - 6$, and find the other two.

(2) Show that $(x - 3)$ is a factor of $2x^3 - 3x^2 - 8x - 3$, and find the other two.

(3) Factorise completely the expression $f(x) = 3x^3 + x^2 - 12x - 4$, and hence solve the equation $f(x) = 0$.

(4) Factorise completely the expression $f(x) = 2x^3 + 7x^2 + 2x - 3$, and hence solve the equation $f(x) = 0$.

(5) Solve the equation $f(x) = x^4 - 29x^2 + 100 = 0$.

(6) Given that $(x - 3)$ is a factor of $f(x) = 5x^3 + ax^2 + bx - 6$, and that the remainder when $f(x)$ is divided by $(x + 2)$ is -40, find a and b, and the other two factors.

(7) Show by using long division that $(3x - 2)$ is a factor of $12x^3 + 4x^2 - 17x + 6$. Show also that this is true by using the Factor Theorem.

(8) Using long division, find the remainder when $6x^3 + 5x^2 - 8x + 1$ is divided by $(2x - 1)$. Check that your answer is correct by using the Remainder Theorem.

3 Relations and functions

We now build on the work of the previous two chapters to introduce functions. These are very important in scientific and engineering applications, and this chapter helps you to understand how they work.

It is split up into the following sections.

3.A Two special kinds of relationship
(a) Direct proportion, (b) Some physical examples of direct proportion, (c) More exotic examples, (d) Partial direct proportion – lines not through the origin, (e) Inverse proportion, (f) Some examples of mixed variation

3.B An introduction to functions
(a) What are functions? Some relationships examined, (b) $y = f(x)$ – a useful new shorthand, (c) When is a relationship a function? (d) Stretching and shifting – new functions from old, (e) Two practical examples of shifting and stretching, (f) Finding functions of functions, (g) Can we go back the other way? Inverse functions, (h) Finding inverses of more complicated functions, (i) Sketching the particular case of $f(x) = (x + 3)/(x - 2)$, and its inverse, (j) Odd and even functions

3.C Exponential and log functions
(a) Exponential functions – describing population growth, (b) The inverse of a growth function: log functions, (c) Finding the logs of some particular numbers, (d) The three laws or rules for logs, (e) What are 'e' and 'exp'? A brief introduction, (f) Negative exponential functions – describing population decay

3.D Unveiling secrets – logs and linear forms
(a) Relationships of the form $y = ax^n$, (b) Relationships of the form $y = an^x$, (c) What can we do if logs are no help?

3.A   Two special kinds of relationship

We start this chapter with some more practical examples of the use of equations. Many physical laws can be described by the two particular sorts of relation which we shall consider next.

3.A.(a)   Direct proportion

This describes a situation in which two quantities are related together so that as one gets bigger the other does also, in the same proportion. If the first quantity is doubled then the second quantity will be doubled also. We could take as an example the number of identical objects bought and the price paid.

The relationship between the number pairs making up the coordinates of the points on the straight line shown in Figure 3.A.1 also fits this description because it passes through the origin.

Fill in the blanks for the points C, D and E yourself.

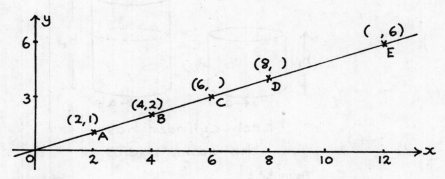

Figure 3.A.1

You should have C is (6,3), D is (8,4) and E is (12,6).

Each fraction y/x gives the gradient of the line because all of them give the relative change of y with respect to x measured from the origin. We have

$$\frac{1}{2} = \frac{2}{4} = \frac{3}{6} = \frac{4}{8} = \frac{6}{12} = \frac{y}{x} = \text{the gradient, } m.$$

For any two general pairs (x_1,y_1) and (x_2,y_2), we have $y_1/x_1 = y_2/x_2 = \frac{1}{2}$. We know from Section 2.B.(f) that the equation of the line through these points is given by $y = \frac{1}{2}x$. The $\frac{1}{2}$ is called the constant of proportionality and tells us the relation between this particular set of ys and xs.

If two quantities x and y vary directly then we can write

$x \propto y$ or $x = ky$ where k is a constant.

The symbol $\propto$ means 'is proportional to'.

3.A.(b)   **Some physical examples of direct proportion**
Here are some examples of physical quantities which are related in this way.

EXAMPLE (1)   *Charles' Law of gases*. This states that the volume, V, of a certain mass of gas is directly proportional to its temperature, T, measured from absolute zero, which is $-273\,°C$. Therefore we can say

$$V \propto T \quad \text{or} \quad \frac{V_1}{T_1} = \frac{V_2}{T_2} \quad \text{etc.} \quad \text{or} \quad V = kT.$$

where k is the constant of proportionality. The numerical value of k will depend upon the units in which we measure V and T.

EXAMPLE (2) The volume, V, of a cylinder of a given cross-section is directly proportional to its height, h. (This is shown with two such cylinders in Figure 3.A.2.)

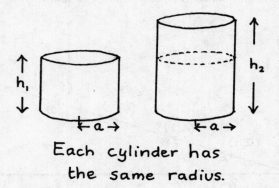

Each cylinder has the same radius.

Figure 3.A.2

We can say $V \propto h$ or $V_1/h_1 = V_2/h_2$ or $V = kh$.

Can you see what k will be this time?

The formula for the volume of a cylinder is $V = \pi a^2 h$, so $k = \pi a^2$.

EXAMPLE (3) For simple tension or compression (so no bending is involved), stress, σ, is directly proportional to strain, ε.

We can say $\sigma \propto \varepsilon$ or $\sigma_1/\varepsilon_1 = \sigma_2/\varepsilon_2$ or $\sigma = E\varepsilon$ where E is the constant of proportionality.

A possible (rather simplified) situation is shown in Figure 3.A.3(a).

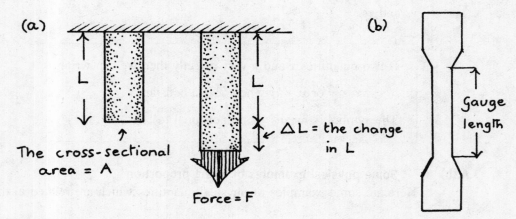

(a) The cross-sectional area = A

$\Delta L =$ the change in L

Force = F

(b) Gauge length

Figure 3.A.3

Figure 3.A.3(b) shows the cross-section of a typical test specimen with a pre-determined gauge length to perform the test on, and large end pieces to enable them to be clamped firmly.

The strain is the fractional change in length, and the stress is the stretching force per unit cross-sectional area. ΔL stands for the change in the original length, L. (The symbol 'Δ' is often used to mean 'the change in'.)

Relations and functions

So we have

$$\varepsilon = \frac{\Delta L}{L} \quad \text{and} \quad \sigma = \frac{F}{A} \quad \text{and therefore} \quad F/A = E\,\frac{\Delta L}{L}.$$

E, the constant of proportionality, is called Young's Modulus of elasticity and is a physical property of the particular material concerned.

Physically, the relationship will only be one of direct proportion, and so represented by a straight line through the origin, up to a certain critical point which will depend upon the properties of the material concerned. When the strain is increased beyond this critical value, deformation takes place and the material behaves differently. The mathematical model of direct proportion only works over a limited physical range.

3.A.(c)   More exotic examples

EXAMPLE (1) The kinetic energy, E, of an object of mass M moving at a speed of v is given by the relation $E = \frac{1}{2}Mv^2$. (Notice that we have used the symbol E to mean different things in this example and the last one. This is because engineers and physicists do commonly use this same letter with these two different meanings. It is very important in any practical application to make sure that you know what the different symbols represent.)

For two objects moving at the same speed, v, the kinetic energies will be directly proportional to the masses of the objects. For example, a lorry of mass 6 tonnes moving at a speed of $10\,\mathrm{m\,s^{-1}}$ has six times the kinetic energy of a car of mass 1 tonne, also moving at $10\,\mathrm{m\,s^{-1}}$.

But how does the kinetic energy of the car compare when it is moving at a speed of $10\,\mathrm{m\,s^{-1}}$ to when it is moving at a speed of $30\,\mathrm{m\,s^{-1}}$?

The speed is now three times greater but the kinetic energy is proportional to the square of the speed. Therefore the kinetic energy is nine times greater.

Here, $E = kv^2$ with this particular k being $\frac{1}{2}$ since the mass of the car is one tonne.

EXAMPLE (2) The area of a circle, A, of radius r is given by $A = \pi r^2$.
What is A directly proportional to?
What is the constant of proportionality?

A is directly proportional to r^2, and the constant of proportionality is π.

The table below shows possible values for A, r and r^2.

A	0	π	4π	9π	16π	25π
r	0	1	2	3	4	5
r^2	0	1	4	9	16	25

Figure 3.A.4(a) shows a sketch of the graph of A against r, and Figure 3.A.4(b) shows a sketch of the graph of A against r^2.

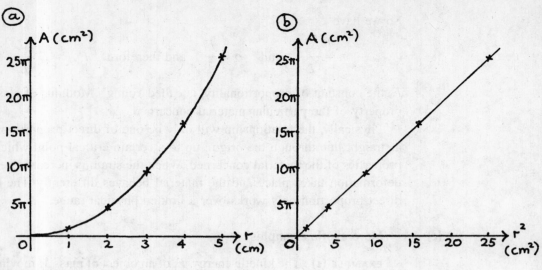

Figure 3.A.4

From these you will see that plotting A against r gives a graph of the same form as $y = x^2$, but plotting A against r^2 gives a straight line through the origin of gradient π.

EXAMPLE (3) The volume, V, of a sphere of radius r is given by $V = \frac{4}{3}\pi r^3$.
What is V directly proportional to?
What is the constant of proportionality?

V is directly proportional to r^3 and the constant of proportionality is $\frac{4}{3}\pi$.

EXAMPLE (4) In Section 2.A.(d), we used the formula $T = 2\pi\sqrt{l/g}$ for the period, T, of a simple pendulum of length l. (g stands for the acceleration due to gravity.)
What is T directly proportional to here?
What is the constant of proportionality?

T is directly proportional to $\sqrt{l}$, the square root of the length, so $T = k\sqrt{l}$. The constant of proportionality is $2\pi/\sqrt{g}$. (This is assuming that the acceleration due to gravity can be taken to be constant when we are making our measurements.) A graph of T against $\sqrt{l}$ will give a straight line through the origin with gradient $2\pi/\sqrt{g}$.

EXERCISE 3.A.1

Try answering these questions yourself. Each question is an example of a relationship involving direct proportion, and you are asked to compare pairs of physical measurements.

(1) Compare the volumes of the cylinders (a) A and B (b) C and D shown in Figure 3.A.5.

(2) Compare the kinetic energy, E_1, of a car moving at a speed of 5 m s⁻¹ with its kinetic energy E_2 when it is moving at 30 m s⁻¹.

(3) Compare the volumes V_1 and V_2 of two spheres if the first sphere has a radius of 2 cm and the second has a radius of 8 cm.

(4) Compare the time of the swing of a simple pendulum of length 9 cm with a pendulum of length 25 cm.

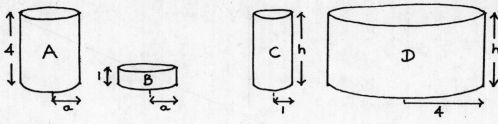

Figure 3.A.5

3.A.(d)   Partial direct proportion – lines not through the origin

We have seen that every direct proportion relationship gives us a straight line graph through the origin.

Can we give any physical meaning to pairs of points lying on a straight line which doesn't pass through the origin?

If we take *any* straight line, so that its equation can be written in the form $y = mx + c$ (Section 2.B.(f)), then y is partly directly proportional to x and partly made up of the constant, c.

An electricity bill is a physical example of such a relationship. This is made up partly of the cost of the number of units of electricity used and partly of a standing charge which is a constant amount added to each bill. (See Figure 3.A.6.)

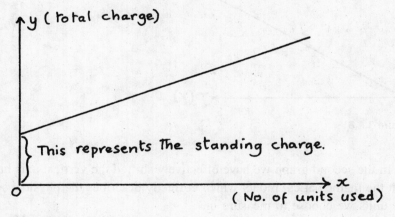

Figure 3.A.6

The equation for a typical electricity bill might read $y = 7.42x + 910$ where the cost in pence per unit used is 7.42 and the standing charge is £9.10.

y, the total cost, is given in pence by this equation.

There are many other physical situations which can be described in a similar way. A second example is given by the relationship between the volume and the temperature of a gas if we *don't* measure the temperature on a scale starting from absolute zero. This is because we can only have zero volume if the temperature is also at absolute zero, so measurements on a temperature scale which starts from here are necessary to make the line pass through the origin.

If the temperature is measured in °C, we shall get a graph like the one shown in Figure 3.A.7.

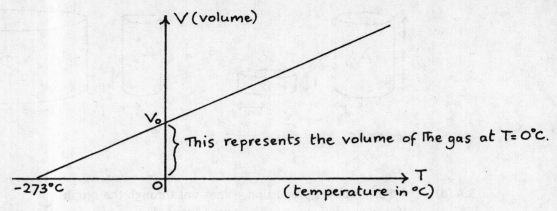

Figure 3.A.7

The equation which relates the volume to the temperature is $V = kT + V_0$ where k (the gradient) $= V_0/273$.

Compare this with the graph of Figure 3.A.8 which shows the simple relationship of direct proportion of volume to absolute temperature, so $V = kT$. (The absolute temperature is measured in degrees Kelvin where 0 K is equivalent to $-273\,°C$.)

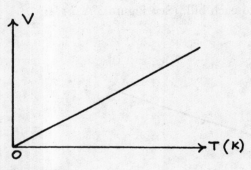

Figure 3.A.8

In the second graph we have effectively shifted the vertical axis back by 273 °C. We see that the mathematical model which correctly describes the physical situation depends upon the units we choose to measure in.

3.A.(e)   **Inverse proportion**

Two quantities are in inverse proportion if, as one gets larger, the other gets proportionally smaller and vice versa.

For example, if 24 apples are to be shared out equally among different numbers of people, we have all the possibilities shown in the table below.

x (number of apples)	1	2	3	4	6	8	12	24
y (number of people)	24	12	8	6	4	3	2	1

Evidently, in each case xy must be equal to 24.

If we plot these pairs of values we no longer get a straight line graph. (The graph we get is shown in Figure 3.A.9(a).

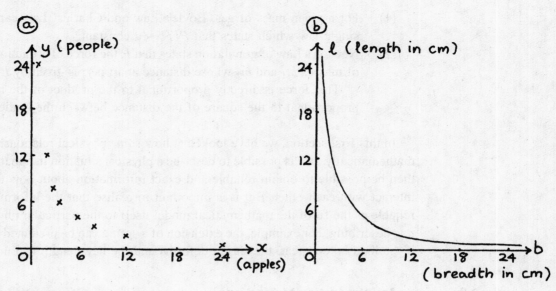

Figure 3.A.9

Nor can we reasonably join the points together to form a curve unless we start dividing up the apples (or, even more alarmingly, the people).

However, if we consider instead the possible variation in the measurements of the length and breadth of a rectangle of a given area of $24 \, \text{cm}^2$, we get exactly the same pairs of values as in the table above but we also get all the intermediate values too, including fractions as in the pair $\frac{1}{2}$ and 48, and irrationals such as $\sqrt{24}$, since $\sqrt{24} \times \sqrt{24} = 24$.

This time, the set of all possible pairs *does* give a smooth curve and this is shown in Figure 3.A.9(b).

Notice what happens at the two ends of this curve.

As we make one measurement smaller, so the other measurement has to become correspondingly larger to give the fixed area of $24 \, \text{cm}^2$. If the rectangle gets very thin it will also have to be extremely long. The points on the curve become closer and closer to the two axes but they can never touch since a zero measurement either way gives a zero area. Lines like this which a curve approaches but never touches are called **asymptotes**.

The relationship here is that $l \times b = 24$ which is a constant.

A relationship of inverse variation can always be written in this form.

If two quantities x and y vary inversely,
then we can write $xy = c$ where c is a constant.

Another physical example of inverse variation is Boyle's Law for gases which states that, for a given mass of gas at a constant temperature, the pressure is inversely proportional to the volume, so $PV = $ a constant.

3.A Two special kinds of relationship

3.A.(f)   **Some examples of mixed variation**

Some physical laws involve a combination of direct and inverse variation. Here are two examples.

(1)   For a given mass of gas, Boyle's Law and Charles' Law can be combined into a single law which states that PV/T = a constant.

(2)   Newton's Law of gravitation states that F, the force of attraction between two bodies of masses m_1 and m_2 whose distance apart is r, is given by $F = k\, m_1 m_2 / r^2$.

        This force is directly proportional to the product of the masses, and inversely proportional to the square of the distance between the bodies.

In this first section, we have looked at how some physical relationships can be expressed mathematically. If it is possible to describe a physical situation in a mathematical way, it will then be possible to obtain reliable and exact information about how the physical variables interact with each other. But it is important to realise that the information will only be as reliable as the fit of the mathematical model itself to the particular physical situation which it is describing. For example, the extension of a spring can be predicted for a known load but, if the load is too great, the spring deforms and the new length can no longer be found.

3.B   An introduction to functions

3.B.(a)   **What are functions? Some relationships examined**

To be able to describe physical situations mathematically, and so to be able to extract detailed information about how they can behave, you need to be confident about handling the necessary maths. This next section is about different kinds of mathematical relationship and how they work. In particular, we shall look at the special relationships which are called functions.

Suppose we consider the four equations:

(a) $y = 2x + 3$,   (b) $y = x^2 - 2x - 3$,

(c) $y = \dfrac{1}{x^2 + 4}$,   (d) $y = (3x + 1)^{1/2}$.

Each of these gives a relationship between x and y from which we could build up a set of ordered pairs or coordinates to draw a graph.

For each of these four in turn, try answering for yourself the following four questions.

(1)   If you feed different values of x into the relationship, is there just one corresponding value of y for each possible value of x?

(2)   Does every new value of x which you feed in give you a correspondingly new value of y, or do you sometimes find that two different values of x lead to the same y value?

(3)   Do you think that you could reasonably choose *any* real number as a value of x to feed into each of the four cases above? (That is, could you choose any number which lies somewhere on the x-axis? Section 1.E gives you a description of all the different kinds of number which can be found here.)

(4)   Finally, if we make the set of x values as large as possible in each case, what happens to the complete set of possible values for y? Is it the same as the set of possible values for x? If not, what is it?

It will very much help your understanding if you think about these four questions carefully yourself and write down what you think is going to happen in each case before you go on to look at my answers.

I will answer the four questions for each example in turn.

(a) $y = 2x + 3$

It is clear that for every value of x which we feed in there is just one possible value of y, and also that each value of y can only come from one possible value of x. Also there is no reason for excluding any real number from the possible values of x if we want to make the choice as wide as we can. Likewise, y can take all real values. We can see this graphically in Figure 3.B.1.

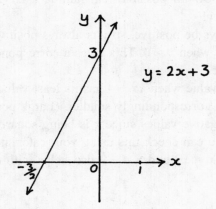

Figure 3.B.1

The arrows indicate that the line is infinitely long in either direction. Imagining this extension, we see that all possible values of x are included, and also all possible values of y. Also, each x value gives only one possible y value, and vice versa.

(b) $y = x^2 - 2x - 3$

This time, for every value of x which we feed in, again there is only one possible value of y.

But what about the other way round? For example, if we put $x = 4$ we get $y = 5$, and if we put $x = -2$ we also get $y = 5$. Similarly, both $x = 3$ and $x = -1$ give $y = 0$, so the answer to question (2) is 'no' for this relationship.

The graph sketch looks like Figure 3.B.2. We also see from this that, while there is no reason why we shouldn't choose any real number for an x value, the possible values for y

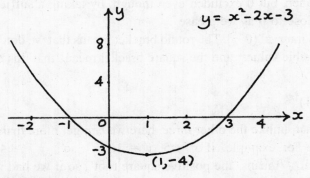

Figure 3.B.2

only go down to the lowest value of the curve. This we can find by completing the square like we did in Section 2.D.(b) in the last chapter.

We have $y = x^2 - 2x - 3 = (x-1)^2 - 1 - 3 = (x-1)^2 - 4$.

The least possible value of y is -4 and this happens when $x = 1$.

We see that the range of possible values for y *is* restricted, because $y \geq -4$.

(c) $\quad y = \dfrac{1}{x^2 + 4}$

Again, it is clear here that each value of x fed in gives only one possible value of y. But, like last time, we can get the same y value from two different values of x.

For example, if $x = +1$ then $y = \frac{1}{5}$ and if $x = -1$ then $y = \frac{1}{5}$ also. Notice that every symmetrical pair of $\pm$ values of x will give the same value for y.

There is no reason not to allow all possible real numbers as values for x, but think carefully about what happens to y!

First of all, $x^2 + 4$ must always be positive, so y is always positive.

The least value of $x^2 + 4$ is 4 when $x = 0$. This gives a corresponding value of $y = \frac{1}{4}$ so the point $(0, \frac{1}{4})$ lies on this curve.

Also, y must have its largest value when $x^2 + 4$ has its least value since $y = 1/(x^2 + 4)$. As x becomes larger, y becomes correspondingly smaller. (Large positive values of x will have the same effect as large negative values since x is being squared.) The graph will be symmetrical about the y-axis. You can check this using your calculator if you like; putting in a few values such as $x = \pm 1$, $x = \pm 2$ and $x = \pm 4$ also helps with drawing the sketch of Figure 3.B.3 below.

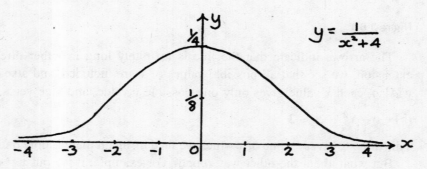

Figure 3.B.3

We see that the possible values of y lie between 0 and $\frac{1}{4}$.

Also, y can have the value of $\frac{1}{4}$, but it never actually reaches 0 although it gets infinitely close to it. We say that the values of y lie in the interval from 0 to $\frac{1}{4}$ on its number line, with the value $\frac{1}{4}$ included, but 0 excluded even though, by taking a sufficiently large value of x, we can get as close to 0 as we please.

We write this interval $(0, \frac{1}{4}]$. The round bracket means that we don't include that end point in the set of possible values; and the square bracket means that this end point is included.

(d) $\quad y = (3x + 1)^{1/2}$

Firstly, we see that, unlike the other three, here we *can* get more than one value of y for just one value of x. For example, if $x = 5$, $y = 16^{1/2}$ so $y = \pm 4$. (Remember that the convention is that $\sqrt{}$ means 'the positive square root', so if we had written $y = \sqrt{3x + 1}$ we would have avoided the complication of double-valued ys.)

However, it does look as though each possible y value can come from only one x value. For example, if $y = -5$, we have $(3x + 1)^{1/2} = -5$ so $3x + 1 = 25$ and $x = 8$.

Can we choose *any* real numbers for our values of x? Not unless we want complications coming from trying to take the square root of negative numbers, which is not something which we can yet do.

We must keep $3x + 1 \geq 0$ so $3x \geq -1$ and $x \geq -\frac{1}{3}$.

The possible y values include all the real numbers, however.

You can see that this will be so from the example which we took of $y = -5$. For any chosen number, we could repeat this process.

Figure 3.B.4(a) shows a sketch of the graph of $y = (3x + 1)^{1/2}$. Figure 3.B.4(b) shows the graph of $y = \sqrt{3x + 1}$. If we always take the positive square root, we just get the top half of (a).

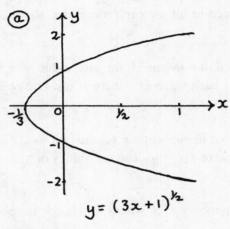
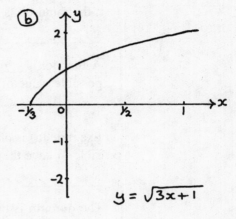

Figure 3.B.4

3.B.(b) $y = f(x)$ – a useful new shorthand

To make explanations simpler, it is often helpful to write what we have so far been calling y as $f(x)$, so that we have $y = f(x)$. (We have already used this notation for cubic equations in Section 2.E.(a).)

This means that y can be found from x according to some rule, in the way that the different ys of (a), (b), (c) and (d) above can be found, for example.

In the case of (a), we would have $y = f(x) = 2x + 3$,

$$\text{so} \quad f(2) = 4 + 3 = 7 \quad \text{and} \quad f(-3) = -6 + 3 = -3 \quad \text{etc.}$$

In case (b),

$$y = f(x) = x^2 - 2x - 3, \quad \text{so} \quad f(0) = -3 \quad \text{and} \quad f(3) = f(-1) = 0 \quad \text{etc.}$$

This notation is particularly useful when we want to talk about specific values, as we have done here. It is also useful for making clear what the variable quantity is.

An example of this is the case of the ball thrown up in the air, given in Section 2.D.(g). There, we used the formula $s = ut - \frac{1}{2}gt^2$ to find s, the distance moved from the thrower's hands. Both u and g are constants, and t gives the changing measurement of time. Therefore, we could write $s = f(t)$ meaning that the distance moved is a function of the time that the ball has been in the air.

A function is a particular form of relationship. Just what makes it particular is the subject of the following section.

When is a relationship a function?

We shall now use the answers which we have just found to the four questions above to lead us to some important definitions.

> If a relationship $y = f(x)$ is a **function** then, for any chosen value of the variable x, there is only one corresponding possible value of y.

Of the four examples from Section 3.B.(a), we found that (a), (b) and (c) are all functions, but (d) is not. However, $y = \sqrt{3x + 1}$ would have been.

Looking at this requirement graphically, we see that any vertical line on the graph must never cut the curve more than once if it is the graph of a function. I call this the raindrop test; the raindrop is only allowed to hit the curve once as it slides down the paper.

> A function $y = f(x)$ is called **one-to-one** if, for each value of y, there is just one possible value of x, and for each value of x there is just one possible value of y.

Example (a) is one-to-one but neither (b) nor (c) are one-to-one since in both cases it is possible to have the same value of $f(x)$ for different values of x.

> The **domain** is the set of numbers from which we choose the possible values of x.

In our four examples we deliberately made this choice as wide as possible, but as we saw in case (d), it may be restricted because of the formula involved. There might be circumstances in which you would choose to restrict the domain yourself. For example, if you were considering a physical problem in which x represented a length, you would require the domain to be restricted to positive numbers.

> The set of all possible values of y is called the **range**.

We found that in (a) this was the complete set of real numbers (any value for y was possible), but in each of (b) and (c) it was restricted in some way. Case (d) is a bit more subtle: if $y = (3x + 1)^{1/2}$ then, as we can see from Figure 3.B.4(a), y can take any value. But, as we also saw there, $y = (3x + 1)^{1/2}$ isn't a function. If we force a function by writing $y = \sqrt{3x + 1}$ then, as we can see from Figure 3.B.4(b), the possible values of y are restricted to $y \geq 0$.

Stretching and shifting – new functions from old

What kinds of effect will we get if we create new functions from old ones by adding or multiplying the first function in various different ways? We will now look at the results obtained from four possible different types of alteration.

(1) Adding a fixed amount to a function

What happens if we go from $f(x)$ to $f(x) + a$, where a is some given constant number? Here are two examples, both taking $a = 3$.

(a) $f(x) = 2x + 1$

so $f(x) + 3 = 2x + 4.$

(b) $f(x) = x^2$

so $f(x) + 3 = x^2 + 3.$

I show sketches of the two pairs of graphs below in Figure 3.B.5(a) and (b).

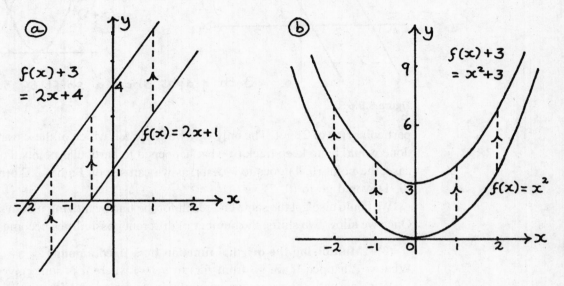

Figure 3.B.5

We see that the effect of adding 3 to $f(x)$, so that $y = f(x) + 3$, is to shift the graph up by 3 units.

(2) Adding a fixed amount to each x value

What will happen if we add a fixed amount to each x value instead, so that we go from $f(x)$ to $f(x + a)$ in each case? Again, we look at two examples, taking $a = 3$.

(a) $f(x) = 2x + 1$

so $f(x + 3) = 2(x + 3) + 1 = 2x + 7.$

(b) $f(x) = x^2$

so $f(x + 3) = (x + 3)^2.$

Notice that, to find $f(x + 3)$ from $f(x)$, we just replace x by $(x + 3)$.

I show sketches of the two pairs of graphs in Figure 3.B.6(a) and (b).

This time, the effect is to slide the whole graph 3 units to the left. Notice that the interesting bits happen 3 units sooner. For example, each contact with the x-axis happens 3 units earlier now.

What actually happens here is not what you might think at first; notice that $f(x + 3)$ is what you get if you slide $f(x)$ three units to the *left*, not to the right.

Because the function of (1) is a straight line, we can get the same effect as this sideways shift by giving the line an upwards shift of 6 units, so making $f(x)$ go to $f(x) + 6$ with our

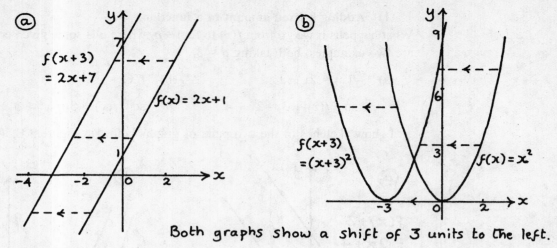

Both graphs show a shift of 3 units to the left.

Figure 3.B.6

particular $f(x)$ of $2x + 1$. The only way we could tell which of these transformations had been done would be to keep track of what happened to particular points. For example, in the first case, the point $(0, 1)$ goes to $(-3, 1)$, as we can see on Figure 3.B.6(a). In the second case, $(0, 1)$ would go to $(0, 7)$.

We could also get the same end result for the line by moving it both sideways and upwards. Once we allow two shifts, the number of different possibilities becomes infinite.

(3) Multiplying the original function by a fixed amount

What will happen if we go from $f(x)$ to $a\,f(x)$ where a is some given constant number?

Working with the same two examples as before, and with $a = 3$ again, we get

(a) $\begin{cases} f(x) = 2x + 1 \\ \text{so} \quad 3f(x) = 6x + 3. \end{cases}$ (b) $\begin{cases} f(x) = x^2 \\ \text{so} \quad 3f(x) = 3x^2 \end{cases}$

Sketches of the two pairs of graphs are shown below in Figure 3.B.7(a) and (b).

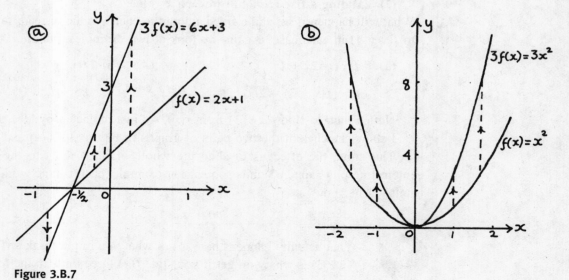

Figure 3.B.7

This time, the whole graph has been pulled away from the x-axis by a factor of 3, so that every point is now three times further away than it was originally. Therefore the only points on the graph which will remain unchanged are those on the x-axis itself.

(4) Multiplying x by a fixed amount

What will happen if we go from $f(x)$ to $f(ax)$?

Taking our same two examples, with $a = 3$, we have

(a) $\begin{cases} f(x) = 2x + 1 \\ \text{so} \quad f(3x) = 2(3x) + 1 = 6x + 1 \end{cases}$ (b) $\begin{cases} f(x) = x^2 \\ \text{so} \quad f(3x) = (3x)^2 = 9x^2. \end{cases}$

Notice that we simply replace x by $3x$ to find $f(3x)$ from $f(x)$.

I show sketches of the two pairs of graphs below in Figure 3.B.8(a) and (b).

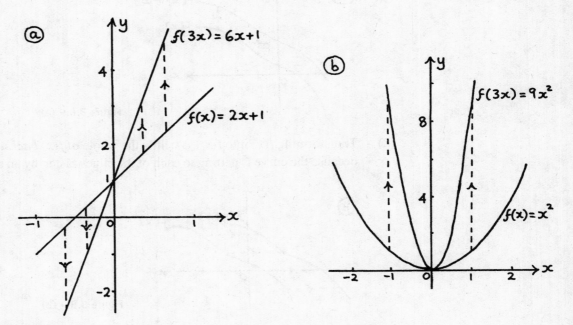

Figure 3.B.8

This time the stretching effect is more complicated because it only affects the part of the function involving x. Any purely number parts remain unchanged. The points which are unaffected by the stretching are those where the graphs cut the y-axis, so $x = 0$.

Notice too that the strength of the effect now depends upon the power of x. Having $(3x)^2$ in example 4(b) gives a more extreme effect than the $3x^2$ in 3(b), since the 3 is also being squared here.

We can relate examples 3(a) and 4 (a) to the real-life situation of the electricity bill graph shown earlier in Section 3.A.(d). The positive parts of the two graphs of 3(a) correspond to a situation of increasing both the standing charge and the cost per unit by a factor of three, while the positive parts of the two graphs of 4(a) could show an increase in the cost per unit of three, but an unchanged standing charge. (In this physical application, negative values of x or y would be meaningless.)

It has been easier in all these descriptions to stick to the same variable, x, for the functions. However, there is no reason why another letter should not be used.

In the physical example in Section 2.D.(g), on the motion of a ball when it is thrown up in the air, we described the distance travelled in terms of t, the time from when it left the thrower's hands.

We used the function $s = f(t) = ut - \frac{1}{2}gt^2$, and the horizontal axis was a t-axis instead of an x-axis.

We have now looked at the four simplest kinds of transformation of functions, and their graphical effects. I will list these for you below.

A summary of some effects of transforming functions

(1)   Transforming $f(x)$ to $f(x) + a$ shifts the whole of $f(x)$ upwards by a distance a. We have

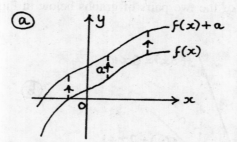

Figure 3.B.9 (a)

(2)   Transforming $f(x)$ into $f(x + a)$ shifts the whole of $f(x)$ *back* a distance a, because the curve is getting to each of its values faster, by an amount a. We have

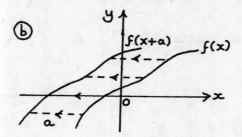

Figure 3.B.9 (b)

Shifts are sometimes called **translations**.

(3)   Transforming $f(x)$ into $af(x)$ stretches out each value of $f(x)$ by a factor a. We have

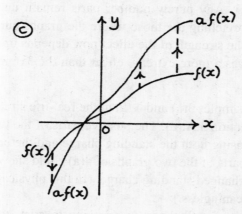

Figure 3.B.9 (c)

(4)   Transforming $f(x)$ into $f(ax)$ has a more complicated effect, since how much a affects each part of $f(x)$ depends on what is happening to x itself in $f(x)$. For example, if $f(x) = x^2 + x + 1$, then $f(ax) = a^2x^2 + ax + 1$. Each term has been affected differently. Therefore it is not possible to show this case on one sketch; the change in shape will depend entirely upon the function concerned.

The following exercise gives you a chance to practise recognising these shifts and stretches for yourself. Although f is the letter most commonly used for functions, it is sometimes more convenient to use other letters to avoid confusion. I do this here, having functions called $g(x)$, $h(x)$ etc.

EXERCISE 3.B.1

This exercise contains four questions, each of which involves one of the following four functions.

(1) $f(x) = 3x - 1$ (2) $g(x) = 2x - 2$ (3) $h(x) = \frac{1}{2}x + 1$ (4) $p(x) = x^2 - 4x + 3$.

Each question shows the original function on the left, followed by two examples of stretching or shifting it beside it. (See Figure 3.B.10 below.)

You have to decide what particular stretch or shift has happened in each case, and then write it in beside its graph. (For example, in Figure 3.B.5(a) earlier, I showed the shift of $f(x)$ to $f(x) + 3$.) Then check in the answers given at the back of the book to see if you have decided correctly. (Don't be tempted to go straight there!)

To make the questions easier for you, the constant number involved in each transformation (its 'a') is always either +2 or −2. This also means that you will be able to tell whether I have shifted my straight lines up or down or sideways to get them to their new positions.

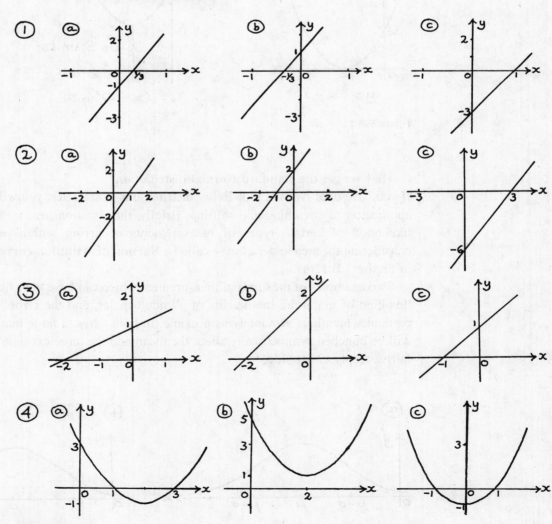

Figure 3.B.10

3.B.(e)   Two practical examples of shifting and stretching
        The method of completing the square
When we do the process of completing the square for a quadratic expression, as we did in Section 2.D.(b), we are actually finding what shift we would need to do to make the curve sit on the x-axis.

For example, if we take the curve $y = x^2 - 4x + 9$, we can use the method of completing the square to rewrite this as $y = (x - 2)^2 - 4 + 9 = (x - 2)^2 + 5$.

The curve $y = (x - 2)^2$, which I have drawn in Figure 3.B.11(a), just touches the x-axis when $x = 2$.

The curve $y = (x - 2)^2 + 5$ is the result of shifting the curve $y = (x - 2)^2$ up by 5 units. I have drawn this in Figure 3.B.11(b).

We can see from this picture that $y = (x - 2)^2 + 5 = x^2 - 4x + 9$ has a minimum value of 5 when $x = 2$.

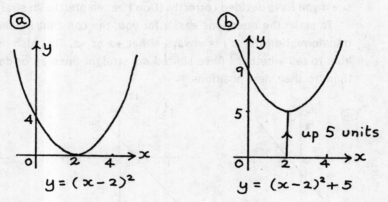

Figure 3.B.11

How we get the standard Normal distribution
If you have used Normal probability distributions in statistics, you will already have met an application of stretching and shifting. Briefly, the situation here is that we can model the likelihood of certain types of measurements occurring within particular intervals by considering the area under a curve called a **Normal distribution curve** which I sketch below in Figure 3.B.12(a).

Two examples of the kinds of measurement which can have their likelihoods modelled by this kind of graph are the heights of all adult males, and the errors made in measuring a particular length as accurately as possible. In both cases, a large number of measurements will be bunched symmetrically about the mean and the more extreme examples will tail off fairly steeply either side.

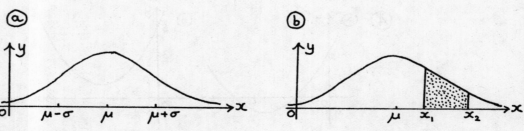

Figure 3.B.12

On the graph sketch, μ represents the mean or average measurement, and σ represents a measure of how spread out these measurements are. The curve flexes itself at a distance σ away from μ either side.

The area under the curve gives the probabilities of measurements lying between certain values. For example, the likelihood of a randomly chosen x lying between x_1 and x_2 is given by the shaded area shown in Figure 3.B.12(b).

These areas are extremely difficult to calculate since the equation of the curve is mathematically complicated, but since they are very frequently needed, tables have been calculated from which the different probabilities can be read off.

There is only one problem: it would be impossible to print the tables for *every* Normal distribution curve, and the tables just give the results for the simplest possible case, which I show in Figure 3.B.13(a). For this curve, $\mu = 0$ and $\sigma = 1$. The variable along the horizontal axis is called the **standard Normal variable**. This is always given the letter z.

Beside the standard Normal distribution curve, I show again the general Normal distribution curve in Figure 3.B.13(b).

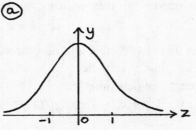 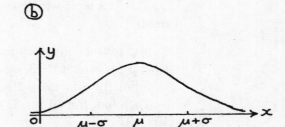

Figure 3.B.13

How can we get from the curve shown in (b) to the curve shown in (a)?

In order to transform (b) into (a) we have to shift the y-axis forwards by μ, so this would make $z = x - \mu$.

But this alone is not sufficient because, in (a), we have also squeezed the x measurements by a factor of $1/\sigma$. So to get from (b) to (a), we put

$$z = \frac{x - \mu}{\sigma}.$$

This is the formula for finding the standard Normal variable, z, which corresponds to a value x in a Normal distribution curve like Figure 3.B.13(b) above with mean μ and standard deviation σ.

To sketch the correct graph, the y measurements have to be stretched by a factor of σ since the total area under the graph remains one unit. (This is because it gives the sum of all the possible likelihoods or probabilities of the measurements concerned.)

The equation of each Normal distribution curve is in terms of its particular μ and σ, and this stretching of the y measurement takes place automatically in the new curve because of the property of unit area.

Instead of having to find the area between x_1 and x_2 shown in Figure 3.B.12(b) above, we can now use the tables to find the area between the corresponding z_1 and z_2 of the standard Normal curve. The tables give the two cumulative areas measured from the left-hand end of the curve up to z_1 and z_2 respectively, and the required area is the difference between these two. Since the total area remains 1, this area is unchanged in the two graphs. It is just a different shape.

There is one other rather neat spin-off from this transformation. Because the standard Normal curve is symmetrically placed about the origin, the tables only have to give values for one side. In practice, this is the right-hand side, and values for the left-hand side are found by using the symmetry of the curve.

3.B.(f)   **Finding functions of functions**

In Section 3.B.(d), we were able to see graphically the effects that some simple changes have on functions. But suppose the changes are more complicated because they have been built up from a number of simple steps. It's not so easy then to work out what is happening geometrically, but it *is* easy to find out what has happened using algebra. We can think of these changes as involving functions of functions.

Suppose we start with the two functions $f(x) = 2x + 3$ and $g(x) = 5x$.

What kind of meaning can we give to the expressions $f(g(x))$ and $g(f(x))$?

Do they mean the same thing?

This is a topic which sometimes makes students nervous, so we will look at it in some detail.

The instruction which $f(x)$ gives us is to 'double and add three', so we will have $f(\text{lump}) = 2 (\text{lump}) + 3$, whatever the 'lump' may be.

Similarly, $g(\text{lump}) = 5(\text{lump})$, whatever that lump may be.

Therefore $f(g(x)) = f(5x) = 2(5x) + 3 = 10x + 3$ and $g(f(x)) = g(2x + 3) = 5(2x + 3) = 10x + 15$.

The two results are different, and in general $f(g(x))$ will *not* be the same as $g(f(x))$. In fact, in this example, $f(g(x))$ is never equal to $g(f(x))$ for *any* value of x since we can't find an x so that $10x + 3 = 10x + 15$.

Notice the order of the operations. The inside function acts on x first, and then the outside function acts on the result.

$$f(g(x)) \quad \text{and} \quad g(f(x))$$

EXERCISE 3.B.2

Try these for yourself.
Find   (a) $f(g(x))$ (b) $g(f(x))$ if
(1) $f(x) = 3x - 5$ and $g(x) = 2x$
(2) $f(x) = x^2$ and $g(x) = 4 - x$
(3) $f(x) = \frac{1}{x}$ and $g(x) = x - 4$.

Similarly, $f(f(x))$, which is the function of the function itself, holds no terrors. We'll look at two examples to prove that this is so.

EXAMPLE (1) $f(x) = 2x + 3$ so $f(f(x)) = 2(f(x)) + 3 = 2(2x + 3) + 3 = 4x + 9$.
We can check that this works by putting $x = 2$, say. Then we can find $f(f(2))$ either by doing f twice, getting $f(2) = 7$ and $f(7) = 17$, or in one step using $f(f(x)) = 4x + 9$ so $f(f(2)) = 8 + 9 = 17$.
Try doing one for yourself before we go on.
If $g(x) = 2x^2 + 3$ what is $g(g(x))$? Check with $x = 1$.

$$g(g(x)) = g(2x^2 + 3) = 2(2x^2 + 3)^2 + 3$$
$$= 2(4x^4 + 12x^2 + 9) + 3 = 8x^4 + 24x^2 + 21.$$

Check: $g(1) = 5$ and $g(5) = 50 + 3 = 53$.
Alternatively, $g(g(1)) = 8 + 24 + 21 = 53$.

EXAMPLE (2) Now we'll find $f(f(x))$ if $f(x) = \dfrac{2x + 3}{3x + 2}$.

To find $f(f(x))$ we simply replace the x of the formula by $f(x)$, so we get

$$f(f(x)) = \frac{2\left(\frac{2x+3}{3x+2}\right) + 3}{3\left(\frac{2x+3}{3x+2}\right) + 2}.$$

We then simplify this unwieldy fraction by multiplying top and bottom by $(3x + 2)$. (Remember that this leaves the value of the fraction unchanged – see Section 1.C.(a) if necessary.) So we have

$$f(f(x)) = \frac{2(2x + 3) + 3(3x + 2)}{3(2x + 3) + 2(3x + 2)} = \frac{13x + 12}{12x + 13}.$$

We must exclude the one value of x for which the function is undefined by saying $x \neq -\frac{13}{12}$. This value would make $12x + 13 = 0$, and so involve us in trying to divide by zero which is impossible. (This is also in Section 1.C.(a).)

Try this very similar example for yourself, because it is also good practice for tidying up fractions within fractions, sort of double-decker fractions. See if you can get right through without referring back to the example above. (You could have another good look at that one first.)

If $f(x) = \dfrac{2x - 5}{4x + 1}$ find (a) $f(3)$, (b) $f(x^2)$, (c) $f(2x + 1)$ and (d) $f(f(x))$.

Here are the answers.
First of all, you wouldn't even consider cancelling the 2 and the 4 in the definition of $f(x)$. If you would, you should return to Sections 1.C.(a) and (b) and go through them again! You should have:

(a) $f(3) = \dfrac{2(3) - 5}{4(3) + 1} = \dfrac{1}{13}$

(b) $f(x^2) = \dfrac{2(x^2) - 5}{4(x^2) + 1} = \dfrac{2x^2 - 5}{4x^2 + 1}$

(c) $f(2x + 1) = \dfrac{2(2x + 1) - 5}{4(2x + 1) + 1} = \dfrac{4x - 3}{8x + 5}$

(d) $f(f(x)) = \dfrac{2\left(\frac{2x-5}{4x+1}\right) - 5}{4\left(\frac{2x-5}{4x+1}\right) + 1}$

$= \dfrac{2(2x-5) - 5\,(4x+1)}{4(2x-5) + (4x+1)}$ (multiplying top and bottom by $(4x+1)$)

$= \dfrac{-16x - 15}{12x - 19}$

$= \dfrac{16x + 15}{19 - 12x}$ (multiplying top and bottom by -1 to make the answer look more tasteful).

3.B.(g)   **Can we go back the other way? Inverse functions**
We have now worked with quite a large number of functions each of which gives us a rule for finding the function from any given starting value of x. We also know that, in order for this relationship to *be* a function, the rule must give just one possible answer for each starting value of x.

Is it possible to go back the other way? If we know a value of $f(x)$ for a particular function can we work out from this what the original value of x must have been?

Can you see any difficulty which we might have?

We can only do the backwards process if each value of $f(x)$ comes from just one possible x. This is why the answer to the second question of Section 3.B.(a) was so important. For example, in the case of function (b) which was $y = f(x) = x^2 - 2x - 3$, we have $f(4) = f(-2) = 5$. Therefore, from knowing that $f(x) = 5$, it is not possible to say what value of x gave this, since it could be either 4 or -2. Since the backwards relation has more than one possible answer, it is not a function.

The function (if it exists) which undoes the effect of $f(x)$ and brings you back to where you started, is called **the inverse function of** x. It is written $f^{-1}(x)$.

- A function can only have an inverse function if it is one-to-one. This means that $f(a) = f(b)$ only if $a = b$.
- If $f^{-1}(x)$ exists, then $f^{-1}(f(x)) = f(f^{-1}(x)) = x$.
- Each of f and f^{-1} undoes the effect of the other.

$f^{-1}(x)$ does *not* mean $1/f(x)$.

You can, if you want, write $1/f(x)$ as $(f(x))^{-1}$. It is just unfortunate that the mathematical way of writing these two very different things looks so similar.

For simple functions, it is often very easy to see what the inverse function must be. Here are two examples.

(1)   If $f(x) = x + 3$, then $f^{-1}(x) = x - 3$ so, for example, $f(4) = 7$ and $f^{-1}(7) = 4$.

(2)   If $g(x) = 5x$ then $g^{-1}(x) = \frac{1}{5}x$ so $g(2) = 10$ and $g^{-1}(10) = 2$.

Graphically, these two examples correspond to shifting x up and then shifting back down by 3 units in the case of (1), and stretching x and then shrinking it back by a factor of 5 in the case of (2). (These graphical effects were looked at in Section 3.B.(d).)

To make clearer what is happening here, it can sometimes be helpful to use an alternative way of writing functions which emphasises the carrying across or mapping of x into the function $f(x)$.

Taking $f(x) = x + 3$ as an example, we can also write this as $f: x \mapsto x + 3$ which means the function f in which x maps to $x + 3$. Then we write the inverse function as $f^{-1}: x \mapsto x - 3$.

Similarly, if $g:x \mapsto 5x$, then $g^{-1}: x \mapsto \frac{1}{5}x$.

Try finding the inverse functions of the following three functions yourself.

(1) $f(x) = x - 2$ (2) $g(x) = 2x$ (3) $p(x) = 6 - x$

You should have (1) $f^{-1}(x) = x + 2$ and (2) $g^{-1}(x) = \frac{1}{2}x$.

Students often find (3) a little bit tricky. Clearly, it *isn't* true that $p^{-1}(x) = 6 + x$ since this doesn't bring us back to where we started.

If you haven't been able to find an answer, try finding $p(1)$, $p(5)$, $p(2)$ and $p(4)$.

You will see that doing $p(x)$ twice brings you back to the original x, so that $p(x)$ is its own inverse function. We can say that $p(p(x)) = x$.

A function which is its own inverse is called **self-inverse**.

If $f(x)$ is self-inverse, then $f^{-1}(x) = f(x)$ so $f(f(x)) = x$.

---

(4) Can you find the inverse function for $q(x) = 12/x$?

Trying the pairs of values for x of 12 and 1, 6 and 2, and 3 and 4, shows us that this function is also self-inverse. These pairs of values are behaving symmetrically with respect to each other.

This is the same kind of relationship as those that we looked at in Section 3.A.(e) on inverse proportion. However, unlike the physical examples of inverse proportion which we looked at there, this function also includes negative pairs such as -3 and -4, and -2 and -6.

I show in Figure 3.B.14 graph sketches for the pairs of functions and their inverses from the four questions above, taking equal scales on the x and y axes.

This is a good place to add colour to the sketches yourself. If you use two colours so that you can highlight each function and its inverse function differently, you will bring

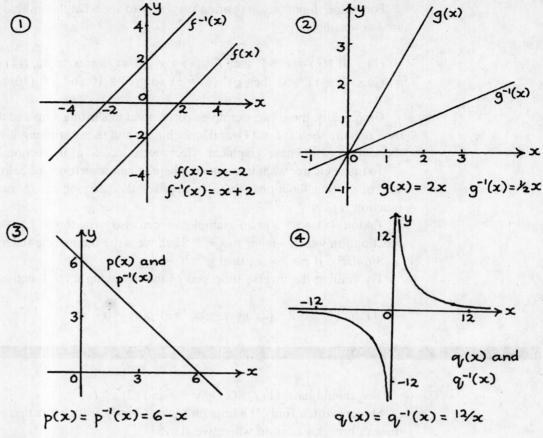

Figure 3.B.14

out two important points. The first is that the two self-inverse functions are the *same* function; they lie on top of one another. The second is that all the four pairs of graphs shown have the same line of symmetry. Try sketching in this line yourself on each of the four graphs.

Each function and its inverse function are symmetrically placed about the diagonal line $y = x$. This symmetry stresses the equal standing of each function with its inverse; each is the inverse of the other. They are mirror images of each other in the line $y = x$ because the original function is taking x to y, and the inverse function takes y back to x. This symmetry means that the domain, the set of all possible x values for the original function, is the same as the range, the set of all possible y values for the inverse function, and the range of the original function gives the domain of the inverse function.

For the two self-inverse functions, the original function is itself symmetrical about the line $y = x$. Each half of the line or curve reflects onto the other half, and therefore we can see geometrically that these functions *must* be their own inverses.

Notice that this symmetry means that it is always possible to sketch an inverse function if we know what the original function looks like. This sketching is easier if equal scales are chosen on the two axes, so that the line $y = x$ is at 45°. A quick sketch is much the easiest way of seeing how an inverse function works.

3.B.(h)   Finding inverses of more complicated functions

How can we find the inverse function if the starting function is more complicated? For example, what is f^{-1} for $f(x) = 2x - 5$ or $f: x \mapsto 2x - 5$?

It's not very easy to write down the answer immediately. (Try it and see, checking with some numbers to see if your answer works.) However, we can work out what it must be in the following way.

We have $y = f(x) = 2x - 5$. This gives the rule or formula for finding y if we know x.

We are looking for the rule which, if we know y, will take us back to the original x. We can find this by rearranging $y = 2x - 5$ to change it to the form $x =$ some rule involving y. This is called changing the subject of the formula to x, and we have already done this for some physical formulas in Section 2.A.(d).

We have $y = 2x - 5$ so $y + 5 = 2x$ so $x = \frac{1}{2}(y + 5)$, so giving us the rule which will take us back from y to the original x.

We can check that it works by doing a numerical test. If $x = 3$ then $y = 6 - 5 = 1$ and if $y = 1$ then $x = \frac{1}{2}(1 + 5) = 3$.

We now use the rule we have found to write the inverse function so that it is itself a function of x. Using the mirror-image property of the function and its inverse about $y = x$, we simply swap x and y getting $f^{-1}(x) = \frac{1}{2}(x + 5)$. The line giving $f^{-1}(x)$ is $y = \frac{1}{2}(x + 5)$.

I show both $f(x)$ and $f^{-1}(x)$ in Figure 3.B.15.

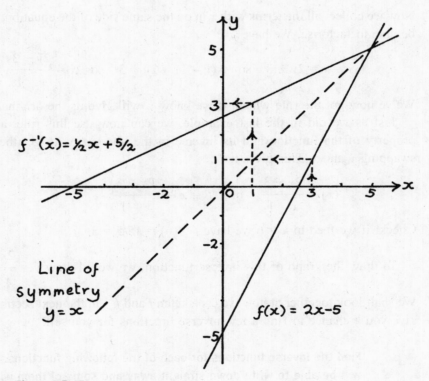

Figure 3.B.15

I have also shown $3 \mapsto 1$ using $f(x)$, and $1 \mapsto 3$ using $f^{-1}(x)$.

Can you work out where the two functions cross over each other?

They cross over where $f(x) = f^{-1}(x)$ so $2x-5 = \frac{1}{2}(x+5)$ giving $4x - 10 = x+5$ so $x = 5$.

Check: $f(5) = 10 - 5 = 5$ and $f^{-1}(5) = \frac{1}{2}(5 + 5) = 5$.

The crossing point is at $(5, 5)$ on the line $y = x$ which checks with what we know must be true geometrically.

We set about finding the inverse function for a function involving a fraction like $f(x) = (x+3)/(x-2)$ in exactly the same kind of way. We have

$$f(x) = \frac{x + 3}{x - 2} \quad \text{or} \quad f: x \mapsto \frac{x + 3}{x - 2}$$

meaning that, under the function f, x maps to $(x+3)/(x-2)$, so, for example, 3 maps to 6. Let

$$y = \frac{x + 3}{x - 2}$$

where y gives the outcome of feeding x into the function, as 6 is the outcome of feeding 3 into the function.

As before, we are looking for a formula which, if we know y, will take us back to the original x, so we change the subject of the formula to x.

$$y = \frac{x + 3}{x - 2} \quad \text{so} \quad y(x - 2) = x + 3 \quad \text{so} \quad xy - 2y = x + 3.$$

Now we collect all the terms with x in on the same side of the equation, because then we will be able to factorise. We have

$$xy - x = 2y + 3 \quad \text{so} \quad x(y - 1) = 2y + 3 \quad \text{so} \quad x = \frac{2y + 3}{y - 1}.$$

We've now got the rule which, if we know y, will give us the original x.

Just as we did in the last example, we can now use this rule, and the mirror-image property of the function and its inverse in the line $y = x$, to get the inverse function by swapping y and x. This gives us

$$f^{-1}(x) = \frac{2x + 3}{x - 1} \quad \text{or} \quad f^{-1}: x \mapsto \frac{2x + 3}{x - 1}.$$

Check: if we feed in $x = 6$ we have $f^{-1}(6) = 15/5 = 3$.

To draw the graph of this inverse function, we would draw $y = \dfrac{2x + 3}{x - 1}$

We shall look together at how we can sketch f and f^{-1} in the next section, but before that I'll give you a chance to find a few inverse functions for yourself.

EXERCISE 3.B.3

Find the inverse functions for each of the following functions. (Some of them you will be able to write down straight away and some of them will need rearranging like the last two examples.)

(1) $f(x) = 5x$ (2) $f(x) = x - 9$ (3) $f(x) = 5x - 9$

(4) $f(x) = 8 - x$ (5) $f(x) = x/4$ (6) $f(x) = 4/x$

(8) $f(x) = \dfrac{x - 3}{x + 2}$ $(x \neq -2)$ (9) $f(x) = \dfrac{2x + 3}{x - 2}$ $(x \neq 2.)$

We say $x \neq -2$ in (8) and $x \neq 2$ in (9) to make it clear that we don't think that we can divide by zero.

3.B.(i)   **Sketching the particular case of $f(x) = (x + 3)/(x - 2)$, and its inverse**

We will now look into how we can set about drawing graph sketches for

$$f(x) = \frac{x + 3}{x - 2} \quad \text{and} \quad f^{-1}(x) = \frac{2x + 3}{x - 1}.$$

Each of these functions is more complicated than any that we have sketched so far, but they have interesting properties that it will be useful for you to see here. Also, if we can draw a sketch for $f(x)$ we shall then be able to reflect this in the line of symmetry $y = x$ to draw the sketch of $f^{-1}(x)$.

In order to sketch $y = f(x)$ we need to find out what it does at all its interesting bits. We do this rather than making a table of values because we might choose the x values badly, so that what we sketched was just a boring bit, such as a piece of curve which is almost a straight line. (Many students panic at this stage, and make it into a completely straight line, so finishing up with a total disaster.)

To investigate the interesting bits, we need to answer the following questions.

(a)   When does $f(x) = 0$?
(b)   What is the value of $f(x)$ when $x = 0$?
(c)   Is there any value of x which we can't have because $f(x)$ would be undefined for this value? If so, what happens to $f(x)$ when x gets near this forbidden value?
(d)   What happens to $f(x)$ when x becomes very large?
        Test your theory with some large positive and negative values of x.

Try answering each of these four questions yourself for the function $f(x)$ above which we want to sketch.

(a) $f(x) = 0$ if $\dfrac{x + 3}{x - 2} = 0$.

This happens if $x = -3$. (Notice that we only have to look at the top of the fraction to answer this question. However many parts something is divided into, if you get none of those parts you've got nothing.)

We now know that $f(x)$ cuts the x-axis at $(-3, 0)$.

(b) $f(x) = -\frac{3}{2}$ when $x = 0$ so $f(x)$ cuts the y-axis at $(0, -\frac{3}{2})$.

(c)   We can't have $x = 2$ because we can't divide by zero.
        If x is very close to 2, say 1.999 or 2.001, then $(x - 2)$ is very small, and dividing by a very small number gives a very large result.
        Just before $x = 2$, $f(x)$ is very large and negative, and just after $x = 2$, $f(x)$ is very large and positive. (You can check this on your calculator if you wish.)
        $f(x)$ becomes closer and closer here to the line $x = 2$. (This line is called a **vertical asymptote**.)

(d)   What happens to $y = f(x) = \dfrac{x + 3}{x - 2}$ as x becomes very large?

The easiest way of seeing what must happen here is to divide the top and bottom of $f(x)$ by x. This gives us

$$f(x) = \frac{x + 3}{x - 2} = \frac{1 + (3/x)}{1 - (2/x)}.$$

Now, as x becomes very large, (either positive or negative), both $(3/x)$ and $(2/x)$ will become extremely small. The larger x becomes, the tinier they get, and indeed we can make them as small as we please by choosing a large enough value of x. (We can't actually make them equal to zero because this would require x to be infinitely large and, as we saw with the two straight lines in Section 1.E.(d), infinite quantities of things behave in strange ways.)

We see that, as x becomes very large, $f(x)$ will become closer and closer to $1/1 = 1$.

This means that we know that the curve of $y = f(x)$ becomes closer and closer to the straight line $y = 1$ as the values of x become larger and larger. (This line is called a **horizontal asymptote**.)

We now have enough information to be able to have a good try at sketching this curve. First, we draw the two axes and mark on them where the curve crosses them using our answers to (a) and (b). Then we draw in the two lines $y = 1$ and $x = 2$ which we know the curve gets closer and closer to. We then sketch in the curve which seems to fit in best with this information. I've done this in Figure 3.B.16.

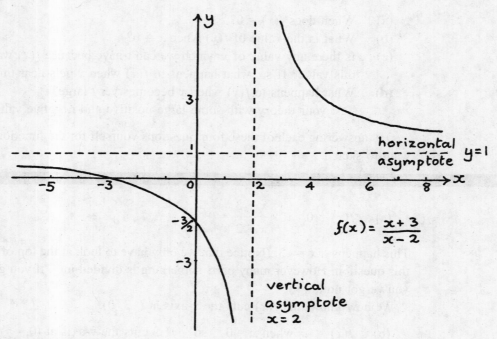

Figure 3.B.16

The only question we can't yet answer is how the slope of the curve changes from point to point. Could it perhaps have some kinks and wiggles that we don't know about? Finding out how slopes change is the subject of Chapter 8, and in Section 8.E.(c) I shall give you a full list of curve-sketching help which will include this. Also, in Section 8.C.(e), we shall show that this particular curve must always have a negative slope (except when $x = 2$).

For this particular curve, it is also possible to show that its slope is always downhill by taking any two points which lie on it which are both either to the left of $x = 2$ or to the right of it. If you then work out the gradient of the straight line joining them, you will find that it is always negative.

This curve is interesting because of another special property. It's only the second one we've met which does this particular thing. Can you see what it is?

It does a jump. This jump, which happens when $x = 2$, is called a **discontinuity**. Because of it, this curve can't be drawn with a continuous pencil line. (The other one like it is example (4) at the end of Section 3.B.(g) – in fact, it is very like it indeed. When we've finished this graph sketch, I shall show you how to turn this one into that one.)

Using the fact that the graph of $f^{-1}(x)$ is the same as the graph of $f(x)$ reflected in the line of symmetry $y = x$, we can now sketch both of these graphs together.

HELPFUL
HINT

If you are sketching an inverse function by this method, the best method for drawing it convincingly is to turn your paper so that the line $y = x$ is vertical. This makes it much easier to get f and f^{-1} symmetrically placed either side of this line.

I show my two graphs in Figure 3.B.17.

The two asymptotes of $y = f(x)$ will also be reflected in the line $y = x$ to give the corresponding pair of asymptotes of $y = f^{-1}(x)$.

Adding your own colours to f and f^{-1} and the two pairs of asymptotes $x = 2$ and $y = 1$, and $x = 1$ and $y = 2$ would help you to see exactly what is going on.

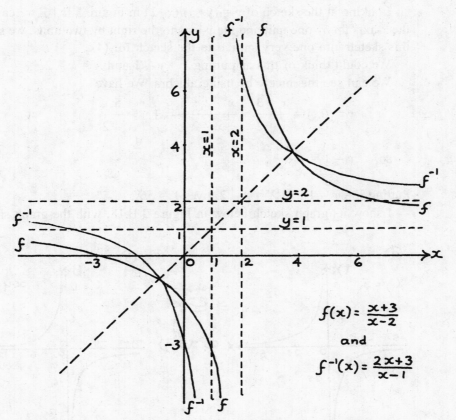

The asymptotes for $f(x)$ are the lines $x = 2$ and $y = 1$.

The asymptotes for $f^{-1}(x)$ are the lines $x = 1$ and $y = 2$.

Figure 3.B.17

3.B An introduction to functions

From this graph sketch, you can see the symmetry of the gaps in the domain and range of $f(x)$ and $f^{-1}(x)$ respectively. The value 2 is excluded from the domain, the set of possible x values for $f(x)$, and also from the range, the set of possible y values for $f^{-1}(x)$, and the value 1 is excluded from the range of $f(x)$ and the domain of $f^{-1}(x)$.

EXERCISE 3.B.4

Using similar methods to those we used together above, find out as much information as you can about the following two functions.

$$(1) \quad g(x) = \frac{x - 2}{x + 4} \qquad (2) \quad h(x) = \frac{2x - 5}{x + 1}$$

Use this information to sketch the graphs of the two functions. (Of course, for all of this sketching you could just use a graph-sketching calculator – but if you answer the questions for each curve like we did in the example, you'll know *why* it does what it does.)

Find also the two inverse functions, $g^{-1}(x)$ and $h^{-1}(x)$.

(3) Sketch the function

$$f(x) = \frac{2x + 3}{x - 2}$$

from question (9) of Exercise 3.B.3 and draw in the line $y = x$ on your sketch.

Now we find out how to turn $y = (x+3)/(x-2)$ into $y = 12/x$ which was (4) at the end of Section 3.B.(g).

Looking at the sketch of $y = (x+3)/(x-2)$ in Figure 3.B.16, we can see that, if we move the x-axis up by one unit and the y-axis to the right by two units, we shall have transformed this sketch into one very similar to the sketch for (4).

We could think of this as putting $Y = y - 1$ and $X = x - 2$.

We can see this nicely by using algebra. We have

$$y = f(x) = \frac{x + 3}{x - 2} = \frac{x - 2 + 5}{x - 2} = 1 + \frac{5}{x - 2}$$

so $\quad y - 1 = \dfrac{5}{x - 2}.$

Putting $Y = y - 1$ and $X = x - 2$ gives $Y = 5/X$.

I show its graph sketch below in Figure 3.B.18, with the graph sketch of $y = 12/x$.

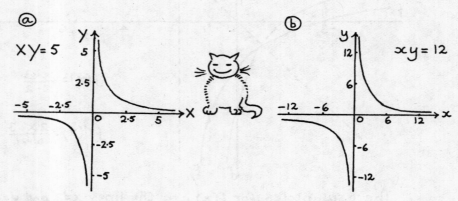

Figure 3.B.18

The only difference now is one of scale. If we shrink (b) by a factor of 5/12, we get the identical graph to (a).

Relations and functions

3.B.(j)   **Odd and even functions**

Make sketches for yourself of the graphs of the following four functions.

(a) $y = x$ (b) $y = x^2$ (c) $y = x^3$ (d) $y = |x|$.

$|x|$ means 'take the positive value whatever the sign of x itself'.
What kinds of symmetry do you see in your sketches? Describe them.

Your four graphs should show two different sorts of symmetry, so giving you examples of what are called **even** and **odd** functions.

Even functions

A function is even if it is symmetrical about the y-axis.
For these functions, $f(x) = f(-x)$ for any value of x.

The functions (b) and (d) above are both examples of this. The standard Normal distribution, which we talked about in Section 3.B.(e), is also an even function, and it is this property which makes it possible to halve the size of the tables needed to work with it.

The sketches for (a) and (c) show a different sort of symmetry. In each case, if we rotate the graph through a half turn about the origin, then it exactly fits onto itself. Put another way, turning the page upside down leaves the graph unchanged.

Odd functions

A function is odd if rotation through a half-turn leaves it unchanged.
This is the same as saying that the function reverses its sign if it is
reflected in the y-axis, so $f(x) = -f(-x)$.

Figure 3.B.19 shows my sketches of the four graphs for (a), (b), (c) and (d).

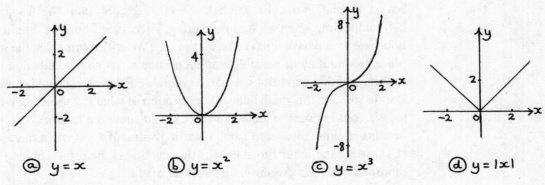

(a) $y = x$ (b) $y = x^2$ (c) $y = x^3$ (d) $y = |x|$

Figure 3.B.19

See if you can decide which of (a), (b), (c) and (d) have inverse functions.

(a) and (c) will each have an inverse function because each value of y is given by only one possible value of x, but (b) and (d) will only have inverse relations.

With (b) for example, if $y = 4$ then x could be $+2$ or -2.

If $y = x^2$ then $x = y^{1/2}$. The inverse relation is $x \mapsto x^{1/2}$, and $x^{1/2}$ can be either $+$ or $-$.

The sketch in Figure 3.B.20(a) shows the graphs of $y = x^2$ and its inverse relation $y = x^{1/2}$.

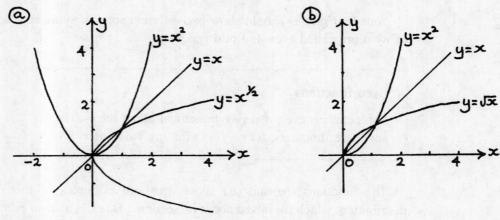

Figure 3.B.20

However, if we say that x cannot be negative, so that we restrict the domain of $y = x^2$ to values of x which are greater than or equal to 0 (which we write as $x \geq 0$), then we shall have a perfectly good inverse function which is $y = \sqrt{x}$. This is shown in Figure 3.B.20(b).

3.C   Exponential and log functions

3.C.(a)   Exponential functions – describing population growth

The functions which we shall look at in this next section are of huge importance to scientists and engineers. This is because they describe many physical situations where there is a smooth rate of growth which depends on how much of the substance is present at any particular time. An example of this is the process by which cell growth takes place through the repeated division of individual cells into two new cells.

To help us to see what is going on in this kind of situation, we'll look at what happens if we have a population of cells which doubles in size every hour. We'll suppose that there are 1000 cells at the time when we start measuring. Then after 1 hour we would have 2000 cells, after 2 hours we would have 4000 cells, and so on. (We will assume that the growth process is taking place as smoothly as possible, so that particular groups of cells don't all double at the same instant, and that conditions remain favourable for this continued growth. When the nutrients start to run out, this mathematical description of what is happening will break down.)

We could make the table shown below to show the number of cells present at particular instants in time, measured from a starting value of $t = 0$ when there are one thousand cells. (I am using the letter t to stand for time as this is the usual choice.) Then x, the number of thousands of cells present, is a function of t.

t (time in hours)	-2	-1	0	$\frac{1}{2}$	1	2	3	4
x (number of cells in thousands)			1		2	4		

I have left some gaps in the table. Try filling in these for yourself, in the following order:

(a) the numbers of thousands of cells which will be present after 3 hours and after 4 hours,

(b) the number of thousands of cells present both 1 hour and 2 hours before the measuring started,

(c) the number of thousands of cells present after half an hour.

(a) For this, you should have 8000 after 3 hours and 16 000 after 4 hours, giving $x = 8$ and $x = 16$. The rule that gives you these answers is $x = 2^t$.

(b) For this, you should have $x = \frac{1}{2}$ when $t = -1$, meaning that there were 500 cells present 1 hour before measuring started, and $x = \frac{1}{4}$ when $t = -2$, meaning there were 250 cells present 2 hours before the measuring started. These numbers fit in with the meanings which we gave to negative powers in Section 1.D.(b).

(c) From Section 1.D.(b), too, we take $2^{1/2}$ as meaning $\sqrt{2}$ so that there will be about 1414 cells after half an hour. You should go through this section now if you are unsure about these last results.

I show in Figure 3.C.1 a sketch of what happens if we plot the first seven of these pairs of values.

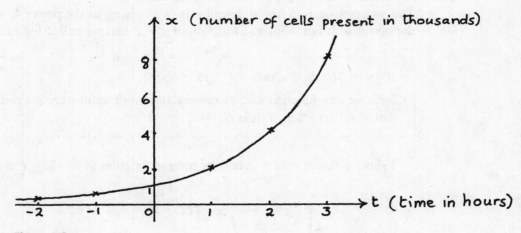

Figure 3.C.1

They appear to form part of a smooth curve, so it would seem reasonable to join them up in this way since it shows very well what is happening physically. We could then use the curve to read off values for 2^t which come between the points which we have plotted. (It's worth mentioning very briefly here that if the process of doubling is *not* smooth, so that it goes in definite steps like the numbers of people involved in a game which starts with one person picking a partner, and then both these people picking partners and so on, then the mathematical description of what is going on will be very different. We shall look at this situation in Section 6.C. Then, later on in Section 8.C.(a), we look at what happens if you start with stepped time intervals, but then make these intervals smaller and smaller, so that you are getting closer and closer to a continuous process – something which is at the heart of the maths of the physical world.)

3.C Exponential and log functions

Now try answering the following questions yourself.

(1)   How many cells will there be after 5 hours?
(2)   How many cells are there after $1\frac{1}{2}$ hours?
(3)   How long is it until there are 16 000 cells?
(4)   How long is it until there are 64 000 cells?

As you answer these four questions, you will probably guess what I'm working towards here. The answers go as follows.

(1)   There will be 32 000 cells after 5 hours (that is, 1000×2^5).
(2)   After $1\frac{1}{2}$ hours there will be approximately 2828 cells (that is, $1000 \times 2^{3/2}$), using a calculator for $2^{3/2}$ and giving the answer to the nearest whole number.
(3)   It takes 4 hours to get 16 000 cells because $1000 \times 2^4 = 16\,000$.
(4)   It takes 6 hours to get 64 000 cells because $1000 \times 2^6 = 64\,000$.

The last two questions are put the other way round from the first two so that, to find the answers, you have to go back from a known x to find the t which gave it. In other words, you are using the inverse function of $x = 2^t$.

So what *is* this inverse function that you are using?

The answer to this question is so important that it needs a section of its own.

3.C.(b)   The inverse of a growth function: log functions

This inverse function has to describe $16 = 2^4$ giving us the power 4, and $64 = 2^6$ giving us the power 6. It is the inverse function of $x = 2^t$ and we call it log to the base 2.

> If $x = f(t) = 2^t$ then $f^{-1}(t) = \log_2 t$.
>
> Because any function and its inverse also work opposite ways round, it is also true that if $f^{-1}(t) = \log_2 t$ then $f(t) = 2^t$.

I show a sketch of $x = 2^t$ and its inverse function of $x = \log_2 t$ in Figure 3.C.2.

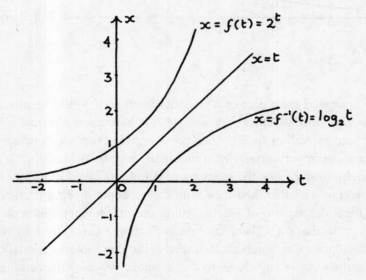

Figure 3.C.2

We know that these curves work well for giving a description of what is happening physically. We can't therefore allow negative roots here, since these would give us points which would not lie on the curve of $x = 2^t$. (For example, we don't want $x = -\sqrt{2}$ when $t = \frac{1}{2}$.) For this reason we only include positive roots, meaning that our inverse function is safe.

This means that we can only have logs of positive numbers.

3.C.(c) Finding the logs of some particular numbers

Many students find logs rather alarming. They are so important in applications that it's important for you not to be scared of them, so now we will look at some particular examples of how they actually work.

We have already seen the particular cases of $\log_2(2^4) = 4$ and $\log_2(2^6) = 6$ from the answers to questions (3) and (4) in the previous section.

We can say that if some number $n = 2^t$ then $t = \log_2 n$.

This means that if we can write any particular number as a power of 2 then it is very easy to write down its log to base 2. Here are two examples.

(1) $128 = 2^7$ so $\log_2(128) = 7$ and   (2) $1/8 = 1/2^3 = 2^{-3}$ so $\log_2(1/8) = -3$.

EXERCISE 3.C.1

Some of the questions in this exercise use the special results for powers from Section 1.D.(b) – you may need to go back to these before you do them.

(1) Try finding the logs to base 2 of the following yourself.
   (a) 4   (b) 8   (c) 2   (d) 1   (e) $\frac{1}{2}$ (f) $\frac{1}{4}$

(2) Logs to other bases work in exactly the same sort of way.
   For example, $27 = 3^3$ so $\log_3 27 = 3$.
   Try finding the logs to base 3 of the following numbers yourself.
   (a) 9   (b) 81   (c) $\frac{1}{27}$ (d) $\frac{1}{3}$ (e) 1   (f) 3   (g) $\frac{1}{9}$ (h) 27   (i) $\sqrt{3}$

(3) Now try finding the logs to base 10 of these numbers.
   (a) 100   (b) 1000   (c) 10   (d) 1   (e) $\frac{1}{10}$ (f) 0.01

Some important points come out of the answers to this exercise. This is the first.

It is always true that $\log_a a = 1$ and $\log_a 1 = 0$ for any base a.

We'll also widen the definition of logs to a general base, here.

If $x = a^t$ then $t = \log_a x$ and   if $t = \log_a x$ then $x = a^t$.

Also, logs to base 10 are given on your calculator, because we count in base 10. This means that you can get the same answers to question (3) above by using your calculator – do this, just to check. You will need to use the key marked 'lg' or 'log'. (The one marked 'ln' or 'log$_e$' will give you a different sort of log which I'll come to in Section 3.C.(e).) Because logs to base 10 are so common, we don't usually bother to write the little 10 below. Your calculator will also give you values for all those in-between points on the smooth curve of $x = \log_{10} t$ where we can't work out the answers in the way we've done the ones above. We can't explain mathematically how it does this yet.

3.C.(d) The three laws or rules for logs

In Section 1.D.(a) we wrote down the three rules for working with powers. These are as follows:

> **Rule (1)** $a^m \times a^n = a^{m+n}$
>
> **Rule (2)** $a^m \div a^n = a^{m-n}$
>
> **Rule (3)** $(a^m)^n = a^{mn}$

We showed there that they worked for whole number powers, and said that they do, in fact, work for any values of m and n provided that $a \neq 0$. We can't yet show that this is true though at least now we have a mental picture of the graph of $x = a^t$ to give us some idea of how the intermediate values work. Our next results come from assuming that the three laws above are indeed true.

The special striking property of these three laws of powers is that they make things easier. They write a multiplication in the form of an addition, a division in the form of a subtraction, and raising to a power in the form of a multiplication.

Because logs are the inverses of powers, they also have this property of making things nicer. Through the three rules for powers, we get the three rules for logs which I have put in a box below.

> **The three rules for working with logs**
>
> **Rule (1)** $\log_a (xy) = \log_a x + \log_a y$
>
> **Rule (2)** $\log_a \left(\dfrac{x}{y} \right) = \log_a x - \log_a y$
>
> **Rule (3)** $\log_a (x^n) = n \log_a x$

I will show you through a numerical example how the first rule for logs comes from the first rule for powers.

Suppose we have $\log_3 (9 \times 81)$.

Then Rule (1) says that $\log_3 (9 \times 81) = \log_3 9 + \log_3 81$.

Can we show by using the first rule of powers that the LHS is equal to the RHS above?

We know that $9 = 3^2$ and $81 = 3^4$ so we can say that $\log_3 9 = \log_3 (3^2) = 2$ and $\log_3 81 = \log_3 (3^4) = 4$.

Therefore the RHS $= \log_3 9 + \log_3 81 = 2 + 4 = 6$.

We can also say that the LHS $= \log_3 (9 \times 81) = \log_3 (3^2 \times 3^4) = \log_3 (3^{2+4}) = 2 + 4 = 6$.

Therefore we have shown that the RHS is equal to the LHS.

In exactly the same way, suppose we have $\log_a (xy)$ and we rewrite each of x and y as powers of a, so that $x = a^m$ and $y = a^n$.

This then means that $m = \log_a x$ and $n = \log_a y$. Then

$$\log_a (xy) = \log_a (a^m \, a^n) = \log_a (a^{m+n}) \quad \text{(from the first rule)}$$

$$= m + n = \log_a x + \log_a y.$$

We can see from what we have just done that it *cannot* be true that $\log_a (x + y) = \log_a x + \log_a y$ (except for the very special case when $xy = x + y$).

We can show similarly that $\log_a (x/y) = \log_a x - \log_a y$.
Again, we start by looking at a numerical example.
Can you show that $\log_2 (32/4) = \log_2 32 - \log_2 4$?

We can say that $\log_2 (32/4) = \log_2 (2^5/2^2) = \log_2 (2^{5-2}) = 5 - 2 = 3$.
Also $\log_2 32 - \log_2 4 = \log_2 2^5 - \log_2 2^2 = 5 - 2 = 3$.
Therefore the LHS above is equal to the RHS.

Now we show in a more general way that

$$\log_a \left(\frac{x}{y} \right) = \log_a x - \log_a y.$$

We rewrite x as a^m and y as a^n as we did before. Then $\log_a x = m$ and $\log_a y = n$. So

$$\log_a \left(\frac{x}{y} \right) = \log_a \left(\frac{a^m}{a^n} \right) = \log_a (a^{m-n}) \qquad \text{(from Rule (2))}$$

$$= m - n = \log_a x - \log_a y.$$

Finally, we look at $\log_a x^n$.
Taking a numerical example first, can you show that $\log_2 (8^4) = 4\log_2 8$?

You can say that $8^4 = (2^3)^4 = 2^{12}$ from Rule (3), so $\log_2 8^4 = \log_2 2^{12} = 12$.
Also, $\log_2 8 = \log_2 2^3 = 3$, so $4 \log_2 8 = 4 \times 3 = 12$.
Therefore, $\log_2 (8^4) = 4 \log_2 8$.

We now show in a more general way that

$$\log_a x^n = n \log_a x.$$

We rewrite x as a^m, so $m = \log_a x$.

Now, we have $\log_a (a^m)^n = \log_a(a^{mn})$ (from Rule (3)) $= mn = n\log_a x$.

A little piece of history
Before calculators were invented, the multiplication and particularly the division of large numbers were very tedious and time-consuming processes. However, it was realised that if the numbers could be written as powers of 10, the processes could be converted into addition instead of multiplication, and, even better, subtraction instead of division. Books with tables of these corresponding powers were published, to use for these calculations.

You can relive the experience of past days by using logs to divide 231.4 by 27.2.

First, find the logs of the two numbers on your calculator, then subtract the second from the first, and finally do INV log or SHIFT log. You get the result 8.5074 to 4 d.p., an answer which you, of course, can obtain far more quickly by simply feeding in the original numbers and pressing the ÷ button. Back in those days, finding the logs from log tables and then subtracting them was vastly preferable to the alternative of long division. Calculators are a great blessing for those faced with complicated arithmetic.

For you, the three rules or laws of logs will be of great importance when you are solving physical problems. They can be used either for splitting expressions up or for combining separate logs together. Being able to rearrange in both directions is important so I will give two examples of each.

In the first two, we split up as far as possible.

EXAMPLE (1) $\log_2 8x^2 = \log_2 8 + \log_2 x^2 = \log_2 2^3 + 2\log_2 x = 3 + 2\log_2 x.$

EXAMPLE (2) $\log_2 (3x^2/y^3) = \log_2 (3x^2) - \log_2(y^3) = \log_2 3 + \log_2 x^2 - \log_2 y^3$
$$= \log_2 3 + 2\log_2 x - 3\log_2 y.$$

In the second two examples, we combine as far as possible.

EXAMPLE (3) $\log_2 3 + 4\log_2 x = \log_2 3 + \log_2 x^4 = \log_2 (3x^4).$

EXAMPLE (4) $\log_{10}(x^2 + 1) - \log_{10}(x^2 - 1) = \log_{10}\left(\dfrac{x^2 + 1}{x^2 - 1}\right).$

You can't split the insides of the brackets here!

EXERCISE 3.C.2

(1) Use the rules of logs to split the following expressions up into separate logs (or numbers) as much as possible.
(a) $\log_3 3x$ (b) $\log_3 27x^2$
(c) $\log_3 (x/y)$ (d) $\log_3 (x^2/y^2)$
(e) $\log_3 (ax^n)$ (f) $\log_3 (9a^x)$ (g) $\log_3 (2x + 3y)$

(2) Combine the logs in the following as far as possible, using the laws of logs.
(a) $\log_{10} x + \log_{10} (x - 1)$ (b) $2\log_{10} x - \log_{10} y$
(c) $\log_{10} (x + 1) - \log_{10} (x - 1)$ (d) $3\log_{10} x + 2\log_{10} y$

3.C.(e) What are 'e' and 'exp'? A brief introduction

In the physical example of cell growth in Section 3.C.(a), the number of cells present at any particular time t was given by the equation $x = 2^t$. Also, the rate of increase of this number of cells was directly proportional to the number of cells present at any particular time. Using the ideas of Section 3.A.(a), we could say that

the rate of increase = k (the number of cells present)

where k is some constant. (We aren't yet in a position to work out the value of this constant – this has to wait until Section 8.F.(d).)

The special and particular property of the number e is that the rate of growth at any instant of a quantity x given by $x = e^t$ is actually equal to x itself. The constant of proportionality, k, is equal to 1, which greatly simplifies many situations. We can't go into what this will mean mathematically until Section 8.B, but because functions involving e are of central importance in describing many physical processes, you are likely to meet them early on in your course. This is why I'm putting in this brief introduction for you here.

The value of e lies between 2 and 3, and its value to 3 d.p. is 2.718. (It is a number like π which cannot be written with an exact numerical value.)

The curve of $x = e^t$ lies between the curves of $x = 2^t$ and $x = 3^t$. I show this in Figure 3.C.3.

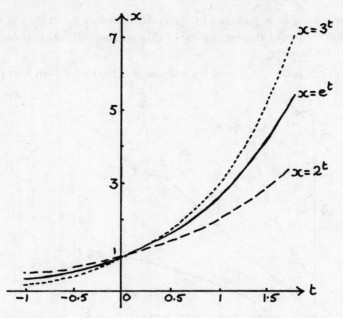

Figure 3.C.3

Notice that all the curves pass through the point (0,1), because $2^0 = e^0 = 3^0 = 1$.

You may sometimes see e^t written as exp(t). (The 'exp' is short for 'exponent'.) This notation is particularly useful if you have a complicated power of e because it makes it much easier to read than the tiny writing of a power.

> The word 'exp' is also sometimes used by calculators when they display very large or very small numbers in scientific notation. For example, 314 000 might be displayed as 3.14 EXP 5, meaning $314\,000 = 3.14 \times 10^5$, or 0.00176 might be displayed as 1.76 EXP −3, meaning $0.00176 = 1.76 \times 10^{-3}$.
>
> When 'exp' is used like this, it is referring to powers of 10 not e.

Calculators also sometimes use a gap instead of putting 'exp' when they are displaying numbers in scientific notation. They may also write the power of 10 raised above the level of the number. It is important for you to know how your own calculator does this. If you are at all unsure, put in $(600\,000)^2$. This is 3.6×10^{11} in scientific notation, and you will be able to see just how your calculator displays the 3.6 and the 11. (Your calculator will display this number in this way because it is too large for the conventional display.)

Logs to base e are written as 'ln' or '$\log_e$'. They are often shown as 'ln' on calculators. Because the behaviour of e^t and therefore of $\ln t$ is so special, these logs are often called **natural logs.** We can say

> if $x = e^t$ then $t = \ln x$
>
> and
>
> if $t = \ln x$ then $x = e^t$.

3.C Exponential and log functions

One example of how e creeps into physical laws is given by the value of the constant k which we referred to at the beginning of this section. We shall show in Section 8.F.(d) that $k = \ln 2$.

I show a sketch of $x = e^t$ and its inverse function of $x = \ln t$ in Figure 3.C.4.

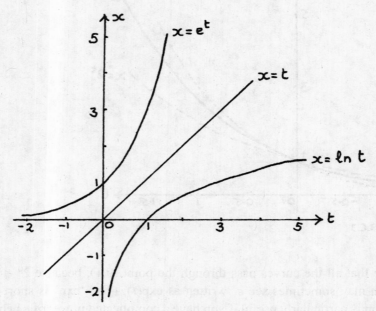

Figure 3.C.4

If you plot the curve of $y = e^x$ as accurately as possible on graph paper, taking values of x between 0 and 4 inclusive, you will be able to see more clearly how the curve builds up. (You can fill in as many intermediate points as you wish, using the e^x button on your calculator. The curve of $y = e^x$ is exactly the same as that for $x = e^t$. We are just using different letters.)

You will see that the steepness of the curve is changing smoothly as the value of x increases. Clearly this is a very different situation from the graphs of straight lines where the steepness, or rate of change of y with respect to x, remains the same, and they have a constant gradient.

Can you think of a way of estimating the steepness or rate of change of the curve of $y = e^x$ when $x = 1.5$, by drawing in a straight line and finding its gradient? (If you choose different scales on the two axes, be careful to allow for this when you find the gradient of the line.)

What answer do you expect to get?

3.C.(f)   Negative exponential functions – describing population decay

The situations represented by the graphs of $x = 2^t$ and $x = e^t$ are examples of what is called **exponential growth**.

What would the graphs of $x = 2^{-t}$ or $x = e^{-t}$ represent?

I show some values for $x = 2^{-t}$ in the table below.

t	-3	-2	-1	0	1	2	3	4
x	8	4	2	1	$\frac{1}{2}$	$\frac{1}{4}$	$\frac{1}{8}$	$\frac{1}{16}$

You will see that the values match those of the table on page 114 except that they have been switched either side of $t = 0$.

I have drawn a sketch of the graphs of $x = 2^{-t}$ and $x = 2^t$ together on the same axes in Figure 3.C.5(a). This shows that they are mirror images of each other in the vertical axis.

In Figure 3.C.5(b), I have sketched the two graphs of $x = e^t$ and $x = e^{-t}$. These, like all similar pairs of equations, also form a pair of mirror images of each other in the vertical axis. These mirror images will always intersect each other at the point $(0,1)$ since $a^0 = 1$ for all non-zero values of a.

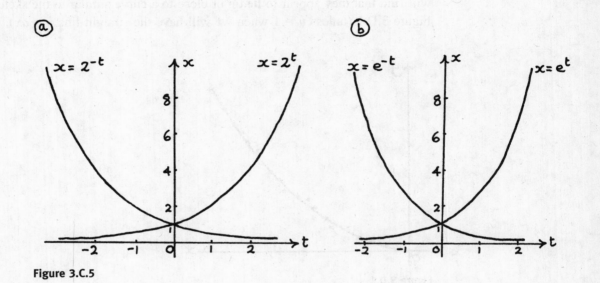

Figure 3.C.5

> **!** Don't confuse the graph of $x = e^{-t}$ with the graph of $x = -e^t$. The second of these is the same as the graph of $x = e^t$ except that every value of x has now become negative. Therefore it is the same as the graph of $x = e^t$ reflected in the horizontal axis.

The graph of $x = 2^{-t}$ could represent the radioactive decay of 1 tonne of a substance with a half-life of one hour. (This means that during each hour the mass of the substance becomes half what it was at the beginning of that hour. The total mass of substance present will probably not change very much since most radioactive elements decay into another element with a very similar mass.) The left-hand side of the graph then shows the mass of the substance present at various times before the instant when we started measuring. These times therefore have negative values.

This graph represents what is called the **exponential decay** of the substance.

We shall look at this kind of situation in more detail in the first example in Section 9.C.(b).

3.D   Unveiling secrets – logs and linear forms

The use of logs gives us an extremely powerful method for analysing experimental results to reveal underlying physical laws of relationship. This section describes how this works. There are some practical applications of these methods to physical examples in Section 9.C.(b), where we look at how we can solve some equations involving rates of change.

3.D.(a)   Relationships of the form $y = ax^n$

Suppose that we have a table of pairs of experimental measurements x and y, and we suspect that there is a relationship between x and y of the form $y = ax^n$, where a and n are two constants which we want to find out.

If our suspicion is correct, and we plot the points given by the pairs on graph paper, we will find that they appear to lie on or close to a curve similar to the sketch I have shown in Figure 3.D.1 (unless $n = 1$ when we will have the straight line $y = ax$).

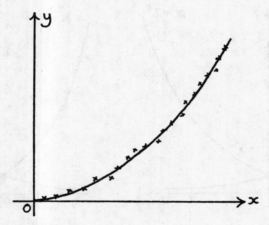

Figure 3.D.1

But this curve will take us no further forward since we can't see from it what its equation is, and so we can't find out from it what a and n are.

However, we know that we *can* get information from a straight line. If we have a straight line with the equation $y = mx + c$ then m is the gradient of the line, and c is its y intercept. (Look in Sections 2.B.(d) and (e) if necessary.)

If we can somehow convert the curve into a straight line, we shall be able to read useful information from it.

How can we do this?

We can take logs of both sides of the equation $y = ax^n$. We do this usually either to base 10, or by finding natural logs (i.e. to base e), since these are the two possibilities given on calculators. In my example, I use logs to base 10.

Then we use the laws of logs to write this new equation in a simpler form.

These three laws or rules of logs come in Section 3.C.(d). As we shall be using them a lot here, I have put them in again for you.

The three laws of logs

$\log(ab) = \log a + \log b$

$\log(a/b) = \log a - \log b$

$\log(a^n) = n \log a$

To fit any of these laws, all the logs involved must be taken to the same base.

 Remember that $\log(a + b)$ is *not* equal to $\log a + \log b$.

If we take logs on both sides of the equation $y = ax^n$, we get

$\log y = \log(ax^n) = \log a + \log(x^n) = n \log x + \log a.$

Now we compare this with the equation of a straight line, $Y = mX + c$.

I've put this in a box for you as it is important.

Finding a linear form for $y = ax^n$

Taking logs gives $\log y = n \log x + \log a$.

Comparing this with $Y = mX + c$ gives

$Y = \log y, \quad X = \log x, \quad m = n \quad$ and $\quad c = \log a$.

So we can now see that if the physical relationship *is* of the form $y = ax^n$ then we should get an approximate straight line if we plot $\log y$ against $\log x$. (I say 'approximate' because if these are experimental values there is likely to be some error in the measurements.)

Drawing a line of best fit through the points will give us something similar to the sketch I have shown in Figure 3.D.2(a). The reason for drawing this line of best fit is that it evens out the inaccuracies as much as possible since it uses *all* the data that we have. Trying to calculate an equation from just two of the pairs of values which we found from taking the logs would be less accurate. Sometimes you may draw this line in by eye, or in some cases you may do the job more accurately by finding a regression line, in which case you will be able to write down the values for $c = \log a$ and $m = n$ immediately from its equation.

If you have drawn a line of best fit by eye, you will now have to use it to find your c and m, so I will explain to you next how you would do this from your graph.

This graph will look similar to my drawing of Figure 3.D.2(a).

In Figure 3.D.2(b), I show a sketch on which I have put some numerical values, so that I can more easily explain to you the process for the next stage.

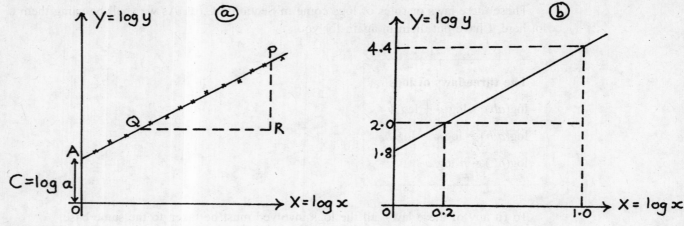

Figure 3.D.2

Firstly, we use the graph to find the value of c. This is given by reading off the value of the Y-intercept. This gives us $c = \log a$ in (a) and $c = \log a = 1.8$ in the numerical example of (b), so $a = 63$ to 2 s.f.

Secondly, because we now have a straight line, we can find its gradient by using any two points lying on the line. (This is explained in Section 2.B.(d).) Because this is a line of best fit, it may be that neither of these points corresponds to an actual pair of plotted measurements.

The gradient is given by PR/QR in (a), and $2.4/0.8 = 3$ giving $n = 3$ in (b). The graph of Figure 3.D.2(b) would give us the result that the pairs of measurements x and y are linked by the relationship $y = 63x^3$.

 Remember that you must take account of the scales that you have used on your horizontal and vertical axes when you work out the gradient of your line. You can't do it simply from the graph paper squares.

Dealing with a possibly tricky situation

In order to make the best use of the pairs of measurements that you have, it is often better to use only the parts of the scales which cover the range of your measurements, rather than showing the entire scale from zero at the origin. The convention for showing that you have done this is to use a zig-zag at the origin as I have done on my X-axis in Figure 3.D.3.

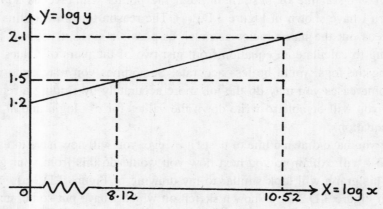

Figure 3.D.3

It's quite easy to find the gradient of the line here, as it is

$$\frac{2.1 - 1.5}{10.52 - 8.12} = \frac{1}{4}.$$

> The tricky bit is finding the Y-intercept correctly. It *isn't* 1.2 because of the break in the x-axis which means that it is not true that $Y = 1.2$ when $X = 0$.
>
> But, since we now know that the gradient of the line is $\frac{1}{4}$, we know that its equation is $Y = \frac{1}{4}X + c$.
>
> We also know that $Y = 1.5$ when $X = 8.12$, so $c = 1.5 - 2.03 = -0.53$.
>
> But $c = \log a$ so $a = 0.295$ and we have the equation linking the measurements as $y = 0.295x^{1/4}$.

3.D.(b)   **Relationships of the form $y = an^x$**

Suppose we have a table of pairs of experimental measurements x and y, and this time we suspect there is a relationship between them of the form $y = an^x$ where, as before, a and n are two constants for which we want to find the value.

Just like last time, if this relationship is true, plotting y against x will give us a curve from which we can obtain no further information except that there does seem to be *some* form of relationship.

Try taking logs of both sides of the equation $y = ax^n$ yourself, and see if you can work out what we should make X and Y be so that we get a straight line when we plot Y against X.

Taking logs of both sides of the equation $y = an^x$, you should have

$$\log y = \log(an^x) = \log a + \log(n^x) = x \log n + \log a.$$

I've put the next part of the working in a box for you, so that it is easy to refer to when you need it. This is what you should have found.

> **Finding a linear form for $y = an^x$**
>
> Taking logs gives $\log y = x \log n + \log a$.
>
> Comparing this with $Y = mX + c$ gives
> $Y = \log y$, $m = \log n$, $X = x$ and $c = \log a$.

Therefore, plotting $Y = \log y$ against $X = x$ should give us a straight line if our suspicion is correct.

Doing this will give us a sketch similar to Figure 3.D.4(a).

Again, I have shown a numerical example in Figure 3.D.4(b).

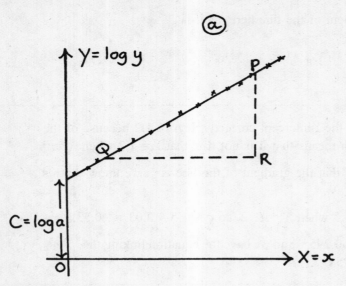

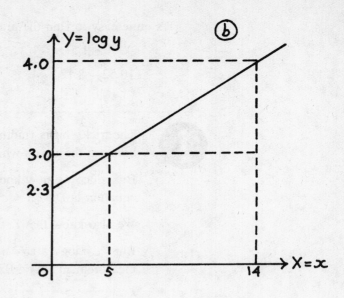

Figure 3.D.4

From Figure 3.D.4(a) we have $c = \log a$ and $m = \log n = PR/RQ$.

From Figure 3.D.4(b) we have $\log a = 2.3$ so $a = 200$ to 2 s.f. and $m = \log n = \frac{1}{9}$ so $n = 1.3$ to 2 s.f.

This would mean the original relationship in this case was $y = 200(1.3^x)$.

If we do not know which of these forms the relationship has, then it would be sensible to try both $\log y$ against $\log x$, and $\log y$ against x, in the hope of getting a straight line.

It is possible to do this by using special log/linear or log/log graph paper, which saves you having to do the logging yourself.

The log scales are in powers of 10 called cycles, so you would choose the number of cycles according to the range of measurements you need to cover. For example, if this range runs from 27 to 1540, then you would need the three cycles 10–100, 100–1000 and 1000–10 000.

3.D.(c)   What can we do if logs are no help?

Unfortunately, it isn't possible to bring all relationships to a linear form by taking logs both sides.

For example, if we suspect a relationship of the form $y = a + bx^2$, taking logs both sides does not help us since $\log(a + bx^2)$ cannot be split up, and so the values of a and b will remain hidden inside the log.

It isn't true that $\log(a + bx^2)$ is the same as $\log a + \log(bx^2)$.

If you think this *should* be true, go quickly back to Section 3.C.(d) and sort out these risky ideas.

All is not lost in the search for the values of a and b.

If you compare $y = a + bx^2$ with $Y = mX + c$, what could you choose for Y and X for the points to lie on a straight line?

How would you then find the values of a and b from this straight line?

Plotting $Y = y$ against $X = x^2$ will give a straight line if the relationship *is* $y = a + bx^2$. In this case, a is the y intercept, and b is the gradient of this line.

This may seem surprising so I will show you that it works by taking the example of $y = 3 + 2x^2$ (which you will recognise gives the left-hand sketch of Figure 3.D.5(a)).

Plotting y against x^2 from the table of values in Figure 3.D.5(b) gives the straight line shown in Figure 3.D.5(c).

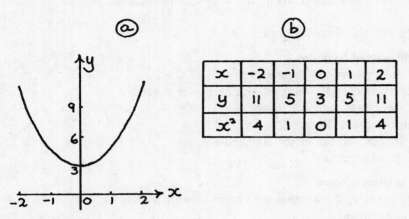

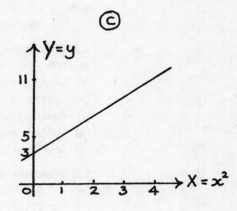

Figure 3.D.5

This straight line has a y intercept of 3 and its gradient is $(11 - 3)/4 = 2$, so $a = 3$ and $b = 2$, giving us the equation we know we should have of $y = 3 + 2x^2$.

If you suspected a relationship of the form (1) $y = a + bx^3$ or (2) $y = a + b\sqrt{x}$ what would you plot in each case in order to get a straight line if your theory is correct?

For (1), you would try plotting values of y against values of x^3.

For (2), you would try plotting values of y against values of $\sqrt{x}$.

You will see that the problem we have here is that, in order to get the straight line, we need to know what power of x is involved. In the first example which we looked at, the logs took care of that problem for us.

4 Some trigonometry and geometry of triangles and circles

This chapter reminds you of what trig is for, and how it works in triangles. It also explains some of the special geometrical properties of triangles and circles, because they may be very useful to you in applications of maths to your own special subject area.

The chapter is divided into the following sections.

4.A Trigonometry in right-angled triangles
(a) Why use trig ratios? (b) Pythagoras' Theorem,
(c) General properties of triangles, (d) Triangles with particular shapes,
(e) Congruent triangles – what are they, and when?
(f) Matching ratios given by parallel lines,
(g) Special cases – the sin, cos and tan of 30°, 45° and 60°,
(h) Special relations of sin, cos and tan

4.B Widening the field in trigonometry
(a) The Sine Rule for any triangle, (b) Another area formula for triangles,
(c) The Cosine Rule for any triangle

4.C Circles
(a) The parts of a circle, (b) Special properties of chords and tangents of circles,
(c) Special properties of angles in circles,
(d) Finding and working with the equations which give circles,
(e) Circles and straight lines – the different possibilities,
(f) Finding the equations of tangents to circles

4.D Using radians
(a) Measuring angles in radians,
(b) Finding the perimeter and area of a sector of a circle,
(c) Finding the area of a segment of a circle,
(d) What do we do if the angle is given in degrees?
(e) Very small angles in radians – why we like them

4.E Tidying up – some thinking points returned to
(a) The sum of interior and exterior angles of polygons,
(b) Can we draw circles round *all* triangles and quadrilaterals?

4.A	**Trigonometry in right-angled triangles**
4.A.(a)	**Why use trig ratios?**

When you began learning trigonometry (often referred to as 'trig'), you will have started by working with right-angled triangles. Since my policy is to make sure of the groundwork for each topic before going further, I will start from here, too.

We begin by looking at the right-angled triangle *ABC* shown in Figure 4.A.1.

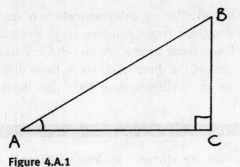

Figure 4.A.1

We will describe the sides of this triangle by their position relative to the angle at A.

BC is the side **opposite** to angle A (opp. for short).

AC is the side **adjacent** to angle A (adj. for short).

(The word 'adjacent' means 'lying next to').

AB is the longest side, opposite to the right angle. It is called the **hypotenuse** (hyp. for short).

Then we give particular names to each of the ratios of the different pairs of sides. We say:

$$\sin A = \frac{BC}{AB} = \frac{\text{opp.}}{\text{hyp.}}, \quad \cos A = \frac{AC}{AB} = \frac{\text{adj.}}{\text{hyp.}}, \quad \tan A = \frac{BC}{AC} = \frac{\text{opp.}}{\text{adj.}}.$$

To do the thing thoroughly, the ratios obtained by turning the above three ratios upside down are also given names as follows:

$$\frac{1}{\sin A} = \frac{AB}{BC} = \text{cosec } A, \quad \frac{1}{\cos A} = \frac{AB}{AC} = \sec A, \quad \frac{1}{\tan A} = \frac{AC}{BC} = \cot A.$$

These three ratios are the **reciprocals** of the first three ratios.

(Sin, cos, tan, cosec, sec and cot are all shortened versions of longer names which are relatively rarely used. They are, in the same order, sine, cosine, tangent, cosecant, secant and cotangent.)

The question now is why did anyone think these different ratios so important that they ought to be given special names? We can see the answer to this by looking at the triangles in Figure 4.A.2 which are nested into each other because they are the same shape. Only their

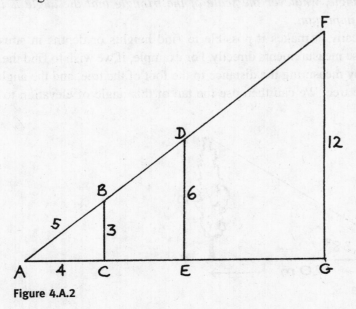

Figure 4.A.2

size is different. Triangles *ADE* and *AFG* are **enlargements** of triangle *ABC*. It is as though triangle *ABC* is stretched out into these larger triangles under a constant pull, so that all the proportions stay the same. (If it is some time since you did any trig, you may find that it helps you to draw in the outlines of the three triangles in three different colours.)

From the lengths shown on the triangles, how long will the sides *AE*, *AD*, *AG* and *AF* be?

Triangle *ADE* has sides which are all twice as long as triangle *ABC*, since it is just a scaled-up version of triangle *ABC*. So *AE* = 8 and *AD* = 10 units long.

Similarly, triangle *AFG* is scaled up by a factor of 4, so *AG* = 16 units and *AF* = 20 units long.

Next, we write down the values of sin *A*, cos *A* and tan *A* in these three triangles.

I have left some blank for you to fill in because you will then see why they are so important.

In $\triangle ABC$,

$$\sin A = \frac{3}{5}, \quad \cos A = \frac{4}{5}, \quad \tan A = \frac{3}{4}.$$

In $\triangle ADE$,

$$\sin A = \frac{6}{10} = \frac{3}{5}, \quad \cos A = \frac{8}{-} = \frac{}{5}, \quad \tan A = \frac{}{8} = \frac{3}{-}.$$

In $\triangle AFG$,

$$\sin A = \frac{12}{20} = \frac{3}{5}, \quad \cos A = \frac{}{-} = \frac{}{-}, \quad \tan A = \frac{}{-} = \frac{}{-}.$$

We see that the fractions or ratios giving the sin, cos and tan of angle *A* remain the same, although the sizes of the triangles are different. *It is this property of remaining constant for a given angle, whatever the scale of the triangle that the angle is in, which makes these ratios so important.*

Practically, it makes it possible to find heights or depths in situations where we can't make these measurements directly. For example, if we wish to find the height of a tree, it can be done by measuring the distance to the foot of the tree, and the angle of elevation *E* to the top of the tree. We can then use the tan of this angle of elevation to find its height.

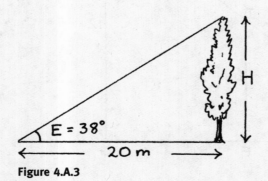

Figure 4.A.3

Some trigonometry and geometry

In the case shown in Figure 4.A.3 we would have:

$$\tan 38° = \frac{H}{20} \quad \text{so} \quad H = 20 \tan 38° = 15.6 \, \text{m to 1 d.p.}$$

There are two standard ways of measuring angles. They can be measured in **degrees**, where 90° is a right angle, as shown in Figure 4.A.4 below. Then 180° is a straight line, and 360° is a full turn.

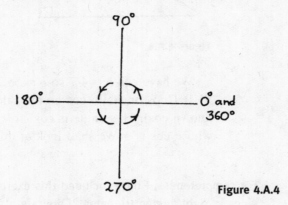

Figure 4.A.4

Angles can also be measured in **radians** which are described later on in this chapter in Section 4.D.(a).

There is a third way of measuring angles on your calculator (called grad), which is very rarely used.

The ratios for any sin, cos or tan are programmed into your calculator so that you can then use them to find either unknown angles, or the lengths of unknown sides of triangles.

Here's a quick revision of how the working out goes, just in case you haven't used it for some time.

EXAMPLE (1) Find the length of *PR* in triangle *PQR*, in which the length of *QR* = 5 cm and the angle *P* is 32°. I show a sketch of this in Figure 4.A.5.

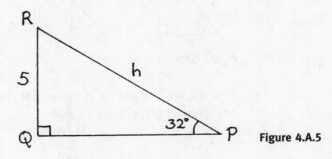

Figure 4.A.5

If we let $PR = h$, we have $\sin P = 5/h = \sin 32°$, so

$$h \sin 32° = 5$$

and

$$h = \frac{5}{\sin 32°} = 9.44 \, \text{cm to 3 significant figures (s.f.)}.$$

EXAMPLE (2) Find angle *b* in triangle *ABC* in Figure 4.A.6, if *AB* = 7 m and
BC = 4 m.

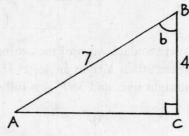

Figure 4.A.6

We have cos *b* = 4/7 so *b* = 55.2° to 1 d.p. (using INV cos or
SHIFT cos or 2nd/F cos on the calculator to find the angle with the
known cos). This angle is cos⁻¹ (4/7), where cos⁻¹ stands for 'the angle
whose cos is'. (We shall look at this in more detail in Section 5.A.(g).)

EXERCISE 4.A.1

For completeness, I have included this exercise on finding angles and lengths of
sides in right-angled triangles. If you are at all unsure that you remember how to
do these, this exercise gives you something to check against.

(A) If the sketches in Figure 4.A.7 all show triangles with lengths given in
centimetres find the lengths of the sides marked with a letter to 2 d.p.

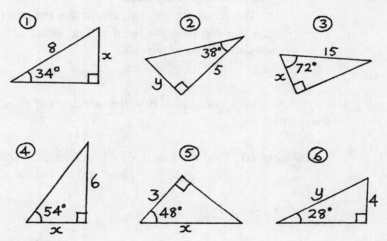

Figure 4.A.7

(B) Find the marked angles in these triangles giving your answers in degrees to
one decimal place (Figure 4.A.8).

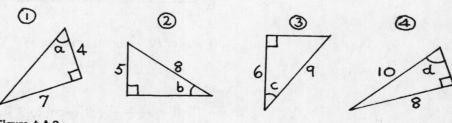

Figure 4.A.8

Some trigonometry and geometry

Comparing the areas of the triangles in Figure 4.A.2

Returning to the three nested triangles of Figure 4.A.2, we know that the lengths of the matching sides go in the ratio of $1:2:4$ as we move from the smallest triangle to the largest triangle.

How do their areas compare? Do they also go $1:2:4$?

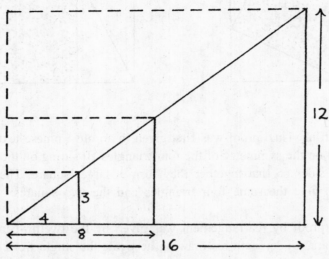

Figure 4.A.9

Each triangle is half a rectangle as you can see from diagram Figure 4.A.9. Using $\triangle$ to stand for 'triangle', we have

$$\triangle ABC = \tfrac{1}{2} \times 4 \times 3 = 6 \text{ square units,}$$

$$\triangle ADE = \tfrac{1}{2} \times 8 \times 6 = 24 \text{ square units,}$$

$$\triangle AFG = \tfrac{1}{2} \times 16 \times 12 = 96 \text{ square units.}$$

The ratio of the areas is given by

$$\triangle ABC : \triangle ADE : \triangle AFG = 6 \ : 24 \ : \ 96$$

$$= 1 \ : \ 4 \ : \ 16$$

$$= 1^2 : \ 2^2 : \ 4^2.$$

The ratio of the areas is the same as the ratio of the lengths squared, which makes sense as the area is found from multiplying two lengths together. So, for example, if each length has been doubled, the area will be four times larger.

4.A.(b)   **Pythagoras' Theorem**

You will almost certainly have recognised the smallest triangle in Figure 4.A.2 as having sides of the smallest whole numbers which fit Pythagoras' Theorem. This says that the square on the longest side (or hypotenuse) of a right-angled triangle is equal to the sum of the squares on the other two sides.

(In this particular case, we have $5^2 = 3^2 + 4^2$.)

The ancient Egyptians knew that they could use a 3, 4, 5 triangle to give them a square corner to true their buildings.

We can see that Pythagoras' Theorem must be true for *any* right-angled triangle from the pair of drawings in Figure 4.A.10.

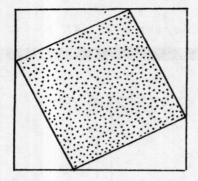 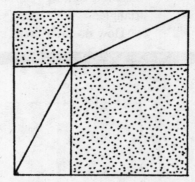

Figure 4.A.10

This beautiful visual proof was first given in an old Chinese text.

It is based on the symmetry of the four triangles all sitting on the sides of the square on their longest sides so that together they form a larger square. The larger square is then rearranged to give the same four triangles and the two squares on each of the shorter sides.

A similar proof by rearrangement was given by the twelfth-century Hindu mathematician, Bhoskara. Underneath his drawing he wrote the single word 'Behold!'.

Two other examples of right-angled triangles in which the sides are whole numbers are given by 5, 12 and 13 units, and 8, 15 and 17 units, because $5^2 + 12^2 = 13^2$ and $8^2 + 15^2 = 17^2$.

Sets of three whole numbers like these are called Pythagorean triples, and there are, in fact, infinitely many of them. In the huge majority of cases, however, the sides of right-angled triangles are not all exact numbers, and therefore involve those irrational numbers like $\sqrt{2}$ which caused Pythagoras such distress. (See Section 1.E.(d).)

Pythagoras' Theorem can be used to calculate the length of the third side of any right-angled triangle if we know the other two.

Here are two examples.

In each of the two triangles in Figure 4.A.11 find the length of the third side.

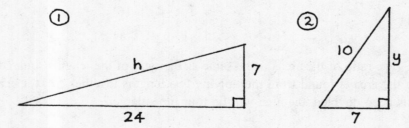

Figure 4.A.11

In (a), $h^2 = 7^2 + 24^2 = 49 + 576 = 625$ so $h = 25$ units.

In (b), $10^2 = y^2 + 7^2$ so $100 = y^2 + 49$ and $y^2 = 51$ so $y = 7.14$ to 2 d.p.

EXERCISE 4.A.2   Find the lengths of the third sides of each of the four triangles from Exercise 4.A.1 part (B).

General properties of triangles

We have just seen that right-angled triangles have a remarkable special property. Do all triangles have special properties regardless of their shape?

The most important property held in common by all triangles is that their interior angles always add up to 180°.

This can be seen from the drawing shown in Figure 4.A.12.

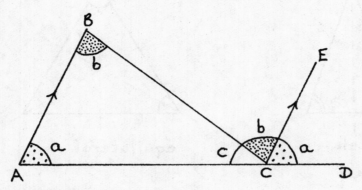

Figure 4.A.12

We start with any triangle *ABC*, and then draw the line *CE* so it is parallel to *AB*. (The two arrows on *AB* and *CE* are to show that these lines are parallel.)

Then the two angles marked *a* exactly slot into each other, and so do the two angles marked *b*.

a + *b* + *c* makes a straight line, and so adds to 180°.

Therefore, the angles of the triangle must also add up to 180°.

We also see from this same diagram that, if we have a triangle with one side extended, then the exterior angle *e* is equal to *a* + *b*, the sum of the two interior opposite angles.

This is shown drawn in on Figure 4.A.13.

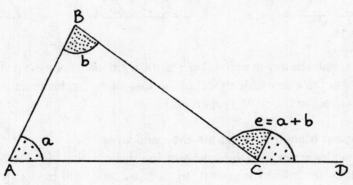

Figure 4.A.13

4.A.(d)   **Triangles with particular shapes**

Triangles can come in an infinite variety of shapes, but there are two particular types which have specific names.

If a triangle has two sides equal then it is called **isosceles** (originally by the Greeks who were very keen on geometry – 'iso' means 'equal' and 'sceles' means 'sides'. 'Trigonometry' also comes from the Greeks – 'trigono' is the Greek word for triangle.)

The two equal sides give these triangles a line of symmetry, so that one half folds exactly on to the other half, and the pair of angles opposite the equal sides are also equal. The line of symmetry divides the triangle into two equal right-angled triangles. (See Figure 4.A.14(a).) The little dashes are there to mark the two equal sides.

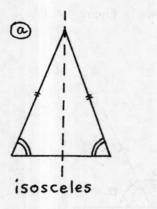
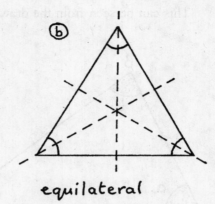

isosceles   *equilateral*

Figure 4.A.14

If a triangle has all three sides equal then it is called **equilateral**. Such a triangle is pictured in Figure 4.A.14(b).

It will have three lines of symmetry as shown, and will fit exactly onto itself three times in a complete turn. Therefore all its angles are equal, and so must be 60° each.

All equilateral triangles can nest into each other, in any chosen corner.

Some are shown here in Figure 4.A.15.

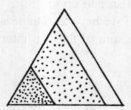
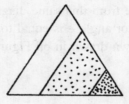
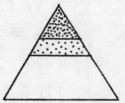

Figure 4.A.15

They are all **similar** to each other. ('Similar' in maths doesn't just mean 'more or less the same as' but 'an exact scale model of' so that all the angles remain the same, and the pairs of sides are all in the same proportion.)

4.A.(e)   Congruent triangles – what are they, and when?

If two triangles are exactly the same size and shape so that they can be fitted onto each other exactly, they are called **congruent**. In this case, they will obviously have three equal pairs of angles and three equal pairs of sides. (It may be necessary to lift one triangle out of the paper, and turn it over, in order to fit it exactly on top of the other one.)

How many measurements (and which ones) do you need to know are the same in order to be sure that two triangles must be congruent?

In general, three pairs of equal measurements will be enough, provided that they are the right pairs. See how many of these you can find – draw little sketches if necessary! (Things are not always what they seem.)

Case (1)   We have already seen that having three pairs of equal angles certainly isn't enough. This would just mean that the triangles were similar.

Case (2)   On the other hand, having three pairs of equal sides is certainly sufficient. The triangles will then exactly match.

Case (3)   If we have two pairs of equal angles, then the third pair of angles *must* be equal since the angles of a triangle add to 180°. Just one pair of equal sides opposite same-sized angles is then enough to tell us that the scale is the same, and so the two triangles are congruent.

Case (4)   If we have two pairs of equal sides and one pair of equal angles, then it all depends where the angle is! You can see the danger in Figure 4.A.16. We are only safe if the angles are between the matching sides (except for one case when it doesn't matter where the matching pair is . . .).

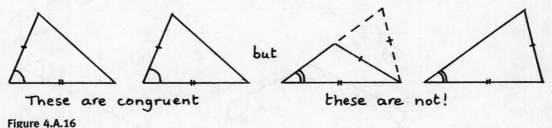

These are congruent   but   these are not!

Figure 4.A.16

Case (5)   This special case is when the two equal angles are both right angles.

In practice, similar and congruent triangles often appear at a slant to each other.

One example of this is shown in Figure 4.A.17 below. The two congruent triangles shown here, with one of them turned through 180° relative to the other one, fit together to form a **parallelogram**.

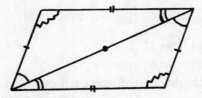

Figure 4.A.17

If the two triangles are isosceles, as shown in Figure 4.A.18(a), then together they make what is called a **rhombus** or diamond.

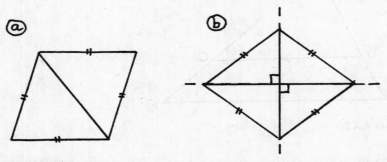

Figure 4.A.18

4.A Trigonometry in right-angled triangles

By showing the two axes of symmetry set horizontally and vertically, we see why this shape is called a diamond, and also that the diagonals cut at right angles.
This is shown in Figure 4.A.18(b).

THINKING POINT

- What do you get if you add up all the interior angles shown in this drawing of a six-sided figure? (See Figure 4.A.19(a)). Does it depend on its shape?
- What is the sum of its exterior angles? (See Figure 4.A.19(b).)
- What would be the sum of the interior angles if the figure had *n* sides? (It would then be what is called an *n*-sided polygon.)
- What would be the sum of its exterior angles?
  See if you can work out the answers yourself to these four questions.
  (I give solutions later on in the chapter for you to check against.)

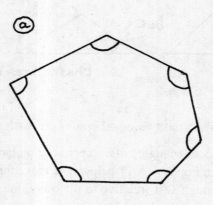
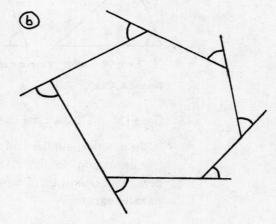

Figure 4.A.19

4.A.(f)   Matching ratios given by parallel lines

Here is another useful property of similar triangles.
Suppose we have two similar triangles nested into each other.
This is shown in Figure 4.A.20.

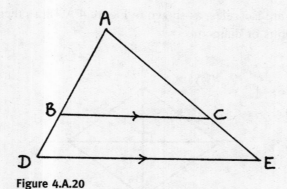

Figure 4.A.20

Then *BC* is parallel to *DE*. This is shown in the diagram by using little arrows.

Because the triangles are similar, corresponding pairs of sides are in the same proportion, so we have

$$\frac{AD}{AB} = \frac{AE}{AC} = \frac{DE}{BC}.$$

But $AD/AB = AE/AC$ can be written as

$$\frac{AB + BD}{AB} = \frac{AC + CE}{AC}.$$

Also

$$\frac{AB + BD}{AB} = 1 + \frac{BD}{AB} \quad \text{and} \quad \frac{AC + CE}{AC} = 1 + \frac{CE}{AC}.$$

Therefore

$$\frac{BD}{AB} = \frac{CE}{AC} \quad \text{or, equally,} \quad \frac{AB}{BD} = \frac{AC}{CE}$$

turning both fractions upside down if we prefer them that way.

You will find that this property of parallel lines cutting off sections with the same ratio is often very useful when working with problems involving similar physical shapes.

4.A.(g)   **Special cases – the sin, cos and tan of 30°, 45° and 60°**

It is often useful to know the ratios of the sides of right-angled triangles which have particularly simple divisions of 90° for the other two angles.

The two most useful ones are as follows:

(a)   the ratios for all triangles which have angles of 90°, 45° and 45°,
(b)   the ratios for all triangles which have angles of 90°, 60° and 30°.

(a)   The 90°, 45°, 45° triangle is isosceles.

The simplest example is the one which has two equal sides of 1 unit, shown in Figure 4.A.21(a).

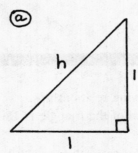 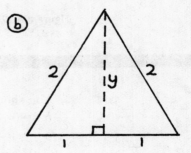

Figure 4.A.21

By Pythagoras, $h^2 = 1^2 + 1^2 = 2$ so $h = \sqrt{2}$ so we have

$$\sin 45° = \cos 45° = \frac{1}{\sqrt{2}} \quad \text{and} \quad \tan 45° = \frac{1}{1} = 1.$$

(b) The 90°, 60°, 30° triangle is half of an equilateral triangle, so if we take 2 units for each side, the base is conveniently divided into 1 unit for each side.

A sketch of this triangle is shown in Figure 4.A.21(b).

Again, we can find the vertical height by using Pythagoras' Theorem. We have $2^2 = 1^2 + y^2$ so $y^2 = 3$ and $y = \sqrt{3}$. This gives us

$$\sin 60° = \cos 30° = \frac{\sqrt{3}}{2} \quad \text{and} \quad \cos 60° = \sin 30° = \frac{1}{2}$$

$$\tan 60° = \sqrt{3} \quad \text{and} \quad \tan 30° = \frac{1}{\sqrt{3}}.$$

You will find that these exact values do also check with the decimal values given on your calculator for these angles. (Make sure of this for yourself.)

4.A.(h)   Special relations of sin, cos and tan

Are there any relationships between the sin, cos and tan of the two angles a and b which will be true in *any* right-angled triangle?

Use the triangle shown in Figure 4.A.22 below to write down the sin, cos and tan of a and b. Then see if you can find any connections between them.

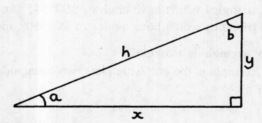

Figure 4.A.22

You should have found the following relationships.

$b = 90° - a$ because the angle sum of the triangle is 180°.

$$\sin a = \frac{y}{h} = \cos b, \quad \cos a = \frac{x}{h} = \sin b, \quad \tan a = \frac{y}{x} = \frac{1}{\tan b}.$$

We see also that

$$\frac{\sin a}{\cos a} = \frac{y/h}{x/h} = \frac{y}{x} = \tan a \quad \text{and} \quad \frac{\sin b}{\cos b} = \tan b.$$

We also find a very nice relationship between the sin and cos of each of a and b which comes directly from Pythagoras' Theorem. We have

$$x^2 + y^2 = h^2 \quad \text{so} \quad \frac{x^2}{h^2} + \frac{y^2}{h^2} = \frac{h^2}{h^2} = 1.$$

But

$$\frac{y^2}{h^2} = \sin^2 a \quad \text{and} \quad \frac{x^2}{h^2} = \cos^2 a$$

$$\boxed{\sin^2 a + \cos^2 a = 1.}$$

This is an enormously useful result and it is worth surrounding its box with bright colour.

It is, of course, equally true that $\sin^2 b + \cos^2 b = 1$. Indeed, all the special relationships which we have shown above will carry through truthfully when we move on to consider general angles instead of just being restricted to angles between $0°$ and $90°$.

NOTE $\sin^2 a$ is the usual way that $(\sin a)^2$ is written. Equally, $\cos^2 a$ means $(\cos a)^2$ etc.

$\sin^2 a$ is **not** the same as $\sin(a^2)$. For example, if $a = 5°$, then $\sin a = 0.0872$ to 3 s.f. and $\sin^2 a = 0.00760$ to 3 s.f. but $\sin(a^2) = \sin 25° = 0.423$ to 3 s.f.

The last result which we found above has two offspring which are also often very useful. We start with

$$\sin^2 a + \cos^2 a = 1. \tag{1}$$

Dividing through by $\cos^2 a$ we get

$$\frac{\sin^2 a}{\cos^2 a} + \frac{\cos^2 a}{\cos^2 a} = \frac{1}{\cos^2 a}$$

so

$$\boxed{\tan^2 a + 1 = \operatorname{cosec}^2 a.} \tag{2}$$

Starting again from (1), and dividing through by $\sin^2 a$, what do you get?

$$\frac{\sin^2 a}{\sin^2 a} + \frac{\cos^2 a}{\sin^2 a} = \frac{1}{\sin^2 a}$$

so

$$\boxed{1 + \cot^2 a = \operatorname{cosec}^2 a.}$$

(3)

It's also worth surrounding (2) and (3) in bright colour.

4.B   Widening the field in trigonometry

4.B.(a)   The Sine Rule for any triangle

We are now in a good position to get trig formulas for *any* triangle, which we will then be able to use to find unknown angles and sides.

We start this process by finding what is called the **Sine Rule**.

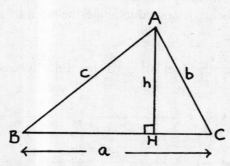

Figure 4.B.1

I have drawn a general-shaped sort of triangle in Figure 4.B.1. I have labelled the sides with the lower case letter corresponding to the capital letter of the opposite angle. (This is the usual way in which such labelling is done.)

I've also drawn in the perpendicular line *AH* (so that we shall have two right-angled triangles to work from!). I have labelled its length *h*.

Then, in $\triangle ABH$,

$$\sin B = \frac{h}{c} \quad \text{so} \quad h = c \sin B.$$

Write down for yourself the same sort of thing for sin C in $\triangle AHC$.

You will have

$$\sin C = \frac{h}{b} \quad \text{so} \quad h = b \sin C.$$

So we can say $c \sin B = b \sin C$. Therefore,

$$\frac{c}{\sin C} = \frac{b}{\sin B}.$$

We could equally have drawn the triangle in such a way that we used A and a.

Therefore, by symmetry, we have

The Sine Rule

$$\frac{a}{\sin A} = \frac{b}{\sin B} = \frac{c}{\sin C}$$

This applies to *any* triangle, and we can use it to calculate the lengths of unknown sides and angles.

Here is an example of this.

In triangle ABC, $\angle B$ is $58°$, $\angle C$ is $40°$ and the side AC is 6 m long. Calculate the lengths of the unknown sides and angles.

We start by drawing a sketch. *A sketch is important in any geometrical or physical problem, because it gives you some idea of what you are looking for.*

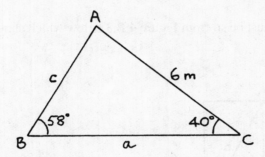

Figure 4.B.2

My sketch is Figure 4.B.2. I have labelled it in the same sort of way that I labelled the original triangle. Also, although it is not accurate, I have tried to make it believable, so that the angles of $58°$ and $40°$ are roughly the right size.

So now we start. What is $\angle A$?

It is $180° - 58° - 40° = 82°$ because the angles of a triangle add to $180°$ (Section 4.A.(c)).

Now, to find a, we have

$$\frac{a}{\sin 82°} = \frac{6}{\sin 58°} \qquad \text{so} \qquad a = \sin 82° \times \frac{6}{\sin 58°} = 7.01 \text{ m to 2 d.p.}$$

To find c, we can say

$$\frac{c}{\sin 40°} = \frac{6}{\sin 58°} \qquad \text{so} \qquad c = 4.55 \text{ m to 2 d.p.}$$

(It is safer not to use the newly found length of a to find c just in case it has a mistake in it.)

Finally, before going on, we look at the sketch to see if our answers seem reasonably convincing for this particular triangle. They do, so we can proceed happily to the next thing, which is an exercise on using the Sine Rule.

Find, if possible, the missing sides and angles in each of the three triangles whose measurements are given below, giving the angles in degrees to 1 d.p. and the sides in centimetres to 2 d.p. In each case, start by drawing a labelled sketch, as I did in the previous example. It's particularly important to do this exercise because things are not always quite as they seem.

(1) Triangle ABC in which $\angle A = 78°$, $\angle B = 65°$ and $AB = 5\,\text{cm}$

(2) Triangle ABC in which $\angle C = 33°$, $BC = 6\,\text{cm}$ and $AB = 4\,\text{cm}$

(3) Triangle ABC in which $\angle C = 40°$, $AB = 9\,\text{cm}$ and $BC = 5\,\text{cm}$

4.B.(b)   Another area formula for triangles

The most usual formula for the area of a triangle is

the area of the triangle = $\frac{1}{2}$ base $\times$ height.

You can see that this must be so from Figure 4.B.3 below which shows the triangle as half a rectangle.

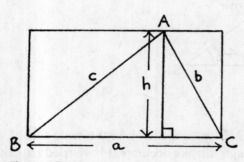

Figure 4.B.3

Sometimes it is useful to be able to write this area in another way. We know that

the area = $\frac{1}{2}\,ah$

but $h = b \sin C = c \sin B$ as we saw when we proved the Sine Rule in Section 4.B.(a), above. So, by symmetry,

the area = $\frac{1}{2}\,ab \sin C = \frac{1}{2}\,ac \sin B = \frac{1}{2}\,bc \sin A$.

In words, we can say

The area of a triangle is equal to one half of any two sides multiplied together and then multiplied by the sine of the angle between them.

Here is an example of the use of this new formula.

Find the area of the equilateral triangle ABC with sides of length 3 cm, shown in Figure 4.B.4.

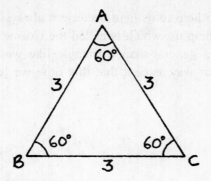

Figure 4.B.4

Instead of having to mess around finding the vertical height, we can say that

$$\text{the area} = \tfrac{1}{2} \times 3 \times 3 \times \sin 60° = \frac{9\sqrt{3}}{4} = 3.90 \, \text{cm}^2 \text{ to 2 d.p.}$$

The new formula is particularly useful for finding the area of triangles enclosed by two radius lengths in circles such as the one I've shown in Figure 4.B.5. I've marked the angle with the Greek letter θ (called theta), since this is often used for angles.

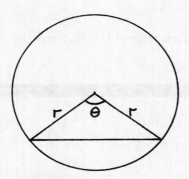

Figure 4.B.5

The area of the triangle is $\tfrac{1}{2} r^2 \sin \theta$.

4.B.(c)   The Cosine Rule for any triangle

Suppose we have a triangle in which we know the lengths of the three sides, and we want to find its angles, like the one in Figure 4.B.6.

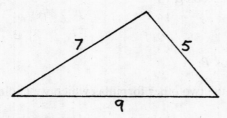

Figure 4.B.6

4.B Widening the field in trigonometry

The Sine Rule will be of no help to us here because it always involves two angles. But there *is* a formula which will help us, which is called the Cosine or Cos Rule.

To get this, we start with a general-shaped triangle like we did with the Sine Rule, and label it in the same sort of way, except that this time we let the length of $BH = x$. (See Figure 4.B.7.)

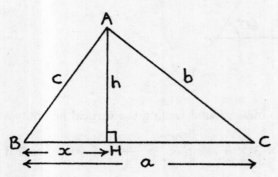

Figure 4.B.7

In triangle ABH, using Pythagoras' Theorem, we have

$$c^2 = h^2 + x^2 \quad \text{so} \quad x^2 = c^2 - h^2.$$

What is the length of CH using the given letters? Use this to write down how Pythagoras' Theorem will go for $\triangle AHC$.

$$CH = a - x.$$

So, in $\triangle AHC$, we have

$$b^2 = h^2 + (a - x)^2 = h^2 + a^2 + x^2 - 2ax.$$

But $x^2 = c^2 - h^2$, so we have

$$b^2 = h^2 + a^2 + c^2 - h^2 - 2ax = a^2 + c^2 - 2ax.$$

In $\triangle ABH$, what is $\cos B$?

We have

$$\cos B = \frac{x}{c} \quad \text{so} \quad x = c \cos B.$$

Therefore, we have

$$b^2 = a^2 + c^2 - 2ac \cos B.$$

Equally, by symmetry, we have the two other formulas which we could have got by labelling the triangle differently.

We now have the Cosine Rule for any triangle.

The Cosine Rule

$$a^2 = b^2 + c^2 - 2bc \cos A \qquad (1)$$

$$b^2 = c^2 + a^2 - 2bc \cos B \qquad (2)$$

$$c^2 = a^2 + b^2 - 2ab \cos C \qquad (3)$$

Notice also that if we put $A = \pi/2$ in (1) above, we get Pythagoras' Theorem for what is now a right-angled triangle.

That is, we get $a^2 = b^2 + c^2$ because $\cos \pi/2 = 0$, so everything connects up as it should do.

Here is an example of using the Cosine Rule to find a side of a triangle. We will use it to find a in $\triangle ABC$ shown in Figure 4.B.8.

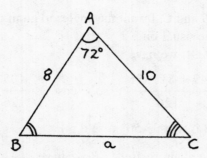

Figure 4.B.8

This triangle is another example of a case in which the Sine Rule will not give us what we want. This is because the known facts slot into it in such a way that every possible equation has two unknowns.

We would have

$$\frac{a}{\sin 72°} = \frac{10}{\sin B} = \frac{8}{\sin C} \quad \text{which is no use.}$$

Using the Cosine Rule, we have $a^2 = b^2 + c^2 - 2bc \cos A$.

Substituting the known values, this gives us $a^2 = 64 + 100 - 160 \cos 72°$ so $a = 10.7$ to 1 d.p.

If we want to find the angles of a triangle using the Cosine Rule, it will pay us to rearrange the three formulas.

For example, we have $a^2 = b^2 + c^2 - 2bc \cos A$ so $2bc \cos A = b^2 + c^2 - a^2$.

Rearranging this gives us

$$\cos A = \frac{b^2 + c^2 - a^2}{2bc}, \qquad (1)$$

$$\cos B = \frac{c^2 + a^2 - b^2}{2ca}, \qquad (2)$$

$$\cos C = \frac{a^2 + b^2 - c^2}{2ab}, \qquad (3)$$

shifting the letters round again in turn to give the other two formulas.

We take the triangle from the beginning of this section to show the use of the Cosine Rule to find its angles. It has sides of 5 cm, 7 cm and 9 cm and I show it again in Figure 4.B.9.

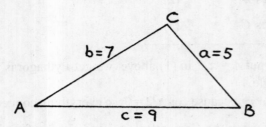

Figure 4.B.9

We will now find the angles A, B and C. I want the angles to go in this way, which is why my lettering of the triangle isn't the usual one.

Using the Cosine Rule to find $\angle A$, we have

$$\cos A = \frac{b^2 + c^2 - a^2}{2bc} = \frac{49 + 81 - 25}{126} \quad \text{so} \quad \cos A = \frac{105}{126}$$

and $\qquad \angle A = 33.6°$ to 1 d.p. $\quad (\angle A = 33.56°$ to 2 d.p.)

Similarly, using the Cosine Rule again to find $\angle B$ we have

$$\cos B = \frac{c^2 + a^2 - b^2}{2ca} = \frac{81 + 25 - 49}{90} = \frac{57}{90} = 50.7(0)° \text{ to 1 d.p.}$$

Working with 2 d.p. to avoid a rounding error in the first decimal place, we can find the third angle using the angle sum of the triangle.

This gives us $\angle C = 180° - 33.56 - 50.70° = 95.7°$ to 1 d.p. which is an angle greater than 90°.

Are we going to have the same problem that we had with the Sine Rule if we are dealing with an angle which might be greater than 90°? Will we be unsure about the shape of the triangle?

If we had used the Cosine Rule to find $\angle C$ we would have got

$$\cos C = \frac{a^2 + b^2 - c^2}{2ab} = \frac{25 + 49 - 81}{70} = -\frac{7}{70}.$$

If you now use your calculator to find $\angle C$ (putting in the fraction complete with its minus sign), you will find that you again get 95.7° to 1 d.p. so it agrees with what we know it should be.

We find, using the Cosine Rule, that angles between 90° and 180° have a negative cos. This means that there can't be any ambiguous cases from using the Cosine Rule – we will know from the sign of the answer whether the angle we have found is less than 90° (acute), or greater than 90° (obtuse).

We saw earlier that, if the angle $A = \pi/2$, then the Cosine Rule for angle A of $a^2 = b^2 + c^2 - 2bc \cos A$ becomes $a^2 = b^2 + c^2$ (that is Pythagoras' Theorem).

If the angle A is acute, we are taking something off $b^2 + c^2$ to get a^2.

If the angle A is obtuse, because $\cos A$ is then negative, we are adding something on to $b^2 + c^2$ to get a^2.

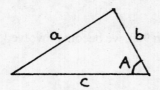
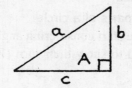
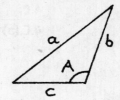

The length of a is the same in all three triangles.

Figure 4.B.10

You can see from the different lengths of *b* and *c* in the three cases which I show in Figure 4.B.10 that this must happen in order that the length of *a* will work out correctly in each case.

If you think that the angle you are finding may be obtuse, it is safer to use the Cosine Rule if possible, rather than the Sine Rule.

I shall explain exactly what we mean by the cos of an angle greater than 90° in Section 5.A.(c).

EXERCISE 4.B.2

Now try the following questions.

(1) Find the sides and angles marked with a question-mark in the three triangles shown in Figure 4.B.11.

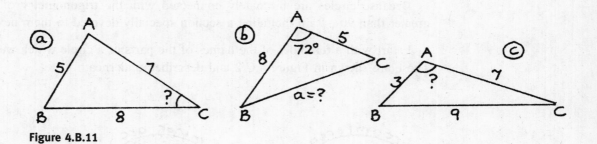

Figure 4.B.11

(2) Figure 4.B.12 shows a triangle formed by joining together the two halves of an equilateral triangle by their shortest sides.

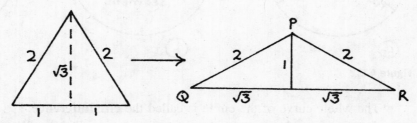

Figure 4.B.12

(a) How large are the angles *Q* and *R*?

(b) How large is ∠*QPR*?

(c) Use the Cosine Rule in △*QPR* to find the cos of ∠*QPR*.

(d) Use the Sine Rule in △*QPR* to find the sin of ∠*QPR*.

4.B Widening the field in trigonometry

4.C   Circles

4.C.(a)   The parts of a circle

Once we start considering angles larger than 90°, we become involved with the circles which are used to show their turn (Figure 4.C.1).

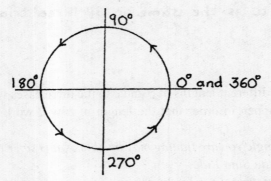

Figure 4.C.1

The convention is that angles are shown turning anticlockwise from the positive *x*-axis, so that angles from 0° to 90° lie in the quarter-circle or quadrant where all measurements are positive. (Bearings are not measured like this; they turn clockwise from a zero position at due north.)

Because circles are intimately connected with the trigonometry of angles which are greater than 90°, I am including a section specially devoted to them next.

I start with a reminder of the names of the parts of a circle which we shall need to use. These are shown in Figure 4.C.2 and described underneath.

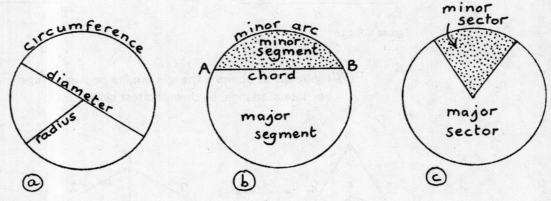

Figure 4.C.2

- The whole curve of the circle is called the **circumference**.
- Any line from the centre to the circumference is called a **radius** (plural: radii). Clearly, from the symmetry of the circle, these are all the same length.
- A line drawn right across a circle through its centre is called a **diameter**.
- A line like *AB* drawn across a circle is called a **chord**, so a diameter is a special case of a chord.
- The curved piece of the circle from *A* to *B* is called an **arc**. The short way round from *A* to *B* is called the minor arc, and the long way round is called the major arc.

- The part of the circle enclosed between the minor arc *AB* and the chord *AB* is called a **minor segment**. The rest of the circle is a **major segment**.
- The shaded piece shown in circle (*c*) is called a **minor sector**. The rest of the circle is called a **major sector**.

To avoid mixing up segments and sectors, you can remember that 'a sector is like a piece of cake because it's got a "c" in it'.

If the radius of the circle is *r*, then the length of the circumference is $2\pi r$, and the area of the circle is πr^2. π is a number which cannot be written exactly as a fraction (though 22/7 is sometimes used as an approximation to it.) To 4 d.p. it is 3.1412. As a decimal, it is non-repeating, and has been calculated to a huge number of decimal places using computers.

If *C* stands for the circumference and *A* stands for the area

$$C = 2\pi r \quad \text{and} \quad A = \pi r^2.$$

4.C.(b) **Special properties of chords and tangents of circles**

The chords and tangents of circles have special properties because any diameter of a circle is a line of symmetry.

(The circle can be folded along any diameter so that the two halves exactly match.)

The most important properties of chords and tangents

- Any line perpendicular to a chord from the centre of the circle divides that chord equally in two (or bisects it).
- If a line from the centre of a circle divides a chord equally in two then it must be perpendicular to that chord.
- Any line which is perpendicular to a chord and bisects it must pass through the centre of the circle.
- If a chord is pushed to the edge of a circle and extended to make a tangent (a line which touches the circle and gives its slope at that point), the tangent is perpendicular to the radius to the point of contact.
- The two tangents to a circle from any outside point must be equal in length.

I show examples of all these properties in Figure 4.C.3.

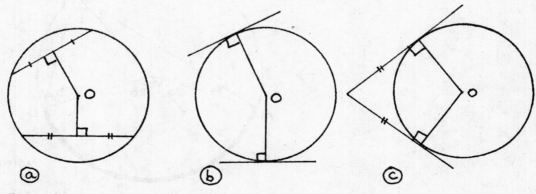

Figure 4.C.3

The matching pairs of little marks show lines which are equal in length.

Draw in the diameters which show the lines of symmetry in colour if it helps you.

4.C.(c)   **Special properties of angles in circles**

We come next to a result which does not come so obviously from the symmetry of the circle.

In Figure 4.C.4, I have shown three angles all standing on the same arc of the circle. This arc is drawn with a thicker line. If you measure these three angles, you will find that they are all equal. Any similar drawings will give other sets of equal angles. Why should this be so?

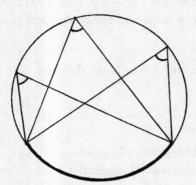

Figure 4.C.4

To find the answer to this, we compare the size of the angle at the centre of the circle with any angle at the circumference which stands on the same arc.

We can do this in the way I have shown in the sequence of drawings in Figure 4.C.5.

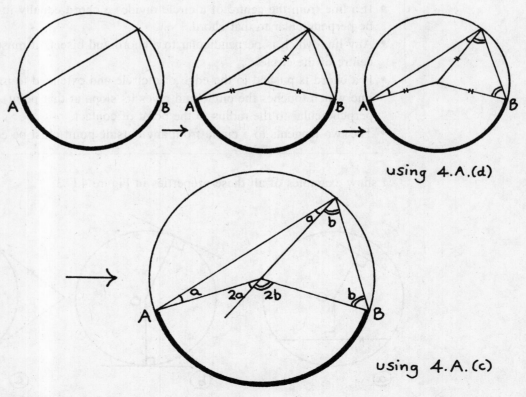

using 4.A.(d)

using 4.A.(c)

Figure 4.C.5

Some trigonometry and geometry

From this, we see that the angle at the centre of the circle is twice the size of the angle at the circumference.

This will be true wherever this angle touches the circumference above *AB*, so long as it is standing on the same arc, so *all* the angles standing on this arc must be equal; an unexpected and beautiful result.

If the angle is *below AB*, as I show in Figure 4.C.6, the angle at the circumference is still half the angle at the centre, but we are looking at the situation upside down, so the angle at the centre is now greater than 180°. (An angle like this is called a **reflex angle**.) The two angles are now standing on the major arc of the circle which I have shown using a thicker line.

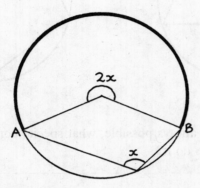

Figure 4.C.6

From these two results we can now deduce a useful special case, which is that the angle in a semi-circle is a right angle.

We can see that this must be so either way round from the two diagrams shown in Figure 4.C.7.

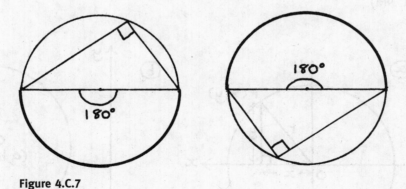

Figure 4.C.7

A summary of special properties of angles in circles

- The angle at the centre of a circle is twice any angle standing on the same arc.

- Angles at the circumference and standing on the same arc are equal.

- The angle in a semi-circle is a right angle.

THINKING
POINT

(a) Is it possible to draw a circle round *any* triangle as in Figure 4.C.8(a)?

(b) Is it possible to draw a circle round *any* four-sided shape (quadrilateral) as shown in Figure 4.C.8(b)?

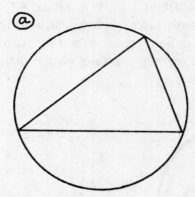

Figure 4.C.8

In each case, if it isn't always possible, what special conditions must you have in order to be able to do it?

4.C.(d)   Finding and working with the equations which give circles

How can we find the equation of the curve which gives a particular circle in terms of x and y?

We will start by considering the simplest case which is a circle of radius r symmetrically placed so that its centre is at the origin. I have drawn a circle like this in Figure 4.C.9(a).

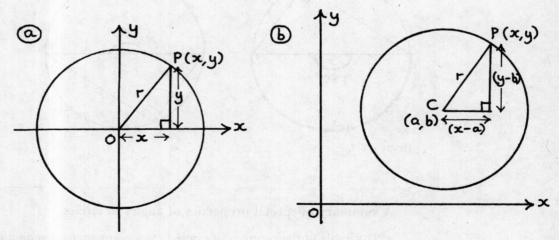

Figure 4.C.9

Any point P on it, with coordinates (x, y), must be a distance r from the origin, so $x^2 + y^2 = r^2$ by Pythagoras' Theorem.

156   **Some trigonometry and geometry**

The equation of any circle with radius r and whose centre is the origin can be written in the form $x^2 + y^2 = r^2$.

For example, if the radius r is 4 units, we get the circle whose equation is $x^2 + y^2 = 3^2$ or $x^2 + y^2 = 9$.

If the centre of the circle is not at the origin, we can still use the property that the distance of any point on the circumference from the centre is equal to the constant length of the radius.

In Figure 4.C.9(b) the length of PC remains constant, and equal to r.

If P has coordinates (x, y), using Pythagoras' Theorem here gives us $(x-a)^2 + (y-b)^2 = r^2$.

The equation of the circle with centre (a,b) and radius r is given by

$$(x - a)^2 + (y - b)^2 = r^2.$$

For example, the circle with a radius of 4 units, and with its centre at the point (6,5), has the equation

$$(x - 6)^2 + (y - 5)^2 = 4^2$$

or $\qquad x^2 - 12x + 36 + y^2 - 10y + 25 = 16$

giving $\quad x^2 - 12x + y^2 - 10y + 45 = 0$.

(These numbers will fit Figure 4.C.9(b) quite nicely. If you are at all unsure about the algebra version of the equation of this circle, feed in the numbers to make yourself an actual example of the algebra working.)

Now we do the same thing of multiplying out with the algebra version of the equation given in the box above.

We have $(x - a)^2 + (y - b)^2 = r^2$.

Multiplying out the brackets gives $x^2 - 2ax + a^2 + y^2 - 2by + b^2 = r^2$

Tidied up, this gives us an alternative form for the equation of this circle.

The equation of the circle with centre (a,b) and radius r can also be written as

$$x^2 - 2ax + y^2 - 2by + c = 0 \text{ where } c = a^2 + b^2 - r^2.$$

For an equation like this to give a circle it must fit the following conditions.

(1)   There must be equal coefficients of x^2 and y^2. The coefficient is the number which tells us how many we've got. The coefficient of $3x^2$ is 3. The coefficient of y^2 is 1. If there are no terms in x, say, then the coefficient of x is zero.

(2)   There must only be, at the most, terms in x^2, y^2, x, y and a number. (We mustn't have any terms with xy, for instance.)

(3)   The value of r^2 must be positive so that we have a physically possible length for the radius.

It's easy to remember that the circle with equation $x^2 - 2ax + y^2 - 2by + c = 0$ has its centre at the point (a, b). But its radius is *not c*.

From above, we have $r^2 = a^2 + b^2 - c$ so $r = \sqrt{a^2 + b^2 - c}$.

This is a very clumsy formula to remember. I think that much the best way of finding the centre and radius of a circle is to complete the two squares. (Completing the square is explained in Section 2.D.(b).)

Here is an example of this, to show you how it works.

Suppose we have the circle whose equation is $x^2 - 4x + y^2 + 6y - 3 = 0$.

Completing the two squares gives us $(x - 2)^2 - 4 + (y + 3)^2 - 9 - 3 = 0$ so $(x - 2)^2 + (y + 3)^2 = 16$.

Therefore the centre of the circle is at $(2, -3)$ and its radius is 4 units.

 Notice that the signs flip to give the coordinates of the centre, just as they do to give the solutions to quadratic equations.

EXERCISE 4.C.1

Find the centre and radius of each of the following circles.
(1) $(x - 1)^2 + (y + 2)^2 = 16$. (2) $x^2 + y^2 - 2x - 4y = 0$.
(3) $x^2 + y^2 - 8x + 7 = 0$. (4) $x^2 + y^2 - 6x + 2y - 6 = 0$.
(5) $x^2 + y^2 - x + y = 0$. (6) $x^2 + y^2 + 3x + 2y + 1 = 0$.
(7) Find the equation of the circle which is concentric with the circle
$x^2 + y^2 + 2x - 4y = 0$ and which has a radius of 5 units.
('Concentric' means 'having the same centre as'.)
(8) Find the equation of the circle which passes through the origin and the points
(3,0) and (0,4), writing it in the form $x^2 - 2ax + y^2 - 2by + c = 0$.
Find also its centre and radius.

4.C.(e)   Circles and straight lines – the different possibilities

What are the three possible relationships between a straight line and a circle? Try sketching them for yourself.

You should have a line which passes through the circle so that it cuts it twice, a line which just touches the circle and so is a tangent, and a line which misses the circle altogether.

How will these three different possibilities show up if we work from the equations of the particular line and circle?

We will go through the following example together, to see what happens.

EXAMPLE (1) Find whether, and if so where, the lines
(a) $y = 2x - 4$ (b) $3y = x + 11$ and (c) $y = 3x + 6$
cut the circle whose equation is $x^2 - 4x + y^2 - 2y - 5 = 0$.
Draw a sketch showing the three lines and the circle.

(a) If the line $y = 2x - 4$ cuts the circle, the values of x and y at the points where it cuts must fit both the equations of the circle and of the line. (In other words, we have two simultaneous equations at these points, but they involve a line and a circle instead of two straight lines like the ones in Section 2.C.)

This means that we can put $y = 2x - 4$ into the equation of the circle to find the possible values of x.

This gives us

$$x^2 - 4x + (2x - 4)^2 - 2(2x - 4) - 5 = 0$$

$$x^2 - 4x + 4x^2 - 16x + 16 - 4x + 8 - 5 = 0$$

$$5x^2 - 24x + 19 = 0$$

$$(5x - 19)(x - 1) = 0$$

$$x = 1 \quad \text{or} \quad x = \tfrac{19}{5}.$$

(You could use the formula for quadratic equations from Section 2.D.(d) to find these two roots if you prefer.)

Substituting these values of x back in the line $y = 2x - 4$ gives us the corresponding two values for y of -2 and $\tfrac{18}{5}$.

So the line $y = 2x - 4$ cuts the circle at the two points with coordinates $(1, -2)$ and $(\tfrac{19}{5}, \tfrac{18}{5})$. Sometimes, the word 'intersects' is used instead of the word 'cuts'.

(b) To find if the line $3y = x + 11$ cuts the circle, we can rewrite its equation as $x = 3y - 11$ and substitute this for x in the equation of the circle.

This gives us

$$(3y - 11)^2 - 4(3y - 11) + y^2 - 2y - 5 = 0$$

$$9y^2 - 66y + 121 - 12y + 44 + y^2 - 2y - 5 = 0$$

$$10y^2 - 80y + 160 = 0$$

$$y^2 - 8y + 16 = 0$$

$$(y - 4)^2 = 0.$$

The two possible cutting points have come together here to give the single point for which $y = 4$ and $x = 12 - 11 = 1$.

This means that the line $3y = x + 11$ just *touches* the circle – it is a **tangent** to it.

The point of contact has the coordinates $(1, 4)$.

(c) This time, we put $y = 3x + 6$ in the equation of the circle.

This gives us

$$x^2 - 4x + (3x + 6)^2 - 2(3x + 6) - 5 = 0$$

$$x^2 - 4x + 9x^2 + 36x + 36 - 6x - 12 - 5 = 0$$

$$10x^2 + 26x + 19 = 0.$$

Using the quadratic formula on this equation, with $a = 10$, $b = 26$ and $c = 19$ gives $b^2 - 4ac = -84$, so we can't find any value for x which will satisfy this equation. This must mean that the line misses the circle completely.

The three different quadratic equations of (a), (b) and (c) have revealed exactly what is happening geometrically.

For the sketch, we need the centre and the radius of the circle.

We have

$$x^2 - 4x + y^2 - 2y - 5 = 0$$

$$(x - 2)^2 - 4 + (y - 1)^2 - 1 - 5 = 0$$

so

$$(x - 2)^2 + (y - 1)^2 = 10.$$

The centre of the circle is at the point (2,1) and its radius is $\sqrt{10}$.

I have drawn a sketch of the three lines and the circle in Figure 4.C.10.

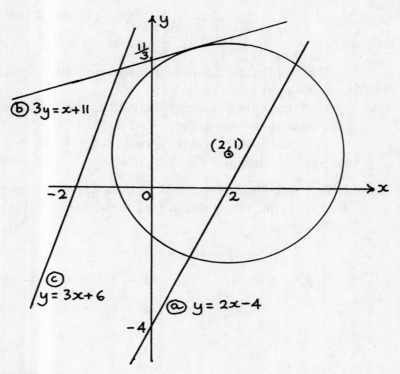

Figure 4.C.10

I've summarised the results which we have just found in the box below for you.

Straight lines and circles

Substituting the equation of the line into the equation of the circle will give you a quadratic equation in x or y.

There are then three possibilities.

- The equation has two roots. This means that the line cuts the circle in two points.
- The equation has one repeated root. This means that the line is a tangent to the circle – it just touches it.
- '$b^2 - 4ac$' is negative, and the equation has no real roots.
  This means that the line misses the circle altogether.

Find whether, and if so where, the lines (a) $3y = x - 5$ (b) $2y = x + 4$ and
(c) $y = 2x + 3$ cut the circle $x^2 - 6x + y^2 - 2y + 5 = 0$.
Draw a sketch showing the three lines and the circle.

4.C.(f)   Finding the equations of tangents to circles

The circle is the first curve for which we can find the steepness or gradient at any point on it. We saw in Section 4.C.(b) that any tangent to a circle must be perpendicular to the radius going to the point of contact. The gradient of the tangent will then tell us the slope or gradient of the circle at this point of contact.

We will look at the following example together to see how these ideas work out in practice.

EXAMPLE (1) Find the equations of the four tangents to the circle

$$x^2 - 6x + y^2 - 4y - 12 = 0$$

with points of contact (a) (7,5), (b) (–1, –1), (c) (8,2) and (d) (3,7).
Draw a sketch showing the circle and these four tangents.
We start by finding the centre and radius of the circle.
We have

$$x^2 - 6x + y^2 - 4y - 12 = 0 = (x - 3)^2 - 9 + (y - 2)^2 - 4 - 12.$$

So the equation of the circle is also given by $(x - 3)^2 + (y - 2)^2 = 25$.
Its centre is at the point (3,2) and its radius is 5 units.
I have drawn a sketch of this circle in Figure 4.C.11 showing the first tangent that we shall find. I think that it will help you in the working which follows if you sketch in how you think the other three tangents will go.

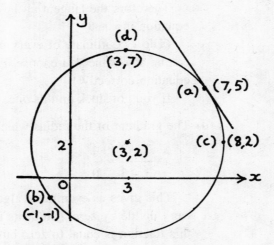

Figure 4.C.11

(a) The first tangent touches the circle at the point (7,5).
    The radius to the point of contact joins (3,2) to (7,5), so its gradient is

$$\frac{y_2 - y_1}{x_2 - x_1} = \frac{5 - 2}{7 - 3} = \frac{3}{4} \quad \text{using Section 2.B.(d).}$$

4.C Circles   161

The tangent is perpendicular to this radius, so its gradient is $-\frac{4}{3}$, using $m_1 m_2 = -1$ from Section 2.B.(h).

It passes through the point (7,5) so its equation is $y - 5 = -\frac{4}{3}(x - 7)$.

(This uses $y - y_1 = m(x - x_1)$ from Section 2.B.(f).)

Tidied up, this gives $3y - 15 = -4x + 28$ or $3y + 4x = 43$.

I have shown this tangent on my sketch on the previous page.

Try finding the other three tangents yourself.

If curious things happen, look at the sketch and see if you can see why.

This is what you should have.

(b) The gradient of the radius which joins (3,2) to (−1, −1) is $\dfrac{-1-2}{-1-3} = \dfrac{3}{4}$.

Therefore, the gradient of tangent (b) is $-\frac{4}{3}$.

The equation of tangent (b) is $y + 1 = -\frac{4}{3}(x + 1)$ or $3y + 4x + 7 = 0$.

You can sketch this tangent yourself, if you haven't already done so. It is parallel to the one which we found in (a).

(c) The gradient of the radius which joins (3,2) to (8,2) is $\dfrac{2-2}{8-3} = 0$.

This gives us a real problem for finding the equation of the tangent by algebra but, when we look at the sketch, everything becomes clear.

The gradient of this radius is zero because it is horizontal.

Therefore the tangent at the point (8,2) is vertical and its equation is $x = 8$.

(The x coordinate of every point on it is 8 while the y coordinate can be any value you choose. Excellent thinking if you got this equation correctly!)

If you got stuck on this one, have another go now at answering (d).

(d) The gradient of the radius which joins (3,2) to (3,7) is given by

$$\frac{7-2}{3-3} = \frac{4}{0}.$$

This gives us even more algebraic trouble since we know we can't divide by zero. (Students in desperation sometimes say that this fraction is equal to zero but this is not true!)

Again, looking at the sketch we see that everything falls into place.

This radius is vertical and the tangent at the point (3,7) is horizontal. Its gradient is zero and its equation is $y = 7$.

Add tangents (c) and (d) to the sketch if you haven't already done so.

Because the circle *is* a curve for which we can find out what is happening with the algebra which we can do now, the example above will be very useful to you when you start working with the slopes of general curves using implicit differentiation in Section 8.F.(a). It will help you to see why things happen in the way that they do.

EXERCISE 4.C.3

Draw a sketch of the circle $x^2 + 16x + y^2 - 4y - 101 = 0$.

Find the equations of the four tangents to this circle with the points of contact

(a) $(4, -3)$,   (b) $(-3, 14)$,   (c) $(-21, 2)$ and   (d) $(-8, -11)$.

Show these four tangents on your sketch.

4.D   Using radians

4.D.(a)   Measuring angles in radians

So far, all the angles to which we have given a size have been measured in degrees. This form of measurement has an arbitrary element about it in that somebody originally decided that 90 would be a nice number of units to have in a right angle. It could equally well have been 100 or 80, say. Had the scale been chosen by Napoleon, it probably *would* have been 100, to fit with his other metric measurements. (Indeed, the mysterious gradians on your calculator are divided so that there are 100 parts to each right angle.)

The special property of the radian is that it does not depend upon any arbitrary choice of number. It *does* depend on that beautiful and symmetrical shape, the circle.

I show how in Figure 4.D.1.

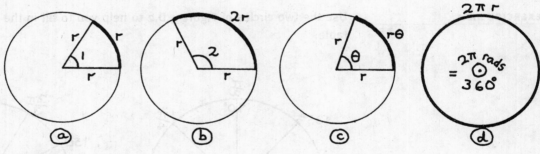

Figure 4.D.1

If we draw an angle as shown in Figure 4.D.1(a), so that the length of the arc is equal to the radius, then this angle is defined to be 1 radian.

If the arc is 2 radius lengths long, the angle is 2 radians (Figure 4.D.1(b)).

From Figure 4.D.1(c), an angle of θ radians gives an **arc length** of $r\theta$.

(θ is the Greek letter theta and is a hot favourite for describing an unknown angle, just as x is for describing general unknown quantities.)

4.D Using radians

163

Since a full turn gives an arc length of the whole circumference of the circle, which is an arc length of $2\pi r$, we see from Figure 4.D.1(d) that a full turn is 2π radians.

This means that 2π radians is the same angle as 360°.

Remembering, too, that π is a bit bigger than 3, we have the following box of results.

Useful rules connecting degrees and radians

- π radians is the same angle as 180°.
  (You can think of π as a symbol for a straight line angle.)
- To convert degrees to radians, multiply by $\pi/180$.
- To convert radians to degrees, multiply by $180/\pi$.
- It is useful to remember that one radian is just slightly less than 60°.

(In practice, you very rarely have to use the conversion from degrees to radians or vice versa, because you will set your calculator in either degree or radian mode depending upon which units you want to work in.)

Because radians come from the structure of the circle, they will slot directly into any working involving angles when we use calculus. If we work with degrees, however, we shall keep having to do a sort of gear change – and it's much nicer not having to worry about that! For this reason you need to be happy working with radians, so it is a good idea now to become familiar with the corresponding radian measurements for the standard divisions of 360°.

EXERCISE 4.D.1

Use the two circles of Figure 4.D.2 to help you to fill in the missing angles in the table.

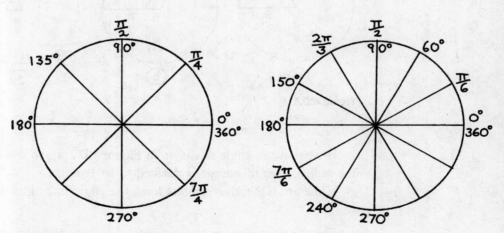

Figure 4.D.2

Degrees	0			60	90		135	150	180		240	270		360
Radians	0	$\frac{\pi}{6}$	$\frac{\pi}{4}$		$\frac{\pi}{2}$	$\frac{2\pi}{3}$				$\frac{7\pi}{6}$			$\frac{7\pi}{4}$	

Some trigonometry and geometry

4.D.(b)   **Finding the perimeter and the area of a sector of a circle**

I have shown the minor sector *AOB* shaded in the circle with radius *r* in Figure 4.D.3.

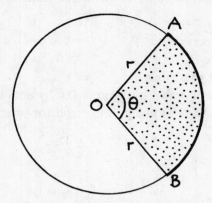

The angle θ is in radians.

Figure 4.D.3

We know from the last section that the arc length *AB* is equal to $r\theta$.

Therefore, the length of the perimeter of the sector *AOB* (that is, the distance round its boundary) is given by $2r + r\theta$.

   Don't forget to include the two radius lengths here.

The perimeter of the sector is $2r + r\theta$.

We can find the area of the sector *AOB* by thinking of it as a fraction of the area of the whole circle (which is πr^2).

The area of the sector *AOB* is given by $\dfrac{\theta}{2\pi} \times \pi r^2 = \frac{1}{2} r^2\theta$.

   Both these formulas are only true if θ is in radians.

Try writing down for yourself what the area of the *major* sector *AOB* is (that is, the area of the rest of the circle).

Subtracting the area of the minor sector AOB from the area of the whole circle gives the result that the area of the major sector $AOB = \pi r^2 - \frac{1}{2}r^2\theta$.

Alternatively, you could say that the angle of the major sector is $2\pi - \theta$.

Therefore its area is given by

$$\tfrac{1}{2}r^2(2\pi - \theta) = \pi r^2 - \tfrac{1}{2}r^2\theta.$$

4.D.(c) Finding the area of a segment of a circle

We can find the area of the segment drawn in Figure 4.D.4 by noticing that it comes from subtracting $\triangle AOB$ from sector AOB. (I'm using $\triangle$ to stand for 'triangle'.)

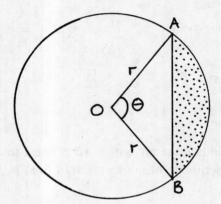

Figure 4.D.4

Again, the angle θ is in radians.

We know from Section 4.B.(b) that the area of $\triangle AOB$ is equal to $\frac{1}{2}r^2 \sin\theta$, so the area of the segment shown (that is, the minor segment), is given by the rule below.

The area of the segment $AOB = \frac{1}{2}r^2\theta - \frac{1}{2}r^2 \sin\theta = \frac{1}{2}r^2 (\theta - \sin\theta)$

(Make sure that your calculator is in radian mode when you find this!)

Now try writing down for yourself the area of the major segment AB (that is, the unshaded part of the circle in Figure 4.D.4).

It is given by $\pi r^2 - \frac{1}{2}r^2 (\theta - \sin\theta) = \frac{1}{2}r^2 (2\pi - \theta + \sin\theta)$.

4.D.(d) What do we do if the angle is given in degrees?

I will call the angle $D°$ to avoid confusing it with the angle θ in radians.

There are two things you can do in this situation.

METHOD (1) Immediately convert the angle $D°$ into radians by multiplying it by $\pi/180$. (See Section 4.D.(a) if necessary.) Then you can use all the rules given above for angles in radians. This is the method I would recommend.

METHOD (2)   Alternatively, you can change the rules that we have already found so that they will be right for working with angles in degrees by replacing θ by $D\pi/180$.

This will then give you, for an angle D measured in degrees,

(1) The arc length is $r \times \dfrac{D\pi}{180} = \dfrac{\pi rD}{180} = \dfrac{\text{angle}}{360} \times$ circumference.

(2) The area of the sector is

$$\frac{1}{2}\,r^2 \times \frac{D\pi}{180} = \frac{\pi r^2 D}{360} = \frac{\text{angle}}{360} \times \text{ the area of the circle.}$$

These rules are more clumsy than the rules for radians because of the arbitrary nature of the choice of 360 for the number of degrees in a full turn.

Because radians use the structure of the circle itself, they give much nicer results.

EXERCISE 4.D.2

Now try these questions, giving your answers correct to 2 d.p. (if they are not exact) in the units used on the drawings.

(1) Using the sketch shown in Figure 4.D.5(a), find
   (a) the minor arc length *AB*,
   (b) the area of △ *AOB*,
   (c) the area of the minor segment *AB*.

(2) Find the shaded area (that is, the major sector) shown in Figure 4.D.5(b).

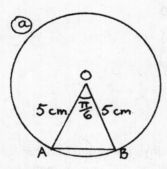
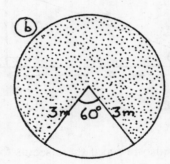
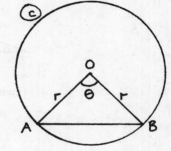

Figure 4.D.5

THINKING POINT

The circle shown in Figure 4.D.5(c) above has a fixed radius of *r* units. What do you think the size of the angle θ should be in order to make triangle AOB have maximum area?

4.D.(e)   Very small angles in radians – why we like them

Radians have a second very special quality, as well as being independent of anyone's particular choice of number.

Suppose we start with an angle of θ radians as shown in Figure 4.D.6.

4.D Using radians

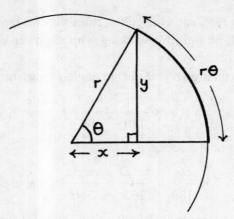

Figure 4.D.6

We know from Section 4.D.(a) that the arc length is $r\theta$, and we also know that

$$\sin \theta = \frac{y}{r}, \quad \cos \theta = \frac{x}{r} \quad \text{and} \quad \tan \theta = \frac{y}{x}.$$

What happens to these trig ratios as θ becomes very small?

Try finding this out yourself experimentally with your calculator. Use radian mode, and put in very small values for the angle, say 0.001 as one possible value. See what values the answers are close to. Can you see why this might be if you look at the drawing of Figure 4.D.7?

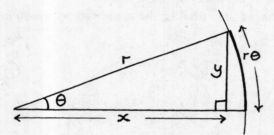

Figure 4.D.7

Look also to see if there seems to be any connection between the size of the angle that you put in and the values for sin, cos and tan that you get out.

 Remember that your calculator must be in radian mode for this experiment. A mistake here will seriously affect your results. (For example, 1° is quite a small angle, but 1 radian is about 60°, so an input of 1 will give you vastly different results depending on which mode your calculator is in.)

You should now have a good experimental idea of what is happening.

We will now look together at why this should be so.

Figure 4.D.7. shows a very small angle θ set inside its circle.

As θ becomes increasingly smaller, x becomes closer and closer to r so $\cos \theta \to 1$.

(The $\to$ symbol I have used above is a mathematical shorthand for saying 'becomes increasingly closer in value to'. It saves a lot of writing!)

Also, y becomes very small indeed, so $\sin \theta \to 0$, and $\tan \theta \to 0$ also.

But you should also have found a more startling result. *Not only are $\sin \theta$ and $\tan \theta$ becoming very small, they are also becoming very close to θ itself, as θ becomes small.*

We can see from the diagram that this must be so.

As y becomes smaller it gets closer and closer in length to the arc $r\theta$. So

$$\sin \theta \to \frac{r\theta}{r}, \text{ that is } \sin \theta \to \theta \text{ as } \theta \to 0.$$

The smaller the angle becomes, the closer these two are. We also see that $\sin \theta$ will always be slightly less than θ because y stays less than $r\theta$. Notice that the arc $r\theta$ will become closer and closer to a straight line as θ becomes smaller.

Now, what happens to $\tan \theta$?

Since $\tan \theta = y/x$, it is clearly going to get smaller and smaller just as $\sin \theta$ does. It looks from the calculator as if it is close to θ too, but a little bit larger.

Will it stay like this?

We can see that it will from Figure 4.D.8.

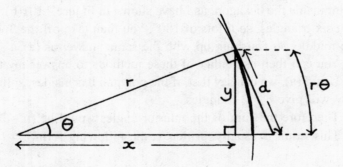

Figure 4.D.8

This uses the fourth property from Section 4.C.(b) to give the right angle between the radius and the tangent. Using this right-angled triangle, $\tan \theta = d/r$, but d is getting closer and closer to $r\theta$ while remaining just slightly larger.

So

$$\tan \theta \to \frac{r\theta}{r}, \text{ that is } \tan \theta \to \theta \text{ also, as } \theta \to 0.$$

But it stays slightly larger than θ while $\sin \theta$ stays slightly smaller.

The fact that when we measure in radians $\sin \theta$ and $\tan \theta$ are approximately the same as θ when θ is very small is of crucial importance when we come to calculus.

4.E.(a)   **The sum of interior and exterior angles of polygons**

At the end of Section 4.A.(e) on congruent triangles, I asked you if you could find the sum of the interior angles of a six-sided figure. (This is called a **hexagon**.)

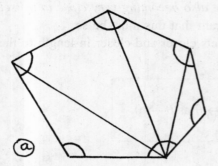

Figure 4.E.1

(a)   One way of answering this question is to split the shape into triangles by joining up to one corner as I have shown in Figure 4.E.1(a).

This gives us four triangles, that is, two fewer triangles than there are sides.

Together they account for all the interior angles.

We see, therefore, that the sum of the interior angles is $4 \times 180° = 720°$.

(b)   You could also have got this answer by joining up each corner (or vertex) to some point inside the hexagon, as I have shown in Figure 4.E.1(b). This would then give you six triangles, so 6 lots of 180°. You then take off the 360° for the full turn in the middle, so finishing up with the same answer as (a).

You can then use either of these methods to answer my third question.

Using (a), we can say that, if the polygon has n sides, splitting it up in the same way will give $n - 2$ triangles.

Therefore the sum of the interior angles would be $(n - 2) \times 180°$.

This result is usually written in the following form.

The sum of the interior angles of an n-sided polygon is equal to $(2n - 4)$ right angles.

The sum of the **exterior** angles will be the same whatever the shape of the hexagon is, so long as we are turning inwards all the while as we go round.

We find this sum by noticing in Figure 4.E.2(a) that we have six straight lines formed by the exterior angles and the interior angles together.

Therefore, the exterior angles together make $6 \times 180° - 720° = 360°$ or a full turn.

We can see that this must be so because if we start at A and travel round the sides of the shape, we will have made a full turn when we come back to A. This full turn is built up from all the small turns made by the exterior angles, as I have shown in Figure 4.E.2(b). Exactly the same thing will happen however many sides the shape has, provided we are always turning inwards as we go round, that is, none of the interior angles is greater than 180°. The exterior angles will always add to four right angles.

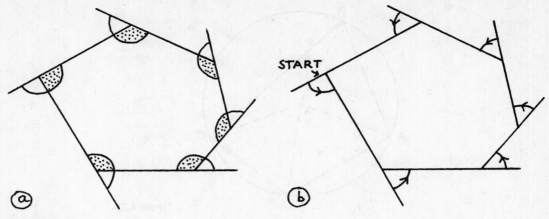

Figure 4.E.2

Indeed, this result is still true if our particular choice of shape means that we *do* sometimes turn outwards, but in this case we must count these outwards turns as negative.

4.E.(b) **Can we draw circles round *all* triangles and quadrilaterals?**
I asked you this question at the end of Section 4.C.(c) on the special properties of circles. The answer is that it *is* always possible to draw a circle round a triangle.

You can see this from the drawings of Figure 4.E.3(a) and (b).

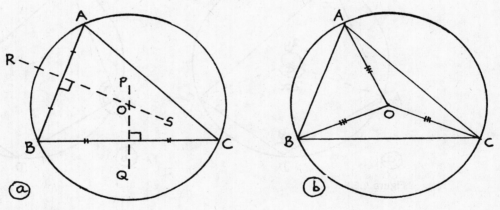

Figure 4.E.3

From (3) in Section 4.C.(b), the centre of the circle would have to lie on the line PQ. (The little marks are to show that PQ divides BC equally in two as well as being perpendicular to it.)

For the same reason, it would have to lie on RS.

But where PQ and RS cross, we have $CO = BO$ and $BO = AO$. So $CO = AO$ too, and O is the centre of the circle which triangle ABC sits inside.

We can also see from this that it *isn't* always possible to draw a circle round a quadrilateral like $ABCD$.

If we have a quadrilateral $ABCD$ sitting inside a circle, as in Figure 4.E.4, then this must be the particular circle which can be drawn round triangle ABC.

But a small adjustment to D, either inwards or outwards, will mean that this point is no longer on the circle which works for A, B and C.

So what particular property must $ABCD$ have for it to be possible to draw a circle through its four corners?

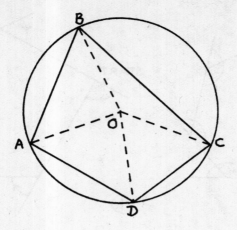

Figure 4.E.4

We can see the answer to this from Figure 4.E.5(a).
Using (1) from Section 4.C.(c), we know that $\angle AOC = 2\angle ABC$.
Looked at the other way up, the other part of $\angle AOC = 2\angle ADC$.
But the two parts together of $\angle AOC$ make 360°, so $\angle ABC + \angle ADC = 180°$.
Also, since $\angle A + \angle B + \angle C + \angle D = 360°$, $\angle A + \angle C = 180°$ too.

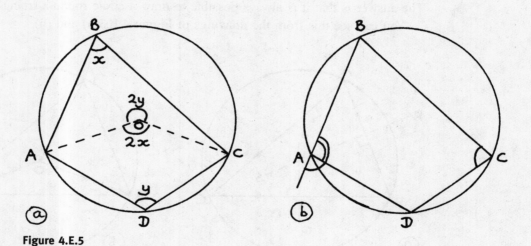

Figure 4.E.5

It is only possible to draw a circle through the four corners of a quadrilateral if its opposite angles add up to 180°. Such a quadrilateral is called **cyclic**.

This is the same as saying that each exterior angle must equal its interior opposite angle. We can see that this must be so from Figure 4.E.5(b) since the two angles at A together make a straight line.

5 Extending trigonometry to angles of any size

This chapter makes it possible for us to use trig ratios with angles of any size, and looks at the graphs of these trig functions. These are very important in many physical applications, so we look at what happens if we shift them and combine them. We also look at methods of handling trig functions and equations.

The chapter is divided into the following sections.

5.A Giving meaning to trig functions of any size of angle
(a) Extending sin and cos, (b) The graph of $y = \tan x$ from 0° to 90°,
(c) Defining the sin, cos and tan of angles of any size,
(d) How does X move as P moves round its circle?
(e) The graph of $\tan \theta$ for any value of θ, (f) Can we find the angle from its sine?
(g) $\sin^{-1} x$ and $\cos^{-1} x$ – what are they?
(h) What do the graphs of $\sin^{-1} x$ and $\cos^{-1} x$ look like? (i) Defining the function $\tan^{-1} x$

5.B The trig reciprocal functions
(a) What are trig reciprocal functions?
(b) The trig reciprocal identities: $\tan^2 \theta + 1 = \sec^2 \theta$ and $\cot^2 \theta + 1 = \csc^2 \theta$,
(c) Some examples of proving other trig identities,
(d) What do the graphs of the trig reciprocal functions look like?
(e) Drawing other reciprocal graphs

5.C Building more trig functions from the simplest ones
(a) Stretching, shifting and shrinking trig functions,
(b) Relating trig functions to how P moves round its circle and SHM,
(c) New shapes from putting together trig functions,
(d) Putting together trig functions with different periods

5.D Finding rules for combining trig functions
(a) How else can we write $\sin (A + B)$?
(b) A summary of results for similar combinations,
(c) Finding $\tan (A + B)$ and $\tan (A - B)$, (d) The rules for $\sin 2A$, $\cos 2A$ and $\tan 2A$,
(e) How could we find a formula for $\sin 3A$?
(f) Using $\sin (A + B)$ to find another way of writing $4 \sin t + 3 \cos t$,
(g) More examples of the $R \sin (t \pm \alpha)$ and $R \cos (t \pm \alpha)$ forms,
(h) Going back the other way – the Factor Formulas

5.E Solving trig equations
(a) Laying some useful foundations, (b) Finding solutions for equations in $\cos x$,
(c) Finding solutions for equations in $\tan x$, (d) Finding solutions for equations in $\sin x$,
(e) Solving equations using $R \sin (x + \alpha)$ etc.

5.A Giving meaning to trig functions of any size of angle

5.A.(a) Extending sin and cos

In the last chapter we discovered that we were able to find the sin and cos of some angles between 90° and 180° by using the Sine and Cosine Rules for any triangle. (In fact, it would be possible, by choosing suitable triangles, to find the sin and cos of *any* angle in this range.)

It seemed, from the results which we got there, that we would need to put sin (180° − x) = sin x and cos (180° − x) = − cos x in order to make the Sine and Cosine Rules work for all triangles. If we use this to draw graphs of y = sin x and y = cos x for values of x from 0° to 180° we will get curves like those in Figure 5.A.1.(a) and (b).

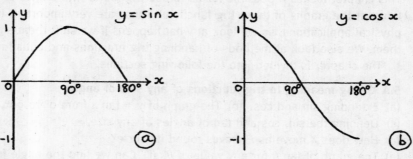

Figure 5.A.1

The shape of these two curves suggests that what we have here is part of a much longer pattern, and that indeed they are parts of the *same* graph which has just been shifted by 90° to the left to give the second case.

This view will seem very reasonable if you have seen, for example, sound waves displayed on an oscilloscope, or the graph of an alternating electric current in a wire, or the waves which you get along a rope if you fix one end and move the other end up and down.

From these physical examples, we will get the pair of graphs shown in Figure 5.A.2.(a) and (b). I have used units of radians here for the angles. I explain how radians work in Section 4.D and if you are at all unsure about them you should go back there now, before going on. This is because they are very important throughout this chapter and for future work, particularly if it involves calculus.

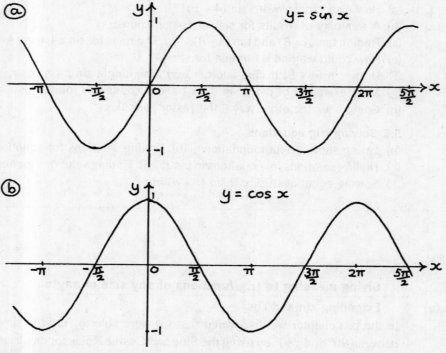

Figure 5.A.2

Clearly, there is no particular reason to stop anywhere, so we imagine the two graphs as extending an infinite distance in both the + and − directions.

How many special distinctive properties can you see in these two graphs?

Make a note of as many as you can.

Here are some of the important particular properties of these two graphs which I hope that you will have noticed.

(1)   The cos graph is symmetrical about the *y*-axis, or the line $x = 0$.
For example, $\cos \frac{\pi}{2} = \cos(-\frac{\pi}{2})$. In fact, $\cos x = \cos(-x)$, whatever *x* is.
A graph like this is called **even**, as we saw in Section 3.B.(j).

(2)   The sin graph exactly fits onto itself if it is rotated through half a complete turn about the origin. If you turn the page upside down, this graph is unchanged.
You could also describe this by saying that the graph of sin *x* reverses sign if it is reflected through the *y*-axis.
$\sin \frac{\pi}{2} = -\sin(-\frac{\pi}{2})$, and $\sin x = -\sin(-x)$ whatever *x* is.
A graph like this is called **odd**. (Again, there were similar ones in Section 3.B.(j).)

(3)   They are the same graph, except that the sin graph must be shifted $\pi/2$ to the left to give the cos graph.
For example, $\sin \frac{\pi}{2} = \cos 0$, $\sin \pi = \cos \frac{\pi}{2}$ and, in general, $\sin (x + \frac{\pi}{2}) = \cos x$. (There are other examples of shifts in Section 3.B.(d).)

(4)   Both of the graphs infinitely repeat themselves, with the length of the unit of repeat being 2π in each case. This is called the **period** of the graph.

(5)   In both cases, the graphs are enclosed in a pair of horizontal lines which are one unit either side of the *x*-axis so the maximum **displacement** of the graph from this axis is one unit.

EXERCISE 5.A.1

We have already found (in Section 4.A.(g)), values for the sin and cos of angles of 0°, 30°, 45°, 60° and 90°.

I have shown these values again set out in the table below, using both radians and degrees.

Angle (x) degrees	−180	−120	−90	−30	0	30	45	60	90	120	180	210	270	315	360
radians	$-\pi$	$-\frac{2\pi}{3}$	$-\frac{\pi}{2}$	$-\frac{\pi}{6}$	0	$\frac{\pi}{6}$	$\frac{\pi}{4}$	$\frac{\pi}{3}$	$\frac{\pi}{2}$	$\frac{2\pi}{3}$	π	$\frac{7\pi}{6}$	$\frac{3\pi}{2}$	$\frac{7\pi}{8}$	2π
sin x					0	$\frac{1}{2}$	$\frac{1}{\sqrt{2}}$	$\frac{\sqrt{3}}{2}$	1						
cos x					1	$\frac{\sqrt{3}}{2}$	$\frac{1}{\sqrt{2}}$	$\frac{1}{2}$	0						

Use these values, and the symmetrical properties of the graphs shown in Figure 5.A.3 (a) and (b), to write down the values of the sin and cos of the other angles listed in the table. Check your values using your calculator.

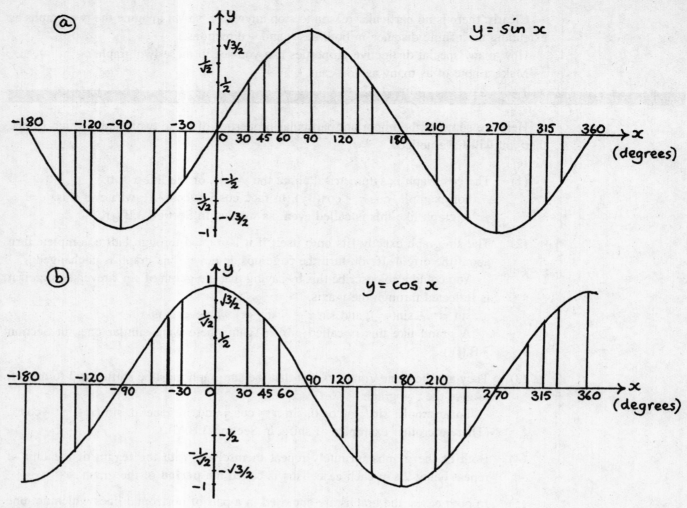

Figure 5.A.3

5.A.(b)   The graph of $y = \tan x$ from $0°$ to $90°$

We have not yet thought about what the graph of $y = \tan x$ will look like. We know from Section 4.A.(g) that

$$\tan 45° = 1, \quad \tan 30° = \frac{1}{\sqrt{3}} = 0.58 \text{ to 2 d.p.} \quad \text{and}$$

$$\tan 60° = \sqrt{3} = 1.73 \text{ to 2 d.p.}$$

We also know, from Section 4.A.(h), that

$$\tan x = \frac{\sin x}{\cos x} \quad \text{so} \quad \tan 0° = \frac{0}{1} = 0 \quad \text{and} \quad \tan 90° = \frac{1}{0} = \text{trouble,}$$

since we can't divide by zero.

Using your calculator, you can see that, the closer the angle gets to $90°$, the larger its tan becomes. (Try this for yourself.) You can also see that this will happen from the three triangles in Figure 5.A.4(a) by finding the tans of the three marked angles. The height of the triangles remains the same but the horizontal measurement becomes smaller, so the fraction which gives the tan is becoming larger.

Using all our known information, we get a sketch for $y = \tan x$ from $0°$ to $90°$ which looks like Figure 5.A.4(b).

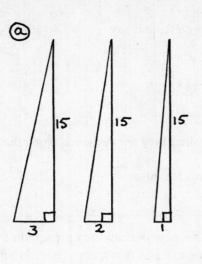

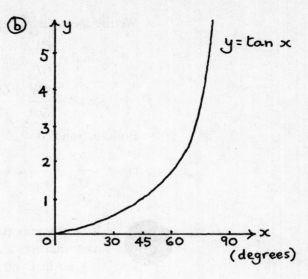

Figure 5.A.4

5.A.(c)   Defining the sin, cos and tan of angles of any size

There is no general Tangent Rule which works for any triangle, like the Sine and Cosine Rules, so we have no simple way to sketch the continuation of the graph for tan x.

It would be good to have a *definition* for the sin, cos and tan of angles of any size so that we wouldn't have to rely on what is apparently happening physically, although, to be useful, any definition would have to fit in with observed wave phenomena.

We shall now do this by using the turn or angle measured out on a circle. (We have already used this method for showing the turn of angles in Figure 4.C.1 in the last chapter.)

We consider the rotation of a unit length through a full turn about the origin, in an anticlockwise direction from the positive x-axis. I have shown this in four separate diagrams which show rotations round to each quadrant or quarter-circle, in turn. The angles of rotation are shown shaded.

You can think of OP as a rod of length one unit which is turning about O.

First quadrant

In the first quadrant, shown in Figure 5.A.5, the definition exactly tallies with the definitions given at the beginning of the last chapter in Section 4.A.(a) for the sin, cos and tan of angles between 0° and 90°. I have used the symbol θ for the angle here, as I want to keep x for the length OX. (θ is the Greek letter theta.)

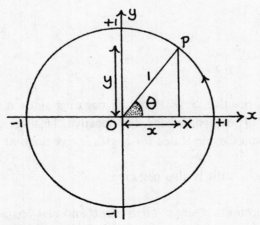

Figure 5.A.5

5.A Trig functions of any size of angle   177

We use the right-angled triangle *OPX*, and say

$$\sin \theta = \frac{PX}{OP} = \frac{PX}{1} = y$$

$$\cos \theta = \frac{OX}{OP} = \frac{OX}{1} = x.$$

Both $\sin \theta$ and $\cos \theta$ are positive since they are measured along the positive x and y axes.

$$\tan \theta = \frac{PX}{OX} = \frac{y}{x} \quad \text{so } \tan \theta \text{ is positive, also.}$$

> **NOTE**
>
> It is very important that this new definition is giving $\sin \theta$ and $\cos \theta$ as measurements along the y- and x-axes respectively – so important that I suggest that you use one colour for $y = \sin \theta$ and another for $x = \cos \theta$ here, and on the following three diagrams.

Second quadrant

The angle we are considering is now between $\pi/2$ and π radians (or 90° and 180°.) Again, we use the right-angled triangle *OPX* for our definitions.

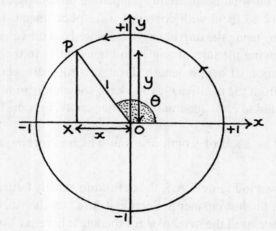

Figure 5.A.6

We say

$$\sin \theta = \frac{PX}{OP} = \frac{PX}{1} = y,$$

$$\cos \theta = \frac{OX}{OP} = \frac{OX}{1} = x.$$

This time, although y is positive, x will now be negative since it is measured along the negative x-axis, so $\sin \theta$ is positive but $\cos \theta$ is negative. This agrees with what we found when we used the Sine and Cosine Rules for angles larger than 90°.

$$\tan \theta = \frac{PX}{OX} = \frac{y}{x} \quad \text{so it is also negative.}$$

We can see from the diagram that $\sin(\pi - \theta) = \sin \theta$ and that $\cos(\pi - \theta) = -\cos \theta$.
$(\pi - \theta) = \angle POX$ in size, so it would come in the first quadrant.

Third quadrant

Again using the right-angled triangle OPX for our definitions, we say

$$\sin \theta = \frac{PX}{OP} = \frac{PX}{1} = y,$$

$$\cos \theta = \frac{OX}{OP} = \frac{OX}{1} = x.$$

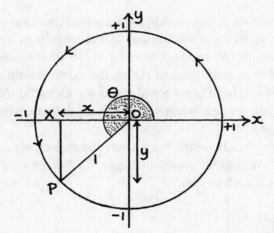

Figure 5.A.7

This time, both $\sin \theta$ and $\cos \theta$ are negative, since they are measured along the negative y and x axes respectively.

$$\tan \theta = \frac{PX}{OX} = \frac{y}{x} \quad \text{so it is positive.}$$

We also see from the diagram that $\sin \theta = -\sin(\theta - \pi)$ and $\cos \theta = -\cos(\theta - \pi)$.

$(\theta - \pi) = \angle POX$ in size, so it would come in the first quadrant.

Fourth quadrant

Again using the right-angled triangle OPX for our definitions, we have

$$\sin \theta = \frac{PX}{OP} = \frac{PX}{1} = y,$$

$$\cos \theta = \frac{OX}{OP} = \frac{OX}{1} = x.$$

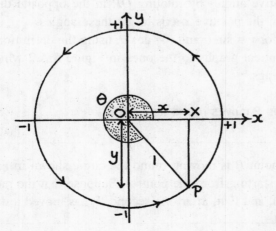

Figure 5.A.8

We see that sin θ is negative, and cos θ is positive, from the positions of y and x on the two axes.

$$\tan \theta = \frac{PX}{OX} = \frac{y}{x} \quad \text{so it is negative.}$$

We also see that $\sin \theta = -\sin(2\pi - \theta)$ and $\cos \theta = \cos(2\pi - \theta)$.

$(2\pi - \theta) = \angle POX$ in size, so it would come in the first quadrant.

You can see from these four diagrams that, by using the right-angled triangle OPX in each quadrant, we have now *defined* the sin and cos of the angle θ in terms of the shadow or projection of the unit length OP on the x-axis for cos θ (the distance shown as x in the diagrams), and the shadow or projection of OP on the y-axis for sin θ (the distance shown as y in the diagrams). If you have highlighted x and y with two different colours on these diagrams, it will emphasise for you, when you look back at them, where the sin and cos are and how they are changing.

The $+$ or $-$ signs automatically follow from where the projections lie on the two axes. You may find it helpful to use the picture shown in Figure 5.A.9 to remember the changing signs for sin, cos and tan in a complete turn.

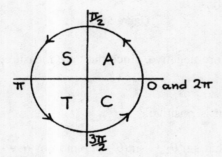

Figure 5.A.9

The letters **A S T C** stand for whatever is positive in that particular quadrant. **A** = 'all', **S** = 'sin', **T** = 'tan' and **C** = 'cos'. This can be remembered by a catch-phrase if you like, such as 'All Silly Tom Cats.'

When OP has turned through an angle of 2π it will have returned to its original position. (It has completed one cycle.) If we then continue to rotate it, the whole identical process will be repeated with each new full turn or cycle.

We can obtain negative angles by rotating OP in the opposite direction, so we would rotate it clockwise from the positive x-axis to get these angles.

Plotting the graphs for $y = \sin \theta$ and $y = \cos \theta$, using the definitions which we have just given, will give us identical graphs to the ones in Figure 5.A.2 which we know describe actual physical happenings.

5.A.(d)   **How does *X* move as *P* moves round its circle?**

THINKING POINT

Suppose the point P is moving round the circle shown in Figure 5.A.10 at a steady speed, starting from the point A. Suppose that the radius of the circle is 1 m (metre), and that, after one second, P has moved a distance of 1 m.

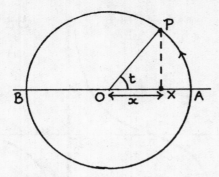

Figure 5.A.10

Try answering the following questions.

(1) What angle (in radians) has the line *OP* turned through after one second? (See Section 4.D.(a) if you need help with radians.)

(2) How long will it take *P* to make a full turn round the circle?

(3) How far is the point *X* from *O* after a time of
(a) 0 seconds, (b) 1 second, (c) 1.5 seconds, (d) $\pi/2$ seconds, (e) π seconds, (f) $3\pi/2$ seconds, and (g) *t* seconds?

(4) As *P* turns round the circle at its steady speed, how is the point *X* moving? Does it also have a constant speed? If not, when do you think it is moving fastest? When is it moving slowest?

These are the answers which I hope you have found.

(1) One radian. We say that the **angular velocity** of *P* is one radian per second.

(2) A full turn is 2π radians, so 2π seconds.

(3) (a) $OX = 1$ m. (b) $OX = \cos t = \cos 1 = 0.54$ m to 2 d.p.
(c) After 1.5 seconds, $OX = \cos 1.5 = 0.07$ m to 2 d.p.
(d) After $\pi/2$ seconds, *X* is at *O*, so $OX = 0$.
(e) After π seconds, the distance *OX* is again 1 m as *P* is now at *B*. We can think of this distance as negative, since it is measured in the opposite direction to *OA*.
(f) After $3\pi/2$ seconds, $OX = 0$.
(g) After *t* seconds, $OX = \cos t$ metres. If we let $OX = x$, we could write the equation giving the position of *X* after time *t* as $x = \cos t$.

(4) *X* is not moving at a constant speed. It moves fastest as it passes through *O* and slowest at the points *A* and *B* when it instantaneously comes to rest before turning back on itself.

The point *X* is moving in what is called simple harmonic motion or SHM.

Surely, if we know the distance or displacement of *X* from *O* at any time, we have enough information to discover its speed exactly? Indeed we have, and we shall be able to do just this in Section 8.A.(e).

5.A.(e)   The graph of tan θ for any value of θ

Using $\tan \theta = y/x = \sin \theta/\cos \theta$ in the four diagrams of Figures 5.A.5–5.A.8, we can now define tan θ for any size of angle θ. We can therefore draw the extended graph of $y = \tan \theta$ which I've done in Figure 5.A.11.

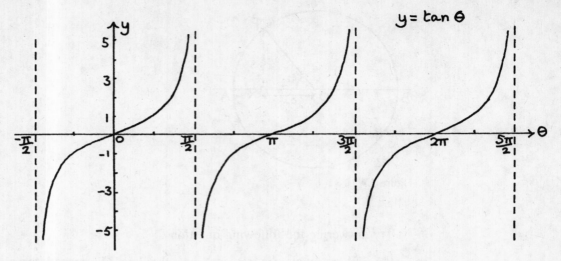

Figure 5.A.11

What special properties does this graph have? Make a note of as many as you can.

The graph shows these special properties.

- It is **periodic**, but the period of repeat this time is π rather than 2π, as it was for $\sin \theta$ and $\cos \theta$.
- It is **odd**, that is, if you rotate it through half a turn about the origin, it fits exactly onto itself, so if you turn the page upside down you get the same graph. Equally you could say that, if you reflect it through the y-axis, it reverses its sign, so

    $$\tan x = -\tan(-x).$$

- The tan of an angle just less than $\pi/2$ (or 90°) is very large and positive.
- The tan of an angle just greater than $\pi/2$ (or 90°) is very large and negative.
- There is a jump or discontinuity in the graph when $\theta = \pi/2$ and we therefore see that the tan of 90° *can't* be given a value, and any calculator asked to display it will give an ERROR message. The same thing happens for all odd multiples of $\pi/2$, so on the graph we see it happening at

    $$-1 \times \frac{\pi}{2}, \quad +1 \times \frac{\pi}{2} \quad \text{and} \quad +3 \times \frac{\pi}{2} \quad \text{and} \quad +5 \times \frac{\pi}{2}.$$

The graph has a **vertical asymptote** for each of these values of θ, just as the graph in Section 3.B.(i) had a vertical asymptote of $x = 2$.

5.A.(f) Can we find the angle from its sine?

In Figure 5.A.12, I show again the graph of $y = \sin x$ for values of x from $-\pi$ to 2π.

From this graph, find x for these values of $\sin x$.

(a) $\sin x = 1$ (b) $\sin x = 0$ (c) $\sin x = -1$ (d) $\sin x = \frac{1}{2}$ (e) $\sin x = -\frac{1}{2}$.

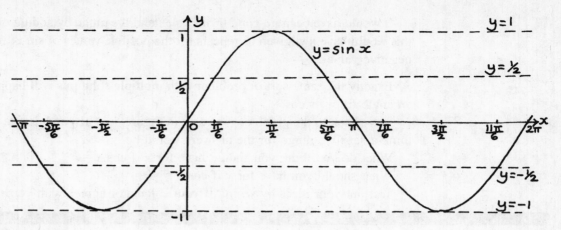

Figure 5.A.12

Here are the answers which you should have found.

(a) $x = \pi/2$.

(b)   As soon as we try this one, we find that we've got a more complicated situation. There are four possible values of x on this graph for which $\sin x = 0$. We can have $x = -\pi$ or $x = 0$ or $x = \pi$ or $x = 2\pi$.

(c)   Similarly, if $\sin x = -1$, from the graph we have $x = -\pi/2$ or $+ 3\pi/2$.

(d)   If $\sin x = \frac{1}{2}$, then from the graph we have $x = \pi/6$ or $5\pi/6$.

(e)   If $\sin x = -\frac{1}{2}$, then from the graph we have $x = -5\pi/6$ or $-\pi/6$ or $7\pi/6$ or $11\pi/6$.

We can see that extending the graph further in either direction would give us more solutions for x for any given value of $\sin x$, and that there are, in fact, an infinite number of possible solutions.

Although this infinitely repeating possibility will be very important in describing some situations, such as those involving waves of one kind or another, in many other circumstances they will just be an awkward embarrassment. If you have $\sin x = 0.6$, for example, and you want to find an angle from this on your calculator, you don't really want it to try to flash up an infinite number of answers for you.

So what do we do?

It would make sense for us to *restrict* the possible angle shown for a given sin to a short range so that we only get one answer, but every possible value for $\sin x$ is included, that is, we have all values of $\sin x$ from -1 to $+1$. If we *do* require further answers, we can then find them using the repeating pattern of the graph. (We shall look into this in more detail later on in Section 5.E.(d).)

We shall want to include $0°$ to $90°$ (or 0 to $\pi/2$ radians) in our range because this is the cradle of civilisation as far as trig is concerned – it all started with right-angled triangles. But this will only give us answers for positive values of $\sin x$, so what should we add to it?

We see from the graph that if we add $-90°$ to $0°$ (or $-\pi/2$ to 0 radians) we shall be all fixed up.

Then if, for example, $\sin x = -0.4$, using INV or SHIFT or 2nd Function Sin on your calculator (in degree mode) should give you an angle lying between $-90°$ and $0°$. Try it and see.

You should get $-23.6°$ to 1 d.p.

It would have been no good trying to extend the range by adding on 90° to 180° because this would have just given us repeats for the positive values of sin x and no solutions for the negative values.

Exactly the same sort of problem with multiple solutions will happen if we want to find an angle from its cos.

Look back to the graph of $y = \cos x$ in Figure 5.A.3(b) and decide for yourself what you think a sensible range for the answers would be.

What do you think you should have for x if $\cos x = \frac{1}{2}$?

What should you have for x if $\cos x = -\frac{1}{2}$?

Test out your ideas by seeing if your calculator agrees with you.

You should have the range from 0° to 180° this time (or 0 to π radians). This then gives you one and only one possible solution for any value of cos x between −1 and +1, and includes those important angles between 0° and 90°.

Using this range gives $x = 60°$, or $\pi/3$ radians, if $\cos x = \frac{1}{2}$, and $x = 120°$, or $2\pi/3$ radians, if $\cos x = -\frac{1}{2}$.

What we have cunningly done here, by restricting the range of values which we will allow for the angle from a given sin or cos, is to give ourselves **inverse functions** to take us back from a known sin or cos to just the one possible angle. (If you need help with inverse functions, you should go back now to Section 3.B.(g).)

We have already dealt with a similar situation to the one which we have here when we were looking for an inverse relation for $f(x) = x^2$ in Section 3.B.(j). There we also found that we could define an inverse function by restricting the possible values for x.

5.A.(g) **$\sin^{-1} x$ and $\cos^{-1} x$: what are they?**

Don't panic! We have just found them.

$\sin^{-1} x$ is the inverse function which takes us back from a value of sin x to an angle with that sin, and $\cos^{-1} x$ is the function which takes us back from a value of cos x to an angle with that cos. The possible values of these angles are restricted in the way we have just decided above will make sense.

With these restrictions, there is only **one** possible value for the angle from a given sin or cos, which is a condition which we must have for a relation to be a function as we saw in Section 3.B.(c).

Two inverse trig functions

$\sin^{-1} x$ is the angle in the range from −90° to +90°
   (or − $\pi/2$ to +$\pi/2$ radians) whose sin is x.

$\cos^{-1} x$ is the angle in the range from 0° to 180°
   (or 0 to π radians) whose cos is x.

$\sin^{-1} x$ is sometimes called arcsin x and $\cos^{-1} x$ is sometimes called arccos x.

 $\sin^{-1} x$ does *not* mean 1/sin x. This would be written as $(\sin x)^{-1}$. It is one of those tricky bits of mathematical notation which make a trap for the unwary.

5.A.(h)   **What do the graphs of sin⁻¹ x and cos⁻¹ x look like?**

We can use the method which we found in Section 3.B.(g) to draw a sketch of these two graphs.

Since the inverse relations take us from the y values back to the original x values, their graphs are mirror images of the original graph in the line $y = x$. The sketches will be easier to draw if we take equal scales on the two axes. We then get graphs as sketched in Figure 5.A.13 and Figure 5.A.14. If you are sketching these graphs for yourself, you may find it helps if you use the helpful hint I suggested for complicated inverse sketches in Section 3.B.(i). If you use equal scales on your two axes, and turn your paper so that the line $y = x$ is vertical, it is much easier to sketch the mirror image of $f(x)$ in the line $y = x$ which gives you the graph of $f^{-1}(x)$.

You can see from Figure 5.A.13(a) that, without the restrictions, the inverse relation is not a function – extending the graph would give an infinite number of solutions to 'y is the angle whose sin is x.' (Remember the raindrop test in Section 3.B.(c).)

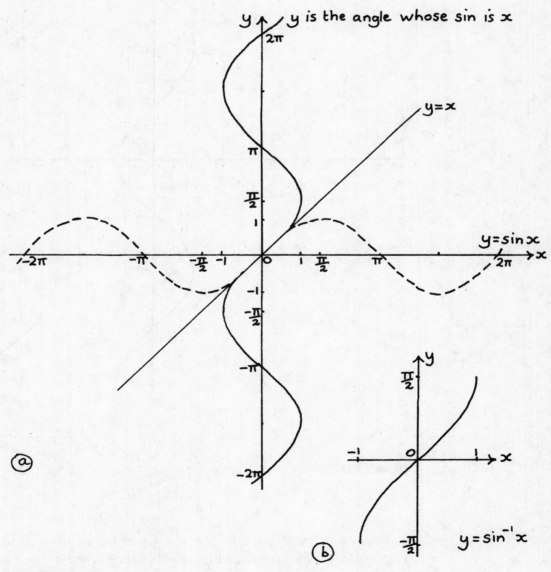

Figure 5.A.13

5.A Trig functions of any size of angle   185

You can also see how we have forced a function from this relation by restricting the range of values which we will accept.

This is shown in the graph in Figure 5.A.13(b) which represents the *function* 'y is the angle in the range from −π/2 to +π/2 radians whose sin is x.' Notice that this function is only *defined* for values of x lying between −1 and +1 inclusive, that is, $-1 \leq x \leq +1$, because this is the range of possible values for sin x.

Similarly, the graph in Figure 5.A.14(a) shows the repeated solutions of 'y is the angle whose cos is x', while Figure 5.A.14(b) shows the *function* 'y is the angle in the range from 0 to π radians whose cos is x', which gives a single solution for y for each x.

Again, $-1 \leq x \leq +1$.

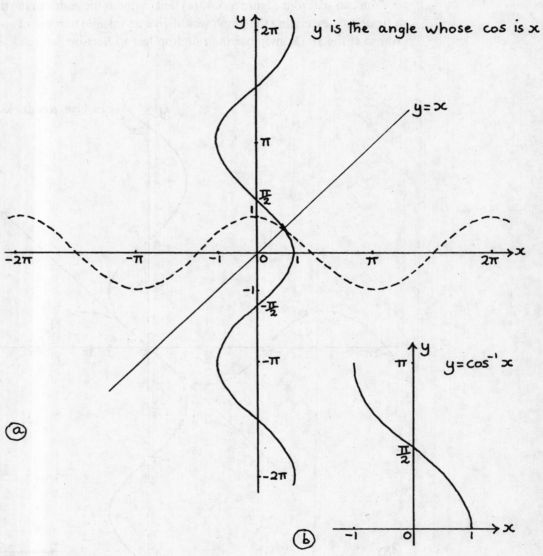

y is the angle whose cos is x

y = x

y = cos⁻¹ x

Figure 5.A.14

I think it will help you a lot here if you put your own two colours on each of the pairs of graphs of y = sin x and y is the angle whose sin is x, and y = cos x and y is the angle whose cos is x. It's much easier then to see which wiggle belongs to which.

5.A.(i) Defining the function tan⁻¹ x

To do this, we need to look at the graph of $y = \tan x$ which I show in Figure 5.A.15.

We see from this graph that, for any given value of $\tan x$, there will be an infinite possible number of angles x which have this tan value. For example, if $\tan x = 1$ then, from the graph, we could have $x = \pi/4$ or $5\pi/4$ or $9\pi/4$. Clearly, there are infinitely many more answers stretching out in both the right-hand and left-hand directions.

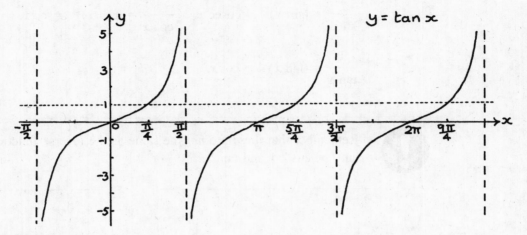

Figure 5.A.15

To define the function $\tan^{-1} x$, we shall again have to restrict the possible range of angles which we will allow. We certainly want to include 0 to $\pi/2$ and we could extend the range so as to go either from $-\pi/2$ to $+\pi/2$, or from 0 to π in order to get just one possible solution for the angle from each possible value of $\tan x$.

The agreed convention is that we take the range from $-\pi/2$ to $+\pi/2$.

I show a sketch of the graph of $y = \tan^{-1} x$ below, in Figure 5.A.16.

I've drawn it by using the reflection in the line $y = x$ of the graph of $y = \tan x$ for values between $-\pi/2$ and $\pi/2$.

Again, using two colours, one for each of $\tan x$ and $\tan^{-1} x$, will make the two graphs stand out more clearly for you.

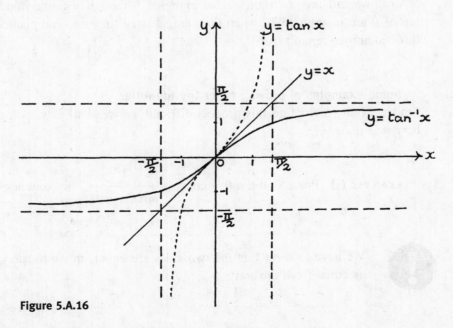

Figure 5.A.16

5.B   The trig reciprocal functions

5.B.(a)   **What are trig reciprocal functions?**

The **reciprocal function** of a function, $f(x)$, is defined as $\dfrac{1}{f(x)}$.

The three trig reciprocal functions are

$$\frac{1}{\sin x} = (\sin x)^{-1} = \operatorname{cosec} x, \qquad \frac{1}{\cos x} = (\cos x)^{-1} = \sec x,$$

$$\frac{1}{\tan x} = (\tan x)^{-1} = \cot x.$$

 Remember that these are *not* the same as the inverse functions, $\sin^{-1} x$, $\cos^{-1} x$ and $\tan^{-1} x$.

5.B.(b)   **The trig reciprocal identities: $\tan^2 \theta + 1 = \sec^2 \theta$ and $\cot^2 \theta + 1 = \operatorname{cosec}^2 \theta$**

In Section 4.A.(h), we used Pythagoras' Theorem to show that the three identities,

$$\sin^2 \theta + \cos^2 \theta = 1,$$
$$\tan^2 \theta + 1 = \sec^2 \theta,$$
$$\cot^2 \theta + 1 = \operatorname{cosec}^2 \theta,$$

are true for any angle θ which is less than 90°.

These three identities will remain true for *any* angle θ since, as we have seen in Section 5.A.(c), we still have the right-angled triangles. Although negative values for the sin, cos and tan of θ are now possible, when they are squared they become positive, and therefore the three identities remain true.

5.B.(c)   **Some examples of proving other trig identities**

Students quite often find this process difficult, so we shall now look at some examples of how it is done.

     EXAMPLE (1)   Prove that $\tan \theta + \cot \theta = \dfrac{1}{\sin \theta \cos \theta}$ for any angle θ.

 We have to *show* that the two sides are equal, so we mustn't write them down as equal from the start.

Instead, we deal with the sides separately. Here,

$$\text{LHS} = \tan\theta + \cot\theta = \frac{\sin\theta}{\cos\theta} + \frac{\cos\theta}{\sin\theta} = \frac{\sin\theta}{\cos\theta} \times \frac{\sin\theta}{\sin\theta} + \frac{\cos\theta}{\sin\theta} \times \frac{\cos\theta}{\cos\theta}$$

$$= \frac{\sin^2\theta}{\sin\theta\cos\theta} + \frac{\cos^2\theta}{\sin\theta\cos\theta} = \frac{\sin^2\theta + \cos^2\theta}{\sin\theta\cos\theta} = \frac{1}{\sin\theta\cos\theta} = \text{RHS.}$$

Just like adding any other fractions, we make it possible to put them over the same denominator in the first line of working above – see Section 1.C.(c) if necessary.

EXAMPLE (2) Try showing that $\sec^2\theta + \operatorname{cosec}^2\theta = \sec^2\theta\operatorname{cosec}^2\theta$ for yourself.

It looks quite an unexpected result!

You could do it like this:

$$\text{LHS} = \sec^2\theta + \operatorname{cosec}^2\theta = \frac{1}{\cos^2\theta} + \frac{1}{\sin^2\theta} = \frac{\sin^2\theta}{\cos^2\theta\sin^2\theta} + \frac{\cos^2\theta}{\sin^2\theta\cos^2\theta}$$

$$= \frac{\sin^2\theta + \cos^2\theta}{\cos^2\theta\sin^2\theta} = \frac{1}{\cos^2\theta\sin^2\theta} = \sec^2\theta\operatorname{cosec}^2\theta = \text{RHS.}$$

I say above 'you could do it like this' because identities can usually be proved in a large number of different ways. This is because the process is a bit like following a maze; you can write down a sequence of true statements starting from one side, but they do not always bring you any closer to the other side. Sometimes, after much effort, they bring you back where you started – at least you know then that what you have written down is true if not helpful.

Usually it is best to start with the more complicated side and show that this can be reduced to the simpler side. In really tough cases, it pays to work on each side separately and bring both of them to some third form. (The example which we have just done can be proved very neatly by using the two relevant identities of Section 5.B.(b) on each side in turn. Try it and see!)

Because there are all these possible branches to follow, you should never spend too long trying to prove an identity in an exam. If it doesn't come out quite quickly, leave it and return to it later if you've got time.

Have a go at the one below too. It is a bit tricky, but you have all the working knowledge and skills to get through it all right. We'll take it in stages.

EXAMPLE (3) Show that $\dfrac{\cos x}{1 - \tan x} + \dfrac{\sin x}{1 - \cot x} = \sin x + \cos x$ for any angle, x.

The LHS is more complicated, so we will work with this and try to show that it is the same as the RHS. It would seem to be a good idea to have the whole of this side in terms of $\sin x$ and $\cos x$. How can we rewrite $\tan x$ and $\cos x$ to do this?

We can put

$$\tan x = \frac{\sin x}{\cos x} \quad \text{and} \quad \cot x = \frac{\cos x}{\sin x}$$

then, at least, everything is in terms of $\sin x$ and $\cos x$. Then

$$\text{LHS} = \frac{\cos x}{1 - \frac{\sin x}{\cos x}} + \frac{\sin x}{1 - \frac{\cos x}{\sin x}}.$$

Now what should we do? (See if you can tidy up what we've now got.)

We get rid of fractions inside fractions by multiplying the first bit top and bottom by $\cos x$, and the second bit top and bottom by $\sin x$. (Try doing this if you didn't already.)

You should get

$$\text{LHS} = \frac{\cos^2 x}{\cos x - \sin x} + \frac{\sin^2 x}{\sin x - \cos x}.$$

Using $\sin x - \cos x = -(\cos x - \sin x)$, how can we rewrite what we've now got?

We can say that

$$\text{LHS} = \frac{\cos^2 x}{\cos x - \sin x} - \frac{\sin^2 x}{\cos x - \sin x} = \frac{\cos^2 x - \sin^2 x}{\cos x - \sin x}.$$

How can we rewrite the top? (Try using a neat factorisation.)

$$\cos^2 x - \sin^2 x = (\cos x - \sin x)(\cos x + \sin x)$$
(using the difference of two squares)

Try to finish it off now.

$$\text{LHS} = \frac{(\cos x - \sin x)(\cos x + \sin x)}{\cos x - \sin x} = \cos x + \sin x = \text{RHS}.$$

NOTE

You may have recognised that $\cos^2 x - \sin^2 x$ could also be written as $\cos 2x$. Although this is true, it would not have helped us here. The trickiest part in proving identities is picking out the possible steps which will also lead you forward in the proof.

Extending trigonometry

5.B.(d)   What do the graphs of the trig reciprocal functions look like?
We start by thinking about how we can draw a sketch of the graph of

$$y = \operatorname{cosec} x = \frac{1}{\sin x}.$$

I show in Figure 5.B.1 a sketch of $y = \sin x$ to work from.

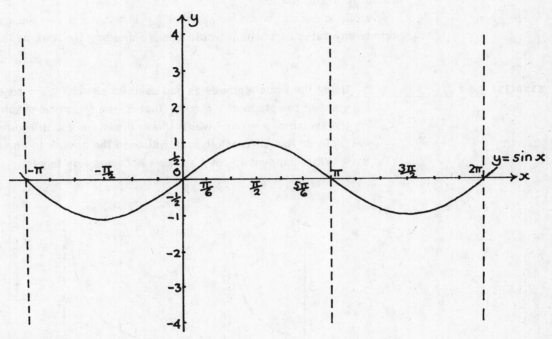

Figure 5.B.1

To help us, we need first to answer the following five questions.

(1)   When $\sin x = 1$, what is $\operatorname{cosec} x$?
(2)   When $\sin x = -1$, what is $\operatorname{cosec} x$?
(3)   Does $\operatorname{cosec} x$ have the same sign as $\sin x$?
(4)   What happens to $\operatorname{cosec} x$ when $\sin x$ is positive but very close to zero?
(5)   What happens to $\operatorname{cosec} x$ when $\sin x$ is negative but very close to zero?

Try answering each of these for yourself.

Here are the answers.

(1) $\operatorname{cosec} x = 1$ (2) $\operatorname{cosec} x = -1$ (3)   Yes it does, since it is just $1/\sin x$.
(4) $\operatorname{cosec} x$ becomes very large and positive.
(5) $\operatorname{cosec} x$ becomes very large and negative.

(When $\sin x = 0$, $\operatorname{cosec} x$ is undefined because we can't divide by zero.)

Using the answers to the five questions above, try to sketch in for yourself the graph of $y = \operatorname{cosec} x$ on the sketch I have already drawn for you of $y = \sin x$. Use pencil so that you can have second thoughts if necessary!

(The sketch is shown in the answers at the back of the book, but it is important to try to draw it yourself before looking.)

Because the functions of $y = \sin x$ and $y = \cos x$ are periodic, so also are the functions of $y = \operatorname{cosec} x$ and $y = \sec x$. The graph of $y = \sec x$ is the same as the one for $y = \operatorname{cosec} x$ shifted by $\pi/2$ to the left.

(Strictly speaking, when we say that $y = \operatorname{cosec} x$ and $y = \sec x$ are functions, we must exclude any value of x which would involve dividing by zero, as this is impossible.)

EXERCISE 5.B.2

Using the same methods as you used for sketching $y = \operatorname{cosec} x$, try sketching for yourself the graph of $y = \cot x$ (that is, the reciprocal graph of $y = \tan x$), using the sketch of $y = \tan x$ which I have drawn for you in Figure 5.B.2.

To do this successfully, you will need the answer to one more question.

What happens to $\cot x$ as $\tan x$ becomes very large?

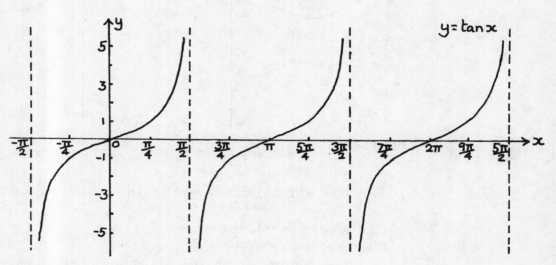

Figure 5.B.2

cot x will become closer and closer to zero, so that when tan x is undefined, say for $x = \pi/2$, cot $x = 0$.

5.B.(e)   Drawing other reciprocal graphs

Drawing and checking the two reciprocal graphs of $y = \operatorname{cosec} x$ and $y = \cot x$ will have shown you many of the basic guidelines to use when drawing reciprocal graphs.

I will summarise these here for you in a box. Then you will be able to have a go at drawing reciprocal graphs for some of the functions which have been mentioned in earlier chapters.

> **Rules for drawing reciprocal graphs**
>
> If we have a function $y = f(x)$, its reciprocal function is $y = 1/f(x)$.
>
> - If the graph of $y = f(x)$ has symmetries (for example, being odd or even or periodic), then the graph of $1/f(x)$ will have the same symmetries.
>
> - If $y = f(x) = 0$ for some value of x, then $1/f(x)$ is undefined. There is a jump or discontinuity in its graph for this value of x.
>   This means that, as $f(x)$ gets close to 0, $1/f(x)$ will become very large in value. Equally, if there is a jump or discontinuity in the graph of $y = f(x)$ for some value of x, then $y = 1/f(x) = 0$ for that value of x.
>
> - If you know a few key values for $y = f(x)$, it is easy to calculate the corresponding values for $y = 1/f(x)$. These can then be used to help you to get the sketch in the right place.

EXERCISE 5.B.3

Using the rules above, try drawing in the reciprocal functions for the six functions shown on my graph sketches. Use any values given on my sketches to write in the corresponding values on the reciprocal sketches.

In case some of these functions are unfamiliar, I have given you a reference back to where I have talked about them earlier in this book.

I suggest that you sketch them first in pencil to allow for second thoughts. When you have got them right, it might help you to use two colours on them (one for the original graph and one for its reciprocal), to emphasise how they depend upon each other.

(1) Sketch $y = \dfrac{1}{x^2 - 2x + 2}$ using my sketch of $y = x^2 - 2x + 2 = (x - 1)^2 + 1$.

(My sketch uses Sections 2.D.(b) and (c) on completing the square and graph sketching.)

(2) Sketch $y = \dfrac{1}{x^2 - 4x + 3}$ using my sketch for $y = x^2 - 4x + 3 = (x - 1)(x - 3)$.

(3) Sketch the graph of $y = 1/x$ using my sketch of $y = x$.

(4) Sketch the graph of $y = 1/x^2$ using my sketch of $y = x^2$.

(5) Sketch the graph of $y = 1/e^x$ using my sketch of $y = e^x$.
You may find that Section 3.C.(f) helps you here.

(6) Sketch the graph of $y = \dfrac{x - 2}{x + 3}$ using my sketch of $y = \dfrac{x + 3}{x - 2}$.

(We drew this sketch in Section 3.B.(i).) See if you can also find the coordinates of the point where this graph and its reciprocal graph cross over each other.)

5.B The trig reciprocal functions

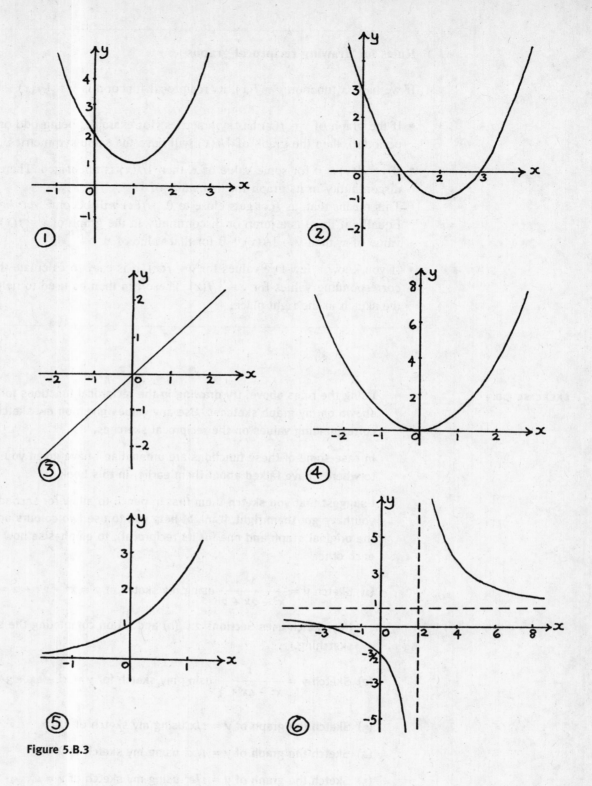

Figure 5.B.3

5.C   Building more trig functions from the simplest ones

5.C.(a)   Stretching, shifting and shrinking trig functions

In Section 3.B.(d), we looked at what happens to functions when we add or multiply them in different ways. You should look back at this section if you haven't yet read it, and make sure that these ideas are familiar to you. I have summarised the effects of the simplest kinds of transformation there.

Because trig functions are periodic, particularly interesting possibilities of combination arise which have profound physical implications. In particular, they are very useful in thinking about mechanically vibrating systems and the behaviour of current and voltage in electric circuits. They can also be used to describe the different qualities of particular notes played on different musical instruments.

We have already seen that, because these functions are periodic, and because of their symmetries, they are very closely related to each other. For example, the cos curve $y = \cos x$ is the *same* as the sin curve $y = \sin x$ except that the sin curve has been shifted $\pi/2$ to the left (Figure 5.C.1).

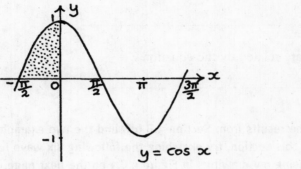 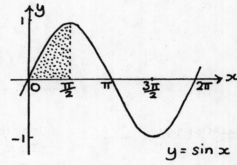

Figure 5.C.1

Using the second result from the summary at the end of Section 3.B.(d), we see that this means that $\sin(x + \pi/2) = \cos x$.

Combinations of sin and cos functions are often used to describe how various kinds of wave motion change with time. In this case we would need to have the horizontal axis in the graphs representing time, and so it is better to use t rather than x for the variable on this axis. The vertical axis is then measuring some displacement, so it is often labelled x, with x being a function of time, t.

Because so many of the different kinds of waves which occur in the natural world can be represented by various combinations of trig functions, these functions are often called wave functions or waveforms.

Using the results summarised in Section 3.B.(d), we can sketch graphs for functions such as $x = 3 \cos t$, or $x = \cos 2t$. I show the sketches for these in Figure 5.C.2(a) and (b). In each case, the graph of $x = \cos t$ is shown by a dashed line.

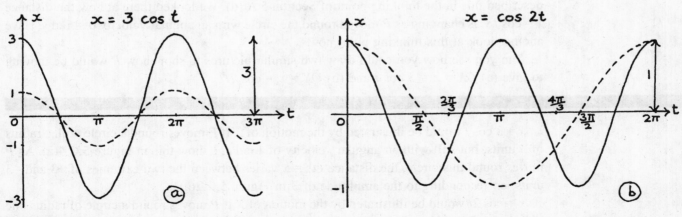

Figure 5.C.2

In the graph of $x = 3 \cos t$, each value of $\cos t$ has been pulled out three times as far from the t-axis.

In the graph of $x = \cos 2t$, each point of the curve as we move out from $t = 0$ is being reached twice as fast. So, if $t = \pi/2$, $\cos \pi/2 = 0$ but $\cos(2 \times \pi/2) = \cos \pi = -1$.

We can now use these two graphs to illustrate some important definitions.

- The maximum displacement or **amplitude**, A, is 3 units in (a), and 1 unit in (b).
- If t is in seconds, the **period**, T, or time taken for each complete cycle is 2π seconds in (a), and π seconds in (b).
- The **frequency**, f, which is the number of cycles per second, is $1/2\pi$ in (a), and $1/\pi$ in (b). The units for frequency are **hertz**, written as Hz.

$$T \text{ and } f \text{ are related by the equation } T = \frac{1}{f}.$$

EXERCISE 5.C.1

Using the results from Section 3.B.(d), and the two examples shown in Figure 5.C.2 in this section, try sketching the following six wave functions for yourself in pencil using my drawings in Figure 5.C.3 on the next page. I have already drawn in the graph of $x = \sin t$ on each of them, to help you.

(1) $x = 2 \sin t$ (2) $x = \sin 2t$ (3) $x = \sin (t/2)$ (4) $x = 1 + \sin t$
(5) $x = \cos t$ (6) $x = \cos (t + \pi/2)$

Also, for each wave function, answer the following questions.

(a) What is its amplitude, A?
(b) What is its period, T?
(c) What is its frequency, f?
(d) Is the function odd or even?

Then check your results against the answers in the back of the book. (If necessary, draw the graph sketches in again so that you have the right version.)

5.C.(b)   Relating trig functions to how P moves round its circle and SHM

We can also think about the two functions whose graphs we sketched in Figure 5.C.2(a) and (b) in the last section by relating them to the motion of X as P moves round its circle. I described this in the thinking point of Section 5.A.(d). We looked there at how the distance $x = \cos t$ was changing as P moved round the circle with an angular velocity of 1 rad/s. Have another look at this thinking point now.

Can you see how you could draw two similar pictures to show how P would be moving to give (a) $OX = x = 3 \cos t$ and (b) $OX = x = \cos 2t$?

$x = 3 \cos t$ would be illustrated by the motion of X if P moves round a circle with a radius of 3 units, but still with an angular velocity of 1 rad/s. I show this in Figure 5.C.4(a). As P moves round this circle, the distance $OX = x$ varies between the two extremes of $+3$ and -3 units, corresponding to the amplitude of 3 in Figure 5.C.2(a).

$x = \cos 2t$ would be illustrated by the motion of X if P moves round a circle of radius one unit, but twice as fast, so its angular velocity is 2 rad/s. I show this on Figure 5.C.4(b).

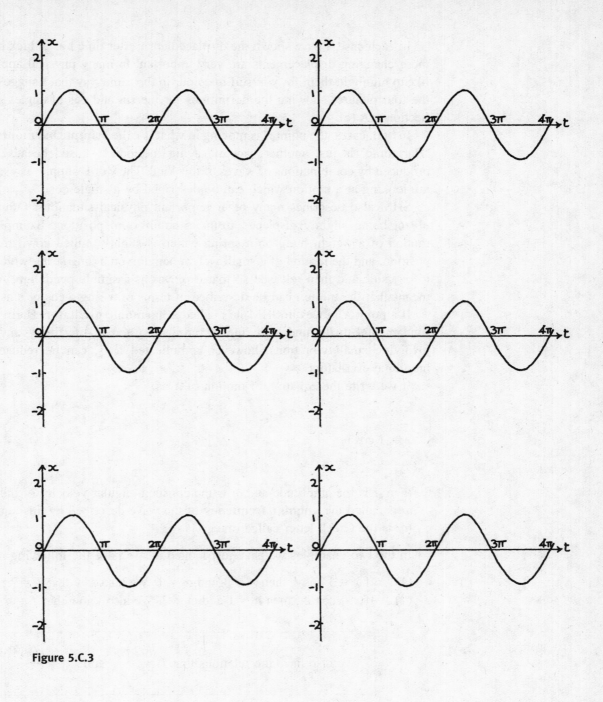

Figure 5.C.3

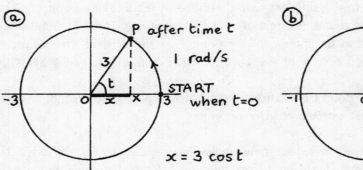

Figure 5.C.4

In each case, I have shown the displacement x after time t as a thick black line. Because these changing displacements are very important in many physical applications, you may like to highlight them for yourself in colour in the same way that I suggested you should for the four pictures showing the definitions for the sin and cos of angles greater than 90° in Section 5.A.(c).

In both cases, the point X is moving in what is called **simple harmonic motion**, or SHM. 'Harmonic' is just another way of saying 'periodic' – used because sound waves are produced by combinations of waves of this kind. The word 'simple' is used here because we are looking at a motion which can be described by a single cos.

SHM also describes many other important physical situations. Often these involve an object being slightly displaced from its equilibrium position. Examples of this are the motion of a weight hung on a spring which is slightly pulled down from its equilibrium position, and the motion of a small weight hanging on a long string which is pulled slightly to one side and then released so that it moves as a simple pendulum. Again, the 'simple' means that the motion can be described in terms of a single cos or sin.

If a point X moves in SHM it is called a **harmonic oscillator**. Harmonic oscillators are fundamental to the understanding of physical systems. Amazingly, any real-life situation involving small vibrations, however complicated it is, can be reduced to a system of harmonic oscillators.

If we write the equation of motion of X as

$$x = A \cos \omega t$$

then A is the amplitude and ω is the constant angular velocity of the point P.
ω is called the **angular frequency** of the wave described by this equation.
(ω is the Greek letter called omega.)

In the two examples we have just looked at, we have the following results.

(1)   If $x = 3 \cos t$, then $A = 3$ and $\omega = 1$. We also saw that $T = 2\pi$ and $f = 1/2\pi$.
(2)   If $x = \cos 2t$, then $A = 1$ and $\omega = 2$. We also know that $T = \pi$ and $f = 1/\pi$.

We also have the relations that $T = \dfrac{2\pi}{\omega}$ and $f = \dfrac{\omega}{2\pi}$.

If, in the simplest case described in the thinking point of Section 5.A.(d), where P is moving round its circle of radius one unit, at a constant angular velocity of 1 rad/s, we had looked at the motion of the point Y on the vertical axis instead, we would have had the equation for OY of $y = \sin t$ (Figure 5.C.5). This is also SHM. Now, when $t = 0$, $y = 0$ also.

The point Y is starting from the central position of its motion, unlike X which started from its most extreme positive position.

These circle diagrams make it much easier to see what is happening with more complicated sin and cos functions. Such functions are very important in physical applications such as describing the voltage and current waveforms in electric circuits. It is

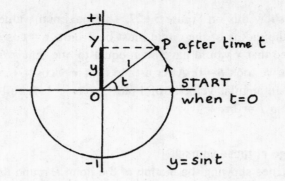

$y = \sin t$ Figure 5.C.5

much simpler to handle them mathematically through the use of complex numbers and the first step in doing this is to become happy with using these circle diagrams.

I have already drawn for you the examples of $x = 3 \cos t$ and $x = \cos 2t$ in Figure 5.C.4, and $y = \sin t$ in Figure 5.C.5. Since I have used x to represent the displacement after time t on all my graph sketches, I shall also use it from now on to show displacements on both the horizontal axis of my circle (which gives a cos function), and on the vertical axis of my circle (which gives a sin function).

Here are two more examples showing this kind of relationship.

EXAMPLE (1) Show the relation of $x = 2 \sin 3t$ to the motion of P round its circle.

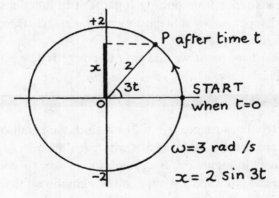

$\omega = 3$ rad /s

$x = 2 \sin 3t$ Figure 5.C.6

I show this on Figure 5.C.6. The maximum value of x is 2, therefore $A = 2$, and the radius of the circle must be 2 units. When $t = 0$, $x = 0$. After a time t, $x = 2 \sin 3t$, so P is moving with an angular velocity of 3 rad/s therefore $\omega = 3$. A full turn or cycle takes $2\pi/3$ s so $T = 2\pi/3$.

EXAMPLE (2) Show the relation of $x = \cos(t + \pi/6)$ to the motion of P round its circle.

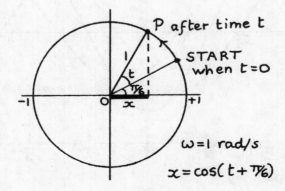

$\omega = 1$ rad/s

$x = \cos(t + \pi/6)$ Figure 5.C.7

I show this on Figure 5.C.7. The maximum value of x is 1, so $A = 1$ and the radius of the circle must be 1 unit. $x = \cos \pi/6$ when $t = 0$. Notice that x would have been equal to one unit $\pi/6$ s *before* the instant when we took $t = 0$. After a time of t, $x = \cos(t + \pi/6)$. P is moving with an angular velocity of 1 rad/s, so $\omega = 1$. A full turn or cycle takes 2π s so $T = 2\pi$.

EXERCISE 5.C.2

Now have a go at these yourself.

Draw sketches showing the motion of the point P round its circle for each of the following:

(1) $x = \cos 3t$ (2) $x = 2 \sin t$ (3) $x = 3 \cos 2t$
(4) $x = 4 \sin(t/2)$ (5) $x = \sin(t + \pi/6)$ (6) $x = \sin(2t + \pi/4)$
(7) $x = 2 \cos(t - \pi/6)$ (8) $x = 5 \sin(3t + \pi/6)$.

Label each sketch in a similar way to my two examples. In each case, you should also give the value of the amplitude, A, and of the angular velocity, ω, and of the period, T. It is very important to actually *do* these sketches yourself; don't just look at my answers.

5.C.(c)   New shapes from putting together trig functions

What happens if we add $\sin t$ to $\cos(t + \pi/2)$? (Have a look at your sketch for question (6) of Exercise 5.C.1.)

What happens if we add $\sin t$ to $\cos t$? Try sketching for yourself what the result would be in each case.

In (6), because $\cos(t + \pi/2) = -\sin t$, the result of adding the two waves is always zero. They are exactly out of phase with each other.

I show in Figure 5.C.8 a sketch for $x = \sin t + \cos t$ drawn from putting together the two curves $x = \cos t$ and $x = \sin t$ and marking in all the easy points such as where one of them is equal to zero, or they are equal to each other and so just double, or they are equal but opposite in sign and so balance out.

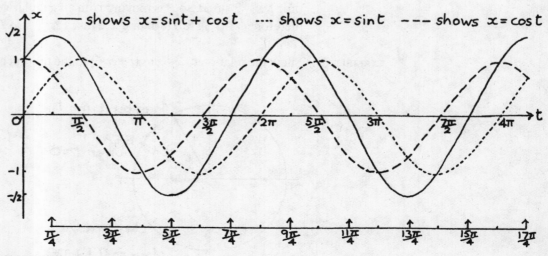

Figure 5.C.8

We see from this sketch that $x = \sin t + \cos t$ has an amplitude of $2 \sin (\pi/4) = 2 \times 1/\sqrt{2}$ $= \sqrt{2}$, and a period of 2π.

It looks as if it might also be sin-shaped. (We shall find out how to show that it *is* a sin curve in Section 5.D.(f).)

Sketching graphs by hand becomes very time-consuming (and difficult if the functions are more complicated), but if you have access to a graph-sketching calculator or computer it would be good to see what happens when you add all the pairs of functions in the six graphs shown in the answers to Exercise 5.C.1.

It is also very interesting to see what happens if you add a sequence of sines. You will see that the shape of the resulting curve gets successively modified to give some remarkable results.

Here are two examples you could try.

(I have used the $\to$ symbol here to mean 'put in the next bit of the sequence and see how it affects your graph.')

$$(1) \quad (\sin t) \to \left(\sin t - \frac{\sin 2t}{2} \right) \to \left(\sin t - \frac{\sin 2t}{2} + \frac{\sin 3t}{3} \right) \to \ldots$$

$$(2) \quad (\sin t) \to \left(\sin t + \frac{\sin 3t}{3} \right) \to \left(\sin t + \frac{\sin 3t}{3} + \frac{\sin 5t}{5} \right) \to \ldots$$

The further you go with these sequences the more interestingly modified the shapes of the graphs become.

By this kind of method it is possible to get graphs which are very close approximations to the ones shown in Figure 5.C.9, both of which are waveforms which can occur naturally in electrical signals.

If you have done the experiments of (1) and (2) above, you will find that you get increasingly good matches except for little overshoots close to the vertical parts of the graph.

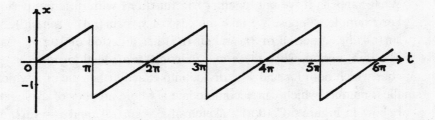

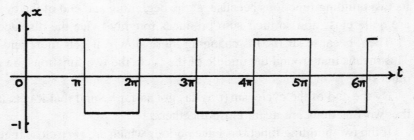

Figure 5.C.9

This is called Gibb's phenomenon and it comes from the problems in accurately representing a graph which is effectively doing a jump at these points.

The fact that these functions can be thought of as sums of sines (or, more generally, to include other cases, as sums of sines and cosines) is of great practical importance. This whole area of what is called harmonic analysis was developed by the French mathematician, Fourier.

Can you see why we couldn't represent *any* periodic functions just by sums of sines of multiples of t as in the two earlier examples I gave you?

The sums of such sines will always give odd functions. If the function we want to represent isn't odd then we shall need also to include cosines of multiples of t to get a correct representation of what is happening.

If the function is made up entirely from cosines of multiples of t it will always be even.

Try the following sequence to see this happening.

$$(\cos x) \to \left(\cos x + \frac{\cos 3x}{3^2} \right) \to \left(\cos x + \frac{\cos 3x}{3^2} + \frac{\cos 5x}{5^2} \right) \to \ldots$$

5.C.(d)   **Putting together trig functions with different periods**

All the examples of putting trig functions together which we have looked at so far in this section have had periods which were the same as at least one of the input functions. For example, both $\sin t$ and $\cos t$ have a period of 2π and $x = \sin t + \cos t$ also has a period of 2π.

$x = \sin t + \frac{1}{3} \sin 3t + \frac{1}{5} \sin 5t$ has the period of 2π belonging to $\sin t$ since all the other functions neatly sit inside this. ($\sin 3t$ has a period of $\frac{1}{3} \times 2\pi$, and $\sin 5t$ has a period of $\frac{1}{5} \times 2\pi$.)

What happens if we put together trig functions with different periods?

For example, suppose we take the case of $x = \sin(t/4) + \sin(t/5)$.

$\sin(t/4)$ has a period of 8π and $\sin(t/5)$ has a period of 10π.

The joint period, when these two functions are added together, is given by the smallest number which both 8π and 10π divide into exactly (their l.c.m.), which is 40π. This is the smallest number which can accommodate a whole number of cycles of both functions.

I show in Figure 5.C.10(a) a sketch of $x = \sin(t/4)$ and $x = \sin(t/5)$ on the same axes. Underneath that, in Figure 5.C.10(b), I show a sketch of the joint function, $x = \sin(t/4) + \sin(t/5)$ so that you can see how it comes from the two functions above.

The complete cycle shown of $x = \sin(t/4) + \sin(t/5)$ has a more complicated shape than its two building functions because, at the beginning and end of the cycle these two functions are quite close and so their sum produces roughly twice the displacement.

Then, because $\sin(t/5)$ is changing more slowly, it gets more and more behind $\sin(t/4)$. This means that around the middle of the cycle the two functions are nearly cancelling each other out.

By the end of the cycle, $\sin(t/5)$ has got so far behind that it gets lapped by $\sin(t/4)$, and the two functions are again close together.

If the two building functions have periods which are *very* close together, then the contrast between the peaking effect at the two ends of each cycle and the level trough near its centre

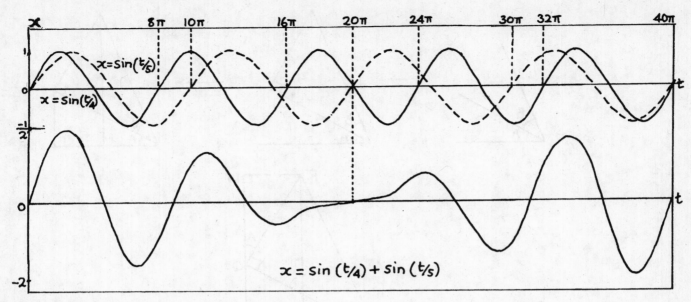

Figure 5.C.10

becomes very much more marked. A physical example of this is what happens if two musical notes, very close to each other in pitch, are played at the same time. The peaks are heard as *beats* which will disappear when the two notes exactly match. This phenomenon is made use of by piano tuners and by other musicians when they tune their instruments.

5.D   Finding rules for combining trig functions

5.D.(a)   How else can we write sin (A + B)?

If A and B are two different angles, is it true that $\sin(A + B) = \sin A + \sin B$?

Test your answer with two examples on your calculator.

Except for some very special cases, such as when $B = 0$, it is *not* true that $\sin(A + B) = \sin A + \sin B$.

 Students sometimes write that $\sin 2A = 2 \sin A$, for example, but from the first two questions of Exercise 5.C.1 earlier it is clearly obvious that $\sin 2t$ and $2 \sin t$ are not at all the same thing.

Can we find a way of writing $\sin(A + B)$ using the sin and cos of A and B?

(As we shall see in Section 5.D.(f) it is often important to be able to do this.)

To show this geometrically, we shall need right-angled triangles to work from.

We start by drawing the tilted triangle for $\angle B$, as this is the trickiest one to get, and then build up the diagram as I show in Figure 5.D.1.

Then we complete this chain by drawing the triangle RNQ. This is because it gives us another right-angled triangle with lengths that we want. $\angle RQN = \angle A$ because NQP is a straight line, and so 180°, and the angles of $\triangle OQP$ also add to 180°.

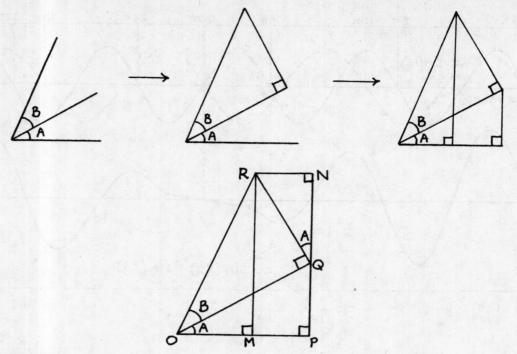

Figure 5.D.1

Then we have:

$$\sin(A + B) = \frac{RM}{OR} = \frac{PQ + QN}{OR} = \frac{OQ \sin A + QR \cos A}{OR}$$

$$= \frac{OQ}{OR} \sin A + \frac{QR}{OR} \cos A = \cos B \sin A + \sin B \cos A.$$

This is more usually written as

$$\boxed{\sin(A + B) = \sin A \cos B + \cos A \sin B.}$$

5.D.(b)   A summary of results for similar combinations

In a very similar way, we can get formulas for $\sin(A - B)$, $\cos(A + B)$ and $\cos(A - B)$. (These can also all be shown to be true for angles larger than 90°.)

These are listed in the box below:

$$
\begin{aligned}
\sin(A + B) &= \sin A \cos B + \cos A \sin B, \\
\sin(A - B) &= \sin A \cos B - \cos A \sin B, \\
\cos(A + B) &= \cos A \cos B - \sin A \sin B, \\
\cos(A - B) &= \cos A \cos B + \sin A \sin B.
\end{aligned}
$$

Notice the + and − signs in the middle of the formulas for $\cos(A + B)$ and $\cos(A − B)$. It makes sense that they should be this way round when you remember that $\cos(60° + 30°) = \cos 90° = 0$ but $\cos(60° − 30°) = \cos 30° = \sqrt{3}/2$.

5.D.(c) **Finding tan(A + B) and tan(A − B)**

How shall we set about getting a formula for $\tan(A + B)$? We can say

$$\tan(A + B) = \frac{\sin(A + B)}{\cos(A + B)} = \frac{\sin A \cos B + \cos A \sin B}{\cos A \cos B − \sin A \sin B}.$$

It would be nicer to have the answer entirely in terms of $\tan A$ and $\tan B$. Can you see what we need to do to the top and bottom of this fraction to make this possible?

If we divide top and bottom by $\cos A \cos B$, and cancel where possible, we shall get

$$\tan(A + B) = \frac{\tan A + \tan B}{1 − \tan A \tan B}.$$

(Remember that each of the four separate chunks in the fraction is getting divided.)
You should now be able to show for yourself that

$$\tan(A − B) = \frac{\tan A − \tan B}{1 + \tan A \tan B}.$$

5.D.(d) **The rules for sin 2A, cos 2A and tan 2A**

These follow immediately from the previous results, putting $B = A$. We get:

$$\sin 2A = 2 \sin A \cos A,$$
$$\cos 2A = \cos^2 A − \sin^2 A,$$
$$\tan 2A = \frac{2\tan A}{1 − \tan^2 A}.$$

In the case of $\cos 2A$, it is possible to write this rule in two other ways, using the identity that $\sin^2 A + \cos^2 A = 1$. We then get:

$$\cos 2A = \cos^2 A − (1 − \cos^2 A) = 2 \cos^2 A − 1,$$
$$\cos 2A = (1 − \sin^2 A) − \sin^2 A = 1 − 2 \sin^2 A.$$

We shall find these alternative versions very useful later on in solving trig equations and for integrating $\sin^2 x$ and $\cos^2 x$. I give you examples of this in Section 5.E.(d) and example (4) of Section 9.B.(c).

5.D.(e)   How could we find a formula for sin 3A?

We can now find a formula for $\sin 3A$ completely in terms of $\sin A$.

We do it by writing $\sin 3A$ as $\sin(A + 2A)$ and then using the $\sin(A + B)$ formula on this. Then we have

$$\sin 3A = \sin(A + 2A) = \sin A \cos 2A + \cos A \sin 2A$$

$$= \sin A(1 - 2\sin^2 A) + \cos A(2 \sin A \cos A)$$

(using the rules for $\sin 2A$ and $\cos 2A$ from the section above)

$$= \sin A - 2\sin^3 A + 2 \sin A \cos^2 A$$

$$= \sin A - 2\sin^3 A + 2 \sin A(1 - \sin^2 A)$$

$$= 3 \sin A - 4 \sin^3 A.$$

You should now be able to find a similar rule for $\cos 3A$ in terms of $\cos A$ for yourself. I have put this pair of rules in the box below for you:

$$\sin 3A = 3 \sin A - 4 \sin^3 A,$$
$$\cos 3A = 4 \cos^3 A - 3 \cos A.$$

5.D.(f)   Using sin (A + B) to find another way of writing 4 sin t + 3 cos t

In Section 5.C.(b), we investigated graphically the effect of adding $\sin t$ to $\cos t$ for each value of t. The result seemed to be a sin curve which had been shifted by some angle from the origin.

There are many physical and mathematical situations where it is much easier to deal with a single sin or cos function rather than having combinations of such functions. Such examples include describing the wave functions for alternating current and voltage, and making it easier to solve certain kinds of trig equation as we shall see in Section 5.E.(e).

I will show you how we can do this conversion to a single function by taking the particular example of $x = 4 \sin t + 3 \cos t$.

We start by noticing that $4 \sin t + 3 \cos t$ looks a little bit like

$$\sin A \cos B + \cos A \sin B, \text{ which is } \sin(A + B)$$

as we saw in Section 5.D.(a). So we try writing

$$4 \sin t + 3 \cos t = R \sin t \cos \alpha + R \cos t \sin \alpha$$

which is $R \sin(t + \alpha)$.

(We need to include the R here to avoid getting into the impossible position of needing a sin or cos greater than 1.)

We now have to find the particular numerical values of R and α which will make this equation be true for *every* value of t, so that each of the two sides is just another way of writing the same thing. This means that the equation is an *identity* and each separate part must match up, just as we matched up the separate terms in the identity in Section 2.D.(h).

Here, the two sides will only be equal for every value of t if we have both the same quantity of $\sin t$ each side, and the same quantity of $\cos t$ on each side.

Matching up the parts with $\sin t$, we get

$$4 \sin t = R \cos \alpha \sin t \quad \text{so} \quad 4 = R \cos \alpha.$$

Matching up the parts with $\cos t$, we get

$$3 \cos t = R \sin \alpha \cos t \quad \text{so} \quad 3 = R \sin \alpha.$$

The easiest way to find R and α is to draw a picture showing the information we now have. I do this here in Figure 5.D.2.

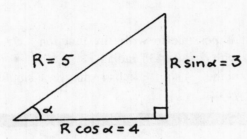

Figure 5.D.2

Using Pythagoras' theorem gives us $R^2 = 3^2 + 4^2 = 25$ so $R = 5$.

We also see that $\tan \alpha = \frac{3}{4}$ so $\alpha = 0.6435$ radians to 4 d.p.

We can now write $x = 4 \sin t + 3 \cos t$ in the alternative form of $x = 5 \sin(t + \alpha)$ with $\alpha = 0.6435$ to 4 d.p. (I shall continue calling this angle α for short.)

What will the graph of $x = 4 \sin t + 3 \cos t = 5 \sin(t + \alpha)$ look like?

(You will find the answer to this question much easier to understand if you did Exercises 5.C.1 and 5.C.2 in Sections 5.C.(a) and 5.C.(b). If you haven't yet done these, you should go back and do them now.)

To help us to sketch the curve of $x = 4 \sin t + 3 \cos t = 5 \sin(t + \alpha)$, we relate this to how the point P moves round its circle. The displacement x will be shown on the vertical axis since it is a sin function. I show this below in Figure 5.D.3(a).

P is moving round its circle of radius 5 units with an angular velocity of one radian per second. It starts at the angle α when $t = 0$.

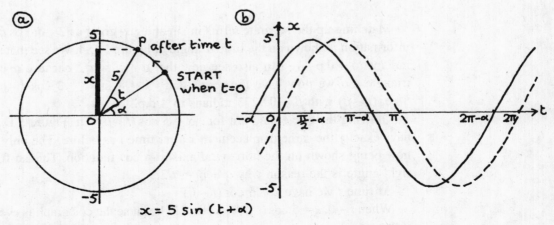

Figure 5.D.3

5.D Rules for combining trig functions

When it has moved through a further angle of t, the displacement x is given by $x = 5 \sin(t + \alpha)$.

We can see from the picture that x will increase first to its maximum value of $+5$ and then decrease through zero to -5.

We can also see that x would have been equal to zero at α or 0.6435 seconds *before* the instant when we are taking $t = 0$.

Using this information we can then draw the sin curve $x = 5 \sin(t + \alpha)$ shown in Figure 5.D.3(b). I have also drawn $x = 5 \sin t$, using a dashed line.

You can see that we have a gap of α between these two graphs. This is called a **phase difference**.

For both graphs, the amplitude $A = 5$, the angular velocity $\omega = 1$, and the period $T = 2\pi$.

We have just seen that it is possible to write the function $x = 4 \sin t + 3 \cos t$ in the form $x = 5 \sin(t + \alpha)$ with $\alpha = 0.6435$ radians.

Would it be possible to combine $4 \sin t + 3 \cos t$ to give a single cos function instead, and if so which rule should we use?

It *is* possible to do this, and we would need to use the rule for $\cos(A - B)$ because this gives us the plus sign in the middle. Doing this will give us

$$3 \cos t + 4 \sin t = R \cos t \cos \beta + R \sin t \sin \beta$$

which is the same as $R \cos(t - \beta)$.

We can see that R will still be equal to 5 here, but I have called the angle β to avoid confusing it with the angle α which we found earlier.

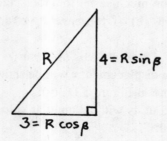

Figure 5.D.4

Matching up the separate terms in sin and cos gives us $3 = R \cos \beta$ and $4 = R \sin \beta$. This information is shown on the little triangle in Figure 5.D.4. We see that $\tan \beta = \frac{4}{3}$ so $\beta = 0.9273$ radians to 4 d.p. We can also see now that $\alpha + \beta = \pi/2$ because α is the top angle in this triangle. So we now have the result that $x = 4 \sin t + 3 \cos t$ can also be written as $5 \cos(t - \beta)$ with $\beta = 0.9273$ radians to 4 d.p.

Drawing the circle diagram for $x = 5 \cos(t - \beta)$ in Figure 5.D.5(a) shows us that we have exactly the same displacement x after time t as before. The only difference is that it is now being shown on the horizontal axis as a cos function. This shift in position through a right angle is the reason why $\alpha + \beta = \pi/2$.

At time t we have $x = 5 \cos(t - \beta)$.

When $t = 0$, $x = 5 \cos(-\beta) = 5 \cos \beta$ because the cos graph is even (see Section 5.A.(a) if necessary).

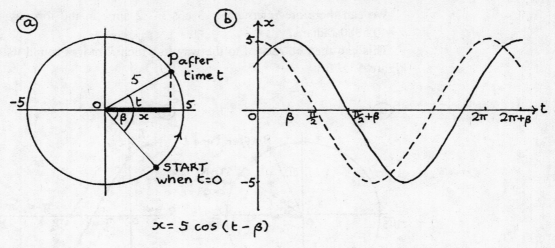

$$x = 5 \cos (t - \beta)$$

Figure 5.D.5

When $t = \beta$, x has its maximum value of $5 \cos (0) = 5$ units.

The graph for $x = 5 \cos (t - \beta)$ is, of course, identical to the graph for $x = 5 \sin (t + \alpha)$ because both represent $x = 4 \sin t + 3 \cos t$.

I have shown it again in Figure 5.D.5(b) with the graph of $x = 5 \cos t$ shown as a dashed line. We see that this time the two graphs have a phase difference of β. The $\alpha + \beta$ together make the $\pi/2$ shift between $x = 5 \cos t$ and $x = 5 \sin t$.

Again, $A = 5$, $\omega = 1$ and $T = 2\pi$ for both graphs.

You can see from Figure 5.D.5(a) that, as P moves round from its starting position, what happens first is that x increases in size to its maximum value of 5 units, and this is what the graph of $x = 5 \cos (t - \beta)$ is also doing.

5.D.(g)   **More examples of the $R \sin(t \pm \alpha)$ and $R \cos(t \pm \alpha)$ forms**

Here is another example, this time involving a minus sign.

Write $x = 3 \cos t - 2 \sin t$ as a single trig function and sketch its curve.

We start by choosing a rule which will fit nicely to what we have this time, including the minus sign in the middle. Which rule should we choose?

$\cos(A + B) = \cos A \cos B - \sin A \sin B$ will give the kind of fit that we want.
We write

$$3 \cos t - 2 \sin t = R \cos (t + \alpha) = R \cos t \cos \alpha - R \sin t \sin \alpha$$

so, matching up the separate parts as before, $3 = R \cos \alpha$ and $2 = R \sin \alpha$.

Using the little triangle in Figure 5.D.6 shows us that $R = \sqrt{13}$ and $\tan \alpha = \frac{2}{3}$ giving $\alpha = 0.5880$ radians to 4 d.p.

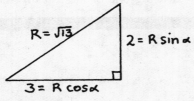

Figure 5.D.6

5.D Rules for combining trig functions   209

We can therefore rewrite $x = 3 \cos t - 2 \sin t$ in the form $x = \sqrt{13} \cos(t + \alpha)$ with $\alpha = 0.5880$ radians.

This can then be related to the way in which P moves round its circle which I show in Figure 5.D.7(a).

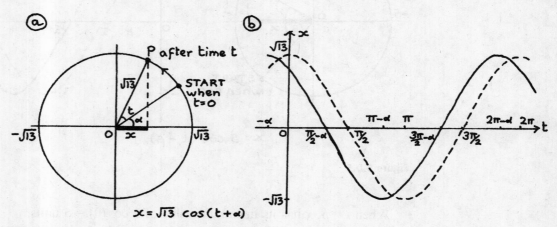

Figure 5.D.7

After time t, the displacement x is given by $x = \sqrt{13} \cos(t + \alpha)$.

When $t = 0$, $x = \sqrt{13} \cos \alpha$.

When $t = -\alpha$ (that is, α seconds before the instant at which we are taking $t = 0$), x will have its maximum size of $\sqrt{13} \cos(0) = \sqrt{13}$.

When $t = \pi/2 - \alpha$, $x = \sqrt{13} \cos(\pi/2) = 0$.

We can now sketch the graph of $x = \sqrt{13} \cos(t + \alpha)$. I show this in Figure 5.D.7(b), with the graph of $x = \sqrt{13} \cos t$ shown as a dashed line. The two graphs have a phase difference of α.

For both the graphs, we have $A = \sqrt{13}$, $\omega = 1$ and $T = 2\pi$.

Each of the circle diagrams which we have drawn shows very nicely how its related graph works. (It's very easy to see on the circle diagram just what effect the shift given by the angle α is having.) But you may be thinking that it is just being perverse to measure time in such a way that we get these shifts to worry about. Surely in the real world we can choose to have $t = 0$ when $\alpha = 0$?

Not necessarily so! There are some physical situations where we have to deal with waves which are out of phase with each other. For example, if we are working with the functions which describe how the voltage and current in an alternating current (a.c.) circuit change with time, and if this circuit includes components with inductance or capacitance, the current will peak after the voltage does, and so the two wave functions describing them will be out of phase with each other.

I'll now give you an example which involves functions of $2t$ instead of t. We'll combine $x = 3 \sin 2t + \cos 2t$ into a single trig function and sketch its graph.

How can we write $3 \sin 2t + \cos 2t$ using one of the rules for combined angles?

We could say either

$$3 \sin 2t + \cos 2t = R \sin(2t + \alpha) = R \sin 2t \cos \alpha + R \cos 2t \sin \alpha$$

or $\quad \cos 2t + 3 \sin 2t = R \cos(2t - \beta) = R \cos 2t \cos \beta + R \sin 2t \sin \beta$.

I shall work with the first of these, but the second would of course give an identical curve. We have

$$x = 3 \sin 2t + \cos 2t = R \sin 2t \cos \alpha + R \cos 2t \sin \alpha.$$

(Notice that everything here is in terms of $2t$ instead of t.)

Now, matching up the separate parts, we have $3 = R \cos \alpha$ and $1 = R \sin \alpha$.

Drawing the little triangle in Figure 5.D.8 shows us that $R = \sqrt{10}$ and $\tan \alpha = \frac{1}{3}$ so $\alpha = 0.3218$ rads to 4 d.p.

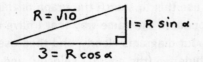

Figure 5.D.8

This gives us

$$x = 3 \sin 2t + \cos 2t = \sqrt{10} \sin (2t + \alpha)$$

with $\alpha = 0.3218$ radians to 4 d.p.

We now know that when $t = 0$, $x = \sqrt{10} \sin \alpha$ and, when $2t + \alpha = \pi/2$, $x = \sqrt{10} \sin (\pi/2)$ $= \sqrt{10}$. This happens when $t = \frac{1}{2} (\pi/2 - \alpha) = 0.624$ seconds to 3 d.p.

As usual, we shall need the circle picture to help us to draw the graph. I show this in Figure 5.D.9(a) below. We shall also use these two diagrams in Section 9.C.(c) when we look at some differential equations which describe SHM.

This time, P is moving at 2 rad/s.

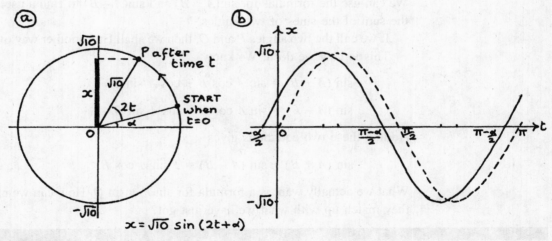

Figure 5.D.9

 From the circle picture, we can see that we shall have to be very careful about labelling the interesting points on the graph sketch this time.

P is moving at 2 rad/s so the period of the function is π seconds. (Each cycle takes π seconds.) Because it is moving at 2 rad/s it would have been at the point A at $\alpha/2$ seconds before the instant when we took $t = 0$.

We also know that x has its first maximum value of $\sqrt{10}$ after $\frac{1}{2}(\pi/2 - \alpha)$ seconds.

Using this information, I have drawn the function $x = \sqrt{10}\sin(2t + \alpha)$ in Figure 5.D.9(b).

I've also sketched $x = \sqrt{10}\sin 2t$ using a dashed line.

This time, the phase difference between the two graphs is $\alpha/2$.

For each graph, $A = \sqrt{10}$, $\omega = 2$ and $T = 2\pi/2 = \pi$.

EXERCISE 5.D.1 **Now try the following questions yourself. Give all your angles in radians, either exactly or to 3 d.p.**

For each question, you should also draw a diagram showing the related motion of P round its circle. Then use this to sketch the graph of the single combined trig function which you have found, in the same way that I have done in my examples. Make sure that you label your diagrams clearly, and then use them to write down the values of A (the amplitude), ω (the angular velocity) and T (the period), of each of your combined trig functions.

(1) Find $x = \sqrt{3}\cos t - \sin t$ in the form $x = R\cos(t + \alpha)$.

(2) Find $x = 5\cos t + 12\sin t$ in the form $x = R\cos(t - \alpha)$.

(3) By choosing a suitable formula, find $x = 15\cos t - 8\sin t$ as a single combined trig function.

(4) By choosing a suitable formula, find $x = 2\cos t - 3\sin t$ as a single combined trig function.

(5) Find $x = \cos 4t + \sin 4t$ in the form $R\cos(4t - \alpha)$.

(6) Write $\sqrt{3}\sin 3t - \cos 3t$ in the form $R\sin(3t - \alpha)$.

5.D.(h)   Going back the other way – the Factor Formulas

We can use the formulas for $\sin(A + B)$ and $\sin(A - B)$ to find a useful new way of writing the sum of the sines of two angles.

If we call the two angles P and Q, then we shall find another way of writing $\sin P + \sin Q$. This is how we do it. We know

$$\sin(A + B) = \sin A \cos B + \cos A \sin B,$$

$$\sin(A - B) = \sin A \cos B - \cos A \sin B.$$

Adding these two equations gives

$$\sin(A + B) + \sin(A - B) = 2\sin A \cos B.$$

What we actually want is a formula for $\sin P + \sin Q$. How can we choose P and Q so that they match up with what we have just got?

We need to put $P = A + B$ and $Q = A - B$. Then we have

$$P + Q = 2A \quad \text{so} \quad A = \frac{P + Q}{2} \quad \text{and} \quad P - Q = 2B \quad \text{so} \quad B = \frac{P - Q}{2}$$

This gives us the result

$$\boxed{\sin P + \sin Q = 2\sin\left(\frac{P + Q}{2}\right)\cos\left(\frac{P - Q}{2}\right).}$$

Similarly, it can be shown that

$$\sin P - \sin Q = 2 \cos \left(\frac{P+Q}{2}\right) \cos \left(\frac{P-Q}{2}\right),$$

$$\cos P + \cos Q = 2 \cos \left(\frac{P+Q}{2}\right) \cos \left(\frac{P-Q}{2}\right),$$

$$\cos P - \cos Q = -2 \sin \left(\frac{P+Q}{2}\right) \sin \left(\frac{P-Q}{2}\right).$$

Notice the minus sign at the start of the rule for $\cos P - \cos Q$.

You can see that it must be there if you put $\angle P = 60°$ and $\angle Q = 30°$, for example. $\cos 60°$ is smaller than $\cos 30°$, but $\sin 45°$ and $\sin 15°$ are both positive.

It is sometimes useful to be able to make use of the midway steps for each of these. We found in the working above that $\sin (A + B) + \sin (A - B) = 2 \sin A \cos B$. The three rules like this one, put together in a box, are:

$$2 \sin A \cos B = \sin(A + B) + \sin(A - B),$$

$$2 \sin A \sin B = \cos(A - B) - \cos(A + B),$$

$$2 \cos A \cos B = \cos(A + B) + \cos(A - B).$$

These two sets of rules are useful to turn adding into multiplying to make it easier to solve certain types of trig equation. I show you an example of this in Section 5.E.(d). They are also useful the other way round, when they turn multiplying into adding, for certain kinds of integral. Example (8) in Section 9.B.(f) shows you how this works.

We have now obtained all the basic trig rules involving two angles, and so have them ready for use whenever we need them.

You might find it helpful now to go through the previous sections highlighting in colour all the boxes with these rules inside, so that you can quickly find them when you need them, and can become familiar with them.

5.E   Solving trig equations

5.E.(a)   Laying some useful foundations

Quite often, students don't like solving trig equations because they find the possibilities of more than one answer confusing. It's in the nature of trig equations that they will have an infinite number of solutions – we only need to look at the repeating graphs of $y = \sin x$ and $y = \cos x$ to see this. (Of course, physical circumstances may limit the number of possible answers; for example, any angle in a triangle must be somewhere between 0° and 180°.)

When infinite numbers of answers *are* possible, we shall use the patterns of how they come to describe them. To do this, we shall need the circle definitions for the trig ratios of angles greater than 90° of Section 5.A.(c). I think you will find that it will help you here if you read through this section again before going on. Then do the following exercise which is based on the results of this section, and which will also give you some particular values which will be useful for solving equations.

EXERCISE 5.E.1

The table below is very similar to the one I gave you for Exercise 5.A.1 in Section 5.A.(a) except that I have only included positive angles here, and I have put in a line for the tan of the angles, too. In that exercise, you worked out the values for the sin and cos of the extra angles by using the graphs of $y = \sin x$ and $y = \cos x$.

Try filling in the blanks again by thinking how each angle will come in the turning circle, and then matching it up with an angle for which I've given you the sin, cos and tan. The values for your angle will then be the same as these except for a change of sign in some cases.

Write your answers in the same form that mine are given in, including $\sqrt{}$ signs if necessary, because you will find when you use these results that exact answers are often easier to work with than strings of decimals. Then check that your answers are right by using your calculator. (It's best to use pencil until you have checked!)

Angles (radians)	0	$\frac{\pi}{6}$	$\frac{\pi}{4}$	$\frac{\pi}{3}$	$\frac{\pi}{2}$	$\frac{2\pi}{3}$	$\frac{3\pi}{4}$	$\frac{5\pi}{6}$	π	$\frac{7\pi}{6}$	$\frac{5\pi}{4}$	$\frac{4\pi}{3}$	$\frac{3\pi}{2}$	$\frac{10\pi}{3}$	$\frac{7\pi}{8}$	$\frac{11\pi}{6}$	2π
Degrees	0	30	45	60	90	120	135	150	180	210	225	240	270	310	315	330	360
sin	0	$\frac{1}{2}$	$\frac{1}{\sqrt{2}}$	$\frac{\sqrt{3}}{2}$	1												
cos	1	$\frac{\sqrt{3}}{2}$	$\frac{1}{\sqrt{2}}$	$\frac{1}{2}$	0												
tan	0	$\frac{1}{\sqrt{3}}$	1	$\sqrt{3}$	U												

U stands for 'undefined'.

We can now start solving trig equations by using the patterns of how these solutions come to give us a way of describing the infinite number of possible answers. This is called giving the **general solution**.

The easiest way for me to explain how to do this is for us to work through some particular examples together. I shall take separate examples for sin, cos and tan with one positive and one negative value in each case, so that we cover all the possibilities. Then we shall use these to build up the rules for the general solutions for each particular case.

When we solve trig equations, we are working back from the sin, cos or tan of the angle to the angle itself. This means that we shall have to use the inverse functions of $\sin^{-1}$, $\cos^{-1}$ and $\tan^{-1}$ (or arcsin, arccos and arctan as they are sometimes known). If you are unsure about these, you should go back now to Sections 5.A(f), (g), (h) and (i) to see how they work.

The angle given by your calculator from a known sin, cos or tan is the angle given by using the inverse function. (Remember that a function gives just *one* possible result for every value fed into it.) We know that for any particular value of sin, cos or tan, there are an infinite number of possible matching angles.

> The angle given by using a trig inverse function is called the **principal value**.

For example, if $\sin x = \frac{1}{2}$, then the principal value for the angle x in radians is $\pi/6$. This is what $\sin^{-1}\left(\frac{1}{2}\right)$ gives you. But other possible solutions to the equation $\sin x = \frac{1}{2}$ are the angles $5\pi/6$, $13\pi/6$, $17\pi/6$, etc. and there are an infinite number of these.

5.E.(b)   **Finding solutions for equations in cos x**

I am starting with cos x because this is the easiest one to write down the patterns for. We'll solve the equation $6\cos^2 x - \cos x - 1 = 0$

  (a)   for the principal values,
  (b)   for all angles between $0°$ and $360°$,
  (c)   for *all* possible angles, giving the answers in degrees.

This is just a quadratic equation like the ones we worked with in Chapter 2. If you like, you can put $\cos x = y$ in the equation, which then gives you $6y^2 - y - 1 = 0$. This factorises to give

$$(2y - 1)(3y + 1) = 0 \quad \text{or} \quad (2\cos x - 1)(3\cos x + 1) = 0$$

replacing y by cos x. You can also factorise straight to this form without bothering with the y if you like.

From this, there are two possible solutions for cos x.

Either $2\cos x - 1 = 0$ so $\cos x = \frac{1}{2}$ and the principal value of x is $60°$, or $3\cos x + 1 = 0$ so $\cos x = -\frac{1}{3}$ and the principal value of x is $109.5°$ to 1 d.p. (This answer is 109.47 to 2 d.p. and I'll use this in any further working to avoid rounding errors.) These two angles give us the answer to (a).

Now we answer (b) by finding all the solutions of the equation between $0°$ and $360°$. It's easiest to see where these must be if we use the two circle diagrams of Figure 5.E.1. From Figure 5.E.1(a) we get a second possible solution of $360° - 60° = 300°$. From Figure 5.E.1(b) we get a second possible solution of $360° - 109.47 = 250.5°$ to 1 d.p. Use your calculator to check that $x = 300°$ and $x = 250.5$ *do* fit the equation which we started with.

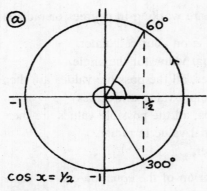
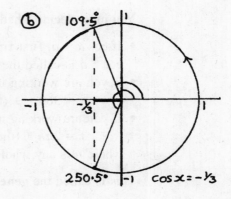

Figure 5.E.1

(c) Now we want to find *all* the possible solutions to the given equation.

Looking at the two circle diagrams of Figure 5.E.1, we can see that each pair of answers is symmetrically placed either side of the horizontal axis.

Adding any number of full turns to each of the four solutions we already have will give further possible solutions.

We can show all these further solutions by writing the ones which we already have in the form

$$x = 360°n \pm 60° \quad \text{and} \quad x = 360°n \pm 109.5°$$

where n is any whole number. (Remember that '$\pm$' means 'plus or minus'.)

The answers which we already have for (ii) could have been found by putting $n = 0$ and $n = 1$ in the two general solutions above and then picking out the ones which come between $0°$ and $360°$. (Try doing this for yourself.)

You can also see that these answers agree entirely with what happens if you use the graph of $\cos x$, by looking at Figure 5.E.2. The answers are given here by the x values at the intersections of $y = \cos x$ with the two lines $y = \frac{1}{2}$ and $y = -\frac{1}{3}$.

We have now seen that the two sets of general solutions are given by

$$x = 360n \pm \text{(the principal value in degrees)}$$

and that this was true whether the principal value was positive or negative.

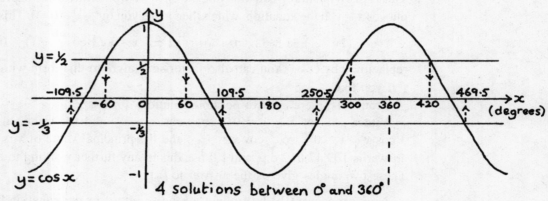

Figure 5.E.2

These are the rules which we now have.

Finding all possible solutions for the angles from a given cos

You must decide whether you are working in degrees or radians before you start.

- If $\cos x = a$, first find $\cos^{-1} a$ on your calculator.
  $\cos^{-1} a$ is called the **principal value** for the angle.
- If you are working in degrees, all the possible values are then given by
  $$x = 360°n \pm \text{(the principal value in degrees)}.$$
- If you are working in radians, all the possible values are then given by
  $$x = 2\pi n \pm \text{(the principal value in radians)}.$$
  where n is any whole number.

This is called the **general solution** of the equation $\cos x = a$.

Never give a mixed answer like $x = 2n\pi \pm 60°$ because this is meaningless. You must work completely either in degrees or in radians. (If you need help with radians, see Section 4.D.)

EXERCISE 5.E.2 Try solving the similar equation $2 \cos^2 x + 3 \cos x + 1 = 0$ for yourself,

(a) for the principal values, (giving your answers in degrees),
(b) for all angles between 0° and 360°,
(c) for all possible angles, that is, the general solution.

5.E.(c)   **Finding solutions for equations in tan x**
We'll use the following example to show how this is done.
Solve the equation $\sec^2 x - \tan x - 3 = 0$

(a)   for the principal values,
(b)   for all angles between 0° and 360°,
(c)   for all possible angles.

We have a difficulty here which is that this equation is partly in terms of $\sec x$ and partly in terms of $\tan x$, and we can't do anything with it as it stands. But we found earlier a relationship between $\sec x$ and $\tan x$ which we can use here.
Can you remember what it is?

We can use the identity $\tan^2 x + 1 = \sec^2 x$ (Section 5.B.(b)).
Substituting for $\sec^2 x$ using this, we now have

$$(\tan^2 x + 1) - \tan x - 3 = 0 \quad \text{so} \quad \tan^2 x - \tan x - 2 = 0$$
$$\text{so} \quad (\tan x - 2)(\tan x + 1) = 0.$$

(a)   Either $\tan x - 2 = 0$ so $\tan x = 2$ and the principal value of x is $63.43 = 63.4°$ to 1 d.p., or $\tan x + 1 = 0$ so $\tan x = -1$ and the principal value of x is $-45°$.

(b)   Now we want all the solutions between 0° and 360°.
Using the definition for the tan of an angle greater than 90° from Section 5.A.(c), we can see where the other two solutions between 0° and 360° must be.
Figure 5.E.3(a) shows the two solutions of $\tan x = 2$, and Figure 5.E.3(b) shows the two solutions of $\tan x = -1$ between 0° and 360°.

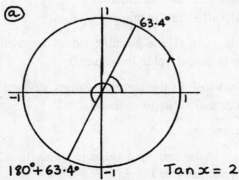
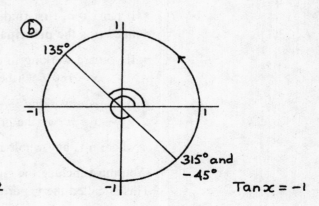

Figure 5.E.3

(c) Adding any number of full turns to the solutions above will give all the possible solutions.

Can you see what pattern these will have? Look particularly at what happens after any number of half turns.

This time, the principal value is always added on to however many half turns have been made.

This adding on takes into account the *sign* of the principal value, so 135° = 180° + (−45°), for example.

The general solution is given by $x = 180°n + 63.4$ and $x = 180°n + 135°$, where n is a whole number (or integer).

You can see how these solutions will also work graphically by looking at Figure 5.E.4 below.

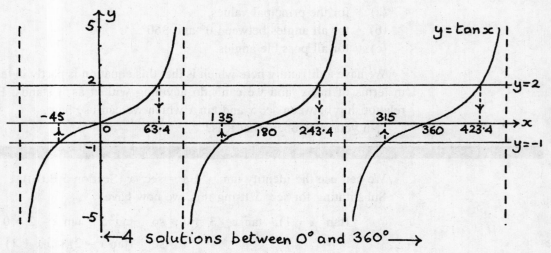

Figure 5.E.4

The solutions are given by the x values at the intersections of $y = \tan x$ with the two lines $y = 2$ and $y = −1$.

These are the rules which we now have.

Finding all possible solutions for the angles from a given tan

- If $\tan x = a$, first find $\tan^{-1} a$ on your calculator.
  $\tan^{-1} a$ is the **principal value** for the angle.

- If you are working in degrees, all the possible values are then given by
  $x = 180°n +$ (the principal value in degrees).

- If you are working in radians, all the possible values are then given by
  $x = n\pi +$ (the principal value in radians)

  where n is any whole number.

(You must include the sign of the principal value in these rules.)
This is called the **general solution** of the equation $\tan x = a$.

Try solving the similar equation of sec² x + 2 tan x – 4 = 0 for yourself
(a) for the principal values, giving your answers in degrees,
(b) for all angles between 0° and 360°,
(c) for all possible angles, that is, the general solution.

5.E.(d)   **Finding solutions for equations in sin x**

We'll use the example of solving the equation $1 + 3 \sin x - 5 \cos 2x = 0$

    (a)   for the principal values,
    (b)   for all angles between 0° and 360°,
    (c)   for all possible angles, giving the answers in degrees.

Again we have a mixed equation. We need to use a trig identity so that we can write it just in terms of sin x.

How else can we write cos 2x?

We can say that $\cos 2x = 1 - 2 \sin^2 x$ from Section 5.D.(d).
Substituting this in the equation gives us

$$1 + 3 \sin x - 5 (1 - 2 \sin^2 x) = 0.$$

From this we get

$$10 \sin^2 x + 3 \sin x - 4 = 0 \quad \text{so} \quad (2 \sin x - 1)(5 \sin x + 4) = 0.$$

    (a)   Either $\sin x = \frac{1}{2}$, which gives the principal value of $x = 30°$, or $\sin x = -\frac{4}{5}$, which gives the principal value of $x = -53.13° = -53.1°$ to 1 d.p.

    (b)   All the possible solutions between 0° and 360° can be seen from the two circle diagrams in Figure 5.E.5.

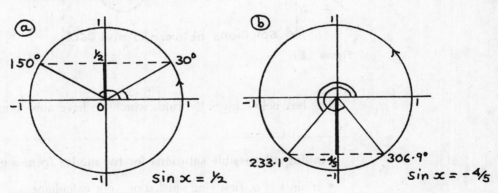

Figure 5.E.5

Circle (a) gives us 30° and 180° – 30° = 150°.
Circle (b) gives us 360° – 53.13° = 306.9° to 1 d.p. and 180° + 53.13° = 233.1° to 1 d.p.

    (c)   The pattern for getting all the possible solutions is a little bit harder to spot this time as the principal value is sometimes being added on and sometimes being taken off. Can you see how to describe this pattern? It might help you if you think about the number of half turns involved as you get to each new solution.

We know that all the possible solutions will be given by adding any number of full turns to the four solutions which we already have.

If we look at Figure 5.E.5(a) first, this gives $360°n + 30°$ and $360°n + 180° - 30°$.

Now $360°n = 2 \times 180°n$, so we can write these two answers as $2 \times 180°n + 30°$ and $2 \times 180°n + 180° - 30°$.

This is the same as $2n (180°) + 30°$ and $(2n + 1) 180° - 30°$.

If the number of half turns is *even*, we add on the 30°.

If the number of half turns is *odd*, we take off the 30°.

These two results can be ingeniously combined by using $(-1)^n$, because $(-1)^n$ gives us +1 if n is even and −1 if n is odd.

All the possible solutions from $\sin x = \frac{1}{2}$ are given by $x = 180°n + (-1)^n 30°$. (The two solutions of (ii) are given by putting $n = 0$ and $n = 1$.)

In just the same way, all the possible solutions of $\sin x = -\frac{4}{5}$ are given by writing $x = 180°n + (-1)^n (-53.1°)$.

You can also see how these solutions are building up in the sketch graph of Figure 5.E.6. They are given by the x values at every intersection of the curve of $y = \sin x$ with the two lines $y = \frac{1}{2}$ and $y = -\frac{4}{5}$ respectively.

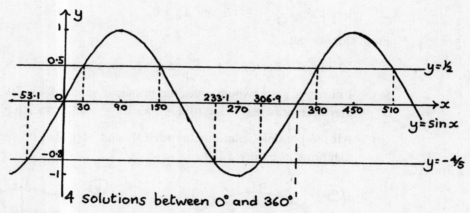

Figure 5.E.6

The box below gives the rules which we have now found.

Finding all possible solutions for the angles from a given sin

- If $\sin x = a$, first find $\sin^{-1} a$ on your calculator.
  $\sin^{-1} a$ is called the **principal value** for the angle.

- If you are working in degrees, all the possible values are then given by
  $x = 180°n + (-1)^n$ (the principal value in degrees).

- If you are working in radians, all the possible values are then given by
  $x = \pi n + (-1)^n$ (the principal value in radians).
  where n is any whole number.

(You must include the sign of the principal value in this rule.)
This is called the **general solution** of the equation $\sin x = a$.

Try solving the equation $\cos^2 x + 2 \sin x = 1$ for yourself
(a) for the principal values (giving your answers in radians),
(b) for all angles from 0 to 2π,
(c) for all possible angles, that is, the general solution.

I will finish this section with an example of a slightly different kind of equation involving $\sin x$.

Suppose we need to solve $\sin 3x = \sin x$ for angles between 0 and 2π.

See how far you can get with this yourself before looking at what I have done.

It's easy to spot that $x = 0$ is one solution of this equation, but how can we set about finding the others?

Figure 5.E.7 shows a snapshot of what's happening graphically.

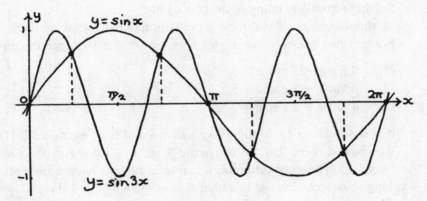

Figure 5.E.7

We can now see that $x = \pi$ and $x = 2\pi$ will also fit, but what values of x will give the other four solutions?

We have $\sin 3x = \sin x$ so $\sin 3x - \sin x = 0$.

Now we use the second of the four factor formulas from Section 5.D.(h)

$$\sin P - \sin Q = 2 \cos\left(\frac{P+Q}{2}\right) \sin\left(\frac{P-Q}{2}\right)$$

and put $3x = P$ and $x = Q$. This gives us

$$2 \cos(2x) \sin x = 0 \quad \text{so} \quad \sin x = 0 \quad \text{or} \quad \cos 2x = 0.$$

From $\sin x = 0$ we get $x = 0$ or π or 2π.

From $\cos 2x = 0$ we get $2x = 2n\pi \pm \pi/2$ so $x = n\pi \pm \pi/4$, giving us the other four solutions of $x = \pi/4, 3\pi/4, 5\pi/4$ and $7\pi/4$.

There is often more than one possible method for solving these equations. For example, we could have done this one by writing $\sin 3x = 3 \sin x - 4 \sin^3 x$ from Section 5.D.(e) and then factorising. Also, in the method above, when we had $\cos 2x = 0$ we could have used $\cos 2x = 1 - 2 \sin^2 x$, giving $\sin^2 x = \frac{1}{2}$ so $\sin x = \pm \frac{1}{\sqrt{2}}$. Sometimes one method is neater than another, but there is no magic 'right way'.

Try solving the following equations which use the whole of Section 5.E so far.

In each case, find (a) the principal value(s), (b) solutions for $0° \leq x \leq 360°$ or $0 \leq x \leq 2\pi$ (I give the units after each question), and (c) the general solution. (Give your answers correct to 1 d.p. for degrees and 2 d.p. for radians.)

I think it is much easier to use the general solutions to find the answers between $0°$ and $360°$ or 0 and 2π. You just need to put in the values for n which give the answers in the desired range. I suggest you try doing this.

(1) $\cos x = \frac{2}{3}$ (deg)   (2) $\tan x = 5$ (deg)   (3) $\cos x = -\frac{1}{2}$ (rad)
(4) $\tan x = -1$ (rad)   (5) $\sin x = 0.4$ (deg)   (6) $6 \sin^2 x + 5 \cos x = 7$ (rad)
(7) $\tan^2 x = \tan x$ (rad)   (8) $3 \sec^2 x + \tan^2 x = 5$ (deg)
(9) $\sin 2x = 3 \cos x$ (rad)

5.E.(e)   Solving equations using $R \sin(x + \alpha)$ etc.

What should you do if you meet a problem like the following one?

Solve, when possible, for angles between $0°$ and $360°$, the three equations

(1) $4 \sin x + 3 \cos x = 6$,
(2) $4 \sin x + 3 \cos x = 5$,
(3) $4 \sin x + 3 \cos x = 2$.

It is not difficult to do this if we use the results of Section 5.D.(f).

We showed there that we can write $4 \sin t + 3 \cos t$ in the form $5 \sin(t + \alpha)$ with $\alpha = \tan^{-1} \frac{3}{4}$. (The only differences here are that we have x instead of t, and that we are working in degrees instead of radians, so $\alpha = 36.87°$ to 2 d.p.)

If you are at all unsure about this, you should go back now to Sections 5.D.(f) and (g), and work through them before going any further. Then see if you can solve the three equations yourself.

This is what I hope you have found.

(1)   There is no possible solution here. We can see this in two ways.
Firstly, if $5 \sin(x + \alpha) = 6$ then $\sin(x + \alpha) = \frac{6}{5}$ which is impossible.
You can also see this by looking at the graph of $y = 5 \sin(x + \alpha)$ which I have sketched in Figure 5.E.8.
You can see here that the line $y = 6$ misses this sine curve completely, so there are no solutions to the equation.

(2)   Again, we can look at this in two ways.
We have $5 \sin(x + \alpha) = 5$ which gives $\sin(x + \alpha) = 1$, so the principal value of $(x + \alpha)$ is $90°$.
From this, we can say that $(x + \alpha) = 180°n + (-1)^n \, 90°$ using the rule for the general solution from Section 5.E.(d).
This then gives us $x = 180°n + (-1)^n \, 90° - \alpha$.
Putting $\alpha = 36.87$ gives us the single solution between $0°$ and $360°$ of $x = 53.1°$ to 1 d.p.

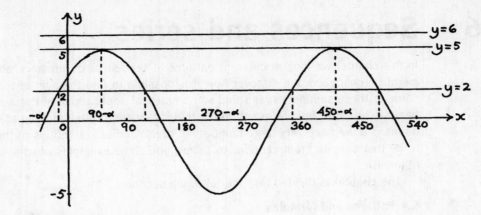

Figure 5.E.8

This answer fits with what we can see is happening graphically. The line $y = 5$ is a tangent to the curve $y = 5 \sin (x + \alpha)$, and only touches it once between $x = 0°$ and $x = 360°$.

(3)   Now we have $5 \sin (x + \alpha) = 2$ so $\sin (x + \alpha) = \frac{2}{5}$ which gives the principal value of $(x + \alpha)$ as $23.58°$ to 2 d.p.

Therefore, the general solution for $(x + \alpha)$ is given by $180°n + (-1)^n (23.58°)$ or $x + 36.87° = 180°n + (-1)^n (23.58°)$, putting $\alpha = 36.87°$.

Putting $n = 0$ gives $x = -13.3°$, $n = 1$ gives $x = 119.6°$ and $n = 2$ gives $x = 346.7°$ all to 1 d.p.

You can see all three of these answers on the sketch graph in Figure 5.E.9. The last two of them give the solutions in the range from $0°$ to $360°$ that we want.

Notice that the answers given by the general solution for (3) are symmetrically placed either side of the answers for (2), and that all these answers have been affected by the sliding along to the left by α of the graph of $y = 5 \sin x$ to give $y = 5 \sin (x + \alpha)$.

The most usual mistake made when solving this sort of equation goes as follows:

The solver gets to $x + \alpha = 23.58°$ correctly and then rearranges this to get the correct answer for x of $-13.3°$.

Then they think 'Curses, I needed a general solution here! Oh well, I'll put $x = 180°n + (-1)^n (-13.3°)$.'

This is not true! The general solution comes from using the graph of $y = 5 \sin (x + \alpha)$ and the solutions must be found taking the whole of $(x + \alpha)$ as I have done.

EXERCISE 5.E.6

Try these two for yourself now.

(1) Solve, when possible, the three equations
    (a) $3 \cos t - 2 \sin t = 4$,
    (b) $3 \cos t - 2 \sin t = \sqrt{13}$,
    (c) $3 \cos t - 2 \sin t = 1$ for $0 \leq t \leq 2\pi$ giving your answers to 2 d.p.
    Show your answers on a sketch graph.

(2) Solve the equation $3 \sin 2t + \cos 2t = 2$ for angles between $0°$ and $360°$.

6   Sequences and series

In this chapter we look at different patterns in sequences of numbers, and how they might be described. We discover how it is possible to find the sum of the terms of some of these sequences, and find some practical applications of these sums. We begin to see how infinite quantities of things behave through looking at what happens if we have very large numbers of them. Endless quantities of things have to be treated with great caution, so I show you some examples of what can happen otherwise.

The chapter is divided into the following sections.

6.A Patterns and formulas
(a) Finding patterns in sequences of numbers,
(b) How to describe number patterns mathematically

6.B Arithmetic progressions (APs)
(a) What are arithmetic progressions? (b) Finding a rule for summing APs,
(c) The arithmetic mean or 'average', (d) Solving a typical problem,
(e) A summary of the results for APs

6.C Geometric progressions (GPs)
(a) What are geometric progressions? (b) Summing geometric progressions,
(c) The sum to infinity of a GP, (d) What do 'convergent' and 'divergent' mean?
(e) More examples using GPs; chain letters, (f) A summary of the results for GPs,
(g) Recurring decimals, and writing them as fractions,
(h) Compound interest: a faster way of getting rich, (i) The geometric mean,
(j) Comparing arithmetic and geometric means,
(k) Thinking point: what is the fate of the frog down the well?

6.D A compact way of writing sums: the Σ notation
(a) What does Σ stand for? (b) Unpacking the Σs,
(c) Summing by breaking down to simpler series

6.E Partial fractions
(a) Introducing partial fractions for summing series,
(b) General rules for using partial fractions, (c) The cover-up rule,
(d) Coping with possible complications

6.F The fate of the frog down the well

6.A   Patterns and formulas

6.A.(a)   Finding patterns in sequences of numbers

We shall start by looking at some lists of numbers for which there is an underlying pattern so that there is some rule for writing down the next number. A list of numbers like this is called a **sequence**. A particular number from a sequence is called a **term** of the sequence.

Here are some examples. In each case, see if you can fill in the next three terms in the sequence, and write down the rule that you are using so that somebody else could continue filling in where you have stopped.

(a)	1, 2, 3, 4, 5, . . .	(b)	1, 3, 5, 7, 9, . . .
(c)	2, 5, 8, 11, 14, . . .	(d)	1, 2, 4, 8, . . .
(e)	1, 2, 4, 7, 11, . . .	(f)	54, 18, 6, 2, $\frac{2}{3}$, . . .
(g)	$\frac{1}{3}, \frac{1}{6}, \frac{1}{12}, \frac{1}{24}, \ldots$	(h)	$\frac{1}{2}, \frac{2}{3}, \frac{3}{4}, \frac{4}{5}, \ldots$
(i)	1, 4, 9, 16, 25, . . .	(j)	1, 2, 3, 5, 8, 13, 21, . . .
(k)	1, 8, 27, 64, . . .	(l)	1, 2, 6, 24, 120, . . .

Here are the answers for you to check yours against.

(a)   6, 7, 8. The counting numbers, or add 1 each time.

(b)   11, 13, 15. The odd numbers. Add 2 each time, starting from 1.

(c)   17, 20, 23. Add 3 each time, starting from 2.

(d)   16, 32, 64. Double each time, starting from 1.

(e)   16, 22, 29. Start by adding 1 to the first term, which is itself 1. Then, for each new term, add 2, 3, 4, etc. so that the number you add is always 1 more than the previous number added.

(f) $\frac{2}{9}, \frac{2}{27}, \frac{2}{81}$. Take one third of the previous term each time, starting with 54.

(g) $\frac{1}{48}, \frac{1}{96}, \frac{1}{192}$. Take one half of the previous term, starting from one third.

(h) $\frac{5}{6}, \frac{6}{7}, \frac{7}{8}$. For each new term, add 1 to both the top and the bottom of the fraction which makes the previous term.

(i)   36, 49, 64. This sequence is formed from the squares of the counting numbers.

(j)   34, 55, 89. After the first two terms, each term is made by adding the previous two terms. This is called a Fibonacci sequence.

(k)   125, 216, 343. These terms are the cubes of the counting numbers.

(l)   720, 5040, 40 320. The terms of this sequence are formed by finding 1, 2 × 1, 3 × 2 × 1, etc. They are called **factorials**, and are written as 1!, 2!, 3!, etc.

6.A.(b)   How to describe number patterns mathematically

It is often useful to be able to write down a rule or formula which will tell us how to find any term we want in a sequence of numbers such as the ones above. To be able do this, we shall need a shorthand system for labelling the terms. We will use the system of calling them $u_1, u_2, u_3, \ldots$ so that u_4 for (b) is 7, and u_5 for (e) is 11. If we don't want to specify a particular number, we can call the term u_n where n is standing for any number which we might later want to choose. We call u_n the **general term**.

 The n in u_n is called a **subscript** and is just a label telling us how far we have gone. Don't confuse it with u^n which means u multiplied by itself n times.

What we now want to do is to find some way of writing a rule which gives the general term or u_n for each of the sequences from (a) to (l).

The easiest way of explaining how we can set about doing this is to take two particular examples.

EXAMPLE (1)   Sequence (c) goes 2, 5, 8, 11, 14, . . .

The description in words for this was 'add 3 each time, starting from 2.'
There are two ways in which we can write this mathematically.
We can say $u_1 = 2$, $u_2 = 2 + 3$, $u_3 = 2 + (2 \times 3)$, $u_4 = 2 + (3 \times 3)$
and so on, so that we are describing each term using the actual numbers
which make it up. We'll call this description (A).

Sticking to the same system, how would you write u_7? How would
you write u_n?

$$u_7 = 2 + (6 \times 3) \quad \text{and} \quad u_n = 2 + ((n-1) \times 3) = 2 + 3n - 3 = 3n - 1.$$

Notice that we needed $(n-1)$ rather than n when we first wrote down the rule for u_n. We
can check this rule by testing it when $n = 5$. We get $3 \times 5 - 1 = 14$ which we know is
correct.

We could also think of this sequence as building up in a chain, each new term coming
from the previous term according to a particular rule. We'll call this description (B).

Description (B) for this sequence would be $u_n = u_{n-1} + 3$. But just knowing this would
not be enough, because, for example, the sequence 1, 4, 7, 10, 13, . . . would also fit this
description. However, if we also give the value of the first term, the sequence is fully
described.

Description (B) is $u_n = u_{n-1} + 3$ and $u_1 = 2$.

EXAMPLE (2)   Sequence (g) goes $\frac{1}{3}, \frac{1}{6}, \frac{1}{12}, \frac{1}{24}$. . . .

The description in words for this was 'take one half of the previous
term starting from one third'.

DESCRIPTION (A)   We can say that $u_1 = \frac{1}{3}$, $u_2 = \frac{1}{2} \times \frac{1}{3}$, $u_3 = \frac{1}{2} \times \frac{1}{2} \times \frac{1}{3} = (\frac{1}{2})^2 \times \frac{1}{3}$
so u_7, say, is $(\frac{1}{2})^6 \times \frac{1}{3}$ and $u_n = (\frac{1}{2})^{n-1} \times \frac{1}{3}$.
Notice that we need a power of $n-1$ here to make u_n work
correctly, not n.

DESCRIPTION (B)   We can say that $u_n = \frac{1}{2}u_{n-1}$ and $u_1 = \frac{1}{3}$.
Just as in the last example, if we don't say what u_1 is, we could
get quite a different sequence. For example, the sequence 24, 12,
6, 3, . . . also fits the description $u_n = \frac{1}{2} u_{n-1}$.

Sometimes both these methods of description are useful when we are considering
particular sequences. Sometimes one is very much easier to find than the other.

EXERCISE 6.A.1
Try finding the following descriptions for yourself now. Keep a special eye out for
sequences which can be described in a similar way to each other because we shall
be looking at some of these in more detail in the next two sections.

(1) Find descriptions (A) and (B) for sequence (a) on page 225.
(2) Find descriptions (A) and (B) for sequence (b).
(3) Find descriptions (A) and (B) for sequence (d).
(4) Find just description (B) for sequence (e).
(5) Find both descriptions (A) and (B) for sequence (f).
(6) Find just description (A) for sequence (h).

(7) Find just description (A) for sequence (i).

(8) Find just description (B) for sequence (j).

(9) Find just description (A) for sequence (k).

(10) Find both descriptions (A) and (B) for sequence (l).

I am giving the answers to this exercise here as we shall be needing some of them in the next two sections.

(1) Description (A) for sequence (a) gives $u_n = n$ and description (B) gives $u_n = u_{n-1} + 1$ with $u_1 = 1$.

(2) For description (A) for sequence (b), we can say that each odd number is one behind the corresponding term in the sequence of even numbers, so $u_n = 2n - 1$.

It is useful to remember this as a formula which must give an odd number. Similarly, $2n + 1$ must also always be an odd number, while $2n$ is always even.

Description (B) for this sequence says $u_n = u_{n-1} + 2$, with $u_1 = 1$.

(3) Description (A) for sequence (d) is $u_2 = 2 \times 1$ and $u_3 = 2^2 \times 1$ etc. so $u_n = 2^{n-1} \times 1 = 2^{n-1}$.

For description (B) we have $u_n = 2u_{n-1}$ with $u_1 = 1$.

(4) Description (B) for sequence (e) is $u_n = u_{n-1} + (n-1)$ with $u_1 = 1$, or you could write this as $u_{n+1} = u_n + n$ with $u_1 = 1$.

It is quite difficult to find a formula for u_n in terms of n here, just by looking at the terms, which is why I didn't ask you to do it.

In fact, the rule for (A) is $u_n = \frac{1}{2}n^2 - \frac{1}{2}n + 1$. Check for yourself that this works for $n = 1, 2$ and 3.

(5) For sequence (f), if we write $u_2 = 18 = (\frac{1}{3})\,54$, and $u_3 = 6 = (\frac{1}{3})^2\,54$, we see that $u_n = (\frac{1}{3})^{n-1}\,54$, so this is description (A).

Notice, here, that the first term uses $(\frac{1}{3})^0 = 1$, which is one of the rules from Section 1.D.(b).

Description (B) is $u_n = \frac{1}{3}(u_{n-1})$ with $u_1 = 54$.

(6) Description (A) for sequence (h) is $u_n = \dfrac{n}{n+1}$.

(7) Description (A) for sequence (i) is $u_n = n^2$.

(8) Description (B) for sequence (j) is $u_n = u_{n-1} + u_{n-2}$ with $u_1 = 1$ and $u_2 = 2$.

The formula for u_n in terms of n is so unlikely that even your wildest guesses would never have produced it.

It is $u_n = \dfrac{1}{\sqrt{5}}\left[\left(\dfrac{1+\sqrt{5}}{2}\right)^{n+1} - \left(\dfrac{1-\sqrt{5}}{2}\right)^{n+1}\right]$.

If you substitute some values for n in this formula, and use a calculator, you will find that you do indeed get the right terms.

(9)   Description (A) for sequence (k) is $u_n = n^3$.

(10)   Description (A) for sequence (1) is $u_n = n!$
    This means that $u_{n-1} = (n-1)!$ But $n! = n(n-1)!$ so description (B) is $u_n = nu_{n-1}$ with $u_1 = 1$.

A formula which describes u_n using the previous terms of the sequence, such as $u_n = u_{n-1} + u_{n-2}$ for the Fibonacci sequence, is called a **recurrence relation** or **difference equation**. Such equations have important applications in electrical engineering.

6.B	**Arithmetic progressions (APs)**

6.B.(a)   What are arithmetic progressions?
The sequences (a), (b) and (c) in Section 6.A.(a) are all examples of arithmetic progressions or APs for short.

If you look back, you will see that in each case each new term is made by adding the same constant number to the previous term.

We can write this type of sequence in the form

$$a, a + d, a + 2d, a + 3d, \ldots$$

where a is the first term (so $u_1 = a$) and d is what is called the **common difference** between each successive pair of terms.

In (a), $a = 1$ and $d = 1$. What are a and d in (b) and (c)?

We would have $a = 1$ and $d = 2$ in (b), and $a = 2$ and $d = 3$ in (c).

The nth term of an AP is given by $u_n = a + (n-1)d$ since we have only added d on $(n-1)$ times.

 It's easy to think that the nth term will be $a + nd$ but this is not so!

If the particular AP which we are considering only has n terms, so that u_n is the last term, we sometimes call this last term l, so then $u_n = l = a + (n-1)d$.

Suppose we have the AP 1, 3, 5, 7, ..., 33.

(The dots in the middle signify that there are a whole lot of other terms here which we do not want to (or even in some cases cannot) list individually. This use of dots is a standard piece of mathematical language.)

How many terms have we got here?

Using $u_n = l = a + (n-1)d$ with $a = 1$ and $d = 2$ gives

$$l = 33 = 1 + (n-1)2 = 1 + 2n - 2 \quad \text{so} \quad 2n = 34 \quad \text{and} \quad n = 17.$$

(Equally, each individual jump is of size 2, and the total jump from 1 to 33 is 32. Therefore, we have 16 jumps and 17 terms. This is like fence-posts and the gaps between them; there is one more post than there are gaps.)

Try these two yourself.

For each of the APs (1) 3, 7, 11, . . . , 79 and (2) 102, 100, 98, . . . , 14 write down the values of a and d. How many terms are there in each series?

You should have these answers.

For (1), $a = 3$ and $d = 4$ which gives $79 = u_n = l = 3 + (n-1)4 = 3 + 4n - 4$ so $80 = 4n$ and $n = 20$.

For (2), $a = 102$ and $d = -2$. (The common difference here is negative.)

We have $u_n = l = 14 = 102 + (n-1)(-2) = 102 - 2n + 2$ so $2n = 104 - 14$ and $n = 45$.

6.B.(b) Finding a rule for summing APs

For practical purposes, we often need the sum of some number of terms of an AP.

When the terms are added together, we call the result a **series**.

The process of actually adding the terms to find their sum is called **summing the series**.

Is there any way in which we can do this without actually having to add on each term separately?

There is a very neat way to do this. Think what happens if we turn the series the other way round, and then add it to itself in the original order. The pairs of terms exactly slot into each other to give the same result, like two staircases fitted opposite ways round.

Figure 6.B.1 shows the steps in adding the first eight terms of an AP as the sums build up term by term.

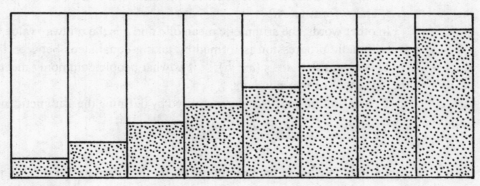

Figure 6.B.1

Turn it upside down and you have the identical situation.

To show how we can use this, we'll take the example of the series (1) which is $3 + 7 + 11 + \ldots + 75 + 79$.

We have just found that it has 20 terms, so we can write, using S for 'sum',

$$S_{20} = 3 + 7 + 11 + \ldots + 75 + 79.$$

Reversing the order, we can also write

$$S_{20} = 79 + 75 + 71 + \ldots + 7 + 3.$$

Adding these two sums, we get

$$2S_{20} = 82 + 82 + 82 + \ldots + 82 + 82$$

and there are 20 lots of 82. Therefore

$$S_{20} = \tfrac{1}{2} \times 20 \times 82 = 820.$$

We can now see how this same system will work for a general AP with a first term of a, a common difference of d and a last term, u_n, of l, by writing

$$S_n = a + (a + d) + (a + 2d) + \ldots + (l - d) + l.$$

Reversing the order, we can also write

$$S_n = l + (l - d) + (l - 2d) + \ldots + (a + d) + a.$$

Adding, we get

$$2S_n = (a + l) + (a + l) + (a + l) + \ldots + (a + l) + (a + l).$$

There are n terms here, so we have

$$2S_n = n(a + l) \quad \text{or} \quad S_n = \tfrac{1}{2}n\,(a + l).$$

Also, since $l = u_n = a + (n - 1)d$, we can say

$$S_n = \tfrac{1}{2}n(a + a + (n - 1)d) = \tfrac{1}{2}n\,(2a + (n - 1)d).$$

The rule for the sum of n terms of an AP is

$$S_n = \frac{n}{2}\left(a + l\right) = \frac{n}{2}\left(2a + (n - 1)d\right).$$

6.B.(c) The arithmetic mean or 'average'

We define the arithmetic mean, A, of two numbers, a and b, to be the number which makes a, A, and b form an AP.

In other words, the arithmetic mean of a and b is the midway value between a and b, since an arithmetic progression is formed by taking equal steps between the terms.

This means that $A = \tfrac{1}{2}\,(a + b)$. A is what people commonly mean when they talk about the 'average' of two numbers.

This definition can also be generalised by defining the arithmetic mean of n numbers to be

$$\frac{a_1 + a_2 + a_3 + a_4 + \ldots + a_n}{n}.$$

Again, this is what is commonly meant by the 'average' of these n numbers.

6.B.(d) Solving a typical problem

Here is an example of a typical problem on APs.

The 7th term of an AP is 23, and the 4th term is 14. Find the sum of the first 20 terms.

First, we must find a and d from the information that we have been given. The 7th term is $a + 6d$, and the 4th term is $a + 3d$, so we have

$$\begin{cases} a + 6d = 23 & (1) \\ a + 3d = 14 & (2) \end{cases}$$

Subtracting equation (2) from (1) gives $3d = 9$ so $d = 3$. Therefore

$$a = 5, \quad \text{and} \quad S_{20} = \frac{20}{2}\,(10 + 19 \times 3) = 670.$$

6.B.(e)   **A summary of the results for APs**

Before asking you to try some similar questions yourself, I will group together all the formulas which we have found for APs.

- We write APs as a, $a + d$, $a + 2d$, ..., where d is called the common difference.
- The nth term is given by $u_n = a + (n - 1)d$. If this is also the last term, we call it l.
- The sum of n terms is given by $S_n = n/2 \, (a + l)$ where l is the last or nth term,

  or $S_n = \frac{n}{2} [2a + (n - 1)d]$.

- The arithmetic mean of two numbers, a and b, is $\dfrac{a + b}{2}$.

- The arithmetic mean of n numbers, a_1, a_2, a_3, ... a_n, is

$$\frac{a_1 + a_2 + a_3 + a_4 + \ldots + a_n}{n}.$$

EXERCISE 6.B.1

Try these questions yourself.

(1) For each of the following APs:
   (i)   write down the values of a and d,
   (ii)   find the number of terms in the series,
   (iii) sum the series.

   (a) $3 + 7 + 11 + \ldots + 79$
   (b) $100 + 95 + 90 + \ldots + 15$
   (c) $6 + 6\frac{1}{4} + 6\frac{1}{2} + \ldots + 17\frac{3}{4}$

(2) (a) Find the sum of the natural numbers from 1 to 100 (that is, find $1 + 2 + 3 + \ldots + 100$).
   (b) Find the sum of the even numbers up to, and including 100, starting with 2.
   (c) Find the sum of the odd numbers up to 100, starting from 1.
   (d) Find the sum of the first n natural numbers.

(3) The first term of an AP is 11 and the sum of the first 18 terms is 1269. What is the common difference?

(4) How many terms must be taken in the series $7 + 11 + 15 + \ldots$ for the sum to be 1375?

(5) An AP is such that the third term equals twice the first term. The sum of the first ten terms is 195. Find the first term and the common difference.

6.C   **Geometric progressions (GPs)**

6.C.(a)   **What are geometric progressions?**

We move on now to consider sequences like those in (d), (f) and (g) in Section 6.A.(a). Each of these is an example of a sequence in which each new term is found by multiplying the previous term by a constant amount. This amount is called the **common ratio**. A sequence like this is called a **geometric progression**, or **GP** for short.

We can write this type of sequence as $a, ar, ar^2, ar^3, \ldots, ar^{n-1}$ where a is the first term, and r is the common ratio.

> The nth term is ar^{n-1}.

(Notice that it isn't ar^n. Again, we are one behind ourselves.) r is called the common ratio because if we divide any term by the previous term, we get r as the answer.

> It is always true for a GP that $\dfrac{u_n}{u_{n-1}} = \dfrac{ar^{n-1}}{ar^{n-2}} = r.$

In other words, the *ratio* between any pair of successive terms is $1 : r$.
It is often helpful to use this property in problems on GPs.
Taking (d) as a numerical example, we have $a = 1$ and $r = 2$, and

$$\frac{2}{1} = \frac{4}{2} = \frac{8}{4} = \frac{16}{8} \text{ etc.} = \text{the common ratio, } 2.$$

6.C.(b)   **Summing geometric progressions**
How can we find $S_n = a + ar + ar^2 + ar^3 + \ldots + ar^{n-1}$?
   It will be no good turning the sum the other way round this time, as the two sums will not slot together nicely as they did for the AP.
   However, if we multiply S_n by r, the whole sequence gets shifted along by one. We get

$$rS_n = \qquad ar + ar^2 + ar^3 + \ldots + ar^n \qquad (1)$$

$$S_n = a + ar + ar^2 + \ldots + ar^{n-1} \qquad (2)$$

Can you see what makes a good next step?

Subtracting (2) from (1) makes nearly everything disappear, and neatly gives us

$$rS_n - S_n = ar^n - a.$$

Factorising, we get $S_n(r - 1) = a(r^n - 1)$, so

> $$S_n = \frac{a(r^n - 1)}{r - 1}. \qquad (\text{G1})$$

   Equally, by multiplying the top and bottom of the previous formula by -1, we can write this as

> $$S_n = \frac{a(1 - r^n)}{1 - r}. \qquad (\text{G2})$$

The working is easier if you use (G2) when r is between -1 and $+1$, and (G1) otherwise.

Here are some typical problems on GPs. (You might like to try having a go yourself first, before looking at how I have done them.)

(1) Sum the following GPs.
   (a) $2 + 6 + 18 + \ldots$ for the first 20 terms.
   (b) $1 - 2 + 4 - 8 + 16 \ldots$ for (i) 10 terms, (ii) 11 terms.

The solutions for this first question are as follows:

(1) (a) We want S_{20} with $a = 2$ and $r = 3$.
      Using formula (G1), we have

$$S_{20} = \frac{2(3^{20} - 1)}{3 - 1} = 3\,486\,784\,398.$$

   (b) We want (i) S_{10}, (ii) S_{11}, with $a = 1$ and $r = -2$.
      Again using (G1), we have

   (i) $$S_{10} = \frac{1((-2)^{10} - 1)}{-2 - 1} = -341$$

   (ii) $$S_{11} = \frac{1((-2)^{11} - 1)}{-2 - 1} = 683.$$

It seems as if, for this series, not only are the terms alternating in sign, but also the sums, as we add on each new term.

6.C.(c) The sum to infinity of a GP

Suppose we have the GP $24 + 12 + 6 + 3 + \ldots$ and we want to find (a) S_4, (b) S_{10} and (c) S_{20}.

We have $a = 24$ and $r = \frac{1}{2}$.

(a) The easiest way to find S_4 is simply to add the first four terms, which gives us 45.
   It is slightly more convenient to use formula (G2) for (b) and (c).

(b) S_{10} is given by

$$S_{10} = \frac{24(1 - (\frac{1}{2})^{10})}{1 - \frac{1}{2}} = 47.953125.$$

(c) Similarly,

$$S_{20} = \frac{24(1 - (\frac{1}{2})^{20})}{1 - \frac{1}{2}} = 47.99995422.$$

We notice here that the difference between the sum of the first four terms and the first ten terms is small. The difference between the sum of the first ten terms and the first twenty terms is very small indeed.

We can see why this is so if we look at the sum of n terms. We have

$$S_n = \frac{24(1 - (\frac{1}{2})^n)}{1 - \frac{1}{2}} = 48(1 - (\frac{1}{2})^n).$$

As n becomes larger and larger, $(\frac{1}{2})^n$ will become smaller and smaller. In fact, by taking a sufficiently large value of n, we can make the value of $(\frac{1}{2})^n$ become as close to zero as we please, although it will never equal zero.

> We can write this mathematically by saying $\lim_{n \to \infty} (\frac{1}{2})^n = 0$.

This means that the limiting value of $(\frac{1}{2})^n$, as n tends to infinity, is zero. The symbol ∞ represents infinity, a boundlessly huge amount.

Since $(\frac{1}{2})^n \to 0$ as $n \to \infty$, we see that the sum to which the series is approaching, is 48.

We call this the **sum to infinity**, and write it as S_∞.

The same kind of thing will happen with *any r* which lies between -1 and $+1$.

The example which we have just looked at could be demonstrated by what happens if you start with a piece of string 48 centimetres long and cut it in half. Lay down the stretched out left-hand piece, and have the right-hand piece. Continue with this process, each time laying the new left-hand piece end to end with the previous pieces, and halving the right-hand piece. The lengths which you have joined end to end are the same as the numbers in the sequence, and your infinite process (mathematicians have no problem in halving infinitely tiny bits of string) brings you closer and closer to your original 48 centimetres of string.

Another way of explaining what conditions r must fit in order for us to have a sum to infinity is to say that we must have $|r| < 1$ where $|r|$ means the **absolute value** of r. This is the value of r taken as positive, whatever the value of r itself, so for example, $|\frac{1}{2}| = \frac{1}{2}$ but $|-3| = 3$. $|r| < 1$ means the same as $-1 < r < +1$.

The sum to infinity of a GP

If $|r| < 1$ and $S_n = \dfrac{a(1 - r^n)}{1 - r}$ then $S_\infty = \dfrac{a}{1 - r}$.

This sum to infinity only exists if $|r| < 1$, so that the values of r^n actually *do* become smaller, as n becomes larger.

For example, if we have the sequence 2, 6, 18, 54, . . . so $a = 2$ and $r = 3$, and we say that

$$S_\infty = 2 + 6 + 18 + 54 + \ldots = \frac{2}{1 - 3} = -1$$

it is clearly absolute nonsense. (It must be, because now r^n is getting larger and larger.)

6.C.(d)   What do 'convergent' and 'divergent' mean?

A series whose sum becomes closer and closer to a definite finite value, S_∞, as we take a larger and larger number of terms, is called **convergent**.

For a convergent series, it must be possible to make the difference $S_n - S_\infty$ as small as we please, by taking a large enough value of n.

If a series is not convergent, then it is called **divergent**.

An AP is always divergent. However tiny we make each individual step, we can always add together enough terms to get an absolute total which is larger than any number we are challenged with, because each step is equal in size.

The different sums that we can find by taking different values of n are called **partial sums**. For example, if we have the series $1 + 2 + 4 + 8 + 16 + \ldots$, then $S_1 = 1$, $S_2 = 1 + 2 = 3$, $S_5 = 1 + 2 + 4 + 8 + 16 = 31$ and each of these are partial sums.

6.C.(e) **More examples using GPs; chain letters**

The following three examples also use GPs.

(1) How many terms of the GP $1 + 2 + 4 + 8 + \ldots$ are required for the sum to be greater than one million?

(2) The third term of a GP is 72, and the sixth term is 243. Find the first term.

(3) The numbers $n + 1$, $n + 5$, and $2n + 4$ are consecutive terms in a GP. (Consecutive terms are terms which come immediately after each other in order.) Find the possible values of n, and of the common ratio. Find also the values of the three given terms in each case.

Have a go at these yourself before looking at what I have done.

Here are my answers.

(1) We have $1 + 2 + 4 + 8 + \ldots$

Suppose we let n be the first number for which $S_n > 1\,000\,000$.

$$a = 1 \quad \text{and} \quad r = 2 \quad \text{so} \quad S_n = \frac{1(2^n - 1)}{2 - 1} = 2^n - 1.$$

$$2^n - 1 > 1\,000\,000 \quad \text{so} \quad 2^n > 1\,000\,001.$$

Taking logs to base 10 both sides, we have

$$\log_{10}(2^n) > \log_{10}(1\,000\,001).$$

Using the third law of logs from Section 3.C.(d), we have

$$n\log_{10}(2) > \log_{10}(1\,000\,001) \quad \text{so} \quad n > \frac{\log_{10}(1\,000\,001)}{\log_{10}(2)}.$$

Therefore $n > 19.93$ to 2 d.p.

The first whole number for which this is true is 20, so $n = 20$.

This series appears in the story of the slave who was offered a reward by a grateful King. Spurning gold, he asked for wheat to be placed on a chess-board, with one grain for the first square, two for the second, and the number of grains doubled for each subsequent square. We have seen that there were already over a million grains by the 20th square. For the 64th square, he had $2^{64} - 1$ grains. This is a seriously large number. If each grain is $\frac{1}{4}$ cm long, and they are placed end to end, they stretch more than one million times round the equator.

Chain letters do not work for the same reason. Suppose you receive a chain letter asking you to post £1 to the sender, and then send off two identical letters yourself. In theory, you end up £1 better off, but, in practice, this is exactly the same situation as the grains of wheat. By the twentieth step in the chain, even with the number of letters only doubling each time, over a million people are involved, and clearly the system must break down. The more letters there are in each step of the chain, the sooner it breaks down. The only people who will safely make money are those near the beginning of the chain. For them, the larger the number of letters the better they do. The system is, in effect, a confidence trick.

(2)   The third term of the GP is 72 so $ar^2 = 72$.

The sixth term is -243 so $ar^5 = -243$. Dividing, we get

$$\frac{ar^5}{ar^2} = -\frac{243}{72}.$$

Because GPs are formed by continued multiplication, dividing is often a technique which works well.

Cancelling down gives us $r^3 = -3.375$.

This can be solved on a calculator by finding the cube root of $+3.375$, by using the '$x^{1/y}$' key.

This gives 1.5, so the cube root of -3.375 is -1.5.

Now, $72 = a(-1.5)^2$, so $a = 32$.

(3)   The ratio from dividing consecutive terms of a GP is constant, so

$$\frac{n+5}{n+1} = \frac{2n+4}{n+5} = \text{the common ratio, } r, \text{ of the series.}$$

We have

$$(n+5)(n+5) = (n+1)(2n+4)$$

so $n^2 + 10n + 25 = 2n^2 + 6n + 4$

which gives

$$n^2 - 4n - 21 = 0.$$

Factorising this, we get

$$(n-7)\,(n+3) = 0 \quad \text{so} \quad n = 7 \quad \text{or} \quad n = -3.$$

Both of these answers are possible.

We substitute back each in turn into $(n+5)/(n+1)$ to find the common ratio.

If $n = 7$, the common ratio is $\frac{12}{8} = \frac{3}{2}$, and the three terms are 8, 12 and 18.

If $n = -3$, the common ratio is $-\frac{2}{2} = -1$, and the three terms are -2, 2 and -2.

6.C.(f)   **A summary of the results for GPs**
• We write GPs as a, ar, ar^2, ..., where r is called the common ratio.
• The nth term is ar^{n-1}.
• The sum of n terms is given by

$$S_n = \frac{a(r^n - 1)}{r - 1} \qquad \text{(best used if } |r| \text{ is greater than 1)} \qquad \text{(G1)}$$

or

$$S_n = \frac{a(1 - r^n)}{1 - r} \qquad \text{(best used if } |r| \text{ is less than 1).} \qquad \text{(G2)}$$

- If $|r| < 1$, then

$$S_\infty = \frac{a}{1 - r} \qquad\qquad \text{(G3)}$$

- $|r| < 1$ means the same thing as $-1 < r < +1$.

EXERCISE 6.C.1

This exercise introduces some very important ideas, so you should do it now as I shall use your answers straight away to show you how things work. Don't be tempted just to look at mine – thinking about your own answers makes an infinite difference to how much you learn.

(1) Which of the following GPs are convergent? If they are convergent, find the sum to infinity in each case.

    (a) $12 + 18 + 27 + \ldots$ (b) $18 + 12 + 8 + \ldots$

    (c) $64 - 48 + 36 - 27 + \ldots$ (d) $16 - 40 + 100 - 250 + \ldots$

    (e) $1 - 1 + 1 - 1 + 1 - 1 + \ldots$ (f) $1 - \frac{1}{2} + \frac{1}{4} - \frac{1}{8} + \frac{1}{16} + \ldots$

(2) The sum of the first two terms of a GP is 30, and the sum of the second and third terms is 20. Find the first term and the common ratio.

(3) The numbers $n + 3$, $3n - 3$, and $5n + 3$ are consecutive terms of a GP. Find the possible values of n and of the common ratio. Find also the values of the three given terms in each case.

(4) (a) Which is the first term of the GP $3 + 12 + 48 + \ldots$ to be greater than $1\,000\,000$?

    (b) How many terms of this GP are required in order to make a **sum** which is greater than 10^{10}?

These are the answers which I hope you will have found.

(1)   (a) $r = \frac{3}{2}$ so $|r| > 1$ and the series is not convergent. In fact, we can easily see that the sums will increase rapidly.

  (b) $r = \frac{2}{3}$ so $|r| < 1$ and the series is convergent.

$$S_\infty = \frac{18}{1 - \frac{2}{3}} = 54.$$

  (c) $r = -\frac{3}{4}$ so $|r| = \frac{3}{4} < 1$ and the series is convergent.

$$S_\infty = \frac{64}{1 - (-\frac{3}{4})} = \frac{256}{7} = 36\frac{4}{7}.$$

  (d) $r = -\frac{5}{2}$ so $|r| > 1$ and the series is not convergent.

  (e) $r = -1$ so $|r| \not< 1$. The symbol '$\not<$' means 'is not less than'. The series is not convergent.

    In fact, a very curious thing happens with (e).

    Normally, if we are adding a string of numbers, we can add them in any order that we please, so for example

$$1 + 2 + 5 + 18 + 24 = (1 + 2) + (5 + 18) + 24 = (1 + 2 + 5) + (18 + 24) \text{ etc.}$$

Here, if we put in brackets to group the terms, we get a very odd result.

    It would appear that it is possible to say

$$S_\infty = (1 - 1) + (1 - 1) + (1 - 1) + \ldots = 0.$$

Also, it would seem reasonable to say

$$S_\infty = 1 + (-1 + 1) + (-1 + 1) + (-1 + 1) + \ldots = 1.$$

Clearly, something is going wrong here.

The fault in the argument is that, by taking the sum to infinity, we are implicitly assuming that the sum of this series is going to get closer and closer to a definite number the further we go. Here, this is not at all true. In fact, if we take an even number of terms the sum is zero, and if we take an odd number of terms the sum is 1, and there is a continual flip-flop between the two. The sum to infinity does not exist and the series is divergent.

At the time when mathematicians were first working on the theory of infinite series, around the beginning of the nineteenth century, this kind of result caused considerable consternation, followed by a big jump forwards in understanding. It is often the cases which behave in peculiar ways which lead to advances in maths, because they make it necessary to look in more detail at what is actually going on. Situations like the one above make it evident that everything is not always as it seems, and that it can be dangerous to jump too soon to conclusions.

It *is* true that we can group together the terms in any way we please in any finite sum of numbers. Also, if all the terms are positive, we can group the terms in any convenient way in an infinite series, because each next term is just another step up in the staircase. Putting some steps together into a larger step will make no difference to the total height of the staircase, whether this height is infinite or not.

(f) Here, $r = -\frac{1}{2}$ so $|r| < 1$ and the series is convergent.

$$S_\infty = \frac{1}{1 + \frac{1}{2}} = \frac{2}{3}.$$

If we calculate some partial sums, that is, sums of different numbers of terms, we find that they are alternately larger and smaller than $\frac{2}{3}$, but getting closer and closer to this value the more terms of the series we take. (Try this for yourself, using a calculator.) By taking a sufficiently large number of terms, we can get as close to $\frac{2}{3}$ as we please. Furthermore, and importantly, any greater number of terms will bring us even closer to $\frac{2}{3}$.

(2) Writing the given information mathematically, we have

$$\begin{cases} a + ar = 30 & (1) \\ ar + ar^2 = 20 & (2) \end{cases}$$

These equations can be solved rather neatly in the following way. Instead of writing equation (2) in the obvious factorisation of $ar(1 + r) = 20$, we write it as $r(a + ar) = 20$. We do this because the $(a + ar)$ exactly matches up with the first equation.

Now we can substitute in this new equation, using equation (1), and we get $30r = 20$ so $r = \frac{2}{3}$. Then, since $a(1 + r) = 30$, $a = 18$.

(3) The ratio of successive terms of a GP is the same, so

$$\frac{3n - 3}{n + 3} = \frac{5n + 3}{3n - 3} = \text{the common ratio.}$$

So

$$9n^2 - 18n + 9 = 5n^2 + 18n + 9.$$

$$4n^2 - 36n = 0 \quad \text{so, factorising, we have}$$

$$4n(n - 9) = 0 \quad \text{so} \quad n = 0 \quad \text{or} \quad 9.$$

If $n = 0$, we get $r = -1$ and the three terms of the series are 3, −3, 3.
If $n = 9$, $r = \frac{24}{12} = 2$ and the three terms are 12, 24 and 48.

(4)   Here, $a = 3$ and $r = 4$.
    (a) Let n be the first number for which u_n is greater than 1 000 000. Then

$$u_n = 3(4)^{n-1} > 1\,000\,000 \quad \text{so} \quad 4^{n-1} > \frac{1\,000\,000}{3}.$$

Taking logs, we have

$$\log_{10} (4^{n-1}) > \log_{10} \left(\frac{1\,000\,000}{3} \right).$$

Now, using the third law of logs, we get

$$(n - 1) \log_{10} (4) > \log_{10} \left(\frac{1\,000\,000}{3} \right)$$

from which $n - 1 > 9.17$ to 2 d.p. So the first possible integer value of n is 11.

    (b) Now let n be the first integer such that $S_n > 10^{10}$.

In the first part of this question, we are looking for the first **term** which is larger than some given value. In the second part, we are looking at the size of the **sum** of all the terms up to that point. Students quite often mix up these two different situations.

We have

$$\frac{3(4^n - 1)}{4 - 1} > 10^{10} \quad \text{so} \quad 4^n > 10^{10} + 1.$$

Taking logs, and using the third law, we have

$$n\log_{10} (4) > \log_{10} (10^{10} + 1)$$

so $n > 16.6$ to 1 d.p. The first possible integer value of n is 17.

6.C.(g)   Recurring decimals, and writing them as fractions

We come next to some applications of GPs.

The first of these gives us a way to convert some decimals to fractions. The strength of the decimal system for writing fractions is that it uses the same system of place values based on powers of 10 as our system of whole numbers uses. This means that decimal fractions are particularly easy to add and subtract and multiply, in just the same way that whole number calculations are straightforward with our number system. If you've ever tried adding or subtracting with Roman numerals, you will appreciate this.

Here are some examples of the place values.

$$0.3 \text{ means } \frac{3}{10}, \quad 0.47 \text{ means } \frac{4}{10} + \frac{7}{100} = \frac{47}{100},$$

$$\text{and } 0.108 \text{ means } \frac{1}{10} + \frac{0}{100} + \frac{8}{1000} = \frac{108}{1000}.$$

(In general, we simply put a zero underneath for every digit on the top.)

 Don't be tempted to say that $\frac{1}{8}$, for example, is 0.8!
In fact, to write $\frac{1}{8}$ as a decimal, we divide the bottom into the top and our number system automatically takes care of the rest so $\frac{1}{8} = 0.125$.

A single-digit repeating decimal, like $\frac{1}{3} = 0.333\ldots$ is written as $0.\dot{3}$.

In a similar way, $\frac{1}{11} = 0.090909\ldots = 0.\overline{09}$, where the line signifies that these two digits are repeated.

Both of these examples are called **recurring** decimals, because the same group of digits is repeated infinitely.

What happens if we want to convert a recurring decimal into fraction form?

For example, suppose we have $0.17171717\ldots$ or $0.\overline{17}$.

It is no use trying to use our rule of zeros underneath for each digit, as this gives us a fraction with an infinitely long top and bottom.

Instead, we use exactly the same device which we used to find the sum of a GP. In other words, we multiply by a number which slides everything along so that it exactly slots for a subtraction to work. Suppose we let

$$F = 0.171717\ldots$$

Then

$$100F = 17.171717\ldots$$

and, subtracting, we get

$$99F = 17 \quad \text{so} \quad F = \frac{17}{99}.$$

You can check this result on your calculator, allowing for the fact that, as it gives a limited number of decimal places, it will round the last digit.

The reason that the same technique works so well is that $0.171717\ldots$ *is* a GP. We can see this by writing it as

$$0.\overline{17} = 0.17171717\ldots$$

$$= \left(\frac{1}{100}\right)(17) + \left(\frac{1}{100}\right)^2 (17) + \left(\frac{1}{100}\right)^3 (17) + \ldots$$

We have

$$a = \frac{17}{100} \quad \text{and} \quad r = \frac{1}{100}.$$

$|r| < 1$, so the sum to infinity of this series exists.

$$S_\infty = \frac{a}{1-r} = \frac{\frac{17}{100}}{1 - \frac{1}{100}} = \frac{17}{100-1} = \frac{17}{99}$$

which agrees with our previous result.

Here is another example. Find in fraction form

$$12.4125125125\ldots \quad \text{or} \quad 12.4\overline{125}.$$

What do you think we should multiply by this time in order to slot everything into the optimum position?

It will need to be 1000. (It is the number of digits which are *repeated* which is important here.)

If we let $F = 12.4\overline{125}$, then we have

$$1000F = 12412.5125125\ldots$$

$$F = 12.4125125\ldots$$

Subtracting, we have $999F = 12400.1$, so

$$F = \frac{12400.1}{999} = \frac{124001}{9990}$$

multiplying top and bottom of this fraction by 10, to tidy it up.

EXERCISE 6.C.2

Try converting the following decimals to fractions yourself.

(1) 0.7 (2) 0.25 (3) 0.401 (4) $0.0\overline{11}$ (5) $0.\dot{7}$

(6) $0.\overline{29}$ (7) $2.5\overline{34}$ (8) $40.2\overline{106}$ (9) $0.\overline{142857}$

6.C.(h) Compound interest: a faster way of getting rich

Another application of GPs is in calculating **compound interest**. If money is invested to obtain compound interest, this means that, in each successive period (usually a year or six months), you not only receive money on the original amount invested (the **principal**) but also on the accumulated interest so far obtained.

With **simple interest**, on the other hand, you receive only the interest on the original capital or principal.

EXAMPLE (1) James invests £800 at 5% compound interest per annum (year). How much money has he at the end of six years?

Compare this with what he would have received if his money was invested at 5% per annum simple interest.

We will look at how much he gets with simple interest first.
At the end of the first year, he receives 5% extra, so he gets

$$\frac{5}{100} \times £800 = £40 \text{ extra.}$$

Exactly the same thing happens in the other five years since he receives no extra interest on his accumulating interest. So at the end of six years he will have

$$£800 + 6 \times £40 = £1040.$$

Under the compound interest system, the result at the end of the first year is unchanged. Writing what happens in detail, we see that he has

$$£800 + \frac{5}{100}(£800) = \left(\frac{105}{100}\right)(£800) = (1.05)(£800) = £840.$$

Now the difference in the two systems starts to show because the interest for the second year is calculated from the total amount of money he now has.

At the end of the second year, he has

$$(1.05)\text{ (the amount now there)} = ((1.05)(1.05)(£800)) = (1.05)^2\,£800.$$

So, at the end of six years, he has $(1.05)^6\,£800 = £1072.08$ to the nearest penny.

We see that he is £32.08 better off with the compound interest.

When James is on a system of simple interest, the steps of his increases form an AP with 'a' = 800 and 'd' = 0.05 × 800 = 40.

When he is on a system of compound interest, the steps of his increases form a GP with 'a' = 800 and 'r' = 1.05.

How much money does James have in total after n years?

If the money was invested at 5% simple interest, he will have $n \times (0.05 \times £800)$ in accumulated interest, giving him a total of $£800 + 0.05n(£800)$.

If his money was invested at 5% compound interest, he would have $(1.05)^n\,£800$ altogether.

Notice that these two formulas give us practical examples of working sequences.

The sequence for his totals with simple interest over periods of a year, in £ units, is the AP which goes:

$$800,\ 840,\ 880,\ 920,\ \ldots,\ [800 + (n-1)(0.05 \times 800)],\ \ldots$$

The nth term of this AP is $800 + (n-1)(0.05 \times 800)$.

This can also be written as a **recurrence relation** or **difference equation**, using the method of description (B) from Section 6.A.(b). We would write

$$u_n = u_{n-1} + (0.05 \times 800) = u_{n-1} + 40 \quad \text{with} \quad u_1 = 800.$$

The sequence for his totals with compound interest form the GP

$$800,\ 840,\ 882,\ 926.10,\ \ldots,\ (1.05)^{n-1}\,800,\ \ldots$$

with $(1.05)^{n-1}\,800$ as its nth term.

It can also be written as a difference equation in the form

$$u_n = (1.05)u_{n-1} \quad \text{with} \quad u_1 = 800.$$

What if James invests the same amount each year with compound interest?

Suppose that he was able to invest £800 at the beginning of each of the six years at the same rate of compound interest of 5%. How much would he have altogether on 2 January of the seventh year, when he has just deposited his most recent £800?

He would have

$$£800 + (1.05)£800 + (1.05)^2\,£800 + \ldots + (1.05)^6\,£800$$

which is a GP with $a = £800$, $r = 1.05$, and $n = 7$. So his total investment is

$$S_7 = \frac{800\,((1.05)^7 - 1)}{1.05 - 1} = £6513.61.$$

6.C.(i)   The geometric mean

We have already seen that the arithmetic mean, A, of two numbers, a and b, is defined as the number A such that a, A and b form an arithmetic progression.

In a similar way, we define the **geometric mean** G, of two positive numbers a and b, to be the number such that a, G, b are in geometric progression.

So a, G, b can also be written as a, ar, ar^2 giving $G = ar$ and $b = ar^2$. Now

$$ab = a(ar^2) = a^2r^2 = G^2 \quad \text{so} \quad G = \sqrt{ab}.$$

For example, suppose we have the pair of numbers 2 and 8.

The **arithmetic mean** of these two numbers is the midway point of 5 (Section 6.B.(c)). This then gives a mini AP of 2, 5, 8 with a common difference of 3.

The **geometric mean** of these two numbers is 4, given by $\sqrt{2 \times 8}$, resulting in the mini GP of 2, 4, 8 with common ratio 2.

The definition of the geometric mean can also be extended to n numbers, provided that they are positive, in the following way.

If the numbers are a_1, a_2, a_3, . . ., a_n then the geometric mean is $\sqrt[n]{a_1 a_2 a_3 \ldots a_n}$.

6.C.(j)   Comparing arithmetic and geometric means

We can also show that the arithmetic mean of any two positive numbers a and b is greater than their geometric mean. We have to show that

$$\frac{a + b}{2} \geq \sqrt{ab}.$$

This can be done rather neatly by putting $a = x^2$ and $b = y^2$. Since we have said that a and b are positive, this is a safe move, and it gets rid of the $\sqrt{\ }$ sign. We now have to show that

$$\frac{x^2 + y^2}{2} \geq xy.$$

Can you see how the rest of the argument will go?

We must show that $x^2 + y^2 \geq 2xy$.

So we must show that $x^2 + y^2 - 2xy \geq 0$, that is, that $(x - y)^2 \geq 0$. But $(x - y)^2$ must be either positive (or zero, if $x = y$), since it is something squared. Therefore $A \geq G$.

6.C.(k)   What is the fate of the frog down the well?

I will finish this section by asking you the following question.

A frog is at the bottom of a well. He finds that he can jump up the side of the well, hanging on briefly between jumps. This procedure is exhausting so he jumps a shorter distance each time, starting with 1 m then $\frac{1}{2}$ m, $\frac{1}{3}$ m, and so on, so that the total height he has reached after n jumps is given by

$$1 + \frac{1}{2} + \frac{1}{3} + \frac{1}{4} + \ldots + \frac{1}{n} \text{ metres.}$$

Obviously, if the well is only 2 metres deep, he will have escaped by his fourth jump. How deep must the well be for him never to escape, or will he always gain his freedom?

It is worth testing your ideas here numerically in any way you can.

You could sum as many terms as you have the patience for on a calculator to get some idea of what is happening.

Even better, if you can write computer programs, you could test any particular depth which you might think would definitely spell the frog's doom, by seeing if there is some number of jumps whose sum would actually come to more than this depth, so that he *does* escape. (I shall return to this puzzle later on in this chapter.)

6.D   A compact way of writing sums: the Σ notation

6.D.(a)   What does Σ stand for?

We have looked fairly thoroughly at APs and GPs because they are relatively easy to sum, and also come up quite often in practical situations. Now we will widen the field by looking at some other kinds of series.

To make this easier, I will show you a neat new method of writing the sum of a series. It is called the Σ notation, from the Greek capital letter S which is written Σ, and pronounced 'sigma'.

$$\text{To write } 1 + \frac{1}{2} + \frac{1}{3} + \ldots + \frac{1}{n} \text{ in this notation, we write } \sum_{r=1}^{n} \frac{1}{r}.$$

What we have done is to write down the sum using the general term of the series. The value of r at the bottom of the Σ gives the first term, and the value (of r) at the top of the Σ gives the last term. You can think of this Σ as meaning 'The sum of all such things as $1/r$ with r going from 1 to n'.

> **NOTE** The letters used need not necessarily be r and n but the general idea will be the same.

Here is another example, which uses n as the letter inside the Σ.

$$\sum_{n=1}^{10} n = 1 + 2 + 3 + \ldots + 10.$$

The r in the first example and the n in the second example are dummy variables with the information about how far they run being written at the bottom and the top of the Σ. Once this information has been filled in, the answer will be purely numerical, and it won't matter what letter we chose to use.

This is similar to some computer programming where, to get repeated steps, you put $n = n + 1$, meaning that you are moving to the next n. The computer then automatically works with this next integer in its next procedure.

EXERCISE 6.D.1

Try writing the following in Σ notation for yourself.

(1) $1 + 4 + 9 + 16 + \ldots + 81$

(2) $\frac{1}{2} + \frac{2}{3} + \frac{3}{4} + \ldots + \frac{11}{12}$

(3) $\frac{1}{1 \times 2} + \frac{1}{2 \times 3} + \frac{1}{3 \times 4} + \ldots + \frac{1}{29 \times 30}$

(4) $-1 + 4 - 9 + 16 - 25 + \ldots - 81$ Be ingenious!

Unpacking the Σs

It will be quite useful for you to get some practice here in unpacking the Σ notation into the separate numerical terms, as sometimes it is necessary to convert back in this way.

Here is an example of this.

Write down the first four terms, and also the nth term and the $(n + 1)$th term, of the series

$$\sum_{r=1}^{n} \frac{1}{r(r + 1)(2r + 1)}.$$

The first four terms are

$$\frac{1}{1(2)(3)} + \frac{1}{2(3)(5)} + \frac{1}{3(4)(7)} + \frac{1}{4(5)(9)}$$

feeding in $r = 1, 2, 3, 4$ in turn. Tidying up, we get

$$\frac{1}{6} + \frac{1}{30} + \frac{1}{84} + \frac{1}{180} = \frac{137}{630}$$

The nth term is

$$\frac{1}{n(n + 1)(2n + 1)}, \quad \text{putting } r = n.$$

For the $(n + 1)$th term, we put $r = n + 1$, and get

$$\frac{1}{(n + 1)(n + 2)\,(2(n + 1) + 1)} = \frac{1}{(n + 1)(n + 2)(2n + 3)}.$$

Students sometimes find this last procedure a bit tricky, but it is well worth practising it now because you will need it if you have to work with more complicated series.

EXERCISE 6.D.2

For each of the following series, write down the first four terms, and then add them together. Also, write down the nth term and the $(n + 1)$th term.

(1) $\sum_{r=1}^{n} (2r + 3)$
  (2) $\sum_{r=1}^{n} 36\left(\frac{1}{3}\right)^{r-1}$
  (3) $\sum_{r=1}^{n} \frac{1}{r!}$

(4) $\sum_{r=1}^{n} \left(\frac{r}{r + 2}\right)(-1)^{r+1}$
  (5) $\sum_{r=1}^{n} \frac{1}{(2r - 1)\,(2r + 1)}$

6.D.(c)   **Summing by breaking down to simpler series**

Sometimes it is possible to sum series by breaking them down into simpler series which have known sums. I will give you some examples of this, using the following three standard sums.

$$1 + 2 + 3 + 4 + \ldots + n = \sum_{r=1}^{n} r = \tfrac{1}{2}n(n + 1) \qquad \text{(S1)}$$

$$1^2 + 2^2 + 3^2 + 4^2 + \ldots + n^2 = \sum_{r=1}^{n} r^2 = \tfrac{1}{6}n(n + 1)(2n + 1) \quad \text{(S2)}$$

$$1^3 + 2^3 + 3^3 + 4^3 + \ldots + n^3 = \sum_{r=1}^{n} r^3 = \tfrac{1}{4}n^2(n + 1)^2 \qquad \text{(S3)}$$

(If not knowing where these have come from worries you, we showed the first one when we did APs in question 2(d) of Exercise 6.B.1. The other two are shown to be true in the next chapter in Section 7.D.)

Here is an example of how they can be used.

Find $\displaystyle\sum_{r=1}^{n} (r+1)(r+2)$.

$$\sum_{r=1}^{n} (r+1)(r+2) = \sum_{r=1}^{n} (r^2 + 3r + 2).$$

This can then be split into separate sums since it makes no difference what order we do the adding in. We say

$$\sum_{r=1}^{n} (r^2 + 3r + 2) = \sum_{r=1}^{n} r^2 + \sum_{r=1}^{n} 3r + \sum_{r=1}^{n} 2.$$

Also,

$$\sum_{r=1}^{n} 3r = 3 \sum_{r=1}^{n} r$$

since multiplying each separate number by 3, and then adding, is the same as adding first and then multiplying the total by 3.

You can see all this actually working if I put $n = 3$.

$$\sum_{r=1}^{3} (r^2 + 3r + 2) = \sum_{r=1}^{3} r^2 + 3 \sum_{r=1}^{3} r + \sum_{r=1}^{3} 2.$$

The LHS of this is $(1^2 + 3 + 2) + (2^2 + 6 + 2) + (3^2 + 9 + 2) = 38$.
The RHS of this is $(1^2 + 2^2 + 3^2) + 3(1 + 2 + 3) + (2 + 2 + 2) = 38$.

 Notice $\displaystyle\sum_{r=1}^{3} 2$ is $2 + 2 + 2$ and not just 2. The 2 is being added in three times.

So we have

$$\sum_{r=1}^{n} (r+1)(r+2) = \sum_{r=1}^{n} r^2 + 3 \sum_{r=1}^{n} r + \sum_{r=1}^{n} 2.$$

Using (S1) and (S2), we find this is the same as

$$\tfrac{1}{6} n(n+1)(2n+1) + 3 \left[\tfrac{1}{2}n(n+1)\right] + 2n.$$

(The 2 is now being added n times.)

Factorising this by taking out $\tfrac{1}{6}n$, we get

$$\tfrac{1}{6}n\left[(n+1)(2n+1) + 9(n+1) + 12\right].$$

(It is good to have the $\tfrac{1}{6}$ out of the way in the front. If you are doubtful about what is inside the bracket, check by multiplying out.) Multiplying out the inside brackets, we have

$$\tfrac{1}{6}n\left[(2n^2 + 3n + 1) + (9n + 9) + 12\right] = \tfrac{1}{6}n(2n^2 + 12n + 22) = \tfrac{1}{3}n(n^2 + 6n + 11)$$

taking out an extra factor of 2, and cancelling. So

$$\sum_{r=1}^{n} (r+1)(r+2) = \tfrac{1}{3}n(n^2 + 6n + 11).$$

Check: If $n = 3$, we have just seen that

$$\text{LHS} = \sum_{r=1}^{3} (r + 1)(r + 2) = 38.$$

Putting $n = 3$ in the answer gives

$$\text{RHS} = \tfrac{1}{3}n(n^2 + 6n + 11) \text{ with } n = 3, \text{ which is } \tfrac{1}{3}(3)(9 + 18 + 11) = 38.$$

EXERCISE 6.D.3 Try these two yourself. Find

$$(1) \sum_{r=1}^{n} (r - 1)(r + 3) \qquad (2) \sum_{r=1}^{n} r(r - 1)(r + 1).$$

In each case, check your answers by putting $n = 3$.

6.E Partial fractions

6.E.(a) Introducing partial fractions for summing series

In the earlier part of this chapter, we found out how to sum APs and GPs. Now we look at a rather ingenious technique which can be used for summing series involving fractions. (This particular technique also has many other uses.)

Suppose we want to find

$$\sum_{r=1}^{n} \frac{1}{r(r + 1)}$$

that is, we want to find

$$\frac{1}{1.2} + \frac{1}{2.3} + \frac{1}{3.4} + \frac{1}{4.5} + \ldots + \frac{1}{n(n + 1)} = \frac{1}{2} + \frac{1}{6} + \frac{1}{12} + \frac{1}{20} + \ldots + \frac{1}{n(n + 1)}.$$

As it stands, there is no simple way of calculating this sum.

However, the fraction $\frac{1}{r(r + 1)}$ looks as if it has come from putting two simpler fractions into one single fraction, as we did in Section 1.C.(c). Suppose we try writing

$$\frac{1}{r(r + 1)} \equiv \frac{A}{r} + \frac{B}{r + 1}$$

where A and B are standing for numbers which we would need to find out. I've used the '$\equiv$' sign here to emphasise that the two sides are just different ways of writing the same thing. What we have here is another example of an **identity**. I explained what this means in Section 2.D.(h).

To find A and B, we get rid of fractions by multiplying through by $r(r + 1)$.
Cancelling where possible, we get $1 \equiv A(r + 1) + Br$.
Since this is just a rewriting, or identity, it must be true for all values of r.
Putting $r = 0$, we get $1 = A$.
Putting $r = -1$, we get $1 = -B$, so $B = -1$.
We can check by putting $r = 1$, say. With these values of A and B, we get the LHS = 1, and the RHS = $2 - 1 = 1$ also.
We now know that we can replace

$$\frac{1}{r(r + 1)} \quad \text{by} \quad \frac{1}{r} - \frac{1}{r + 1}.$$

Will this help us? We can say

$$\sum_{r=1}^{n} \frac{1}{r(r+1)} \equiv \sum_{r=1}^{n} \left(\frac{1}{r} - \frac{1}{r+1}\right)$$

$$= \sum_{r=1}^{n} \left(\frac{1}{r}\right) - \sum_{r=1}^{n} \left(\frac{1}{r+1}\right)$$

$$= \left(1 + \frac{1}{2} + \frac{1}{3} + \frac{1}{4} + \ldots + \frac{1}{n}\right)$$

$$- \left(\frac{1}{2} + \frac{1}{3} + \frac{1}{4} + \ldots + \frac{1}{n} + \frac{1}{n+1}\right),$$

and we see that it does indeed help us.

The second bracket is almost exactly the same as the first bracket. It has the same number of terms, but everything has been slid one place to the right.

When we do the subtraction, we are left with just $1 - 1/(n+1)$ so

$$\sum_{r=1}^{n} \frac{1}{r(r+1)} = 1 - \frac{1}{n+1}.$$

You can check that this actually works by putting $n = 2$. This gives a LHS of $\frac{1}{2} + \frac{1}{6}$ and a RHS of $1 - \frac{1}{3}$, so the two sides do come out the same.

What will happen as n becomes very large? Will this series have a sum to infinity? In other words, is it convergent?

The larger n gets, the closer $1/(n+1)$ becomes to zero, so the sum of the series will get closer and closer to 1.

The series *is* convergent, with a sum to infinity of 1. We can say

$$\sum_{r=1}^{\infty} \frac{1}{r(r+1)} = 1.$$

Now have a go at using the same method yourself to find the sum of the series

$$\frac{2}{3} + \frac{2}{8} + \frac{2}{15} + \frac{2}{24} + \ldots + \frac{2}{n(n+2)} = \sum_{r=1}^{n} \frac{2}{r(r+2)}.$$

Check how you got on.

$\dfrac{2}{r(r+2)}$ can be split up into two simpler fractions as $\dfrac{A}{r} + \dfrac{B}{r+2}$.

Then, multiplying by $r(r+2)$ to get rid of fractions, we have

$$2 \equiv A(r+2) + Br.$$

Putting $r = -2$ gives $2 = -2B$, so $B = -1$.
Putting $r = 0$ gives $2 = 2A$, so $A = 1$.
Checking, by putting $r = 1$, we have the LHS = 2 and the RHS = 3 − 1 = 2.

We can therefore say

$$\frac{2}{r(r+2)} \equiv \frac{1}{r} - \frac{1}{r+2},$$

and we now have

$$\sum_{r=1}^{n} \frac{2}{r(r+2)} \equiv \sum_{r=1}^{n} \frac{1}{r} - \sum_{r=1}^{n} \frac{1}{r+2}$$

$$= \left(1 + \frac{1}{2} + \frac{1}{3} + \frac{1}{4} + \ldots + \frac{1}{n}\right)$$

$$- \left(\frac{1}{3} + \frac{1}{4} + \ldots + \frac{1}{n} + \frac{1}{n+1} + \frac{1}{n+2}\right).$$

(The last three terms in the second bracket come from putting $r = n - 2$, $n - 1$, and n respectively.)

This time, it is as though the right-hand bracket has been slid along two places instead of just one, as it was in the previous example.

Subtracting all the overlapping parts, we are left with

$$\sum_{r=1}^{n} \frac{2}{r(r+2)} = \left(1 + \frac{1}{2}\right) - \left(\frac{1}{n+1} + \frac{1}{n+2}\right) = \frac{3}{2} - \frac{1}{n+1} - \frac{1}{n+2}.$$

Both $\frac{1}{n+1}$ and $\frac{1}{n+2}$ will become very small as n becomes large. We can say that

$$\frac{1}{n+1} \to 0 \quad \text{and} \quad \frac{1}{n+2} \to 0 \quad \text{as } n \to \infty$$

so we see that the sum of the series is getting closer and closer to $\frac{3}{2}$.

The series $\sum_{r=1}^{n} \frac{2}{r(r+2)}$ *is* convergent, and its sum to infinity is $\frac{3}{2}$.

The number $\frac{3}{2}$ forms a barrier beyond which the sum cannot go, however many extra terms we add, although we can get as close to it as we please if we take a sufficiently large number of terms. (We never *quite* get there, though! We are always a tiny bit less than it since all the terms of the series are positive.)

6.E.(b) General rules for using partial fractions

When we summed the series

$$\sum_{r=1}^{n} \frac{1}{r(r+1)} \quad \text{and} \quad \sum_{r=1}^{n} \frac{2}{r(r+2)},$$

we split up the complicated fraction into two simpler fractions, in each case.

This technique of rewriting complicated fractions in the form of separate simpler fractions is called the method of **partial fractions**. It is often extremely useful, not only for summing series as we have already used it, but also in integration, as you will see in Section 9.B.(e).

Because it is such an important technique, we shall look at it now in more detail. The two examples which we have already met both had two factors underneath. If the fraction has more factors underneath, it is simply split into more fractions.

So, for example,

$$\frac{6}{(x - 1)(x + 1)(2x + 1)} \quad \text{is written as} \quad \frac{A}{x - 1} + \frac{B}{x + 1} + \frac{C}{2x + 1},$$

where A, B and C are standing for numbers which we have to find.

Getting rid of fractions as before, by multiplying by $(x - 1)(x + 1)(2x + 1)$ and cancelling where possible, we get

$$6 \equiv A(x + 1)(2x + 1) + B(x + 1)(2x + 1) + C(x - 1)(x + 1).$$

Putting $x = 1$ gives $6 = 6A$, so $A = 1$.
Putting $x = -1$ gives $6 = 2B$, so $B = 3$.
Putting $x = -\frac{1}{2}$ gives $6 = -\frac{3}{4}C$, so $C = -8$.

Notice that we cunningly choose values of x so that two parts get knocked out each time, and we can easily find the value of the remaining letter.

Then it is sensible to check the values we have found, by putting $x = 0$, say, with these values, and making sure that the two sides balance.

Here, the LHS = 6, and the RHS = $A - B - C = 1 - 3 + 8 = 6$.

Often, finding the partial fractions is only a small part of the complete problem, so it is wise to check that nothing has gone wrong at this stage.

6.E.(c)   **The cover-up rule**

In a case like the above, it is also possible to find A, B and C by what is known as the **cover-up rule**.

To do this, we choose each of the three values of x in turn which gives a zero in the denominator of

$$\frac{6}{(x - 1)(x + 1)(2x + 1)}$$

(that is, we choose the same three values which we used in the previous working).

Suppose we start with $x = 1$. Then we cover up the bracket $(x - 1)$, and feed $x = 1$ into the rest of the fraction.

This gives $6/6 = 1$ as A, the number over $(x - 1)$.

Similarly, covering up $(x + 1)$, and feeding in $x = -1$ to the rest of the fraction, gives $B = 6/2 = 3$.

Finally, covering up $(2x + 1)$, and feeding in $x = -\frac{1}{2}$ to the rest of the fraction, gives $C = -8$.

You can use whichever method you prefer.

EXERCISE 6.E.1   Use whichever method you find most convenient to write the following as partial fractions.

$$\text{(1)} \quad \frac{4}{(x + 2)(x + 3)} \qquad \text{(2)} \quad \frac{6}{(2y - 1)(2y + 1)} \qquad \text{(3)} \quad \frac{10}{x(x - 1)(x + 4)}$$

6.E.(d)   **Coping with possible complications**

Unfortunately, sometimes complications arise. These can be split into three types and I'll describe each of them in turn.

Repeated factors

Suppose we have the fraction

$$\frac{4}{(x+1)(x-1)^2}.$$

Can we say

$$\frac{4}{(x+1)(x-1)^2} \equiv \frac{A}{x+1} + \frac{B}{(x-1)^2} \; ?$$

We'll see what happens when we try to find A and B.

Getting rid of fractions, we have $4 \equiv A(x-1)^2 + B(x+1)$.

Putting $x = 1$ gives $4 = 2B$ so $B = 2$.

Putting $x = -1$ gives $4 = 4A$ so $A = 1$.

Now check with $x = 0$. The LHS $= 4$ and the RHS $= 1 + 2 = 3$.

Clearly, something has gone wrong!

If we think what fractions we could have put together to give the original fraction then we see that there could have been a hidden one extra to the two which we wrote down above. Can you see what this extra one is?

There could also have been the fraction

$$\frac{C}{x-1}.$$

If we now write

$$\frac{4}{(x+1)(x-1)^2} \equiv \frac{A}{x+1} + \frac{B}{(x-1)^2} + \frac{C}{x-1}$$

and get rid of fractions by multiplying by $(x+1)(x-1)^2$, cancelling where possible, we get

$$4 \equiv A(x-1)^2 + B(x+1) + C(x-1)(x+1). \quad (1)$$

 You need to think carefully here about the cancelling down. If you try to get rid of the fractions on autopilot, you will almost certainly go wrong.

Now, putting $x = 1$ we get $4 = 2B$ so $B = 2$ as before.

Putting $x = -1$ gives us $4 = 4A$ so $A = 1$, also as before.

To find C, we can apply the very useful technique which we employed when we were factorising cubic equations in Section 2.E.(a).

The way to do this is as follows.

Since equation (1) above is an identity, the coefficients of each separate power of x on each side of it must match up. For example, there must be the same number of x^2 terms on each side; this is the only way that (1) can be true for all values of x.

Looking at the terms in x^2, we have $0 = Ax^2 + Cx^2$ so $C = -A$ so $C = -1$.

Now we check again, putting $x = 0$.

This time, the LHS = 4 and the RHS = 1 + 2 + 1 = 4, which is a much better state of affairs. Our final result is

$$\frac{4}{(x+1)(x-1)^2} \equiv \frac{1}{x+1} + \frac{2}{(x-1)^2} - \frac{1}{x-1}.$$

The rule for dealing with repeated factors

If there is a repeated factor underneath, we must put in extra fractions to make up the whole power. For example,

$$\frac{1}{(x+1)(x+3)^3} \equiv \frac{A}{x-1} + \frac{B}{x+3} + \frac{C}{(x+3)^2} + \frac{D}{(x+3)^3}.$$

EXERCISE 6.E.2

Try these two for yourself. Find partial fractions for

(1) $\dfrac{5}{(x-2)(x+3)^2}$,   (2) $\dfrac{2}{y^2(y-1)}$.

Non-linear factors

Suppose we have   (1) $\dfrac{3}{(x+1)(x^2-4)}$ and   (2) $\dfrac{3}{(x+1)(x^2+4)}$.

How could we split up (1) to find its partial fractions?

We could use the difference of two squares (again!) on $x^2 - 4$, and write

$$\frac{3}{(x+1)(x^2-4)} \equiv \frac{3}{(x+1)(x-2)(x+2)} \equiv \frac{A}{x+1} + \frac{B}{x-2} + \frac{C}{x+2}.$$

Finish this for yourself. You should get

$$\frac{3}{(x+1)(x^2-4)} \equiv \frac{-1}{x+1} + \frac{\frac{1}{4}}{x-2} + \frac{\frac{3}{4}}{x+2}.$$

However, when we come to (2), we can't split up $x^2 + 4$ into two linear factors. (A **linear** factor is one like $(x+2)$ where, if we plotted $y = x + 2$, we would get a straight line.)

Now, if we are dividing by $x^2 + 4$, the remainder can have xs in, as well as numbers, so we have to split (2) up into partial fractions as follows:

$$\frac{3}{(x+1)(x^2+4)} \equiv \frac{A}{x+1} + \frac{Bx+C}{x^2+4}.$$

Getting rid of fractions, $3 \equiv A(x^2 + 4) + (Bx + C)(x + 1)$.
Putting $x = -1$ gives $3 = 5A$, so $A = \frac{3}{5}$.
Putting $x = 0$ gives us $3 = 4A + C$, so $C = \frac{3}{5}$.
Matching the terms in x^2 gives us $0 = Ax^2 + Bx^2$, so $B = -A = -\frac{3}{5}$.
Checking with $x = 1$ gives the LHS = 3, and the RHS = 3 + 0 = 3.

So

$$\frac{3}{(x+1)(x^2+4)} \equiv \frac{\frac{3}{5}}{x+1} + \frac{(-\frac{3}{5}x+\frac{3}{5})}{x^2+4} = \frac{3}{5}\left(\frac{1}{x+1} - \frac{x-1}{x^2+4}\right)$$

taking out the factor of $\frac{3}{5}$. Notice carefully the *signs* in the two forms of writing this answer. Remember that the line of the fraction acts as a bracket. (See, if necessary, Section 1.C.(e) on subtracting fractions.)

The rule for dealing with non-linear factors

If one of the factors on the bottom of a fraction has an x^2 term, and this factor won't itself factorise any further, then we need both xs and numbers on the top, like the $Bx + C$ above.

Similarly, if we had a factor underneath with an x^3 term, and this factor wouldn't itself factorise, we would need to have $Ax^2 + Bx + C$ on the top, and so on.

EXERCISE 6.E.3

Try finding partial fractions for

(1) $\dfrac{14}{(x^2+3)(x+2)}$,   (2) $\dfrac{4}{y(y^2+1)}$.

Top-heavy fractions

Consider these four examples.

(1) $\dfrac{x^2+3x-5}{x^2+2x-8}$ (2) $\dfrac{x^2+4x-2}{x^2+5x+6}$ (3) $\dfrac{x^2+1}{x^2-9}$ (4) $\dfrac{x^3+3x^2+2x-3}{(x+2)(x-1)}$

Each of these fractions is top-heavy. By this I mean that the highest power of x on the top is greater than, or equal to, the highest power of x on the bottom.

If we have this situation, it is necessary to *divide* before finding partial fractions for the rest of the expression.

(This division is exactly the same process that we use in writing the fraction $\frac{19}{8}$ as $2\frac{3}{8}$. The arithmetical fraction $\frac{19}{8}$ is top-heavy.)

Fortunately, quite often this dividing can be done without using the full long-division process.

(1)   In this example, we can cunningly rewrite the top of the fraction as follows:

$$\frac{x^2+3x-5}{x^2+2x-8} \equiv \frac{x^2+2x-8+x+3}{x^2+2x-8}.$$

This can then be written as

$$1 + \frac{x+3}{x^2+2x-8}.$$

Now we find partial fractions for

$$\frac{x+3}{x^2+2x-8}.$$

This factorises to

$$\frac{x + 3}{(x + 4)(x - 2)}$$

giving partial fractions of

$$\frac{\frac{1}{6}}{x + 4} + \frac{\frac{5}{6}}{x - 2}.$$

(Check this for yourself.)

The complete solution is then given by

$$\frac{x^2 + 3x - 5}{x^2 + 2x - 8} \equiv 1 + \frac{\frac{1}{6}}{x + 4} + \frac{\frac{5}{6}}{x - 2}.$$

 It's very easy to forget to include the 1 here.

(2)   Can you see how to rewrite the top of the fraction in example (2) to make the division easy?

We can say

$$\frac{x^2 + 4x - 2}{x^2 + 5x + 6} \equiv \frac{x^2 + 5x + 6 - x - 8}{x^2 + 5x + 6}.$$

This can then be written as

$$1 - \frac{x + 8}{x^2 + 5x + 6}.$$

Notice the signs again! The line of the fraction is acting as a bracket.
Now, find partial fractions for

$$\frac{x + 8}{x^2 + 5x + 6}.$$

You should have

$$\frac{x + 8}{x^2 + 5x + 6} = \frac{x + 8}{(x + 3)(x + 2)} \equiv \frac{A}{x + 3} + \frac{B}{x + 2}$$

so $x + 8 \equiv A(x + 2) + B(x + 3)$.

Putting $x = -2$ gives $6 = B$.
Putting $x = -3$ gives us $5 = -A$.
Notice that, in this example, it is necessary to substitute for x on the LHS too.
So the complete solution is

$$\frac{x^2 + 4x - 2}{x^2 + 5x + 6} \equiv 1 - \left(\frac{-5}{x + 3} + \frac{6}{x + 2} \right) \equiv 1 + \frac{5}{x + 3} - \frac{6}{x + 2}.$$

There are two things to remember here: we must include the 1 like last time, and we also have to remember the minus sign in front of the big bracket.

(3) Try doing this example for yourself.

━━━

You should have

$$\frac{x^2 + 1}{x^2 - 9} \equiv \frac{x^2 - 9 + 10}{x^2 - 9}$$

$$= 1 + \frac{10}{x^2 - 9} \equiv 1 + \frac{10}{(x - 3)(x + 3)}.$$

$10/(x - 3)(x + 3)$ can then be easily split into partial fractions, giving a final complete answer of

$$1 + \frac{\frac{5}{3}}{x - 3} - \frac{\frac{5}{3}}{x + 3}.$$

(4) Here, we shall have to have recourse to the full long-division process. I explained how to do this in Section 2.E.(b). We have

$$\frac{x^3 + 3x^2 + 2x - 3}{x^2 + x - 2},$$

so we find

$$
\begin{array}{r}
x + 2 \\
x^2 + x - 2 \overline{) x^3 + 3x^2 + 2x - 3} \\
\underline{x^3 + x^2 - 2x} \\
2x^2 + 4x - 3 \\
\underline{2x^2 + 2x - 4} \\
2x + 1
\end{array}
$$

Since $x^2 + x - 2 = (x + 2)(x - 1)$, we now have

$$\frac{x^3 + 3x^2 + 2x - 3}{(x + 2)(x - 1)} \equiv x + 2 + \frac{2x + 1}{(x + 2)(x - 1)}.$$

You should check for yourself that this comes to

$$\frac{x^3 + 3x^2 + 2x - 3}{x^2 + x - 2} \equiv x + 2 + \frac{1}{x + 2} + \frac{1}{x - 1}$$

remembering to include the $x + 2$ in the final answer.

┌───┐

The rule for dealing with top-heavy fractions

If the fraction is top-heavy, that is, if the highest power of x on the top is greater than or equal to the highest power of x on the bottom, then we must divide out first, and find partial fractions for the remaining fraction.

└───┘

We shan't need to use partial fractions which are as complicated as these for summing series, but you will need them for integration, and you are now set up for dealing with them when this happens.

EXERCISE 6.E.4

The following questions involve a mixture of the complications we have just been looking at. In each case, find suitable partial fractions.

(1) $\dfrac{4}{(x + 3)(x - 1)^2}$

(2) $\dfrac{3p + 1}{(2p - 1)(p + 2)^2}$

(3) $\dfrac{7x - 14}{(2x + 1)(x^2 - 6x + 9)}$

(4) $\dfrac{10y}{(y - 1)(y^2 + 9)}$

(5) $\dfrac{10x}{(x - 1)(x^2 - 9)}$

(6) $\dfrac{r^2 + 1}{r^2 - 1}$

(7) $\dfrac{x^4 + 1}{x^4 - 1}$

(8) $\dfrac{u^2 - 1}{u^2(2u + 1)}$

(9) $\dfrac{x^2 + 1}{(x + 2)(x + 4)}$

(10) (a) Write down the first four terms of the series $\displaystyle\sum_{r=1}^{n} \dfrac{2}{4r^2 - 1}$.

(b) Factorise $4r^2 - 1$ and then use this to find partial fractions for $\dfrac{2}{4r^2 - 1}$.

(c) Now use these to find $\displaystyle\sum_{r=1}^{n} \dfrac{2}{4r^2 - 1}$.

(d) What is the sum to infinity for this series?

6.F   The fate of the frog down the well

In this last section, we return to the series $1 + \frac{1}{2} + \frac{1}{3} + \frac{1}{4} + \ldots$ which describes the attempts of the frog to escape from the well in the thinking point of Section 6.C.(k). What I was really asking you there was whether this series is convergent or divergent. If it is divergent then, however deep the well, the frog will eventually escape. If it is convergent, then it must be possible to find a depth D so that anything deeper than this spells his doom. (D wouldn't necessarily have to be the sum to infinity of the series – this could well be tricky to find. It's like the headroom of a bridge: if a lorry crashes into it we know that anything higher than the lorry certainly won't get through, and we know this without having measured the exact headroom of the bridge.) Even if this series *is* convergent, there will be some depths which the frog can escape from, just like most cars can probably go safely under the bridge.

We know that four jumps are sufficient to escape from a well which is 2 metres deep. Adding up the terms on a calculator, it is quite easy to discover that 31 jumps are sufficient if the well is 4 metres deep. We also know that each individual jump is getting smaller and smaller the more jumps the frog makes.

Is knowing this sufficient for us to say that this series must converge towards some particular sum? (We know from Section 6.C.(c) that it *would* be enough in the case of a GP because, if the terms get smaller, then its common ratio must be less than 1 and therefore it will have a sum to infinity.)

Might it help us here if we find the ratio of successive terms? We can see that, as n becomes large, there will be very little difference between $1/n$ and $1/(n + 1)$, although each of them separately is also becoming very tiny. We can say that

$$\frac{u_{n+1}}{u_n} = \frac{1/(n + 1)}{1/n} = \frac{n}{n + 1} = \frac{1}{1 + 1/n}.$$

(We did this same sort of thing when we were graph-sketching in Section 3.B.(i).)

Now, since $\frac{1}{n}$ becomes closer and closer to zero the larger n becomes, this ratio gets closer and closer to 1. This still leaves us in a bit of a quandary. The terms are getting more and more equal but they are also getting exceedingly tiny. Which will win?

Mathematicians have actually shown that, if the terms of a series are positive, and if the ratio of successive terms gets closer and closer to some number less than 1, then the series is convergent. If this ratio gets closer and closer to a number greater then 1 then the series is divergent. But if the ratio is *equal* to 1, we need to do more investigation.

Figure 6.F.1 gives a picture of what is happening as the number of jumps increases. I have laid them out sideways to fit them into the space better. The full height travelled is what we get if we place all these lines on top of each other, including the ones which will be too small to see, but which go on for ever.

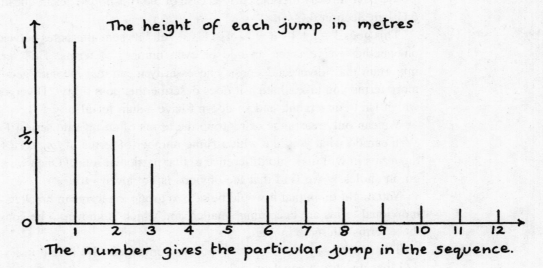

The height of each jump in metres

The number gives the particular jump in the sequence.

Figure 6.F.1

There is a very neat way of showing what happens in the case of this series. It goes like this:

Since all the terms are positive, we can reasonably group them in any way we please, because where we add bits on makes no difference to the total result. Every term you add on is moving you in the same positive direction, so each of these forward steps will have the same effect wherever it is placed.

So we can say

$$1 + \frac{1}{2} + \frac{1}{3} + \frac{1}{4} + \frac{1}{5} + \frac{1}{6} + \frac{1}{7} + \frac{1}{8} + \dots$$

$$= 1 + \frac{1}{2} + \left(\frac{1}{3} + \frac{1}{4}\right) + \left(\frac{1}{5} + \frac{1}{6} + \frac{1}{7} + \frac{1}{8}\right) + \dots$$

$$> 1 + \frac{1}{2} + \left(\frac{1}{4} + \frac{1}{4}\right) + \left(\frac{1}{8} + \frac{1}{8} + \frac{1}{8} + \frac{1}{8}\right) + \dots$$

that is, $\quad > 1 + \frac{1}{2} + \frac{1}{2} + \frac{1}{2} + \dots$

Clearly, this second series is divergent since we can make the sum as large as we like by taking enough terms. Therefore, the first series must also be divergent, and the frog *does* eventually escape. Actually, although mathematically his escape is assured, practically his situation is not very rosy. After 1000 jumps he has still only gone about $7\frac{1}{2}$ metres. This series is very close to the convergence/divergence divide. Its true name is the **harmonic series**. Each term is related to a different mode of oscillation of a stretched string, with 1 corresponding to the fundamental mode or first harmonic. Oscillation modes are important in all oscillating systems including the strings of musical instruments, which explains the use of the word 'harmonic'.

In working out what happened in the case above we were able to compare the series we got by grouping the terms of the original series with the behaviour of a known series. Such comparisons make a very good method of attack on series which we can't easily sum, but we have to be very pernickety about when we can rearrange or regroup the terms of a series.

We have already met the curious case of the flip-flop series in question (1)(e) of Exercise 6.C.1 in Section 6.C.(f).

This goes $1 - 1 + 1 - 1 + 1 - 1 + 1 - \ldots$ and its sum alternates between 0 and 1 depending on whether we've taken an odd or even number of terms. This series is divergent. It's important that 'divergent' doesn't necessarily mean that the sum gets larger and larger the more terms you take, though it does describe this possibility. 'Divergent' means any series which isn't convergent, and so doesn't have a sum to infinity.

We can only rearrange or regroup the terms of an infinite series if they are all positive. (You can do what you like with a finite number of terms of *any* series – the order you add the terms in will make no difference to that particular total.) Once we start letting the series go on endlessly we find that the obvious is not always true.

You might think that it would be safe to group the terms in brackets in a series where the individual terms are becoming smaller, and which is known to be convergent, even though these terms alternate in sign.

The series $1 - \frac{1}{2} + \frac{1}{3} - \frac{1}{4} + \frac{1}{5} - \frac{1}{6} + \ldots$ is convergent. We'll find in Example (4) of Section 8.G that its sum is equal to $\ln 2$.

Now have a look at the following apparently plausible steps of working.

$$\ln 2 = 1 - \tfrac{1}{2} + \tfrac{1}{3} - \tfrac{1}{4} + \tfrac{1}{5} - \tfrac{1}{6} + \tfrac{1}{7} - \tfrac{1}{8} \ldots$$

$$= 1 - \tfrac{1}{2} - \tfrac{1}{4} + \tfrac{1}{3} - \tfrac{1}{6} - \tfrac{1}{8} + \tfrac{1}{5} - \tfrac{1}{10} - \tfrac{1}{12} + \ldots \qquad \text{well, why not?}$$

$$= (1 - \tfrac{1}{2}) - \tfrac{1}{4} + (\tfrac{1}{3} - \tfrac{1}{6}) - \tfrac{1}{8} + (\tfrac{1}{5} - \tfrac{1}{10}) - \ldots \qquad \text{hmm} \ldots$$

$$= \tfrac{1}{2} - \tfrac{1}{4} + \tfrac{1}{6} - \tfrac{1}{8} + \tfrac{1}{10} - \tfrac{1}{12} \ldots$$

$$= \tfrac{1}{2}(1 - \tfrac{1}{2} + \tfrac{1}{3} - \tfrac{1}{4} + \tfrac{1}{5} - \tfrac{1}{6} \ldots = \tfrac{1}{2} \ln 2. \qquad \text{a minefield!}$$

It is because of unexpected and curious results like this that mathematicians have had to investigate what actually happens so carefully. Since series are deeply involved in many practical applications, knowing what can and can't be done with them is very important. For these purposes, it may often only be necessary to consider what happens when you take a limited number of terms, but you need to know when it is safe to do this. It is the difference between taking a permitted liberty and sailing ahead without noticing the warning signs. Mathematically, as well as socially, this can lead to disaster.

7 Binomial series and proof by induction

In this chapter we find out how to do binomial expansions, and see how they can describe some real-life situations. We also look at a new method of proving mathematical statements.

The chapter is divided into the following sections.

7.A Binomial series for positive whole numbers
(a) Looking for the patterns, (b) Permutations or arrangements,
(c) Combinations or selections, (d) How selections give binomial expansions,
(e) Writing down rules for binomial expansions,
(f) Linking Pascal's Triangle to selections, (g) Some more binomial examples

7.B Some applications of binomial series and selections
(a) Tossing coins and throwing dice,
(b) What do the probabilities we have found mean?
(c) When is a game fair? (Or are you fair game?)
(d) Lotteries: winning the jackpot . . . or not

7.C Binomial expansions when *n* is not a positive whole number
(a) Can we expand $(1 + x)^n$ if *n* is negative or a fraction? If so, when?
(b) Working out some expansions, (c) Dealing with slightly different situations

7.D Mathematical induction
(a) Truth from patterns – or false mirages?
(b) Proving the Binomial Theorem by induction,
(c) Two non-series applications of induction

7.A   Binomial expansions for positive whole numbers

7.A.(a)   Looking for the patterns

The first half of this chapter describes what are called binomial series. I have given them so much space because they have many applications. For this reason it is important that you should be able to do binomial expansions correctly and happily. The word 'binomial' comes from the two quantities put together in a bracket which we start from. Binomial expansions are what we get when we raise these brackets to different powers and then multiply the brackets together to find the result. In this first section all these powers will be positive whole numbers.

Here are some examples.

$$(a + b)^1 \text{ is just } a + b$$

$$(a + b)^2 = (a + b)(a + b) = a^2 + 2ab + b^2.$$

The $2ab$ comes from the two middle terms of ab which add together because it doesn't matter what order we multiply a and b in.

Next comes

$$(a + b)^3 = (a + b)(a + b)(a + b) = a^3 + 3a^2 b + 3ab^2 + b^3.$$

We find the answer by picking one letter from each bracket in every possible way and then multiplying these choices together.

There is only one way of getting a^3 and b^3.

The a^2b term comes in three ways, as we can choose the b from any of the three brackets, and then multiply it with the a terms in the other two brackets. Similarly, ab^2 can be made in three possible ways.

What happens with

$$(a + b)^4 = (a + b)(a + b)(a + b)(a + b)?$$

There will be just one a^4 and just one b^4. There will also be some numbers of terms for each of a^3b, a^2b^2 and ab^3.

Because the a and the b are symmetrically placed in the brackets, there must be the same number of terms in a^3b as there are in ab^3.

There will be four of each since we can pick either a single b or a single a in four different ways from the four brackets.

The six possibilities for a^2b^2 are given by $aabb$, $abba$, $abab$, $baab$, $baba$ and $bbaa$.

We see that by multiplying the four brackets together, we get

$$(a + b)^4 = a^4 + 4a^3b + 6a^2b^2 + 4ab^3 + b^4.$$

Now we ask two questions.

Firstly, is there an easier way than this of finding, for example, the $6a^2b^2$ term?

Secondly, is there a general pattern building up from these results?

If we write down how many we have of each possible combination of as and bs for all the brackets which we have multiplied out so far, we get the four lines of numbers written out below, which make a kind of blunt-topped triangle.

```
      1   1
    1   2   1
  1   3   3   1
1   4   6   4   1
```

These numbers give the **coefficients** for the different combinations of as and bs.

Can you see what the next line of it will be?

It is

```
1   5   10   10   5   1
```

Each number in each row is found by adding the two numbers nearest in the line above. If it is at the end of a row, the single number closest to it is used.

We can use the row which we have just worked out to write down the expansion of $(a + b)^5$. It is

$$(a + b)^5 = a^5 + 5a^4b + 10a^3 b^2 + 10a^2b^3 + 5ab^4 + b^5.$$

This triangle, which gives the various different sets of binomial coefficients, is called **Pascal's Triangle**, after the French mathematician who first observed it, Blaise Pascal. Provided the power is not too high, it is the easiest way of working out what the coefficients will be.

Write down, by extending this triangle, the expansions of
(1) $(a + b)^6$ (2) $(a + b)^7$

I've put the answers in straight away because they show something important. You should have

(1) $a^6 + 6a^5 b + 15a^4b^2 + 20a^3b^3 + 15a^2b^4 + 6ab^5 + b^6$

(2) $a^7 + 7a^6 b + 21a^5b^2 + 35a^4b^3 + 35a^3b^4 + 21a^2b^5 + 7ab^6 + b^7$.

Notice how the power of a moves down by 1 and the power of b up by 1 for each new term. The powers together add up to 6 for (1) and 7 for (2).

We will now get some practice in the mechanics of binomial expansions in which the 'a' and the 'b' are replaced by more complicated expressions. (These often form part of the working of longer problems, and it is important that you should be able to do them confidently and accurately.)

We'll work out $(2x + 3y)^6$ as an example.

Here, the 'a' is $2x$, and the 'b' is $3y$, and $n = 6$.

We get the binomial coefficients by using the sixth line of Pascal's Triangle. This is

$$1 \quad 6 \quad 15 \quad 20 \quad 15 \quad 6 \quad 1. \qquad \text{(P6)}$$

I've labelled it (P6) so I can easily refer back to it.

The expansion goes

$$(2x + 3y)^6 = (2x)^6 + 6(2x)^5 (3y) + 15(2x)^4 (3y)^2$$
$$+ 20(2x)^3 (3y)^3 + 15(2x)^2 (3y)^4 + 6(2x)(3y)^5 + (3y)^6.$$

Notice again the pattern of the powers. They move down by 1 each time for the 'a' and up 1 each time for the 'b' of the expansion.

Added together, they always give n, the overall power we are calculating.

Multiplying out, we have

$$(2x + 3y)^6 = 64x^6 + 576x^5y + 2160x^4y^2 + 4320x^3 y^3 + 4860x^2 y^4 + 2916xy^5 +$$
$$243y^6.$$

> Don't forget the part of each coefficient which comes from the 'a' and the 'b' raised to the various different powers. Students very frequently make mistakes here. It is safer always to put brackets round the whole of the 'a' and the 'b' as I have done above.

Try expanding these for yourself.

(1) $(x - 2y)^6$ (2) $(2x^2 - y^2)^5$ (3) $\left(2x - \dfrac{1}{x}\right)^4$ (4) $\left(\dfrac{3}{x} + 4x^2\right)^3$

7.A.(b)   Permutations or arrangements

The pattern shown in Pascal's Triangle is very neat and, as we have seen, is very useful for writing down the answers for binomial expansions when the power is not too large. It would, however, be rather tedious to have to go much further than (P7) and we look now at how we can find a general rule to give us these results. (This will also explain why we get this pattern in the first place.)

To do this, we will look at the numbers of different possibilities of choosing some objects from a larger number of objects. We know that when we multiply out the brackets the order of the letters doesn't matter, so, for example, both *aba* and *baa* count as a^2b. It's actually easier to find a general rule for what happens when the order of choice *does* matter, so we'll look at some examples of this first.

Because it can make it easier to see what is happening if we look at it pictorially, and because the total number of choices quite quickly becomes amazingly large as we increase the possibilities, we will start with a relatively simple situation.

Let's consider the number of possible choices of three counters from four differently shaped counters, and let's also suppose that the order of choice matters. Then the first counter can be chosen in four ways. The second one can be chosen in three ways from the three which are now left, and the third counter can then be chosen in two ways. This gives us a grand total of $4 \times 3 \times 2 = 24$ choices.

All the possibilities are shown in Figure 7.A.1.

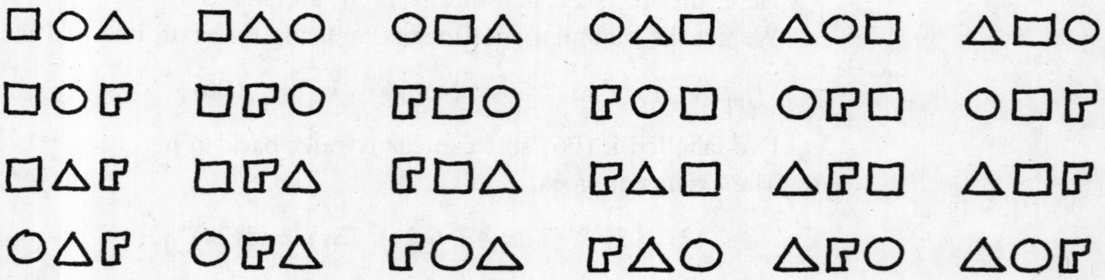

Figure 7.A.1

Here is another example.

Suppose there is a class of ten children and six of them will be given a prize. It is not allowed for any child to have more than one prize, and six different books have been bought for the purpose. We'll also suppose that these prizes are being handed out randomly – no awards for merit here!

The child who gets the first book may be chosen in ten ways. For each of these ten choices, there are nine ways of choosing the child to get the second book. Then, for each of these choices, there are eight ways of choosing the third child. The total number of choices of the six fortunate children is given by $10 \times 9 \times 8 \times 7 \times 6 \times 5 = 151\,200$ which is a surprisingly large number. The order of choice of the children matters because the books are all different so the same six children chosen in a different order will count as a different choice, since they would each then get different books.

We can use the fact that the numbers are running down by 1 each time to write the total number of ways of distributing the prizes in a very neat compact form. We let the top run right down to 1 and then divide this by the extra part on the bottom (so that cancelling would bring us back to the original multiplication).

We can then say that this total number is

$$10 \times 9 \times 8 \times 7 \times 6 \times 5 = \frac{10 \times 9 \times 8 \times 7 \times 6 \times 5 \times 4 \times 3 \times 2 \times 1}{4 \times 3 \times 2 \times 1}$$

$$= \frac{10!}{4!}.$$

The symbol ! is used for multiplications like these. The 10! above is called 'ten factorial'. (Factorials came in also when we looked at series (l) in Section 6.A.(a).)

The expression 10!/4! gives the number of **permutations or arrangements** of six objects (or people) chosen from ten objects (or people).

We can see that it must be 4! on the bottom by noticing that

4 = 10 (the total number we chose from) − 6 (the number of choices we are making).

> For permutations or arrangements, the order of choice matters. A different order gives a different arrangement.

> The number of **permutations** or **arrangements** of r objects from n objects is given by
>
> $$\frac{n!}{(n-r)!}.$$

7.A.(c) **Combinations or selections**

How much difference will it make if we have a situation in which we don't care what order the choices are made in?

Returning first of all to the example of choosing three counters from four differently shaped counters, if the order of choice isn't important, how many different possibilities are there?

There are only four.

These are shown in Figure 7.A.2. (Any order would have done equally well.)

☐○△   ☐○▽   ☐△▽   ○△▽

Figure 7.A.2

If you now look back at the 24 possibilities shown in Figure 7.A.1. you will see that these are the four different possibilities shown in the left-hand column. Each row is then made up of the different arrangements of that particular choice of three counters, and there are six of each because each possible set of three counters was shown there in all its different orders. So there were three different choices for the first counter, two for the second and just one for the third, giving $3 \times 2 \times 1 = 3! = 6$ for each group of three counters.

The total number of choices of three counters from four counters, if we don't care about the order of choice, is given by

$$\frac{24}{6} = \frac{4!}{1!\,3!}.$$

We have to divide the total of 24 by 6 or 3! to get rid of all the different internal arrangements of each group of three counters, which we aren't interested in this time.

We can take a second example by looking again at the different ways in which the children can receive their prizes.

Suppose this time that six identical copies of the same book had been bought for the prizes. The order of choice of the children no longer makes any difference because all six are getting the same book anyway.

The number of *different* choices is now given by the number of different groups of six children. To find these, we no longer need to take account of the order in which any particular group was chosen.

So we must divide our previous total of 10!/4! by 6! to get rid of all these unwanted internal different orderings.

This gives us that the number of **combinations** or **selections** (that is, choices in which the *order* of choice doesn't matter) of six people from ten people, is

$$\frac{10!}{6! \, 4!}.$$

This is sometimes called 'ten pick six' or 'ten choose six'.

 NOTE

For combinations or selections, the order of the choices made does not matter. If the same objects are chosen, it makes no difference which one was chosen first, which second, etc.

The number of **combinations** or **selections** of r objects from n objects is given by

$$\frac{n!}{r! \, (n-r)!}.$$

This is sometimes written as ^nC_r or $\binom{n}{r}$.

 SPECIAL CASES

The number of ways of picking n objects from n objects if the order of choice doesn't matter, is just 1. Using the rule above, we would have

$$\frac{n!}{n! \, 0!} = 1.$$

In order to make this rule work we say that $0! = 1$.

7.A.(d) How selections give binomial expansions

We now link the work we have just done on selections back to what we saw was happening with binomial expansions. The procedure in these expansions is that we are choosing one of two possibilities from each bracket, then multiplying these choices together and finally grouping together all the similar results.

For example, we look again at finding $(a + b)^4 = (a + b)(a + b)(a + b)(a + b)$.

It's easy to see that all the as can be chosen in only one way, giving a^4.

Similarly, all the bs can be chosen in only one way, giving b^4.

Three as and one b can be chosen in four ways since the single b can be chosen from any of the four brackets and the other three will then necessarily be as. This gives us $4a^3b$.

Similarly, three bs and an a can be chosen in four different ways, giving us $4ab^3$.

Finally, in how many different ways can we choose two as?

We are choosing two as from four as and the order of choice doesn't matter, so this can be done in $4!/2!\, 2! = 6$ ways.

We have found the 6 without using either Pascal's Triangle or having to draw the six possibilities.

In exactly the same way, suppose we want to find the term in a^5b^{11} in the expansion of $(a + b)^{16}$.

The power here is of such a size that we wouldn't really want to have to extend Pascal's Triangle this far. (Besides, we only want one term.)

We think of the term we want as giving the number of ways of choosing five as from 16 as if the order of choice doesn't matter.

$$\text{This is given by } \frac{16!}{5!\, 11!} = \frac{16 \times 15 \times 14 \times 13 \times 12}{5 \times 4 \times 3 \times 2 \times 1} = 4368.$$

Since we must choose one letter from each bracket, choosing five as means that we must also have 11 bs so, equally, we could have said that this term would be given by the number of ways of choosing 11 bs from 16 bs. This is

$$\frac{16!}{11!\, 5!} = 4368 \text{ as before.}$$

In each case, once a certain number of one letter has been chosen, we know that the gaps *must* be filled by the other letter, so we don't have to worry about making choices for that.

EXERCISE 7.A.3

We have just found that the coefficient of the term in a^5b^{11} in the expansion of $(a + b)^{16}$ is $16!/5!\, 11! = 4368$ so the term is $4368a^5b^{11}$.

Find the coefficients of the following terms in the same expansion, giving your answers both in factorial form and as numbers.

(1) a^{16} (2) $a^{15}b$ (3) $a^{14}b^2$ (4) $a^{12}b^4$ (5) a^8b^8

(6) a^4b^{12} (7) a^2b^{14} (8) b^{16} (9) a^rb^{16-r}

In each case, say also what the actual term would be.

7.A.(e)   **Writing down rules for binomial expansions**

We can use the results which we have found in this exercise to write down the whole expansion of $(a + b)^{16}$ as follows:

$$(a + b)^{16} = a^{16} + 16a^{15}b + \frac{16.15}{2!}a^{14}b^2 + \ldots + \frac{16!}{r!(16 - r)!}a^{16-r}b^r + \ldots + b^{16}.$$

(The ... stands for missing terms in the same way that we used it in Chapter 6.)

We could also use the Σ notation which we met in Section 6.D, and write

$$(a + b)^{16} = \sum_{r=0}^{16} \frac{16!}{r!(16 - r)!}a^{16-r}b^r.$$

Notice that we start with $r = 0$ so that we have a^{16} and $b^0 = 1$ in the first term.

If n is a positive whole number, we can write down this rule for the binomial expansion of $(a + b)^n$:

$$(a + b)^n = a^n + na^{n-1}b + \frac{n(n-1)}{2!}\,a^{n-2}b^2 + \frac{n(n-1)(n-2)}{3!}\,a^{n-3}b^3 + \ldots$$

$$+ \frac{n!}{r!(n-r)!}\,a^{n-r}b^r + \ldots + b^n. \qquad \text{(B1)}$$

If you put $n = 16$, you will get the example of $(a + b)^{16}$ which we have just done.

I have always found it best to remember the binomial expansion in the way in which I give it here, with the first three terms in their cancelled down form, because this is the easiest form to feed into, if you want to work out just the first few terms of a particular expansion.

Have a go at one yourself, now.
Try using the rule above to write down the expansion of $(a + b)^5$.
You will need to put $n = 5$.

You should get:

$$(a + b)^5 = a^5 + 5a^4b + \frac{5(4)}{2!}\,a^3b^2 + \frac{5(4)(3)}{3!}\,a^2b^3 + \frac{5(4)(3)(2)}{4!}\,ab^4 + b^5$$

so $\qquad (a + b)^5 = a^5 + 5a^4b + 10a^3b^2 + 10a^2b^3 + 5ab^4 + b^4$

which gives the same result as using Pascal's Triangle.

In many circumstances, it happens that the first term in the bracket (which we called a above) is 1.

Then, putting $a = 1$ and $b = x$ to avoid confusion between the two forms, we get:

$$(1 + x)^n = 1 + \frac{n}{1!}\,x + \frac{n(n-1)}{2!}\,x^2 + \ldots + \frac{n!}{r!(n-r)!}\,x^r + \ldots + x^n. \qquad \text{(B2)}$$

I've included the 1! in the second term to keep the pattern of the factorials running through. We'll need this later on in Section 8.B.(a) when we take another look at e.

Notice also that the second term has x and the third has x^2, so

the term $\dfrac{n!}{r!(n-r)!}\,x^r$ is actually the $(r + 1)$th term.

Similarly, in (B1), the general term

$$\frac{n!}{r!\,(n-r)!}\,a^{n-r}b^r \text{ is actually the } (r+1)\text{th term.}$$

When we wrote the series using Σ we made the sum run from *zero* to n, so there are $n + 1$ terms altogether.

Here is an example which uses the formula (B1).
Write down the first four terms of the expansion of $(2x - \frac{1}{2}y)^{12}$.
The value of n here is so large that it would be tedious to continue Pascal's Triangle as far down as we would need.
Instead, we use form (B1), putting 'a' = $2x$, 'b' = $-\frac{1}{2}y$ and $n = 12$.

 Remember that the minus sign must be included as part of 'b'.

Substituting in these values, we have for the first four terms of the expansion

$$(2x)^{12} + 12(2x)^{11}\left(-\tfrac{1}{2}y\right) + \frac{12 \times 11}{2 \times 1}(2x)^{10}\left(-\tfrac{1}{2}y\right)^2 + \frac{12 \times 11 \times 10}{3 \times 2 \times 1}(2x)^9\left(-\tfrac{1}{2}y\right)^3.$$

Tidying up these first four terms, we get

$$4096x^{12} - 12288x^{11}y + 16896x^{10}y^2 - 14080x^9y^3.$$

EXERCISE 7.A.4

Now try these for yourself.
Write down and simplify the first four terms in the expansions of
(1) $(2x - y)^{12}$ (2) $(1 - 2x)^{18}$ (3) $(1 + x^2)^{10}$ (4) $(\tfrac{1}{2}x + 3y)^{16}$

7.A.(f)   **Linking Pascal's Triangle to selections**
We are now in a position to be able to see comfortably how the links work between Pascal's Triangle and the selections which give the coefficients, using formula (B2). We use (B2) because it makes it a bit easier to see what is going on, but (B1) would work in exactly the same way.
We begin by writing down the eighth row of Pascal's Triangle, giving the coefficients in the expansion of $(1 + x)^8$. I have labelled it (P8). It is:

$$1 \quad 8 \quad 28 \quad 56 \quad 70 \quad 56 \quad 28 \quad 8 \quad 1 \qquad (P8)$$

Try answering the following questions, and then we'll look at them together.

(1)   Use (P8) to write down the next row of the triangle, giving the coefficients for the expansion of $(1 + x)^9$. Label it (P9).
(2)   Using (P8), write down the coefficients of (a) x^4 and (b) x^5 in the expansion of $(1 + x)^8$.

(3)   In factorial form, the coefficient of x^4 in the expansion of $(1 + x)^8$ is 8!/4! 4!. Write down the coefficient of x^5 in factorial form.

(4)   Using (P9), write down the coefficient of x^5 in the expansion of $(1 + x)^9$.

(5)   Now write down the coefficient of x^5 in this expansion in factorial form.

Here are the answers.

(1)   1   9   36   84   126   126   84   36   9   1.   (P9)

(2)   The coefficient of x^4 in (P8) is 70. The coefficient of x^5 is 56.

(3)   The coefficient of x^5 from $(1 + x)^8$ in factorial form is 8!/5! 3!.

(4)   From (P9), the coefficient of x^5 in the expansion of $(1 + x)^9$ is 126.

(5)   The coefficient of x^5 in this expansion in factorial form is 9!/5! 4!.

Now we try answering this question.

We used $70 + 56$ in (P8) to get 126 in (P9).

Obviously this must also be true written in factorials, so

$$\frac{8!}{4!\ 4!} + \frac{8!}{5!\ 3!} \quad \text{must equal} \quad \frac{9!}{5!\ 4!}.$$

We now show that this must be true by factorising and tidying up the first two fractions. We have

$$\frac{8!}{4!\ 4!} + \frac{8!}{5!\ 3!} = \frac{8!}{4!\ 3!}\left(\frac{1}{1 \times 4} + \frac{1}{5 \times 1}\right).$$

(Check this step for yourself by multiplying it back. You'll need to use $4 \times 3! = 4!$ and $5 \times 4! = 5!$)

$$= \frac{8!}{4!3!}\left(\frac{5 + 4}{4 \times 5}\right) = \frac{8! \times 9}{(4! \times 5)(3! \times 4)} = \frac{9!}{5!\ 4!}.$$

(This step involves adding fractions as we did in Section 1.C.(c).)

We can also see that this must happen if we think of $(1 + x)^9$ as coming from $(1 + x)(1 + x)^8$. Then the term with x^5 in $(1 + x)^9$ comes from $1 \times$ the term in x^5 from $(1 + x)^8 + x \times$ the term in x^4 from $(1 + x)^8$.

EXERCISE 7.A.5

With the above example to look back at, you should be able to answer the following three questions yourself.

You first have to fill in the gaps marked with asterisks (*), and then combine the factorials.

(1) The coefficient of x^3 in the expansion of $(1 + x)^9$ is $\dfrac{9!}{3!\ 6!}$.   (a)

The coefficient of x^4 in the expansion of $(1 + x)^9$ is $\dfrac{*!}{*!\ *!}$.   (b)

The coefficient of x^4 in the expansion of $(1 + x)^{10}$ is $\dfrac{10!}{*!\ *!}$.   (c)

Show, by factorising and tidying up, that (a) + (b) = (c).

(2) The coefficient of x^3 in the expansion of $(1 + x)^{12}$ is $\dfrac{*!}{3!\ 9!}$.   (a)

The coefficient of x^4 in the expansion of $(1 + x)^{12}$ is $\dfrac{*!}{*!\ *!}$.   (b)

The coefficient of x^4 in the expansion of $(1 + x)^{13}$ is $\dfrac{*!}{*!\ *!}$.   (c)

Show, by factorising and tidying up, that (a) + (b) = (c).

(3) The coefficient of x^{r-1} in the expansion of $(1 + x)^k$ is $\dfrac{k!}{(r-1)!\ (k-r+1)!}$.   (a)

The coefficient of x^r in the expansion of $(1 + x)^k$ is $\dfrac{*!}{*!\ *!}$.   (b)

The coefficient of x^r in the expansion of $(1 + x)^{k+1}$ is $\dfrac{*!}{*!\ *!}$.   (c)

Show, by factorising and tidying up, that (a) + (b) = (c).

7.A.(g)   Some more binomial examples

Here are three more examples showing ways in which we can pick out particular terms.

EXAMPLE (1)   Write down the term containing (a) p^6, (b) q^6, in the expansion of $(p-2q)^{14}$.

To do this, we can use the expression for the general term in form (B1). This is

$$\frac{n!}{r!\ (n-r)!}\, a^{n-r}b^r.$$

(Remember that this is the $(r + 1)$th term of the series, not the rth term.)

Here, $n = 14$, 'a' $= p$, 'b' $= -2q$ and the term in p^6 is given when $n - r = 6$ so $r = 8$.

The term in p^6 is $\dfrac{14!}{8!\ 6!}\, p^6\, (-2q)^8 = 768768p^6q^8$.

The term in q^6 is $\dfrac{14!}{6!\ 8!}\, p^8\, (-2q)^6 = 192192p^8q^6$.

Notice the symmetry of the binomial coefficients:

$$\frac{14!}{8!\ 6!} = \frac{14!}{6!\ 8!}.$$

EXAMPLE (2)   Find the constant term in the expansion of $\left(4x^2 + \dfrac{3}{x}\right)^{12}$.

This is the one term in the expansion which is purely a number, and so doesn't depend upon the value of x for its size. It happens because the powers of x in this expansion are cancelling each other out to some extent on each term.

Can you work out for yourself when it will be that they will cancel out exactly?

The term we want will involve $(4x^2)^4 \left(\frac{3}{x}\right)^8$, so it is

$$\frac{12!}{8!\ 4!}(4x^2)^4 \left(\frac{3}{x}\right)^8 = 831\,409\,920.$$

EXAMPLE (3) Find the term in x^{11} in the expansion of $(1-x)^8\,(3+2x)^5$.

The complication here is that the term in x^{11} arises from three different multiplications of pairs of terms, because x^{11} can come from $x^8 \times x^3$ and $x^7 \times x^4$ and $x^6 \times x^5$.

Any other combinations are impossible from this particular pair of brackets.

We need to write down the terms of these separate multiplications fully in order to work out the complete term in x^{11}. We get

$$\left[(-x)^8\right]\left[\frac{5!}{2!\ 3!}(3)^2(2x)^3\right] + \left[8(-x))^7\right]\left[\frac{5!}{1!\ 4!}(3)(2x)^4\right]$$

$$+ \left[\frac{8!}{2!\ 6!}(-x)^6\right]\left[(2x)^5\right].$$

Each separate part of the three terms we have added together here is enclosed in square brackets to make it easier for you to see how each bit has been worked out.

Now, tidying up the above working, we get

$$720x^{11} - 1920x^{11} + 896x^{11} = -304x^{11}.$$

EXERCISE 7.A.6

Try these questions yourself.

(1) Find the term in x^6 in the expansion of

   (a) $(2-3x)^{11}$ (b) $(2x-y)^8$ (c) $(y^2-2x^2)^{10}$

(2) Find the constant terms in the expansions of

   (a) $\left(2x-\dfrac{3}{x}\right)^{10}$ (b) $\left(x+\dfrac{1}{x^2}\right)^9$ (c) $\left(2x^3+\dfrac{1}{x}\right)^{16}$

(3) Find the term in x^{10} in the expansion of $(1+x)^7\,(2-3x)^5$.

7.B   Some applications of binomial series and selections

7.B.(a)   Tossing coins and throwing dice

Binomial expansions can be applied very neatly to describe the likelihoods of the different possible outcomes to some events involving chance. When you do a binomial expansion, you are making a free choice of which of two terms to pick in each of the equal brackets, and then writing down all the different possible results. This fits any real-life situation in which there are repeated events, each of which has just two possible outcomes, and where the outcome of one event doesn't have any effect on subsequent events.

For example, suppose you toss a fair coin. The likelihood or probability of getting a head is $\frac{1}{2}$. ('Fair' here means that it is equally likely to fall heads or tails.)

What will be the likelihood or probability of each of the different outcomes if you toss the coin three times instead?

We can show all these probabilities by writing the binomial expansion

$$(\tfrac{1}{2}T + \tfrac{1}{2}H)^3 = (\tfrac{1}{2}T)^3 + 3\,(\tfrac{1}{2}T)^2\,(\tfrac{1}{2}H) + 3\,(\tfrac{1}{2}T)(\tfrac{1}{2}H)^2 + (\tfrac{1}{2}H)^3.$$

I have used H and T as markers for heads and tails, and the two halves in the first bracket stand for the probabilities of each of these on a single toss. Tidied up, we get

$$\tfrac{1}{8}T^3 + \tfrac{3}{8}T^2\,H + \tfrac{3}{8}TH^2 + \tfrac{1}{8}H^3.$$

This carries all the information on the possible outcomes of the three trials, that is,

- a probability of $\tfrac{1}{8}$ of getting three tails,
- a probability of $\tfrac{3}{8}$ of getting two tails and one head,
- a probability of $\tfrac{3}{8}$ of getting one tail and two heads,
- a probability of $\tfrac{1}{8}$ of getting three heads.

This idea can be extended to situations where the outcomes on each trial aren't equally likely. Suppose you throw three dice and you want to know the probabilities of getting the different possible numbers of sixes. The probability of getting a six on a single throw of a fair die is one sixth because there are six possible equally likely outcomes, and only one of them gives a six. The probability of *not* throwing a six is $\tfrac{5}{6}$. If I use markers of P (for success in throwing a six) and Q (for throwing a different score) then I can show the probabilities for all the different outcomes by writing

$$(\tfrac{5}{6}Q + \tfrac{1}{6}P)^3 = (\tfrac{5}{6}Q)^3 + 3(\tfrac{5}{6}Q)^2\,(\tfrac{1}{6}P) + 3(\tfrac{5}{6}Q)\,(\tfrac{1}{6}P)^2 + (\tfrac{1}{6}P)^3$$

$$= \frac{125}{216}\,Q^3 + \frac{75}{216}Q^2P + \frac{15}{216}\,QP^2 + \frac{1}{216}\,P^3.$$

So

the probability of getting three sixes is $\tfrac{1}{216}$,
the probability of getting two sixes is $\tfrac{15}{216}$,
the probability of getting one six is $\tfrac{75}{216}$,
and the probability of getting no sixes is $\tfrac{125}{216}$.

Notice that all the probabilities added together give $\tfrac{216}{216} = 1$. We are certain that the dice will fall in one of these ways. (This makes a useful check on the arithmetic.)

I only listed the probabilities of the outcomes of three trials in each of my examples. It wouldn't be too hard to work these out by drawing tree diagrams or listing all the possible equally likely outcomes (remembering that, for example, you can get just one tail in *three* different ways because there are three coins). The strength of the binomial expansion is that it works equally well for some huge number of dice where it would be hideously tedious to write down all the possible outcomes. It would also work equally well in forecasting the likelihoods of the numbers of faulty items off a production line in batches of a given size, provided the probability of any one item being faulty remained constant. Once you understand the mathematical structure of a model, you can apply it in a vast range of situations which are similar mathematically, though physically they are very different.

7.B.(b)   What do the probabilities we have found mean?

What does it actually mean when we say, for example, that the probability of getting two sixes if we throw two dice is $\tfrac{1}{36}$?

It does *not* mean that if we throw two dice 36 times then there will be exactly one double six. We know from our own experience that this can't be so. What it does mean is that if we throw two dice a very large number of times then the *proportion* of double sixes will be roughly 1 in 36. (It will get closer to 1 in 36 the larger the number of trials we make; yet another example of tending to a limit!)

It is important that, in all these examples, what we have found are only theoretical probabilities which give us the likely ratio of the different outcomes in a very large number of trials.

It is possible, for example, to get 12 heads in a row if you toss a coin, but both common sense and the theoretical probability of $(\frac{1}{2})^{12}$ of this happening, tell you that it is very unlikely. You would begin to suspect that you might have a double-headed coin.

Usually, the study of statistics tells us not whether something is possible or impossible, but how *likely* it is. Also, as we have just seen, these likelihoods can be found exactly. If the observed outcomes are, for example, much more frequent than their theoretical probability we are warned that further investigation is sensible. Perhaps all is not as it seems.

These ideas are developed further in the study of statistics, in which such arguments (leading to tests of significance) can be made on a precise mathematical basis, rather than woolly feelings that something is wrong. These feelings may well be correct but a careful statistical test can make it possible to argue the case backed up by sound mathematical reasoning.

7.B.(c)   When is a game fair? (Or are you fair game?)

This is a good point at which to introduce the idea of a 'fair' game. If a game is fair in the mathematical sense then it must be designed so that, over a very large number of goes, none of the contestants is expected to make a profit over the others. So, for example, if we toss a coin with you paying me £1 for a head, and me paying you £1 for a tail, then on average we will end up with neither of us gaining from the other. We have an equal probability of winning overall, even though, on three goes say, I may be lucky with three heads in a row. However, I can't play this game *expecting* to win money from you.

But casinos and lotteries aren't fair in this sense. Clearly, they can't be, because they make profits for the people who run them. The probabilities are built in to be unequal from the start, and they are only fair in the sense that each contestant other than the banker or owner has an equal chance of winning on each attempt.

7.B.(d)   Lotteries: winning the jackpot . . . or not

Let's now consider one other practical application of these ideas before we go on to the next section.

Suppose that the rules of a lottery say that in order to win the big prize or jackpot six numbers must be chosen correctly in the range from 1 to 49.

What is the probability of actually doing this?

There are 49 equal choices which can be made for the first number. Each number in the range can only be chosen once, so although the first choice is made from 49 numbers, the next is from the remaining 48, and so on. It is exactly the same kind of situation as when we were giving out the six identical prizes in Section 7.A.(c). The total number of choices is given by

$$\frac{49!}{6!43!} = 13\,983\,816.$$

(We are using combinations here rather than permutations because the order of choice does not matter. For example, one person might choose 42 first, and another person, with the identical final choice of six numbers, might have had 42 as his second chosen number.)

So the probability of winning the jackpot in this lottery would be 1/13 983 816. In an astronomical number of tries, you could expect to win it roughly once in every fourteen million attempts.

EXERCISE 7.B.1 **Try answering the following questions.**

(1) **Choose six numbers in the range from 1 to 49 as randomly as you can without using any help like the random number generator on a calculator. Now repeat this nine more times. Use squared paper to show your choices on a grid which is 49 squares wide and 10 squares deep. Do you think your choices look really random? Feel free to alter them if you want to.**

(2) **In a lottery like the one described in the previous section, which of these three choices of six numbers would be most likely to win you the jackpot?**
    (a) 1, 2, 3, 4, 5, 6   (b) 2, 14, 21, 29, 33, 45   (c) 44, 45, 46, 47, 48, 49

(3) **Would there be any good reason for picking one group rather than the other two?**

(4) **What would be the probability of guessing at least one number correctly in a lottery like this? Write down what you think it might be, and then work out how near your estimate is to the true answer. *Hint:* work out how many ways there are of choosing all six numbers completely wrongly.**

7.C   Binomial expansions when *n* is not a positive whole number

7.C.(a)   Can we expand $(1 + x)^n$ if *n* is negative or a fraction? If so, when?

All the arguments we have used to justify the binomial series have depended on having a factor multiplied by itself a whole number of times.

It would be interesting and useful if we could extend this. Can we make any sense of something like an expansion of $(1 + x)^{-1}$, for example?

We certainly can't give it the same kind of meaning which we could when we had a positive whole number power; then, we could actually lay out the brackets to make our choices. However, we'll persevere and see what would happen in an experimental kind of way, taking the particular case of $(1 + x)^{-1}$.

We know that we can certainly write $(1 + x)^{-1}$ as $1/(1 + x)$. Now let's see what happens if we try using the (B2) expansion from Section 7.A.(e) on $(1 + x)^{-1}$, putting the *n* of this formula equal to –1. We shall get

$$(1 + x)^{-1} = 1 + \frac{(-1)}{1} x + \frac{(-1)(-2)}{2 \times 1} x^2 + \frac{(-1)(-2)(-3)}{3 \times 2 \times 1} x^3 + \frac{(-1)(-2)(-3)(-4)}{4 \times 3 \times 2 \times 1} x^4 + \dots$$

The first thing that we notice is that the countdown on the top of the fractions isn't going to come to a natural end like it does when *n* is a positive whole number.

$(1 + x)^{-1}$ is giving us an infinite series. We've seen examples in Chapter 6 of the dangers connected with summing infinite series.

Try tidying up this one yourself and see if you recognise what you get. Then you should be able to say whether this expansion works. If so, will this depend in any way on what value *x* has?

Tidying up what we have above for the expansion of $(1 + x)^{-1}$, we get:

$$(1 + x)^{-1} = \frac{1}{1 + x} = 1 - x + x^2 - x^3 + x^4 - \dots$$

This is a GP with '*a*' = 1 and '*r*' = – *x*, and $1/(1 + x)$ is its sum to infinity.

So far, so good, but we know from Section 6.C.(c) that a GP only *has* a sum to infinity if its common ratio lies between -1 and $+1$. So we can say that, in this particular case, the expansion *does* work provided $|-x| < 1$. Now $|-x|$ is the same as $|x|$ since we are taking the positive value whatever the sign. So we must have $|x| < 1$, or $-1 < x < 1$, writing it another way.

You can see for yourself that we will be in trouble if we don't stick to this. For example, suppose $x = 2$. This would give us

$$\frac{1}{1 + 2} = 1 - 2 + 4 - 8 + \ldots$$

The problem here is that successive terms are getting bigger. These terms alternate in sign and so do the partial sums obtained by adding in each new term. Each of these is larger than the previous one in absolute size, so this series can't be getting closer and closer to $\frac{1}{3}$ as we add more and more terms.

> It has been shown by mathematicians that $(1 + \spadesuit)^n$ can be expanded using (B2) if n is either negative or a fraction or both, provided that the $\spadesuit$ fits the requirement that $|\spadesuit| < 1$.
>
> ($\spadesuit$ stands for whatever we have in this position in the bracket.)

7.C.(b) **Working out some expansions**

Now we'll practise the mechanics of how these expansions go, because this process is just an extension of what we have been doing with binomial expansions for positive whole number powers, and it will be useful for you later on to be able to do this.

Here are two examples of such expansions.
Expand as far as the term in x^3, stating the restrictions on the value of x in each case:

(1) $(1 + 3x)^{-2}$ (2) $(1 - x/2)^{1/2}$

For (1), $n = -2$ and $\spadesuit = 3x$. We must have $|\spadesuit| < 1$, so we want $|3x| < 1$, which means $-1 < 3x < 1$, so $-\frac{1}{3} < x < \frac{1}{3}$.
In order for the expansion to be possible, x must lie somewhere in this interval.
If x does fit this requirement, we can say:

$$(1 + 3x)^{-2} = 1 + (-2)(3x) + \frac{(-2)(-3)}{2 \times 1}(3x)^2 + \frac{(-2)(-3)(-4)}{3 \times 2 \times 1}(3x)^3 + \ldots$$

$$= 1 - 6x + 27x^2 - 108x^3 \text{ as far as the fourth term.}$$

For (2), $n = \frac{1}{2}$ and $\spadesuit = -x/2$, so we want $|-x/2| < 1$. But $|-x/2| = |x/2|$, since we are taking the positive value whatever the sign.
So we must have $-1 < x/2 < 1$ which means $-2 < x < 2$.
Provided x fits this requirement, we can write:

$$\left(1 - \frac{x}{2}\right)^{1/2} = 1 + \left(\tfrac{1}{2}\right)\left(-\frac{x}{2}\right) + \frac{(\tfrac{1}{2})(-\tfrac{1}{2})}{2 \times 1}\left(-\frac{x}{2}\right)^2 + \frac{(\tfrac{1}{2})(-\tfrac{1}{2})(-\tfrac{3}{2})}{3 \times 2 \times 1}\left(-\frac{x}{2}\right)^3 + \ldots$$

$$= 1 - \frac{x}{4} - \frac{x^2}{32} - \frac{x^3}{128} \quad \text{as far as the fourth term.}$$

Now, in each of the above cases, substitute $x = 0.001$ and see how closely the two sides match up, as you add in the extra terms on the RHS. You will find that, because x is small, you very quickly get close to the LHS, and indeed are beginning to find an answer accurate to more decimal places than your calculator is giving you, in the second case. This possibility of being able to replace an infinite series by a fast numerical equivalent to any desired degree of accuracy is often important in practical applications.

EXERCISE 7.C.1

Try expanding the following three examples yourself, as far as the term in x^3, stating in each case the restrictions on x for the expansion to be valid.

(1) $(1 + 2x)^{-3}$ (2) $(1 - 3x)^{-1}$ (3) $(1 + \frac{1}{3}x)^{-2}$

7.C.(c)   Dealing with slightly different situations

What should we do if we want to find the expansion of $(2 + 3x)^{-2}$? We can't any longer use the (B2) formula to expand this.

I think that in such a case the simplest method is to rearrange the bracket so that it *is* in $(1 + \spadesuit)$ form. Doing this simplifies the arithmetic quite a bit, as it avoids complicated and changing powers of 'a'.

So we write:

$$(2 + 3x)^{-2} = \left[2 \left(1 + \frac{3x}{2} \right) \right]^{-2} = 2^{-2} \left(1 + \frac{3x}{2} \right)^{-2} = \frac{1}{4} \left(1 + \frac{3x}{2} \right)^{-2}.$$

It is important that the factor which we take out of the bracket was part of this bracket, and so it is raised to the same power as the bracket itself.

Remember, too, that if you are taking out a factor, it applies to the *whole* bracket, so we must write $3x/2$, and not leave the $3x$ unchanged.

For the expansion to be possible, what interval must x lie in?

We must have

$$\left| \frac{3x}{2} \right| < 1 \quad \text{so} \quad -1 < \frac{3x}{2} < 1 \quad \text{so} \quad -2 < 3x < 2 \quad \text{giving} \quad -\frac{2}{3} < x < \frac{2}{3}.$$

Expanding, using (B2), we get that

$$\frac{1}{4} \left(1 + \frac{3x}{2} \right)^{-2} = \frac{1}{4} \left[1 + (-2) \left(\frac{3x}{2} \right) + \frac{(-2)(-3)}{2 \times 1} \left(\frac{3x}{2} \right)^2 + \frac{(-2)(-3)(-4)}{3 \times 2 \times 1} \left(\frac{3x}{2} \right)^3 + \ldots \right]$$

$$= \frac{1}{4} - \frac{3x}{4} + \frac{27x^2}{16} - \frac{27x^3}{8} \ldots$$

This step needs to be done quite carefully if you are not to lose any bits! Try doing it yourself as a check. Remember to square and cube the $\frac{3}{2}$ when necessary.

Here is another situation which you may meet.

Suppose you need to find the expansion of

$$y = \frac{1}{(2-x)(1+2x)}$$

up to the term in x^3, also finding the interval in which x must lie for the expansion to be valid.

There are two ways of doing this.

METHOD (1)   We write

$$y = (2-x)^{-1}(1+2x)^{-1} = 2^{-1}(1-x/2)^{-1}(1+2x)^{-1}$$

$$= \frac{1}{2}\left[1 + \frac{x}{2} + \frac{x^2}{4} + \frac{x^3}{8} + \ldots\right][1 - 2x + 4x^2 - 8x^3 \ldots]$$

using the rules I gave at the end of the answer to question (2) of Exercise 7.C.1 to speed up the working inside these two brackets.

Now we do the multiplying. This is not as bad as it might at first sight seem since we only want terms up to x^3.

I shall multiply the second bracket by each of the terms of the first bracket, only writing down the terms I need. This gives me

$$\frac{1}{2}\,[1 - 2x + 4x^2 - 8x^3$$
$$+ \tfrac{1}{2}x - \;\; x^2 + 2x^3$$
$$+ \tfrac{1}{4}x^2 - \tfrac{1}{2}x^3$$
$$+ \tfrac{1}{8}x^3\,]$$

$$= \tfrac{1}{2}\,[1 - \tfrac{3}{2}x + \tfrac{13}{4}x^2 - \tfrac{51}{8}x^3]$$

$$= \tfrac{1}{2} - \tfrac{3}{4}x + \tfrac{13}{8}x^2 - \tfrac{51}{16}x^3.$$

METHOD (2)   This avoids the multiplication by finding partial fractions for y. (Partial fractions are explained in Section 6.E.) Doing this gives

$$y = \frac{\tfrac{1}{5}}{2-x} + \frac{\tfrac{2}{5}}{1+2x}$$

$$= \tfrac{1}{5}(2-x)^{-1} + \tfrac{2}{5}(1+2x)^{-1} = \tfrac{1}{10}\left(1 - \frac{x}{2}\right)^{-1} + \tfrac{2}{5}(1+2x)^{-1}$$

$$= \tfrac{1}{10}\left(1 + \frac{x}{2} + \frac{x^2}{4} + \frac{x^3}{8} + \ldots\right) + \tfrac{2}{5}(1 - 2x + 4x^2 - 8x^3 + \ldots)$$

$$= \tfrac{1}{2} - \tfrac{3}{4}x + \tfrac{13}{8}x^2 - \tfrac{51}{16}x^3$$

writing down the first four terms.

Finally, whichever method we used, we must find the interval in which x must lie for the expansion to be valid. Both methods involved the same two expansions, so we look at each of these in turn.

For the first, we want $|-x/2| < 1$ so $|x/2| < 1$ and $-2 < x < 2$.

For the second, we must have $|2x| < 1$ so $-\tfrac{1}{2} < x < \tfrac{1}{2}$.

So, to fit *both* requirements, we must take the tighter of the two restrictions, so $-\frac{1}{2} < x < \frac{1}{2}$. This is the same situation as a lorry driving down a road which successfully makes it under the first bridge, but the headroom of the second bridge is lower. Disaster will strike unless the lorry is also lower than this second bridge.

EXERCISE 7.C.2

Try these for yourself.

(1) Expand each of the following as far as the term in x^3. In each case, find the interval in which x must lie for the expansion to be valid.

(a) $(1 - 3x)^{1/3}$ (b) $\left(1 + \dfrac{x}{2}\right)^{2/3}$ (c) $(16 - 3x)^{1/4}$

(d) $(4 + x)^{-1/2}$ (e) $(-2 + x)^{-2}$ (f) $(27 - 4x)^{-2/3}$

You may need to look back at the rules for powers in Section 1.D.(b) for help with the tidying up.

(2) Expand $(3 - 2x)^{-1} (1 + 3x)^{-1}$ as far as the term in x^2, and find the interval in which x must lie for this expansion to be valid.

7.D   Mathematical induction

7.D.(a)   Truth from patterns – or false mirages?

If we find a particular pattern, how can we discover if this pattern will always be true or if it was just a lucky chance that it was true for the cases which we looked at?

To answer this question, we will start by looking at the following pair of series.

(a) $1 + 2 + 3 + 4 + \ldots + n = \displaystyle\sum_{r=1}^{n} r$

(b) $1^3 + 2^3 + 3^3 + 4^3 + \ldots + n^3 = \displaystyle\sum_{r=1}^{n} r^3$

An interesting thing happens if we compare the two sets of partial sums of these series, as they build up.

If we take $n = 1$, so we are just comparing the first terms, we get

S_1 for (a) $= 1$ and S_1 for (b) $= 1$.

Summing the first two terms of each series, we get

S_2 for (a) $= 3$ and S_2 for (b) $= 9$.

Find S_3 and S_4 for each series yourself and see if you can suggest an experimental pattern for what is happening.

You will have

S_3 for (a) is 6, S_3 for (b) is 36,

S_4 for (a) is 10, S_4 for (b) is 100.

It rather looks as though, if we square the sum of (a) for any given number of terms, we get the corresponding sum for (b), for that number of terms.

In other words, it looks as though, if n is any number of terms we might choose to pick, then

$$(S_n \text{ for (a)})^2 = S_n \text{ for (b)}.$$

(Because n is counting the number of terms, it must be a positive whole number or natural number as these counting numbers are sometimes called.)

Now, we have already found a formula for the sum of n terms of series (a) in Exercise 6.B.1. 2(d). We found

$$S_n = \frac{n(n + 1)}{2}$$

Is it true that S_n for (b) is $n^2(n + 1)^2/4$ whatever n is?

We shall prove that this *is* true by using the following process.

Mathematical induction: how to do it

(1) We first show that a statement is true for the case in which $n = 1$.

(2) We then show that

if the statement is true when n is given some particular value, k,

then it must also be true if $n = k + 1$.

We can then argue that, since we know it is true when $n = 1$, it must also be true for $n = 2$, and therefore also for $n = 3$, etc. through all the counting numbers.

We have already done step (1) for this first example.

Now we go to step (2).

We will suppose that the statement

$$\sum_{r=1}^{n} r^3 = \frac{n^2 (n + 1)^2}{4}$$

is true when n is given the particular value, k, so that

$$1^3 + 2^3 + 3^3 + 4^3 + \ldots + k^3 = \frac{k^2 (k + 1)^2}{4} \qquad \text{(This is St[k].)}$$

The St[k] at the right-hand edge of the line above is a convenient shorthand meaning 'the statement of the formula when $n = k$'.

We then show, that *if* St[k] is true, *then* the formula is also true when $n = k + 1$, so that St[$k + 1$] is true.

Here, we must show that if St[k] is true then

$$1^3 + 2^3 + 3^3 + 4^3 + \ldots + k^3 + (k + 1)^3 = \frac{(k + 1)^2 (k + 2)^2}{4}. \qquad \text{(This is St[$k+1$].)}$$

We have added in the extra term on the left-hand side, and replaced k by $k + 1$ in the formula on the right-hand side.

The LHS of St[$k + 1$] can be written as

$$(1^3 + 2^3 + 3^3 + 4^3 + \ldots k^3) + (k + 1)^3 = \frac{k^2 (k + 1)^2}{4} + (k + 1)^3$$

using St[k] to replace $1^3 + 2^3 + 3^3 + \ldots + k^3$ with $k^2(k + 1)^2/4$.

Now we factorise this, by taking out the common factor of $(k + 1)^2$.

It will also pay us here to take out a factor of $\frac{1}{4}$, as it is more convenient to have the fractions at the front, out of the way.

This then gives

$$\frac{k^2(k + 1)^2}{4} + (k + 1)^3 = \tfrac{1}{4}(k + 1)^2 \, (k^2 + 4(k + 1)).$$

Notice the 4 *inside* the bracket, to make it multiply out correctly to give what we started with. But

$$\tfrac{1}{4}(k + 1)^2(k^2 + 4k + 4) = \tfrac{1}{4} \, (k + 1)^2(k + 2)^2$$

so now we have

$$1^3 + 2^3 + 3^3 + \ldots + (k + 1)^3 = \tfrac{1}{4}(k + 1)^2 \, (k + 2)^2 = \text{RHS of St}[k + 1].$$

Therefore we have shown that *if* St[k] is true, *then* St[$k + 1$] is true.

But we know that St[1] is true, so St[2] is true, and so on, for $n = 3, 4, \ldots$ through all the counting numbers.

Here is a second example of proof by induction. Prove that

$$\sum_{r = 1}^{n} r^2 = \tfrac{1}{6}n(n + 1) \, (2n + 1).$$

You may notice a rather serious disadvantage here! The method of mathematical induction is only going to be any use when we have somehow come to what the formula might be by some other route. It won't find an appropriate formula for us.

Working with the formula we have been given here, we first check that it works for $n = 1$, that is, that St[1] is true.

Always start with this; if it is not true there is no point in proceeding any further, and if it is true, showing this is part of the chain of the proof.

If $n = 1$, the LHS = 1, and the RHS = $\tfrac{1}{6}(1)(2)(3) = 1$ so it is true in this case.

Next, we suppose that the formula is true for $n =$ a particular value, k, that is, we suppose

$$1^2 + 2^2 + 3^2 + \ldots + k^2 = \tfrac{1}{6}k(k + 1)(2k + 1)$$ St[k]

We then have to show that this would mean that the formula is also true for $n = k + 1$, that is, we must show that, if St[k] is true, then

$$1^2 + 2^2 + 3^2 + \ldots + k^2 + (k + 1)^2 = \tfrac{1}{6}(k + 1)(k + 2)(2k + 3)$$ St[$k + 1$]

adding in the extra term on the LHS, and replacing k by $k + 1$ on the RHS.

The LHS of St[$k + 1$] can then be rewritten as

$$\tfrac{1}{6}k(k + 1)(2k + 1) + (k + 1)^2 \quad \text{using St}[k] \text{ to replace } 1^2 + 2^2 + 3^2 + \ldots + k^2.$$

Factorising in a similar way to the last example, we have

$$\tfrac{1}{6}k(k + 1)(2k + 1) + (k + 1)^2 = \tfrac{1}{6}(k + 1)(k(2k + 1) + 6(k + 1)).$$

If you are at all doubtful about your factorising at this stage, check by multiplying back that it agrees with the previous step.

Tidying up, we get

$$\tfrac{1}{6}(k + 1)(k(2k + 1) + 6(k + 1)) = \tfrac{1}{6}(k + 1)(2k^2 + 7k + 6)$$

$$= \tfrac{1}{6}(k + 1)(k + 2)(2k + 3) = \text{RHS of St}[k + 1].$$

Therefore, *if* St[k] is true, *then* St[$k + 1$] is true. But we know that St[1] is true, so therefore the statement is true for $n = 2, 3, 4, \ldots$ all through the counting numbers.

It is important that St[k] is shorthand for a *statement*.
It is not a function or part of an equation.
In the example above, you can't say St[k] $= 1^2 + 2^2 + 3^2 + \ldots + k^2$

or St[k] $= \tfrac{1}{6}k(k + 1)(2k + 1)$.

St[k] is the statement that $1^2 + 2^2 + 3^2 + \ldots + k^2 = \tfrac{1}{6}k(k + 1)(2k + 1)$.
St[$k + 1$] is the statement that

$$1^2 + 2^2 + 3^2 + \ldots + k^2 + (k + 1)^2 = \tfrac{1}{6}(k + 1)(k + 2)(2k + 3)$$

St[$k + 1$] is exactly the same as St[k] except that k has been replaced by $k + 1$.

EXERCISE 7.D.1

Try these similar questions yourself.

(1) When we were working on APs, we found in Exercise 6.B.1 question 2(d) that

$$1 + 2 + 3 + 4 + \ldots + n = \tfrac{1}{2}n(n + 1).$$

See if you can prove this, using mathematical induction.

(2) First, see if you can spot a way of finding the sum of n odd numbers by looking at what you get for the first four sums, that is:

(a) $S_1 = 1$ (b) $S_2 = 1 + 3$ (c) $S_3 = 1 + 3 + 5$ (d) $S_4 = 1 + 3 + 5 + 7$.

Then, if you have guessed a formula, see if you can prove it is true by mathematical induction.

(3) Show, using mathematical induction, that

$$(1 \times 2) + (2 \times 3) + (3 \times 4) + \ldots + n(n + 1) = (n/3)(n + 1)(n + 2).$$

7.D.(b)   **Proving the Binomial Theorem by induction**

As the summit of our ambition for this section, we will now prove the Binomial Theorem using induction.

We have already done the only hard bit when we showed in question (3) of Exercise 7.A.5 that

$$\frac{k!}{r! \, (k - r)!} + \frac{(k + 1)!}{r! \, (k + 1 - r)!} = \frac{(k + 1)!}{r!(k - r + 1)!}.$$

So we now set out to show that

$$(1 + x)^n = 1 + \frac{n}{1!} x + \frac{n(n - 1)}{2!} x^2 + \ldots + \frac{n!}{r!(n - r)!} x^r + \ldots + x^n$$

where n is a positive whole number, by using mathematical induction.

We first have to check that the statement is true when $n = 1$ (that is, that St[1] is true).

If $n = 1$, we get $(1 + x)^1 = 1 + x$ so St[1] is true.

Now we have to show that *if* the formula is true when $n =$ a particular value, k, *then* it must also be true when $n = k + 1$. (That is, we show that, *if* St[k] is true, *then* St[$k + 1$] is also true.)

To write down St[k], we must replace n by this particular value k. So St[k] says

$$(1 + x)^k = 1 + \frac{k}{1!} x + \frac{k(k - 1)}{2!} x^2 + \ldots + \frac{k!}{(k - 1)!(k - r + 1)!} x^{r - 1} +$$

$$\frac{k!}{r! \, (k - r)!} x^r + \ldots + x^k.$$

Notice that we have included the term with x^{r-1} as well as the one with x^r. Can you see why?

St[$k + 1$] states that

$$(1 + x)^{k+1} = 1 + \frac{(k + 1)}{1!} x + \frac{(k + 1)(k)}{2!} x^2 + \ldots + \frac{(k + 1)!}{r! \, (k + 1 - r)!} x^r + \ldots + x^{k+1}.$$

(To write this down, we just replaced 'n' by '$k + 1$' in (B2).)

But $(1 + x)^{k+1} = (1 + x)(1 + x)^k$.

We need to show that the term in x^r resulting from this multiplication is the same as the term in x^r in St[$k + 1$].

7.D Mathematical induction   **281**

But, just as in the examples we have already looked at, the term in x^r in $(1 + x)(1 + x)^k$ comes from

$$1 \times (\text{the term in } x^r \text{ from } (1 + x)^k) + x \times (\text{the term in } x^{r-1} \text{ from } (1 + x)^k).$$

So we have to show that

$$\left(\frac{k!}{r!\,(k-r)!} + \frac{k!}{(r-1)!\,(k-r+1)!} \right) x^r = \left(\frac{(k+1)!}{r!\,(k+1-r)!} \right) x^r$$

but this is exactly what we have already shown in question (3) of Exercise 7.A.5.

So we know that, *if* St[k] is true, *then* St[$k + 1$] is also true.

But St[1] is true, and therefore St[2] is true, and St[3] and so on through all the counting numbers, and the theorem is proved.

7.D.(c)   Two non-series applications of induction

The method of mathematical induction is not just restricted to proving results for series. Here are two examples of other ways in which it can be used.

EXAMPLE (1)   Show that, if n is a positive integer, $9^n - 1$ is always divisible by 8.

As always, we test first by putting $n = 1$.

Doing this gives $= 9^1 - 1 = 9 - 1 = 8$ so the statement is true for $n = 1$.

Now we assume that $9^n - 1$ is divisible by 8 when $n =$ a particular value, k. We can show this by writing $9^k - 1 = 8M$ where M stands for some positive whole number.

Stating that $9^k - 1 = 8M$ is St[k].

We have to show now that, *if* St[k] is true, *then* St[$k + 1$] is also true, that is, that $9^{k+1} - 1$ is *also* divisible by 8.

Now $9^{k+1} - 1 = 9(9^k) - 1 = 9(8M + 1) - 1$ using St[k] to replace 9^k by $8M + 1$.

So $9^{k+1} - 1 = 72M + 9 - 1 = 72M + 8 = 8(9M + 1)$.

Therefore $9^{k+1} - 1$ *is* divisible by 8.

We have shown that, *if* St[k] is true, *then* St[$k + 1$] is true.

But St[1] is true so therefore St[2] is true, and so on through all the counting numbers.

HELPFUL HINT The juggling of the powers which we used above by writing 9^{k+1} as $9(9^k)$ so that we could substitute for 9^k is typical of what works for this type of question.

EXAMPLE (2)   Suppose that we have an infinite flat sheet of paper (which is the same as a **plane** in geometry). We then draw straight lines on it so that no two lines are parallel and no new line cuts through a point where two previous lines cross each other.

How is the number of crossing points related to the number of lines?

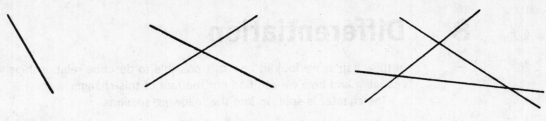

Figure 7.D.1

For example, from the sketches in Figure 7.D.1,

- one line has no crossing points,
- two lines have one crossing point,
- three lines have three crossing points etc.

Draw separate sketches for four and five lines (remembering that you must extend the lines sufficiently far so that all possible crossing points are counted).

Now see if you can find a relationship between the number of lines and the number of crossing points.

You should have got six crossing points for four lines and ten crossing points for five lines.

If you had trouble spotting a relationship, doubling the number of crossing points may help you to see the pattern.

You should then get a possible rule that n lines have $\frac{1}{2}n(n-1)$ crossing points.

But *we do not know* that this is always true; further checking from sketches will only show it to be true for as many sketches as we draw. (Sometimes the most apparently beautiful patterns break down when n is quite large even though they have seemed fine until then.)

However, we can now show that this formula *is* always true by induction.

We know that it is true for $n = 1$.

Suppose that it is true for $n = k$ so that k lines do cut each other in $\frac{1}{2}k(k-1)$ crossing points. This is St[k].

Now St[$k + 1$] states that $k + 1$ lines would cut each other in $\frac{1}{2}(k + 1)(k)$ crossing points. Does this follow from St[k]?

The $(k + 1)$th line cuts all the previous k lines in k extra points, so drawing in this $(k + 1)th$ line gives us a total of $\frac{1}{2}k(k-1) + k$ cutting points. But

$$\tfrac{1}{2}k(k-1) + k = \tfrac{1}{2}k((k-1) + 2) = \tfrac{1}{2}k(k+1)$$

so, if St[k] is true, then St[$k + 1$] is also true.

But St[1] is true, and therefore St[2] is true, and so on for any possible number of lines.

8 Differentiation

In this chapter we look at how it is possible to describe relationships which are changing and how we can find out the rate of this change.

The chapter is split up into the following sections.

8.A Some problems answered and difficulties solved
(a) How can we find a speed from knowing the distance travelled?
(b) How does $y = x^n$ change as x changes?
(c) Different ways of writing differentiation: dx/dt, $f'(t)$, $\dot{x}$, etc.,
(d) Some special cases of $y = ax^n$,
(e) Differentiating $x = \cos t$ answers another thinking point,
(f) Can we always differentiate? If not, why not?

8.B Natural growth and decay – the number e
(a) Even more money – compound interest and exponential growth,
(b) What is the equation of this smooth growth curve?
(c) Getting numerical results from the natural growth law of $x = e^t$,
(d) Relating ln x to the log of x using other bases,
(e) What do we get if we differentiate ln t?

8.C Differentiating more complicated functions
(a) The Chain Rule, (b) Writing the Chain Rule as $F'(x) = f'(g(x))g'(x)$,
(c) Differentiating functions with angles in degrees or logs to base 10,
(d) The Product Rule, or 'uv' Rule, (e) The Quotient Rule, or 'u/v' Rule

8.D The hyperbolic functions of sinh x and cosh x
(a) Getting symmetries from e^x and e^{-x}, (b) Differentiating sinh x and cosh x,
(c) Using sinh x and cosh x to get other hyperbolic functions,
(d) Comparing other hyperbolic and trig formulas – Osborn's Rule,
(e) Finding the inverse function for sinh x,
(f) Can we find an inverse function for cosh x?
(g) tanh x and its inverse function tanh^{-1} x,
(h) What's in a name? Why 'hyperbolic' functions?
(i) Differentiating inverse trig and hyperbolic functions,

8.E Some uses for differentiation
(a) Finding the equations of tangents to particular curves,
(b) Finding turning points and points of inflection,
(c) General rules for sketching curves, (d) Some practical uses of turning points,
(e) A clever use for tangents – the Newton–Raphson Rule

8.F Implicit differentiation
(a) How implicit differentiation works, using circles as examples,
(b) Using implicit differentiation with more complicated relationships,
(c) Differentiating inverse functions implicitly,
(d) Differentiating exponential functions like $x = 2^t$,
(e) A practical application of implicit differentiation,

8.G Writing functions in an alternative form using series

What kinds of things can differentiation tell us? I find that sometimes students know some rules but don't really know what use these rules are. We start this chapter by looking at some examples based on earlier thinking points. In these, we wanted to find answers to what is happening in particular physical situations.

If you see how we can use differentiation to help us here, you will understand better what kinds of things it can do for you.

8.A.(a)   How can we find a speed from knowing the distance travelled?

Suppose somebody is walking at a steady speed of 3 miles per hour (m.p.h.). Then the distance travelled for different lengths of time can be shown on a graph sketch like the one in Figure 8.A.1.

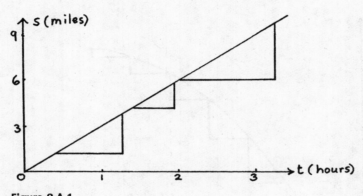

Figure 8.A.1

Since equal distances are covered in equal intervals of time, the speed is represented by the gradient of the line, and this can be found by using any of the triangles I have drawn in; the size does not matter.

Any two points (x_1, y_1) and (x_2, y_2) on the line will give its gradient, using the formula

$$m = \frac{y_2 - y_1}{x_2 - x_1} \quad \text{from Section 2.B.(d).}$$

Each of these triangles will give a gradient of 3. This represents the constant rate of change of distance travelled, or steady speed, of 3 m.p.h.

But how can we find the speed if the rate at which the distance is covered is continually changing?

This question first came up at the end of the thinking point of Section 2.D.(g), in which we looked at how the motion of a ball thrown up in the air changes as time passes.

Look at this again so that we can use it together now.

Because of the pull of gravity, the speed of the ball is changing all the while. It is moving fastest when it leaves the thrower's hands and when it returns to them; and slowest when it comes instantaneously to rest at the highest point of its motion. (We can say that it does this because there is an instant in its motion when, rather like the Grand Old Duke of York, it is neither moving up nor down.) Between these two extremes, its speed is changing smoothly, so that the graph of the distance travelled against the time that this has taken is a curve.

The last question I asked you in this thinking point was whether you could think of a way of estimating the ball's speed one second after it has been thrown up in the air. Surely since we can find how far it has travelled at any instant we should be able to do this?

We used the equation $s = ut - \frac{1}{2}gt^2$ to give us the distance s in metres (m), travelled by the ball after a time of t second(s), if it is thrown up at a speed of u metres per second ($\mathrm{m\,s^{-1}}$).

In our example, the ball was thrown up at a speed of $14\,\mathrm{m\,s^{-1}}$, and we took g, the acceleration due to gravity, as 9.8 metres per second per second ($\mathrm{m\,s^{-2}}$).

This then gave us the equation of $s = 14t - 4.9t^2$ for the curve.

I have drawn a new sketch graph, in Figure 8.A.2.(a), showing how the height of the ball changes with time over the first 1.4 seconds of the motion.

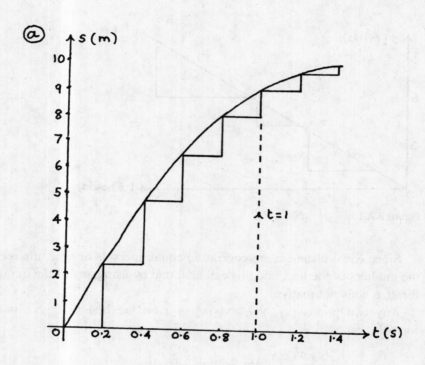

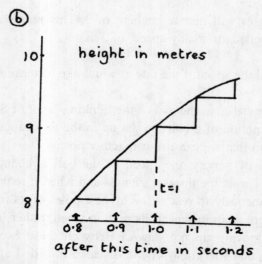

Figure 8.A.2

Differentiation

I have drawn in the separate changes in height for each 0.2 second interval on this graph, to give a picture of how the speed is changing. You can see the inaccuracy in this by drawing in the slant sides of the triangles yourself. The slopes or gradients of these slant sides are giving the average speeds over each 0.2 second interval, but they only give an approximation to the actual shape of the curve. It seems reasonable to think that, at any point where two adjacent triangles touch it, the steepness of the curve will be somewhere between the steepness of the slant sides of these two triangles.

Taking the equation of the curve as $s = 14t - 4.9t^2$, we can make the table below for the different values of s.

(a)

t	0	0.2	0.4	0.6	0.8	1.0	1.2	1.4
s	0	2.60	4.82	6.64	8.06	9.10	9.74	10.00

(b)

t	0.8	0.9	1.0	1.1	1.2
s	8.06	8.63	9.10	9.47	9.74

We now use the triangles either side of $t = 1$ to get estimates of the speed when $t = 1$. I will call the change in height Δs and the corresponding change in time Δt.

(Δ is the Greek capital D, pronounced 'delta'. It is often used to mean 'the change in'. We have used it this way already in Section 3.A.(b).)

The left-hand triangle gives

$$\frac{\Delta s}{\Delta t} = \frac{9.10 - 8.06}{0.2} = 5.2\,\mathrm{m\,s^{-1}}.$$

The right-hand triangle gives

$$\frac{\Delta s}{\Delta t} = \frac{9.74 - 9.10}{0.2} = 3.2\,\mathrm{m\,s^{-1}}.$$

From Figure 8.A.2(a), we believe that $5.2\,\mathrm{m\,s^{-1}}$ is an over-estimate and $3.2\,\mathrm{m\,s^{-1}}$ is an under-estimate of the speed when $t = 1$.

Next, we try taking smaller time intervals either side of $t = 1$. I have done this in table (b), and I show the separate changes in height on this small section of curve in Figure 8.A.2(b). Again, you should draw in the slant sides yourself.

Taking the two triangles on either side of $t = 1$ again, the left-hand triangle gives

$$\frac{\Delta s}{\Delta t} = \frac{0.47}{0.1} = 4.7\,\mathrm{m\,s^{-1}}$$

and the right-hand triangle gives

$$\frac{\Delta s}{\Delta t} = \frac{0.37}{0.1} = 3.7\,\mathrm{m\,s^{-1}}.$$

We see that we are getting closer to an agreement between the estimates.

Infilling again in the same kind of way gives us the table below.

t	0.90	0.95	1.00	1.05	1.10
s	8.63	8.88	9.10	9.30	9.47

8.A Some problems answered

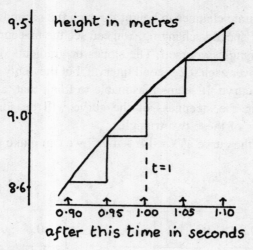

Figure 8.A.3

I have shown again, in Figure 8.A.3, a magnified picture of the small part of the curve which we are considering here. If you now draw in the slant sides of the triangles, you will find that they are almost indistinguishable from the curve itself.

Since the differences are now becoming very small, it seems a good idea to show this by labelling them in a slightly different way. I shall use δ, which is the small Greek letter d, and call the changes δs and δt. δ is very commonly used in maths to mean 'a small change in'.

Now, looking at the two small triangles either side of $t = 1$ shown in Figure 8.A.3, the left-hand triangle gives

$$\frac{\delta s}{\delta t} = \frac{0.22}{0.05} = 4.4 \, \text{m s}^{-1}$$

and the right-hand triangle gives

$$\frac{\delta s}{\delta t} = \frac{0.20}{0.05} = 4 \, \text{m s}^{-1}.$$

So, coming from the left and from the right, we have two sets of approximations which are getting closer and closer to the speed at the instant when $t = 1$. We have

$$5.2 \to 4.7 \to 4.4 \quad \text{and} \quad 4 \leftarrow 3.7 \leftarrow 3.2$$

This system looks very promising. We can see that the smaller the differences are the better the approximation is, so perhaps we should focus on making the differences extremely small and see what happens?

We don't want to specify exactly how small since, for any given interval, we know we could always halve that and so get a better approximation.

So what we will do is to look at what happens to $\delta s/\delta t$, just making the proviso that we are letting δt become smaller and smaller. We are snuggling the little triangles in closer and closer to $t = 1$ from both sides.

Also, it would be much nicer if we could get a rule for finding the speed which works for different initial speeds, u, and for the slightly different possible values of g as we travel over the earth's surface, so that we don't have to recalculate every time these are different. So, instead of taking particular values, we will work with u and g.

We start with $s = ut - \frac{1}{2}gt^2$ and then see what happens to this equation at the nearby time of $t + \delta t$.

If the time has changed by a small amount δt then the distance s will also have changed by a correspondingly small amount δs. So we will have

$$s + \delta s = u(t + \delta t) - \tfrac{1}{2}g(t + \delta t)^2.$$

Now,

$$(t + \delta t)^2 = t^2 + 2t(\delta t) + (\delta t)^2.$$

So

$$s + \delta s = ut + u(\delta t) - \tfrac{1}{2}gt^2 - gt(\delta t) - \tfrac{1}{2}g(\delta t)^2 \qquad (1)$$

But, at time t,

$$s = ut - \tfrac{1}{2}gt^2 \qquad\qquad\qquad\qquad (2)$$

Subtracting (2) from (1) gives

$$\delta s = u(\delta t) - gt(\delta t) - \tfrac{1}{2}g(\delta t)^2$$

so

$$\frac{\delta s}{\delta t} = u - gt - \tfrac{1}{2}g(\delta t).$$

But, if we now let δt get closer and closer to zero, it will become so small that we can ignore the $-\tfrac{1}{2}g(\delta t)$.

Because δs is also becoming very small, the fraction $\delta s/\delta t$ continues to give the slope of the slant side of the little triangle. The smaller this triangle becomes, the closer this slope gets to the slope of the curve itself at the point (t, s).

As δt gets smaller and smaller, $\delta s/\delta t$ will become closer and closer in size to $u - gt$.

We write this mathematically by saying that the limit of $\delta s/\delta t$ as $\delta t \to 0$ is $u - gt$.

The limit of $\dfrac{\delta s}{\delta t}$ as $\delta t \to 0$ is called $\dfrac{ds}{dt}$.

In this particular example, we have $ds/dt = u - gt$. We now have a rule to tell us the speed at any point on the path of the ball.

The value of ds/dt tells us the rate of change of s with respect to t for any chosen value of t while the ball is still in motion.

The line with gradient ds/dt which touches the curve at this particular value of t, showing its steepness there, is called the **tangent** to the curve at this point.

Returning to the particular case of $u = 14$ and $g = 9.8$, we can now work out the speed of the ball one second after it has been thrown into the air.

It is given by $v = ds/dt = u - gt = 14 - 9.8 = 4.2$, so the speed is $4.2 \, \text{m s}^{-1}$. I show this on Figure 8.A.4(a). I also show again, in Figure 8.A.4(b), the little sketch of the actual path of the ball, which is straight up and straight down. The graph of Figure 8.A.4(a) shows how its distance from the ground changes with time.

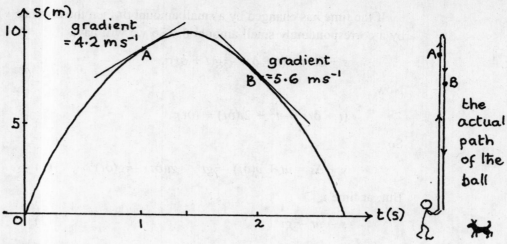

Figure 8.A.4

The gradient of the curve at A, that is, of its tangent there, is 4.2. The speed of the ball after half a second is $4.2 \, \mathrm{m \, s^{-1}}$ vertically upwards.

Similarly, if $t = 2$, $ds/dt = -5.6$. The gradient of the curve, given by the gradient of its tangent at B, is negative. The speed of the ball is $5.6 \, \mathrm{m \, s^{-1}}$ vertically downwards.

Taking the vertically upwards direction as positive, we can say that the **velocity** of the ball (which describes the direction of its motion as well as its speed) is $4.2 \, \mathrm{m \, s^{-1}}$ at A and $-5.6 \, \mathrm{m \, s^{-1}}$ at B.

NOTE

When you first looked at this thinking point, because the acceleration is constant, you may have used the formula $v = u + at$ to find the speed when $t = 1$, putting $u = 14$ and $a = -9.8$. This also gives $v = 4.2$. This method works very well in this particular example, but the method we have just been looking at above is enormously more powerful because it can cope with situations of non-constant acceleration (and much else besides).

8.A.(b)   **How does $y = x^n$ change as x changes?**

We can now answer this question provided that n is a positive whole number.

(I am putting in just enough examples here of where these formulas come from to show you how they link back to past work, and to justify using them in their hundreds of applications.)

We will look at what kind of small change, δy, we will get in y if we change x by the small amount δx. We have

$$y = x^n \quad \text{so} \quad y + \delta y = (x + \delta x)^n.$$

Now, we can expand $(x + \delta x)^n$ using Rule (B1) from Section 7.A.(e). This gives us

$$y + \delta y = x^n + nx^{n-1}(\delta x) + \frac{n(n-1)}{2!} x^{n-2}(\delta x)^2$$

$$+ \text{terms with higher powers of } \delta x.$$

Putting $y = x^n$, and tidying up, gives us

$$\delta y = nx^{n-1}(\delta x) + \frac{n(n-1)}{2!}x^{n-2}(\delta x)^2 + \text{other terms with higher powers of } \delta x$$

so

$$\frac{\delta y}{\delta x} = nx^{n-1} + \frac{n(n-1)}{2!}x^{n-2}(\delta x) + \text{other terms with higher powers of } \delta x.$$

If we now let $\delta x \to 0$, everything except nx^{n-1} becomes so small that we can ignore it, and we have

> The limit of $\dfrac{\delta y}{\delta x}$ as $\delta x \to 0$ is nx^{n-1}.
>
> If $y = x^n$ then $\dfrac{dy}{dx} = nx^{n-1}$.

We know that this result is true if n is a positive whole number because we showed that the Binomial Theorem is true in this case.

Mathematicians have shown that this result is still true if n is *any* real number, and we will use this widened version.

Multiplying by a constant, a, will just have the effect of multiplying the answer by a. This gives us the following general rule.

> If $y = ax^n$ then $\dfrac{dy}{dx} = nax^{n-1}$.

Doing this process is called **differentiating** (with respect to x if the function is in terms of x, or with respect to t if it is in terms of t etc.).

If we have a string of terms similar to this which are added or subtracted, we can go through differentiating term by term in order to find the total rate of change, so, for example, if $y = 3x^2 + 2x$, then $dy/dx = 6x + 2$.

8.A.(c) Different ways of writing differentiation: dx/dt, $f'(t)$, $\dot{x}$, etc.

There is another way of writing dy/dx, dx/dt, etc. which emphasises more that we are doing the process of differentiation to functions.

In Chapter 3, we used $f(x)$, $g(x)$, $f(t)$ and so on to talk about functions of x and t.

If we have $y = f(x)$, then $\dfrac{dy}{dx}$ is also sometimes written as $f'(x)$.

If we have $x = f(t)$, then $\dfrac{dx}{dt}$ can also be written as $f'(t)$.

Writing $x = f(t)$ stresses that x is a function of the variable t.

The dash in $f'(t)$ means that the function $f(t)$ has been differentiated with respect to this variable.

In the particular circumstances when $x = f(t)$ is a function of time, sometimes the **dot notation** is used.

In this notation dx/dt is written as $\dot{x}$.

$$\text{If } x = f(t) \quad \text{then} \quad \frac{dx}{dt} = \dot{x} = f'(t).$$

Historically, the ideas of calculus were developed separately but in parallel by eminent (but rivalrous) mathematicians.

The notation dx/dt was used by the German mathematician Leibnitz.

The notation $\dot{x}$ was used by the English mathematician and physicist Newton.

Here are some examples, using the two most usual notations.

(1) If $y = f(x) = 3x^4 + 2x^3$ then $\dfrac{dy}{dx} = f'(x) = 12x^3 + 6x^2$.

(2) If $s = f(t) = 2t + \frac{1}{2}t^3$ then $\dfrac{ds}{dt} = f'(t) = 2 + \frac{3}{2}t^2$.

(3) If $x = f(t) = 5t + 4t^{1/2}$ then $\dfrac{dx}{dt} = f'(t) = 5 + 4 \times \frac{1}{2} \times t^{-1/2} = 5 + 2t^{-1/2}$.

(4) If $y = f(x) = 5x + \dfrac{2}{x^2} = 5x + 2x^{-2}$ then $\dfrac{dy}{dx} = f'(x) = 5 - 4x^{-3} = 5 - \dfrac{4}{x^3}$.

(If you are unsure about the use of powers here, see Section 1.D.)

EXERCISE 8.A.1

Try these for yourself. Differentiate with respect to whatever letter the function is written in on the right-hand side.

(1) $y = 7x^2 + 3x^4$ (2) $x = 5t - \frac{1}{2}t^3$ (3) $y = 3 - 2/x^3$ (4) $x = 2t^{1/2} + 3t^{-1/2}$.

(5) (a) Show, by thinking about what happens when x is increased by a small amount δx, that if $y = x^3$ then $dy/dx = 3x^2$.

(5) (b) Check what happens at each stage of your working numerically by taking the particular case of $x = 2$ and $\delta x = 0.001$.

8.A.(d)   Some special cases of $y = ax^n$

Students sometimes have difficulty linking the rule for differentiating $y = ax^n$ back to these two particular cases, so I have put in two examples here to show how this works.

(1) If $n = 1$ then $y = ax^n$ is the straight line $y = ax$. (This is using $x^1 = x$ from Section 1.D.(b).)

For example, if $y = 3x$ then, using the rule above, we get $dy/dx = 3x^0 = 3$ since $x^0 = 1$. (This is also in Section 1.D.(b).)

The result of using the rule agrees entirely with what we know to be the gradient of the line. (See Figure 8.A.5(a).)

(2) If $n = 0$ then we have a very particular kind of straight line of the form $y = a$ where a is some number.

For example, if $y = 4$ then we can say $y = 4x^0$.

Now using the rule gives us $dy/dx = 0 \times 4x^{-1} = 0$.

Again, this fits in with what we can see to be true in Figure 8.A.5(b).

The line $y = 4$ is horizontal and its gradient is zero.

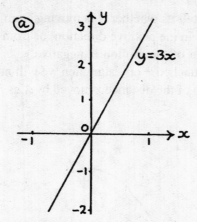

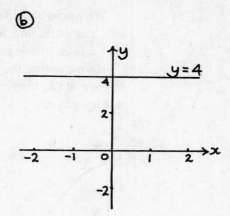

Figure 8.A.5

Two special cases

If $y = ax$ then $\dfrac{dy}{dx} = a.$

If $y = a$ then $\dfrac{dy}{dx} = 0.$

(a stands for any constant number.)

8.A.(e)   **Differentiating $x = \cos t$ answers another thinking point**

In Section 5.A.(d), we looked at how the point X moves on the line AB as P moves round a circle of unit radius at 1 rad/s. You should go back to this now, and answer the questions there, if you haven't already done so. Because this particular kind of motion is of enormous importance in physics and engineering applications, I will use it as a last example of how we can find a rate of change by considering what happens over smaller and smaller time intervals. After this, we will use these results as we need them without specifically proving any further ones.

I show the diagram again here in Figure 8.A.6. The final question of this thinking point was to find the speed of X after a time interval of t seconds, knowing that the distance OX is given by $OX = x = \cos t$.

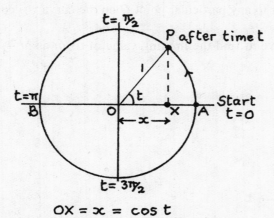

$$OX = x = \cos t$$

Figure 8.A.6

We would also like the answer to tell us whether X is moving from left to right, in which case x is increasing and the motion is in the positive direction; or from right to left, in which case x is decreasing and the direction of the motion is negative.

If we can find the speed with its attached + or – sign then we will have found the **velocity** of the point X. I have shown the graph of the distance x moved by X as P goes round its circle in Figure 8.A.7.

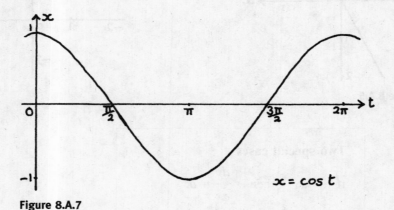

Figure 8.A.7

We know $x = \cos t$.

How does x change as t changes?

We saw in the thinking point that X moves fastest as it passes through O and instantaneously comes to rest every time it gets to either A or B because then it turns back on itself.

Also, when $t = 0$, it starts by moving in the negative direction towards O. Its velocity will be negative for the first π seconds of its motion. Also, its velocity changes regularly with time just like its distance from O does.

Do you have any idea how you could write this velocity in terms of t?

Could it be that the rate of change of X's distance from O with time, that is dx/dt, is equal to $-\sin t$? To answer this question, we shall look how x changes if we change t by a small amount δt.

We are again looking at the gradients of the slanted sides of the little triangles as they tuck in closer and closer to any particular point Q on the curve $x = \cos t$. I show a possible pair in Figure 8.A.8.

To find dx/dt, we have to find the limiting value of $\delta x/\delta t$ as $\delta t \rightarrow 0$.

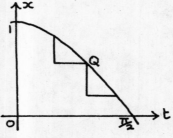

Figure 8.A.8

If the time changes by a small amount δt, so that the distance changes by a correspondingly small amount δx, we have $x + \delta x = \cos(t + \delta t)$. Which of the formulas from Chapter 5 can we use here on the RHS?

We can use $\cos(A + B) = \cos A \cos B - \sin A \sin B$ (Section 5.D.(b)). So then we have

$$x + \delta x = \cos t \cos(\delta t) - \sin t \sin(\delta t).$$

Now comes the step which only works *because* we are measuring the angle turned through by P in radians.

In Section 4.D.(e), we looked at some special properties of very small angles measured in radians. (Have another look at this section now.)

We found there that, for a very small angle θ, $\cos \theta \to 1$ as $\theta \to 0$, and $\sin \theta \to \theta$ as $\theta \to 0$. So here, $\cos(\delta t) \to 1$ and $\sin(\delta t) \to \delta t$ as $\delta t \to 0$. Therefore

$$\text{as } \delta t \to 0, \quad x + \delta x \to \cos t - (\delta t) \sin t$$

but

$$x = \cos t \quad \text{so} \quad \delta x \to -(\delta t) \sin t \quad \text{so} \quad \frac{\delta x}{\delta t} \to -\sin t \quad \text{as } \delta t \to 0.$$

Therefore we have the following result.

$$\boxed{\text{If } x = \cos t \quad \text{then} \quad \frac{dx}{dt} = -\sin t.}$$

So the velocity of the point X after time t is $-\sin t$.

If the radius of the circle is 1 metre then, when X passes through O on its way to B, it has a velocity of $-\sin \pi/2 = -1 \, \text{m s}^{-1}$.

The corresponding result for the curve $y = \sin t$ can be shown in a very similar way. Try doing this for yourself. This is what you should get.

$$\boxed{\text{If } y = \sin t \quad \text{then} \quad \frac{dy}{dt} = \cos t \quad \text{or} \quad \frac{d}{dt}(\sin t) = \cos t.}$$

We are now able to get a very interesting result for the motion of X.

The rate of change of velocity with time is acceleration. But

$$\frac{d}{dt}(-\sin t) = -\cos t = -x.$$

So the acceleration of the point X is always towards O and equal in magnitude to the distance of X from O.

This means that, if X is a particle of unit mass, then the force on X which would make it move in this way is also equal in size to the distance of X from O, and always acts towards O.

These last two results will be unchanged for a larger circle but, if the speed of P is different, the relationship will be altered by some constant factor depending on the new speed.

The point X is moving in what is called **simple harmonic motion** (SHM).

A physical example of this is the motion of the bob of a simple pendulum.

The joint effects of the force of gravity and the tension in the string on the bob produce a force on it which gives it an acceleration of the kind we have described.

We said above that acceleration is the rate of change of velocity with time.

Also, velocity is the rate of change of distance with time.

So acceleration is the rate of change of a quantity which is itself a rate of change.

If we call the velocity v and the acceleration a, then

$$v = \frac{dx}{dt} \quad \text{and} \quad a = \frac{dv}{dt} \quad \text{so} \quad a = \frac{d}{dt}\left(\frac{dx}{dt}\right).$$

This is written as

$$\frac{d^2x}{dt^2}.$$

So, here, we can say that

$$\frac{d^2x}{dt^2} = -x.$$

This is an example of what is called a **differential equation**. A differential equation is an equation which includes terms like dx/dt or d^2x/dt^2.

We know the solution of this particular example of this equation, which is that $x = \cos t$. We'll look at some more equations like this in section 9.C.(c).

$\dfrac{dx}{dt}$ is called the **first derivative** of x with respect to t.

$\dfrac{d^2x}{dt^2}$ is called the **second derivative** of x with respect to t.

This is what happens with the other notations.

$$\text{If } x = f(t) \text{ then } \frac{dx}{dt} = f'(t) = \dot{x} \text{ and } \frac{d^2x}{dt^2} = f''(t) = \ddot{x}.$$

EXERCISE 8.A.2

(1) Do we get the same kind of results if we look at the motion of the point Y on the vertical axis described near the end of Section 5.C.(b)?

The distance OY is given by $y = \sin t$.

Find for yourself the velocity dy/dt and the acceleration d^2y/dt^2 of Y.

Can you link up d^2y/dt^2 and y by an equation?

(2) What happens if we have an object moving so that its distance from the origin can be described as a combination of $\sin t$ and $\cos t$?

For example, what would happen if we had $x = 3 \cos t + 4 \sin t$?

Find dx/dt and d^2x/dt^2, and see if you can find a linking equation between x and d^2x/dt^2.

8.A.(f)   Can we always differentiate? If not, why not?

In all the examples which we have looked at, we have been using the same process of tucking the little triangles in closer and closer to the point we are considering on the curve, to get better and better approximations to its steepness and so to its rate of change at that point.

Is it always possible to do this?

If we have some relationship giving y in terms of x, can we always go ahead and find dy/dx?

What kinds of thing might happen which would mean that we could not differentiate y with respect to x?

The graph sketches in Figure 8.A.9 may suggest some potential problems to you.

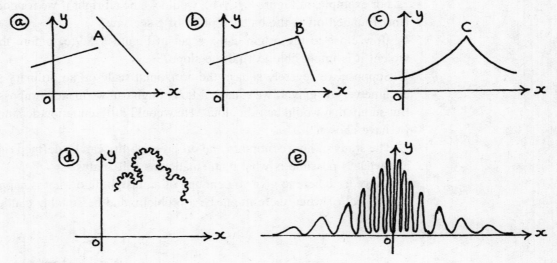

Figure 8.A.9

Also, suppose we can no longer draw the small triangles near some point on a graph because tiny differences in x give rise to huge differences in y? Can you think of such an example on any of the graphs which we have already sketched in this book?

Make a list for yourself of all the circumstances which you think will spell trouble for the process of differentiating.

I hope that you will have thought of some of these possibilities.

In order to differentiate successfully, we must have the following conditions.

(1)   There must be no breaks or **discontinuities** as at A in Figure 8.A.9(a).
       There is no meaning to the slope at the point where the break is.

(2)   It must be true that, moving in from either side with the little triangles, we get the *same* slope for the tangent that we are considering. The left-hand limiting value must be equal to the right-hand limiting value.
       For example, we can't find dy/dx at the points B and C in Figure 8.A.9(b) and (c).

(3)   The graph cannot be infinitely wiggly like a fractal curve where, however small the scale you take, the outline is still very similar to the one I have drawn in Figure 8.A.9(d). A coastline looks much the same in whatever detail you look at it, with smaller and smaller inlets being revealed. For a curve like this, it is impossible to define the slope at *any* point on it.

(4)   It must be true that there is a limiting value to be found, so tiny changes in x don't give uncontrollably huge changes in y. This is what happens, for example, as we get closer and closer to $x = \pi/2$ in the graph of $y = \tan x$.

Any graph which has some value of x for which the function is undefined because it is impossible to divide by zero will give a discontinuity like this.

Another example is the function $f(x) = (x + 3)/(x - 2)$ which we drew in Figure 3.B.16 in Section 3.B.(i). It has a discontinuity like this when $x = 2$. dy/dx does not exist for this value of x.

Unfortunately, it isn't possible to produce watertight definitions of the problems just by using pictures.

For example, in Figure 8.A.9(b), would we be all right if we rounded off the sharp point? How rounded off is the balance point of a see-saw?

How close to the origin can we get in Figure 8.A.9(e) before the wiggles become so violent it is impossible to find the slope?

Suppose we severely squash the horizontal scale on an ordinary sin graph. It will then become very wiggly. If we squash it far enough can we make it impossible to find the slope? But surely that would be ridiculous? How could differentiation depend on the personal scale we have chosen?

The study of how continuity and differentiability can be defined rigorously to make clear just what is possible is what mathematicians call analysis.

I have tried here to give you enough of an insight into what is happening so that you will have a feel of when there might be a problem, and be suitably cautious.

8.B   Natural growth and decay – the number e

I have found that many students regard e as something of a mystery – something that obviously matters a lot in calculus because it is always being used, but why? You will know, if you are studying science or engineering, that e is involved in many of the equations which describe the physical relationships which are important in your subject. This next section sets out to give you at least some of the reasons why e is so important. If you are in a hurry, you can leave the reading of it until later, but you should go through highlighting all the boxes of important results, both so that you can use them now and also to pinpoint them for yourself if you want to do more investigation later on.

I have already described some relationships of natural growth in Section 3.C. If you want to understand how e works, you should start by having another look at this before going on.

8.B.(a)   Even more money – compound interest and exponential growth

In Section 6.C.(h) we looked at how it is possible to make invested money grow faster by using a system of compound interest so that the new interest is calculated as a percentage not only of the original amount of money invested, but also of the interest which has so far been accumulated.

I said there that this updating of interest is usually done either yearly or six-monthly. Would the shorter time interval make very much difference? We would expect it to make some difference because there will be some interest at the end of six months. At the end of the year, you would receive interest on this interest as well as the interest on the original amount of money which you invested.

If this case, how much better would it be to have an even shorter time interval, say three monthly?

This is an important question to answer because rates of growth which depend on how much of a quantity is present at any particular time are very important in many real-life physical situations.

Rather than returning to the situation of Section 6.C.(h), we will look at a slightly different picture. It turns out to be particularly interesting to start from the special case of what happens when the extra amount or interest received at the end of a unit time interval is equal to the amount originally saved.

Unfortunately, this is an unlikely arrangement for a bank to make, so we shall look at the following example instead.

Suppose there is a group of cousins who each receive £100 from their wealthy uncle one Christmas. So strongly does he feel about the virtues of prudence and thrift that he says he will arrange things so that their savings increase at an equal rate to the amount saved, so that if the £100 is saved until the following Christmas, he will then add a further £100 to it.

All five cousins decide that they will save their £100.

The first cousin is happy to look forward to receiving the extra £100 the next Christmas, which will then give him a total of £200.

The second cousin decides to capitalise on her uncle's offer by suggesting that he increase her savings by a system of compound interest. She will split the year into two halves. Her uncle will give her £50 at the end of the first half-year, so she will have £150.

Since she will then be saving £150 instead of £100, at the end of the second half-year she will get an extra £75 instead of just £50, so giving her a total of £225 at the end of the year.

Her uncle agrees, so we can write this in the same form which we used with compound interest in Section 6.C.(h).

We have

Start   Mid-year   End of year

$$£100 \rightarrow £100 + \tfrac{1}{2}(£100) \rightarrow [(£100 + \tfrac{1}{2}(£100)) + \tfrac{1}{2}(£100 + \tfrac{1}{2}(£100))]$$

or

$$£100 \rightarrow (1 + \tfrac{1}{2})£100 \rightarrow [(1 + \tfrac{1}{2})£100 + \tfrac{1}{2}(1 + \tfrac{1}{2})£100]$$

which tidies up as

$$£100 \rightarrow (1 + \tfrac{1}{2})£100 \rightarrow (1 + \tfrac{1}{2})^2 £100 = £225.$$

The two steps in her savings are given by the second and third terms of a geometric progression (GP) which has a first term of £100 and a common ratio of $(1 + \tfrac{1}{2}) = \tfrac{3}{2}$.

The third cousin, seeing this calculation, considers that having the interest updated quarterly would be even more beneficial.

The pattern for her quarterly updates will go

$$£100 \rightarrow (1 + \tfrac{1}{4})£100 \rightarrow (1 + \tfrac{1}{4})^2 £100 \rightarrow (1 + \tfrac{1}{4})^3 £100 \rightarrow (1 + \tfrac{1}{4})^4 £100$$

| Start | 1st quarter | 2nd quarter | 3rd quarter | End of year |

giving her a total at the next Christmas of £244.14 to the nearest penny.

Again, the four steps in the savings are given by a GP, wth a common ratio this time of $(1 + \tfrac{1}{4}) = \tfrac{5}{4}$.

How much would the fourth cousin (who negotiates monthly updates) get by the end of the year?

He would get $(1 + \tfrac{1}{12})^{12}£100 = £261.30$ to the nearest penny.

This time, the twelve steps of the savings are given by a GP which has a common ratio of $(1 + \tfrac{1}{12}) = \tfrac{13}{12}$.

The fifth and youngest cousin is keen to see how much she can negotiate to get.

Try estimating for yourself how much you think she might get. What do you think her best arrangement would be?

She decides to go for the most extreme position and says

'If I am saving as you want, could we not consider that, over the year, the money that you will give me becomes more and more mine, and so it can really be considered as feeding in continuously to become part of *my* savings as the year goes by. And then I shall be getting a rate of increase equal to the total amount I have saved all the while. Since we are reckoning here on infinitely small time intervals, I shall do infinitely better than any of my other cousins!'

Is she right?

If we look at what happens as the time intervals become shorter, we find the following:

Weekly updates would give her a final total of $(1 + \tfrac{1}{52})^{52}£100 = £269.26$.

Daily updates would give her a final total of $(1 + \tfrac{1}{365})^{365}£100 = £271.46$.

Hourly updates would give her $(1 + \tfrac{1}{8760})^{8760}£100 = £271.81$.

See for yourself what happens if the interest is updated every minute.

The amounts *are* increasing, but more and more slowly.

Now we know that the increases for the first four cousins are all coming in definite steps, and we saw that each scheme was described by a different GP.

The increases given by updates every minute are still described by a GP, this time with 525 600 steps, and a common ratio of

$$1 + \frac{1}{525\,600} = \frac{525\,601}{525\,600}.$$

The steps are now exceedingly tiny, but they are still there. This GP would give a grand total at the end of the year of £271.82 to the nearest penny.

When the youngest cousin gets what she wants, the steps will have been smoothed out to give a continuous growth curve. We know that her £100 will have been multiplied by a factor of about 2.7182 by the end of the year.

What *is* this number which is equal to about 2.7182?

To find the answer to this, we'll now look at the pattern of her increases as the time intervals get shorter and shorter. These go

$$£100 \rightarrow \left(1 + \frac{1}{n}\right) £100 \rightarrow \left(1 + \frac{1}{n}\right)^2 £100 \rightarrow \ldots \rightarrow \left(1 + \frac{1}{n}\right)^n £100$$

where n is as large a number as we care to think of, as she is breaking her year into infinitely short time intervals. So she finishes up with

$$\left(1 + \frac{1}{n}\right)^n £100 \quad \text{as} \quad n \rightarrow \infty.$$

Now, we can do a binomial expansion on

$$\left(1 + \frac{1}{n}\right)^n.$$

We use the formula (B2), from Section 7.A.(e), which starts

$$(1 + x)^n = 1 + \frac{n}{1!} x + \frac{n(n-1)}{2!} x^2 + \frac{n(n-1)(n-2)}{3!} x^3 + \ldots$$

We have to put $x = 1/n$, where n is a positive whole number, but a very large one indeed. We get

$$1 + \frac{n}{1!} \left(\frac{1}{n}\right) + \frac{n(n-1)}{2!} \left(\frac{1}{n}\right)^2 + \frac{n(n-1)(n-2)}{3!} \left(\frac{1}{n}\right)^3 + \ldots$$

and, as n becomes larger and larger, $n-1$, $n-2$, etc. are all relatively close to n.

We are getting nearer and nearer to the series

$$1 + \frac{1}{1!} + \frac{1}{2!} + \frac{1}{3!} + \frac{1}{4!} + \ldots$$

as the amount by which we must multiply the £100 to find her total savings.

As we go further and further in summing this series, we find that the running sum gets closer and closer to a value of about 2.71828, so she gets £271.83 to the nearest penny, doing the best of the cousins, but not dramatically better than her next cousin.

This number, to which the pretty series

$$1 + \frac{1}{1!} + \frac{1}{2!} + \frac{1}{3!} + \frac{1}{4!} + \ldots$$

converges is extremely important mathematically, and is indeed the famous e.

You can see its value to as many places as your calculator will allow, by putting in 1 and then pressing e^x.

We now have this important result.

$$\boxed{\text{As } n \rightarrow \infty, \left(1 + \frac{1}{n}\right)^n \rightarrow 1 + \frac{1}{1!} + \frac{1}{2!} + \frac{1}{3!} + \ldots = e.}$$

We have found in this section that, when the interest is updated at the end of equal time intervals, so that the total amount of money is increasing in separate jumps, then these increasing amounts of money form the terms of a GP (with a different GP for each set of equal time intervals).

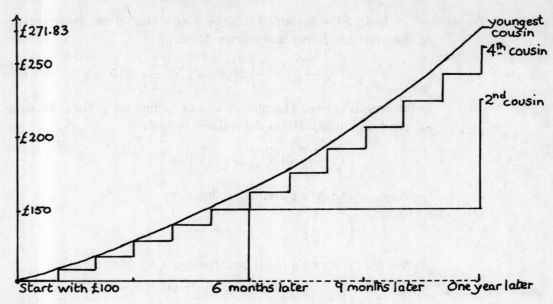

Figure 8.B.1

However, when the interest is updated continuously, so that the amount of money saved is increasing smoothly all the while, the result is no longer described by the steps of a GP but by a smooth growth curve.

You can see these differences in Figure 8.B.1 where I show the growth in the savings of the second, fourth and youngest cousin.

8.B.(b)   **What is the equation of this smooth growth curve?**
In order to be able to apply the mechanism of this smooth growth curve to other situations, we need to know what its equation is.

It becomes easier to see what this must be if we look at how the differences between the graphs are building up at an intermediate point.

For example, after six months we have the following totals.

The first cousin still has £100.

The second cousin has $(1 + \frac{1}{2})$ £100 = £150.

The third cousin has $(1 + \frac{1}{4})^2$ £100 = £156.25.

The fourth cousin has $(1 + \frac{1}{12})^6$ £100 = £161.65.

We can emphasise that we are considering a half-yearly interval here by writing

$$(1 + \tfrac{1}{4})^2 = [(1 + \tfrac{1}{4})^4]^{1/2} \quad \text{and} \quad (1 + \tfrac{1}{12})^6 = [(1 + \tfrac{1}{12})^{12}]^{1/2}$$

so, for example, the fourth cousin has

$$[(1 + \tfrac{1}{12})^{12}]^{1/2} \text{ £100} = \text{£161.65.}$$

It then seems reasonable to say that, at the end of the half-year, the fifth cousin would have

$$[(1 + \tfrac{1}{n})^n]^{1/2} \text{ £100} = e^{1/2} \times \text{£100 since, as } n \to \infty, (1 + \tfrac{1}{n})^n \to e.$$

Now, the accumulating totals for the first four cousins increase in definite jumps, but the total for the fifth cousin is increasing smoothly, so it would seem reasonable to say that, after a time interval of *any* length t, where t is measured in years, she would have a total of $e^t \times$ £100.

Her smooth growth curve has the equation $x = 100\,e^t$ where t represents the time interval along the horizontal axis and x represents her total savings in £ s.

Because the rate of increase of e^t is equal to e^t itself for any value of t, it must be true that

$$\frac{d}{dt}(e^t) = e^t.$$

This property of e^t that its rate of change is always equal to e^t itself makes it very special.

If you tried drawing the sketch in the thinking point of Section 3.C.(e), you should have found that the gradient of the tangent when $x = 1.5$ was about the same as the height of the curve for that value of x.

8.B.(c)   **Getting numerical results from the natural growth law of $x = e^t$**

I have taken the simplest possible form of the natural growth law here, leaving out the 100 which we included for the £100 earlier, to make this section simpler.

Starting from $x = e^t$, see if you can answer the following questions.

(1)   What is x if (a) $t = 2$ (b) $t = \frac{1}{3}$?

(2)   What is t if (a) $x = 1$ (b) $x = 2$ (c) $x = 4.5$?

To help you, I have shown these questions in Figure 8.B.2. (You will need to use your calculator to get the answers.)

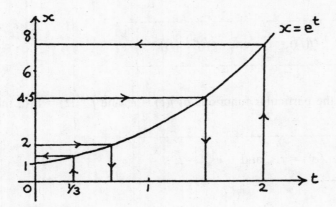

Figure 8.B.2

(1)   This is straightforward. Using $x = e^t$, we have

(a) $x = e^2$ so $x = 7.3891$ to 4 d.p. using a calculator
(b) $x = e^{1/3}$ so $x = 1.3956$ to 4 d.p.

The first answer corresponds to the amount of money, measured in units of £100, which the fifth cousin would have after two years (if her uncle leaves the system of growth unchanged). This would be £738.91. The second answer corresponds to the amount she would have after $\frac{1}{3}$ of a year or 4 months. This is £139.56.

(2) This question is a bit more tricky because we want to go back the other way. We need to use the inverse function which will take us back from x to t.

Because of the way it was obtained, the growth curve is smooth and has no gaps, so there will be a value of x for any particular value of t.

We define the inverse function by introducing the natural log and saying

---

if $x = e^t$ then $t = \log_e x$ or $\ln x$.

(Natural logs, that is logs to the base e, are usually written as ln rather than $\log_e$.) This now gives us the answer for question (2)(a) that $t = \ln 1 = 0$ so $1 = e^0$ which agrees with the meaning we gave to the power 0 in Section 1.D.(b).

It also agrees with the starting amount of money of $1 \times £100$ when $t = 0$.

The answer for question (2)(b) is $t = \ln 2 = 0.693$ to 3 d.p. using a calculator.

The fifth cousin would have £200 after $0.693 \times 12 = 8.3$ months approximately.

Question (2)(c) has the answer of $\ln(4.5) = 1.504$ to 3 d.p., giving the fifth cousin £450 after approximately $1\frac{1}{2}$ years.

If we have a function $x = f(t)$, then we write its inverse function (if it exists) in the form $x = f^{-1}(t)$.

Here, we have $f(t) = e^t$ and $f^{-1}(t) = \ln t$.

Since doing the function followed by doing the inverse function brings you back to where you started, we have

---

$f^{-1}(f(t)) = t$ and $f(f^{-1}(t)) = t$.

For the particular functions of $f(t) = e^t$ and $f^{-1}(t) = \ln t$, this gives us

---

$\ln(e^t) = t$ and $e^{\ln t} = t$.

These equations are extremely useful and are worth surrounding in bright colour.

I have sketched $x = f(t) = e^t$ and $x = f^{-1}(t) = \ln t$ in Figure 8.B.3.

Notice the following points here.

- The sketch includes negative values of t. If t represents time, then these represent times before we started doing the measuring.
- The value of e^t is always greater than zero although, for large negative values of t, it gets infinitely close to zero.
- We can only find the natural log of a positive quantity. (This is true for any log.) This agrees with 2^{-3}, say, being $1/2^3 = 1/8$. We can't actually *get* a negative from a power.

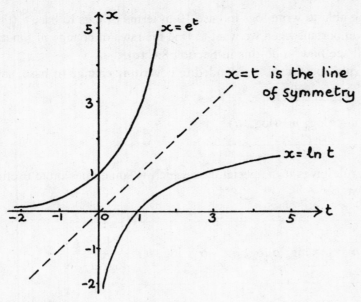

Figure 8.B.3

8.B.(d)   **Relating ln x to the log of x using other bases**

Starting from a similar situation in Section 3.C.(b), we defined the inverse function of $f(x) = 2^x$ as $f^{-1}(x) = \log_2 x$.

It will now be of great practical importance to us to find a rule which will tell us how to write logs to other bases (in particular, base 10) in terms of logs to base e (or natural logs).

To find this rule, we will start with some number a and suppose that $\log_{10} a = y$ and $\ln a = \log_e a = x$. (In this section on changing bases, I will write the natural logs as $\log_e$ rather than ln to emphasise that these are logs to the base e.)

If $\log_{10} a = y$ then $a = 10^y$ and if $\log_e a = x$ then $a = e^x$. (This is what 'base 10' and 'base e' mean.)

But it must also be possible to write 10 itself as a power of e.

Let's say that $10 = e^c$. This means that we can say that $c = \log_e 10$.

(Using my calculator gives me $c = 2.302\,585\,093$ but this is only an approximation to nine decimal places. Any further rounding off will make it even more inexact so we'll carry on calling it c for short.)

Now we say that $a = 10^y = (e^c)^y = e^{cy}$.

But also $a = e^x$ so now we have $e^x = e^{cy}$ so $x = cy$.

Putting back what x, y and c are in terms of logs gives us

$$\log_e a = (\log_e 10)(\log_{10} a) \quad \text{or} \quad \ln a = (\ln 10)(\log_{10} a).$$

It is also worth surrounding this in bright colour.

We now have a rule which makes it possible for us to change a log to base 10 into a log to base e. (One way of remembering it is to think of it as sort of 'cancelling' the 10.) Try choosing some particular values for a and than check on your calculator that the rule does work.

Being able to write logs to base 10 in terms of logs to base e (that is, natural logs) will be very important when we want to find the rates of change of functions of logs to base 10. We shall see how to do this in Section 8.C.(c).

The rule above can be extended to cover *any* change of base, say from m to n.

$$\log_n a = (\log_n m)\,(\log_m a).$$

This rule gives us a special case which is sometimes quite useful. If we put $n = a$ and $m = b$, we get

$$\log_a a = 1 = \log_a b \, \log_b a \quad \text{so} \quad \log_a b = \frac{1}{\log_b a}.$$

We have now seen how logs to other bases can be converted into natural logs. It is possible to define all other logs and powers in terms of logs and powers of e, and this is done in the rigorous approach of mathematical analysis. It is then possible to give a meaning to such unnerving quantities as 2^{π}, for example. Doing this properly is a slow and careful process. In this book I try to give you enough examples of places where you need to be careful, to help you to understand why this detailed analysis is done.

8.B.(e)   **What do we get if we differentiate ln t?**

What is the rate of change of $x = \ln t$ with respect to t?

That is, what is dx/dt?

$$\text{If } x = \ln t \quad \text{then} \quad t = e^x \quad \text{so} \quad \frac{dt}{dx} = e^x.$$

But it seems reasonable in general to say that

$$\frac{dx}{dt} = \frac{1}{dt/dx}$$

since we can say that the fraction

$$\frac{\delta x}{\delta t} = \frac{1}{\delta t/\delta x}.$$

Provided that none of the problems talked about in Section 8.A.(f) is present, then when $\delta t \to 0$, $\delta x \to 0$ also, so this step is justified. Now here we have

$$\frac{dt}{dx} = e^x \quad \text{so} \quad \frac{dx}{dt} = \frac{1}{e^x} = \frac{1}{t}.$$

This gives us the enormously important result that

$$\text{if} \quad x = \ln t \quad \text{then} \quad \frac{dx}{dt} = \frac{1}{t}.$$

This is another box worth surrounding with bright colour.

I should point out here that the letters we use are not important in themselves; they are just names or tags.

So it is equally true, for example, that

$$\text{if } y = f(x) = \ln x \text{ then } \frac{dy}{dx} = f'(x) = \frac{1}{x}.$$

8.C   Differentiating more complicated functions

Before we start looking at ways of how we can do this, I will collect together in a box all the functions we can now differentiate. Remember that the letters of the variables can be changed as you wish. (I have used y, x, t and θ for mine.)

> **Rates of change we already know**
>
> (1) If $y = f(x) = ax^n$ then $dy/dx = f'(x) = nax^{n-1}$.
>
>   So if $y = ax$ then $dy/dx = a$ and if $y = a$ then $dy/dx = 0$ (a stands for any constant number).
>
> (2) If $x = f(t) = \sin t$ then $dx/dt = f'(t) = \cos t$.
> (3) If $x = f(t) = \cos t$ then $dx/dt = f'(t) = -\sin t$.
> (4) If $x = f(t) = e^t$ then $dx/dt = f'(t) = e^t$.
> (5) If $x = f(t) = \ln t$ then $dx/dt = f'(t) = 1/t$.

I have used the letter f for a function here, all through, but of course you can use other letters if you want.

Students sometimes mix up the minus sign in (2) and (3). There are two ways you can use to remember that the minus sign comes when you differentiate a cos.

- Remember the shape of the first bit of the sin and cos graphs.
  The cos graph is going downhill here, so d/dt (cos t) must be $-\sin t$.
- **S**in **D**ifferentiates **P**lus so **S**olve **D**amn **P**roblem.

8.C.(a)   The Chain Rule

It is often necessary to be able to find the rate of change of functions which have been built up from simpler ones. For example, we might have $x = f(t) = \sin 3t$ or $y = f(x) = (3x^2 + 2)^5$ or $y = f(\theta) = \sin^3 2\theta$ etc. The Chain Rule gives us a way of dealing with all of these.

I will explain how this works by showing you the following four examples.

(1) $y = (3x^2 + 5)^5$ (2) $x = \sin(3t + \pi/2)$ (3) $y = e^{x^2+2}$ (4) $y = \ln(2t^2 + 3t)$

Each of these is built up from functions which we can easily differentiate.

We can show this in the following way.

(1) $y = (3x^2 + 5)^5$ becomes $y = X^5$ if we put $X = 3x^2 + 5$.

(2) $x = \sin(3t + \pi/2)$ becomes $y = \sin X$ if we put $X = 3t + \pi/2$.

(3) $y = e^{x^2+2}$ becomes $y = e^X$ if we put $X = x^2 + 2$.

(4) $y = \ln(2t^2 + 3t)$ becomes $y = \ln X$ if we put $X = 2t^2 + 3t$.

In each of these, X stands for a whole lump or chunk which makes a second function.

Taking example (1), we have y as a function not just of x but also of this X which is itself a function of x.

It is for this reason that the Chain Rule is also known as 'function of a function'.

Being able to write y in this way makes the finding of dy/dx very much simpler because we can split it into two easy steps.

We justify this by going back to the stage of the very small changes, and saying

$$\frac{\delta y}{\delta x} = \frac{\delta y}{\delta X}\frac{\delta X}{\delta x}$$ just using the ordinary rules of fractions.

Now, provided none of the potential difficulties which we talked about in Section 8.A.(f) are present at any of the points we are interested in, so that as δx gets very small we also have δX getting very small, we can say that

$$\text{as } \delta x \to 0, \quad \frac{\delta y}{\delta x} \to \frac{dy}{dx}, \quad \frac{\delta y}{\delta X} \to \frac{dy}{dX} \quad \text{and} \quad \frac{\delta X}{\delta x} \to \frac{dX}{dx}.$$

This gives us the following result.

The Chain Rule

If y is a function of X, and X is a function of x, then $\dfrac{dy}{dx} = \dfrac{dy}{dX}\dfrac{dX}{dx}$.

Using this now in each of the four examples which we had above, and changing the letters when necessary, we get the following results.

(1) $\dfrac{dy}{dx} = \dfrac{dy}{dX}\dfrac{dX}{dx} = (5X^4)\,(6x) = 30x\,(3x^2 + 5)^4$

Notice that I have given the final answer in terms of the original x. You should always do this.

(2) $\dfrac{dx}{dt} = \dfrac{dx}{dX}\dfrac{dX}{dt} = (\cos X)(3) = 3\cos(3t + \pi/2)$

 Remember here that $\pi/2$ is a constant, and so gives zero when it is differentiated.

---

(3) $\dfrac{dy}{dx} = \dfrac{dy}{dX}\dfrac{dX}{dx} = (e^X)(2x) = 2xe^{x^2+2}$

Using the Chain Rule also gives us the result that $d/dt(e^{-t}) = -e^{-t}$. This describes a process of decay where the rate of change of the substance present at time t is equal to minus the amount of the substance present at that time. The minus sign shows that this rate of change is *negative*, and the amount of the substance present is *decreasing*.

You will avoid a lot of mistakes if you remember that if e^X is differentiated with respect to X then the answer is e^X.

So if you have $e^{\text{something complicated}}$, then $e^{\text{the same something complicated}}$ must be part of your answer when you differentiate.

$$(4) \quad \frac{dy}{dt} = \frac{dy}{dX}\frac{dX}{dt} = \left(\frac{1}{X}\right)(4t+3) = \frac{4t+3}{2t^2+3t}$$

EXERCISE 8.C.1

Try these questions for yourself now.

It is very important to be able to do these differentiations quickly and reliably because they will be the basic step of many further processes. (In particular, when you come to use partial differentiation, which involves having functions of more than one variable, you need to be able to do this process without any worries.)

For this reason, I start off with easy questions and build them up gradually so that you can get really confident with them.

I think you will find that quite quickly you can work with the X in your head, just writing down the two multiplied bits and then tidying them up for the final answer.

Differentiate each of these functions with respect to the letter used in their description.

(1) $y = (2x^2 + 3)^4$ (2) $x = (t^3 + 2)^5$ (3) $y = (3x^2 - 2x)^4$
(4) $x = (3t + 4)^{1/2}$ (5) $y = 3e^{4x}$ (6) $x = e^{t^2+1}$
(7) $y = 2e^{x^2+x}$ (8) $x = \cos(4t + \pi/3)$ (9) $x = \sin t + \sin 2t$
(10) $y = \sin(x^2)$ (11) $y = \sin^2 x$, which means $(\sin x)^2$. Hint: let $X = \sin x$.
(12) $y = \cos^3 x$ (13) $y = \ln 4x$ (14) $y = \ln(3x + 1)$ (15) $x = \ln(2t^2 + 1)$
(16) $y = \cos(5x^2 + \pi)$ (17) $x = \sin(2t^2 + 3)$ (18) $y = \ln(\sin x)$

The next step is to be able to use the Chain Rule more than once in the same question. With practice on the easy ones (which are then often built into more complicated ones), you will find this no problem.

Try these ten quickies now, doing the X part in your head.

EXERCISE 8.C.2

Differentiate each of the following with respect to x.
(1) $y = e^{5x}$ (2) $y = e^{-2x}$ (3) $y = e^{x^2}$ (4) $y = \ln(2x + 3)$
(5) $y = \ln(1 + x)$ (6) $y = \ln(1 - x)$ (7) $y = \sin 7x$ (8) $y = \cos^4 x$
(9) $y = \sin(2x + \pi)$ (10) $y = \cos(3x + 4)$

Now we are ready to do the functions of functions of functions. (In fact, you can chain together as many as you like, with them all folded inside each other like a set of Russian dollies.)

Here are two examples.

EXAMPLE (1) Find dy/dx for $y = \sin^3 (4x)$ (which means, of course, $y = (\sin(4x))^3$.

We think of this first as $y = X^3$, with $X = \sin 4x$, and write
$$dy/dx = (3X^2)(4 \cos 4x) = 12 \cos 4x \sin^2 4x.$$

The second use of the Chain Rule, on the $\sin 4x$, has become so automatic that you hardly notice that you are doing it.

EXAMPLE (2) Find dy/dt if $y = \ln(\sin 3t)$.

Thinking of this as $\ln X$, with $X = \sin 3t$, we can write
$$dy/dt = \left(\frac{1}{X}\right) (3 \cos 3t) = \frac{3 \cos 3t}{\sin 3t} = 3 \cot 3t.$$

EXERCISE 8.C.3

Try these now for yourself. Differentiate each function with respect to the letter used in their description.

(1) $y = \cos^5 2x$ (2) $y = \sin^3 (4x + 1)$ (3) $x = \ln(\sin(2t + 3))$
(4) $x = (2 \cos 2\theta + 5)^3$ (5) $y = \ln(1 + \cos x)$ (6) $x = \ln(3t + \sin^2 3t)$
(7) $x = \ln(2 + e^{t^2+1})$ (8) $y = \sin(\cos 4x)$ (9) $y = (1 + \sin^2 t)^{1/2}$
(10) $y = \ln[(1 + \sin^2 t)^{1/2})]$ (This is easier than it looks. Think!)

8.C.(b) **Writing the Chain Rule as $F'(x) = f'(g(x)) \, g'(x)$**

You may come across the Chain Rule written in the dash notation for functions as above. It means exactly the same thing as what you have just been doing. I will show you how this is so by taking an example.

Suppose we want to differentiate $y = (3x^2 + 2)^4$ with respect to x.

Because we are using function notation, we need to label the three functions involved here with different letters.

I shall let $y = F(x) = (3x^2 + 2)^4$.

Now y is also a function of $(3x^2 + 2)$. (This is what we have been calling X.)

I shall let $X = 3x^2 + 2 = g(x)$, to show that it also is a function of x.

Since y is a function of X, we can also write $y = f(X) = f(g(x))$. (In this particular example, $y = f(3x^2 + 2)$.)

Next, it is important to be sure what the dash notation means for a function.

$f'(x)$ means the function $f(x)$ differentiated with respect to x,

$f'(t)$ means the function $f(t)$ differentiated with respect to t.

$f'(X)$ means the function $f(X)$ differentiated with respect to X, even though X is itself a function of x, so $f'(g(x))$ means the function $f(g(x))$ differentiated with respect to $g(x)$.

It corresponds to what we have called dy/dX.

So, in this particular example, $g'(x) = 6x$ and $f'(g(x)) = 4(3x^2 + 2)^3$. So we have

$$F'(x) = [4(3x^2 + 2)^3] \, [6x] = 24x \, (3x^2 + 2)^3.$$

Use whichever notation you prefer.

8.C.(c) **Differentiating functions with angles in degrees or logs to base 10**

When we showed that

$$\frac{d}{dx} (\cos t) = -(\sin t)$$

in Section 8.A.(e) everything ran smoothly because the angle t was in radians.

How could we find the slope of the graph of $x = \cos \theta$ if θ is in degrees? (We know that we can draw the graph of $x = \cos \theta$. The only difference is that the horizontal scale will be in degrees instead of radians.)

In order to find $dx/d\theta$ from $x = \cos \theta$ we shall first have to convert θ to radians.

From Section 4.D.(a) we have

$$\theta° = \left(\frac{\pi}{180}\right) \theta \text{ radians}, \quad \text{so} \quad x = \cos\left(\frac{\pi\theta}{180}\right)$$

with the angle now in radians.

We also know from the Chain Rule that, if a is some constant number, and $x = \cos(a\theta)$, then $dx/d\theta = -a \sin \theta$.

Exactly the same principle works here with $a = \pi/180$.

$$\text{If } x = \cos\left(\frac{\pi\theta}{180}\right) \quad \text{then} \quad \frac{dx}{d\theta} = -\frac{\pi}{180} \sin\left(\frac{\pi\theta}{180}\right) = -\frac{\pi}{180} \sin \theta$$

writing the angle again in degrees.

The $\pi/180$ is the gearing mechanism or scale factor which lets us have the horizontal scale in degrees instead of radians.

We can use a similar process to differentiate a function in terms of logs to base ten (or any other base, but ten is the one you are most likely to want to use).

To do this, we go back to the relationship between logs to base e and logs to base 10 which we found in Section 8.B.(d). This says that

$$\ln a = (\ln 10)(\log_{10} a).$$

So, for example, if we want to find dy/dx for the function $y = \log_{10} x$, we rewrite this as

$$y = \frac{\ln x}{\ln 10}.$$

Now $1/\ln 10$ is just a number, so we have

$$\frac{dy}{dx} = \frac{1}{\ln 10}\left(\frac{1}{x}\right) = \frac{1}{x \ln 10}.$$

Here, the $1/(\ln 10)$ is acting as a gearing mechanism or scaling factor which makes the differentiation work in the slightly altered circumstances of a different base.

We now have the following two rules.

> To differentiate functions involving degrees, convert first to radians.
> To differentiate functions involving other logs, convert first to natural logs.

8.C.(d) The Product Rule, or 'uv' Rule

The Product Rule moves us a further step on in being able to differentiate functions which are built up from simpler functions. It is therefore another technique we will need for practical applications.

As its name suggests, it gives us a way of dealing with two functions which are multiplied together to give a third function.

For example, suppose we have $y = f(x) = 3x^2 \sin 2x$.

The function $f(x)$ is made up of two functions, $u(x) = 3x^2$ and $v(x) = \sin 2x$, which are multiplied together. So we can say $y = uv$.

If we alter x by a small amount δx then y will also alter by a small amount δy. Also the two components, u and v, of y will each alter by small amounts since they are also functions of x. (We are assuming here that none of the complications of Section 8.A.(f) is present.) So we can say that u alters by the small amount δu and v alters by the small amount δv.

This gives us

$$y + \delta y = (u + \delta u)(v + \delta v) = uv + v\,(\delta u) + u\,(\delta v) + (\delta u)(\delta v).$$

But $y = uv$, so

$$\delta y = u\,(\delta v) + v(\delta u) + (\delta u)(\delta v).$$

Dividing all through by δx gives

$$\frac{\delta y}{\delta x} = v\,\frac{\delta u}{\delta x} + u\,\frac{\delta v}{\delta x} + \frac{(\delta u)(\delta v)}{(\delta x)}.$$

Now, if we make δx become smaller and smaller, so $\delta x \to 0$, then δu and δv will also become very small.

This means that,

$$\text{as } \delta x \to 0, \qquad \frac{(\delta u)(\delta v)}{(\delta x)} \to 0 \text{ also.}$$

Two very small things multiplied together and then divided by one very small thing give a very small result. This result will become closer and closer to zero as δx itself becomes closer and closer to zero, so we now get the result that

$$\text{the limit of } \frac{\delta y}{\delta x} \text{ as } \delta x \to 0 \text{ is } \frac{dy}{dx}.$$

This gives us the following.

The Product Rule

$$\frac{dy}{dx} = v\,\frac{du}{dx} + u\,\frac{dv}{dx}$$

In the particular example that we started with, $du/dx = 6x$ and $dv/dx = 2\cos 2x$ so we have

$$dy/dx = (\sin 2x)(6x) + (3x^2)\,(2\cos 2x)$$

$$= 6x \sin 2x + 6x^2 \cos 2x = 6x\,(\sin 2x + x \cos 2x).$$

The Product Rule can also be written in function notation as

---

if $y = uv$ then $y' = vu' + uv'$.

This covers the case of y, u and v all being functions of x, or all being functions of any other letter which it might be convenient to work with.

Try these for yourself, tidying up all the answers as far as possible. Find dy/dx for each of the following.

(1) $y = 7x^2 \cos 3x$ (2) $y = e^{3x} \sin 2x$ (3) $y = 4x^5 (x^2 + 3)^3$

Find dx/dt for each of the following.

(4) $x = e^{2t+1} \cos (2t + 1)$ (5) $x = 7t^2 \ln (2t - 1)$ (6) $x = (t^2 + 1)^{1/2} \sin (2t + \pi)$

(7) Find dy/dx if $y = (x^2 + 1)^5 e^{3x} \cos 2x$.

If you have three functions multiplied together like this, there is no special new Product Rule which you should use. You just bunch any two of the functions together and then use the Product Rule twice.

Here, you could say $y = [(x^2 + 1)^5] [e^{3x} \cos 2x]$ and go on from there.

In the following questions, you will need to remember that

$$\frac{d^2x}{dt^2} \text{ means } \frac{d}{dt}\left(\frac{dx}{dt}\right)$$

so to find d^2x/dt^2 you differentiate twice.

These questions are included here not just as practice in differentiating but because, if you have seen them working this way round, they will then be easier for you to solve when you come to do the opposite process in real-life physical applications. There you will be starting with the differential equation (that is the equation which has the terms in d^2x/dt^2 and dx/dt) and finding a solution which fits it.

(8) If $x = (2 + t)e^{3t}$ find (a) $\dfrac{dx}{dt}$ and (b) $\dfrac{d^2x}{dt^2}$.

Show that $\dfrac{d^2x}{dt^2} - 6\dfrac{dx}{dt} + 9x = 0$.

(9) If $x = e^{kt}$ where k stands for some constant number, find (a) dx/dt and (b) d^2x/dt^2.

If $\dfrac{d^2x}{dt^2} - 2\dfrac{dx}{dt} - 3x = 0$ find the two possible values of k.

(10) If $x = Ae^{3t} + Be^{-t}$, where A and B are standing for constant numbers, show that

$$\frac{d^2x}{dt^2} - 2\frac{dx}{dt} - 3x = 0.$$

(There is a very quick way to do this one; look at your answer to the previous question.)

(11) If $x = e^{-t} \ln (1 + e^t)$ show that $\dfrac{dx}{dt} + x = \dfrac{1}{1 + e^t}$.

8.C.(e)   The Quotient Rule or 'u/v' Rule

This rule gives us a good way of differentiating a function which is made up of two simpler functions written as a fraction.

We start with a function $y = f(x) = u/v$ where u and v are both functions of x.

The following result can then be shown by a very similar argument to the one we used for the Product Rule in Section 8.C.(d).

The Quotient Rule

$$\text{If } y = \frac{u}{v} \text{ then } \frac{dy}{dx} = \frac{v\,(du/dx) - u\,(dv/dx)}{v^2}$$

 Notice the minus sign in the middle of the Quotient Rule. Because of this it matters what order the top two bits are written in. This is why I wrote the Product Rule in the same order. Then 'v comes virst' for both.

 NOTE Because the Quotient Rule automatically tidies up the answer by putting it over a common denominator, I think that it is easier to use it for a function like $y = f(x) = 2x/(3x - 1)$, rather than writing this as $2x(3x - 1)^{-1}$ and then using the Product Rule.

Here are two examples of using the Quotient Rule.

EXAMPLE (1) We can use it to find out what the answer is if we differentiate $y = \tan x$ with respect to x. We write

$$y = \tan x = \frac{\sin x}{\cos x} \quad \text{so} \quad u = \sin x \quad \text{and} \quad v = \cos x.$$

So

$$\frac{dy}{dx} = \frac{(\cos x)(\cos x) - (\sin x)(-\sin x)}{\cos^2 x} = \frac{\cos^2 x + \sin^2 x}{\cos^2 x}.$$

But

$$\cos^2 x + \sin^2 x = 1 \quad \text{so} \quad \frac{dy}{dx} = \frac{1}{\cos^2 x} = \sec^2 x.$$

Therefore, we have

$$\frac{d}{dx}(\tan x) = \sec^2 x.$$

EXAMPLE (2) We will use the Quotient Rule to find dy/dx if $\quad y = \dfrac{x + 3}{x - 2}$.

$u = x + 3 \quad$ and $\quad v = x - 2$ so we get

$$\frac{dy}{dx} = \frac{(x - 2)(1) - (x + 3)(1)}{(x - 2)^2} = -\frac{5}{(x - 2)^2}.$$

This is undefined when $x = 2$, but otherwise it will always be negative since $(x - 2)^2$ must be positive.

The value of dy/dx at any particular point of a curve is telling us the slope of the curve at that point. You can see how it tallies with the shape of the curve for this particular function because we sketched it in Section 3.B.(i). We thought there, from the information that we then had, that this curve should always have a negative slope except where $x = 2$ when y itself was undefined. Now we see that this is indeed true!

Knowing dy/dx gives us a rule for finding the slope at any particular point of the curve. We can see this here by taking a couple of examples of points on this curve, say A, $(3,6)$, and B, $(-4, -\frac{1}{6})$.

We get at A $\quad \dfrac{dy}{dx} = -5$ $\quad$ and at B $\quad \dfrac{dy}{dx} = -\dfrac{5}{36}$.

These gradients agree well with the sketch; we can see that the tangent at B would be much less steep than the tangent at A.

The Quotient Rule can also be written in function notation like this.

$$\text{If} \quad y = \frac{u}{v} \quad \text{then} \quad y' = \frac{vu' - uv'}{v^2}.$$

EXERCISE 8.C.5

Try these questions yourself now.

(1) By writing $\cot x = \dfrac{\cos x}{\sin x}$ show that $\dfrac{d}{dx}(\cot x) = -\operatorname{cosec}^2 x$.

(2) By writing $\sec x = \dfrac{1}{\cos x}$ or $(\cos x)^{-1}$, show that $\dfrac{d}{dx}(\sec x) = \sec x \tan x$.

(3) Show similarly that $\dfrac{d}{dx}(\operatorname{cosec} x) = -\operatorname{cosec} x \cot x$.

(4) Find $\dfrac{dx}{dt}$ if $x = \dfrac{\sin t}{1 + \cos t}$.

(5) Show that $\dfrac{d}{dx}(\ln(\sec x + \tan x)) = \sec x$.

(6) Find $\dfrac{dy}{dx}$ if $y = \dfrac{e^x - e^{-x}}{e^x + e^{-x}}$.

(7) Find $\dfrac{dy}{dx}$ if $y = \ln\left(\dfrac{1 - x}{1 + x}\right)$.

(Think how you can make this one simpler to do!)

(8) Find $\dfrac{dy}{dx}$ if $y = \dfrac{x^2 \sin x}{(3 - x)}$.

(9) Find $\dfrac{dy}{dx}$ if $y = \dfrac{3x - 2}{2x + 3}$ with $x \neq -\frac{3}{2}$.

(10) Find $\dfrac{dy}{dx}$ if $y = \dfrac{ax + b}{cx + d}$

where a, b, c and d are all constant numbers, and $x \neq -d/c$. Are there any values of x which make $dy/dx = 0$?

Here is a summary of the new useful results we now have. (We also have the box of results at the beginning of Section 8.C.)

More rates of change we now know

- If $y = \tan x$ then $dy/dx = \sec^2 x$.
- If $y = \cot x$ then $dy/dx = -\operatorname{cosec}^2 x$.
- If $y = \sec x$ then $dy/dx = \sec x \tan x$.
- If $y = \operatorname{cosec} x$ then $dy/dx = -\operatorname{cosec} x \cot x$.
- If $y = \ln(\sec x + \tan x)$ then $dy/dx = \sec x$.

It is worth highlighting this box because, when you come to do the process of differentiation the opposite way round in the next chapter, being able to spot these will be very helpful to you.

USEFUL
RESULTS

8.D   The hyperbolic functions of sinh *x* and cosh *x*

Now that we know the Chain, Product and Quotient Rules for differentiation we are able to look at an interesting extension of the two graphs of $y = e^x$ and $y = e^{-x}$.

8.D.(a)   Getting symmetries from e^x and e^{-x}

The graph of $y = e^x$ is not symmetrical and neither is the graph of $y = e^{-x}$, and yet the two graphs shown together have a striking mutual symmetry which is clear from Figure 8.D.1. This is because each is the mirror image of the other in the y-axis.

Can we exploit this?

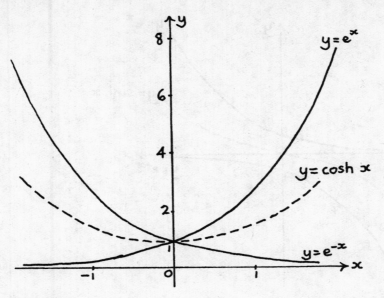

Figure 8.D.1

If we create a new function by taking the average value of e^x and e^{-x} for each value of x, we shall get the function which I have shown by the dashed line in the sketch. It is called $y = \cosh x$. The reason for this is that it behaves in many ways like $\cos x$, curious though this may seem at first sight. Its equation is given by

$$y = \cosh x = \frac{e^x + e^{-x}}{2}.$$

This function is even, that is, $\cosh(-x) = \cosh x$ for any particular value of x.

It describes the curve in which a heavy uniform chain hangs under its own weight. It also describes the sag in a metal tape measure when it is extended, and was used to correct for this before the invention of electronic measuring devices.

If $y = \dfrac{e^x + e^{-x}}{2}$ gives an interesting result, what about $y = \dfrac{e^x - e^{-x}}{2}$?

We can think of this as finding the average value of e^x and $-e^{-x}$ for each value of x.

This gives us the curve shown as a dashed line in Figure 8.D.2.

This function is called $\sinh x$ and it is odd. That is, $\sinh x = -\sinh(-x)$ for any particular value of x.

We now have the pair of definitions

$$\cosh x = \frac{e^x + e^{-x}}{2} \quad \text{and} \quad \sinh x = \frac{e^x - e^{-x}}{2}.$$

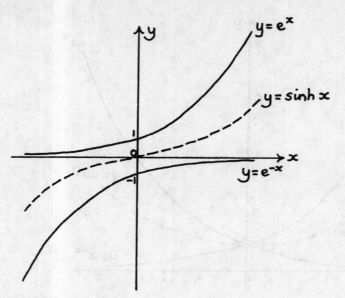

Figure 8.D.2

Remembering from the rules for powers that

$$e^x \times e^x = e^{x+x} = e^{2x}$$

$$e^{-x} \times e^{-x} = e^{-x-x} = e^{-2x}$$

and $\quad e^x \times e^{-x} = e^{x-x} = e^0 = 1,$

we have

$$\cosh^2 x = (\cosh x)^2 = \left(\frac{e^x + e^{-x}}{2}\right)^2 = \frac{e^{2x} + e^{-2x} + 2}{4}.$$

> $\cosh^2 x$ is the way in which mathematicians write $(\cosh x)^2$. It does *not* mean $\cosh x^2$, which is more safely written as $\cosh(x^2)$. In fact
>
> $$\cosh(x^2) = \frac{e^{x^2} + e^{-x^2}}{2}.$$
>
> It is just the same as $\cosh x$ except that the x is replaced by x^2.

We also have

$$\sinh^2 x = \left(\frac{e^x - e^{-x}}{2}\right)^2 = \frac{e^{2x} + e^{-2x} - 2}{4}.$$

So

$$\cosh^2 x - \sinh^2 x = 1.$$

This is true whatever value we choose for x on the x-axis, so it is an example of an **identity**. I described some examples of identities in Section 2.D.(h).

Try showing for yourself that $\cosh^2 x - \sinh^2 x = 1$, without looking at my working, to make sure you can do it.

We begin to see now just why $\cosh x$ and $\sinh x$ have been named in this way. The above relationship is curiously like the trig identity of $\cos^2 x + \sin^2 x = 1$.

8.D.(b)   **Differentiating sinh *x* and cosh *x***
We know that $d/dx\,(e^x) = e^x$ and $d/dx\,(e^{-x}) = -e^{-x}$.

What do we get if we differentiate (a) $y = \sinh x$ and (b) $y = \cosh x$ with respect to x? Have a go at doing this for yourself.

This is what you should have.

$$d/dx(\sinh x) = d/dx\,(\tfrac{1}{2}\,(e^x - e^{-x})) = \tfrac{1}{2}\,(d/dx\,(e^x) - d/dx(e^{-x}))$$

$$= \tfrac{1}{2}\,(e^x + e^{-x}) = \cosh x$$

and, similarly,

$$d/dx(\cosh x) = \sinh x.$$

Again we see that $\sinh x$ and $\cosh x$ are behaving very similarly to $\sin x$ and $\cos x$, though not quite identically since $d/dx(\sin x) = \cos x$ but $d/dx(\cos x) = -\sin x$.

This seems very strange just now, because we have completely different graphs for these two pairs of functions. The mystery of this curious set of links becomes solved later on, in Section 10.C.(b).

Also, just as we did with sin and cos, we can use the Chain Rule to differentiate slightly more complicated functions involving sinh and cosh. For example,

$$\frac{d}{dx}\,(\sinh 3x) = 3\cosh 3x \quad \text{and} \quad \frac{d}{dx}\,(\cosh (x^2 + 1)) = 2x \sinh (x^2 + 1).$$

8.D.(c)   **Using sinh *x* and cosh *x* to get other hyperbolic functions**
Because of the similarities which we have already seen, it makes sense to define further hyperbolic functions to correspond to the other trig functions, so we say

$$\tanh x = \frac{\sinh x}{\cosh x} = \frac{e^x - e^{-x}}{e^x + e^{-x}},$$

$$\operatorname{cosech} x = \frac{1}{\sinh x}, \quad \operatorname{sech} x = \frac{1}{\cosh x}, \quad \coth x = \frac{1}{\tanh x}.$$

Dividing $\cosh^2 x - \sinh^2 x = 1$ by $\cosh^2 x$ gives us

$$1 - \tanh^2 x = \operatorname{sech}^2 x$$

and dividing $\cosh^2 x - \sinh^2 x = 1$ by $\sinh^2 x$ gives us

$$\coth^2 x - 1 = \operatorname{cosech}^2 x$$

again similar but not identical results to the two trig rules of

$$\tan^2 x + 1 = \sec^2 x \quad \text{and} \quad \cot^2 x + 1 = \operatorname{cosec}^2 x.$$

We can now use the Quotient Rule to find d/dx (tanh x). Writing

$$\tanh x = \frac{\sinh x}{\cosh x},$$

we get

$$d/dx(\tanh x) = \frac{(\cosh x)(\cosh x) - (\sinh x)(\sinh x)}{\cosh^2 x} = \frac{1}{\cosh^2 x} = \text{sech}^2 x.$$

(You can get this same result by working directly with tanh x written in terms of e^x and e^{-x} but this is longer. It was question (6) in Exercise 8.C.5.)

Show for yourself that the following three rules are true.

---

(1) $\dfrac{d}{dx}$ (sech x) $= -$ sech x tanh x

(2) $\dfrac{d}{dx}$ (cosech x) $= -$ cosech x coth x

(3) $\dfrac{d}{dx}$ (coth x) $= -$ cosech^2 x

(The working for these is very similar to the working for the corresponding trig functions which came in Exercise 8.C.5.)

EXERCISE 8.D.1

(1) If $e^x = 2$, find the values of (a) sinh x, (b) cosh x and (c) tanh x by using their definitions in terms of e^x and e^{-x}.

(2) If $x = 0$, find the values of (a) sinh x and (b) cosh x. Check that your answers are believable by looking at the graph sketches of these two functions. What is tanh x when $x = 0$?

(3) Differentiate the following with respect to x.

(a) $y = \cosh 2x$ (b) $y = \sinh (3x + 5)$ (c) $y = e^{2x} \sinh 3x$
(d) $y = \tanh 5x$ (e) $y = \ln (\cosh x)$ (f) $y = \cosh^2 3x$

8.D.(d)   Comparing other hyperbolic and trig formulas – Osborn's Rule

In this section, we look at whether some other rules which are true for trig functions are also true for hyperbolic functions.

(1)   In Section 5.D.(d), we showed that $\sin 2A = 2 \sin A \cos A$.
Is it true that $\sinh 2x = 2 \sinh x \cosh x$?

We look at the more complicated side first and see whether it will simplify to give the other side. Doing this gives us

$$2 \sinh x \cosh x = 2 \left(\frac{e^x - e^{-x}}{2} \right) \left(\frac{e^x + e^{-x}}{2} \right) = \frac{e^{2x} - e^{-2x}}{2} = \sinh 2x,$$

so this *is* another rule which transfers exactly.

(2)   Investigate for yourself whether the trig rule of $\cos 2A = \cos^2 A - \sin^2 A$ has the corresponding rule for hyperbolic functions of $\cosh 2x = \cosh^2 2x - \sinh^2 2x$. Indeed, could this be so?

I hope you will have seen straight away that it couldn't be so since we know that $\cosh^2 x - \sinh^2 x = 1$. Try finding for yourself what $\cosh^2 x + \sinh^2 x$ is equal to.

You should have

$$\cosh^2 x + \sinh^2 x = \left(\frac{e^x + e^{-x}}{2}\right)^2 + \left(\frac{e^x - e^{-x}}{2}\right)^2$$
$$= \tfrac{1}{4}((e^{2x} + 2 + e^{-2x}) + (e^{2x} + e^{-2x} - 2))$$
$$= \tfrac{1}{2}(e^{2x} + e^{-2x}) = \cosh 2x.$$

This time we have the two rules

$$\cos 2x = \cos^2 x - \sin^2 x \quad \text{and} \quad \cosh 2x = \cosh^2 x + \sinh^2 x.$$

The different results of (1) and (2) are examples of **Osborn's Rule** which says that the trig rules match the corresponding hyperbolic rules exactly, unless the working somewhere involves two sins or two sinhs multiplied together. In this case, there is a sign change there.

8.D.(e)   Finding the inverse function for sinh x

We look now at the function $y = \sinh x$ to see whether we can find a function that will take us back the other way. We'll start by considering a numerical example so that we can see what is happening here.

Suppose we know that $\sinh x = 2$. What value of x would give this result?

I show this question pictorially in Figure 8.D.3.

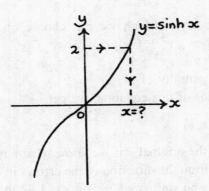

Figure 8.D.3

We say that $x = \sinh^{-1} 2$ meaning that x is the number whose sinh is equal to 2.

 $\sinh^{-1} x$ does *not* mean $1/\sinh x$. This would be written as $(\sinh x)^{-1}$.

Using a sequence like INV-HYP-SIN on your calculator should give you the answer of $x = 1.44$ to 2 d.p. but how can we show this process actually happening? We have

$$\sinh x = \frac{e^x - e^{-x}}{2} = 2 \quad \text{so} \quad e^x - e^{-x} = 4.$$

Multiplying through by e^x gives $e^{2x} - 1 = 4e^x$ so $e^{2x} - 4e^x - 1 = 0$.
This is actually a quadratic equation in e^x, which we can see by putting $e^x = m$.
This gives us $m^2 - 4m - 1 = 0$. We now use the formula to get

$$m = \frac{4 \pm \sqrt{16 + 4}}{2} = \frac{4 \pm 2\sqrt{5}}{2} = 2 \pm \sqrt{5}.$$

Now, $e^x = 2 - \sqrt{5}$ is not a possible solution, because e^x is always positive.
Therefore we have $e^x = 2 + \sqrt{5}$ so $x = \ln(2 + \sqrt{5}) = 1.44$ to 2 d.p.

Having seen what happens with this particular example, we will now see how we can find a general rule for $y = \sinh^{-1} x$.
We use exactly the same method that we did with the numerical example. We start with

$$y = \sinh x = \frac{e^x - e^{-x}}{2} \quad \text{so} \quad 2y = e^x - e^{-x} \quad \text{so} \quad e^x - 2y - e^{-x} = 0.$$

Multiplying through by e^x gives

$$e^{2x} - 2y\,e^x - 1 = 0.$$

Again, this is a quadratic equation. We see this very nicely by putting $m = e^x$.
Then we have $m^2 - 2y\,m - 1 = 0$, and using the formula gives

$$m = \frac{2y \pm \sqrt{4y^2 + 4}}{2} \quad \text{so} \quad m = \frac{2y \pm 2\sqrt{y^2 + 1}}{2} = y \pm \sqrt{y^2 + 1}.$$

Replacing m by e^x gives us

$$e^x = y \pm \sqrt{y^2 + 1}.$$

Now, e^x is always positive for every x which we can choose on the x-axis. However, $y - \sqrt{y^2 + 1}$ is always negative since $\sqrt{y^2 + 1} > y$.
Therefore we cannot have $e^x = y - \sqrt{y^2 + 1}$.
This gives us just the single possibility of $e^x = y + \sqrt{y^2 + 1}$.
Taking natural logs of both sides of this equation, we get

$$\ln(e^x) = x = \ln(y + \sqrt{y^2 + 1}).$$

We now have the rule for finding the original x if we know what y is, but it is giving us x as a function of y. We can see this from the direction of the arrows in Figure 8.D.4(a) which shows $\sinh x = 1$ giving $x = 0.88$, and $\sinh x = 3$ giving $x = 1.82$ to 2 d.p.
We want a rule which will give us y as a function of x so we interchange x and y.
This gives us the inverse function of

$$\boxed{y = \sinh^{-1} x = \ln(x + \sqrt{x^2 + 1}).}$$

Try feeding in $x = 1$ and $x = 3$ to this, so that you can see it actually working.
I show a sketch of this function in Figure 8.D.4(b).

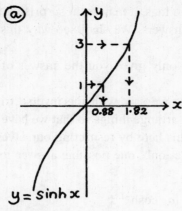

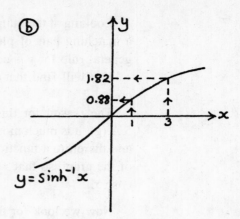

Figure 8.D.4

The interchanging of x and y means that, as for every function and its inverse, the graphs of $y = \sinh x$ and $y = \sinh^{-1} x$ are symmetrical about the line $y = x$.

If you draw your own sketch, showing both $y = \sinh x$ and $y = \sinh^{-1} x$ together, you can see this symmetry.

We can also see graphically in Figure 8.D.4(a) that $y = \sinh^{-1} x$ must be a function because there is only one value of x which can give a particular value of $\sinh x$, so there will be no ambiguity when we want to go back the other way.

EXERCISE 8.D.2

To extract as much information as possible from the two graphs above, and from Section 8.D.(b), try answering the following questions yourself.

(1) What is the gradient of the curve $y = \sinh x$ at the origin?
(2) From your answer to (1), what special property does the line $y = x$ have?
(3) From the symmetry of the two graphs, what is the gradient of the curve
     $y = \sinh^{-1} x$ at the origin?

8.D.(f)   **Can we find an inverse function for cosh x?**

Again we start by looking at a numerical example.

If $\cosh x = 2$, what value of x could have given this result?

We see immediately from Figure 8.D.5 that there will be *two* possible values of x. This is because $\cosh(x) = \cosh(-x)$ for all values of x.

Doing the working in exactly the same way as we did for $\sinh x = 2$ in Section 8.D.(e), we find that $e^x = 2 \pm \sqrt{3}$. (Do this for yourself.)

Both these possibilities are positive so they are both possible solutions.

If we take $e^x = 2 + \sqrt{3}$ we get $x = \ln(2 + \sqrt{3}) = 1.32$ to 2 d.p.

If we take $e^x = 2 - \sqrt{3}$ we get $x = \ln(2 - \sqrt{3}) = -1.32$ to 2 d.p.

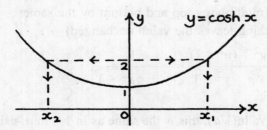

Figure 8.D.5

8.D The hyperbolic functions

Looking at the numbers in these two logs, it may seem surprising to you that they *do* give a matching pair of plus and minus answers. We shall see why this is so when we find a general rule for $y = \cosh^{-1} x$.

You will find that your calculator only gives you the answer of $x = 1.32$ to 2 d.p. for $\cosh^{-1} 2$.

The reason for this is that, just as we saw with the inverse trig functions in Section 5.A.(g), it is much more convenient to arrange things so that we have a single-valued answer and therefore a function. We can do this here by restricting ourselves to the right-hand side of the graph so that $x \geq 0$. We then get only one possible answer for x from each value of $\cosh x$.

Now we look for the general rule for $\cosh^{-1} x$

The procedure is very similar to what we did for $\sinh^{-1} x$ in the last section.

See how far you can get by yourself.

You should have

$$y = \frac{e^x + e^{-x}}{2} \quad \text{so} \quad 2y = e^x + e^{-x} \quad \text{so} \quad e^x - 2y + e^{-x} = 0$$

so

$$e^{2x} - 2y\, e^x + 1 = 0 \quad \text{so} \quad m^2 - 2y\, m + 1 = 0 \quad \text{putting } m = e^x$$

so

$$m = \frac{2y \pm \sqrt{4y^2 - 4}}{2} = \frac{2y \pm 2\sqrt{y^2 - 1}}{2} = y \pm \sqrt{y^2 - 1} = e^x.$$

Both of these possibilities are positive, so we find that we are getting two possible solutions. We have

$$e^x = y \pm \sqrt{y^2 - 1}$$

so, taking natural logs,

$$x = \ln(y \pm \sqrt{y^2 - 1}).$$

It is a nuisance having a general formula with this $\pm$ in the middle of the log where we can't easily get at it, so now we use a cunning trick involving the difference of two squares to put it somewhere better.

It goes like this:

$$y - \sqrt{y^2 - 1} = (y - \sqrt{y^2 - 1}) \times \frac{(y + \sqrt{y^2 - 1})}{(y + \sqrt{y^2 - 1})}$$

(multiplying top and bottom by the same
thing leaves the value unchanged)

$$= \frac{y^2 - (y^2 - 1)}{y + \sqrt{y^2 - 1}} = \frac{1}{y + \sqrt{y^2 - 1}}.$$

Why is this any better?

It is because, if we have $\ln(1/a)$, this is the same as $\ln 1 - \ln a$, using the second rule of logs. These rules are listed in Section 3.C.(d).

But $\ln 1 = 0$ because $e^0 = 1$. You can see that this agrees with Figure 8.D.1. So

$$\ln(1/a) = -\ln a \quad \text{and} \quad \ln\left(\frac{1}{y + \sqrt{y^2 - 1}}\right) = -\ln(y + \sqrt{y^2 - 1}).$$

This gives us the two solutions that $x = \pm \ln(y + \sqrt{y^2 - 1})$ and we see now why $\ln(2 - \sqrt{3}) = -\ln(2 + \sqrt{3})$ in the numerical example earlier.

We now have the two possible values for x from a given y value.

Interchanging x and y so that we can write this as a relation for y in terms of x, we have

$$y = \pm \ln(x + \sqrt{x^2 - 1})).$$

If we restrict the x values by saying $x \geq 0$, we have the inverse *function* of

$$y = \cosh^{-1} x = \ln(x + \sqrt{x^2 - 1})).$$

This is called the **principal inverse function for cosh**.

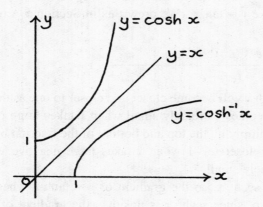

Figure 8.D.6

I show the two functions, $y = \cosh x$ and $y = \cosh^{-1} x$ for $x \geq 0$ in Figure 8.D.6. Just as with any inverse pair of functions, they are symmetrical about the line $y = x$.

8.D.(g)   **tanh x and its inverse function tanh^{-1} x**

What will the graph of $y = \tanh x$ look like? It is not possible to get this one quite so simply from the graphs of $y = e^x$ and $y = e^{-x}$. We have

$$y = \tanh x = \frac{\sinh x}{\cosh x} = \frac{e^x - e^{-x}}{e^x + e^{-x}}.$$

Try answering the following questions yourself.

(1)   What is $\tanh(0)$?

(2)   Can you work out the connection between $\tanh(-x)$ and $\tanh(x)$?
         What will this mean for the graph sketch?

(3)   Multiply the top and bottom of the fraction $(e^x - e^{-x})/(e^x + e^{-x})$ by e^{-x}.
         From your answer to this, can you see what happens to the values of $\tanh x$ when x takes very large positive values?

Now try multiplying the top and bottom of the original fraction by e^x. Can you see what will happen to the value of tanh x when x takes large negative values?

(You could check that your ideas are right by choosing some particular large positive and negative values for x and using your calculator.)

(4)   What is the gradient of the curve $y = \tanh x$ at the origin?

(You may need to look at Section 8.D.(c) to answer this question.)

(5)   See if you can use all the information from your answers to the previous questions to draw a sketch of the graph for $y = \tanh x$.

You should have the following answers.

(1)   tanh $(0) = 0$ because sinh $(0) = 0$.

(2)   Replacing x by $-x$ gives

$$\tanh (-x) = \frac{e^{-x} - e^x}{e^{-x} + e^x} = - \left(\frac{e^x - e^{-x}}{e^x + e^{-x}} \right) = - \tanh x$$

so the left-hand side of the graph will be given by reflecting the right-hand side of the graph in the y-axis and then turning it upside down. $y = \tanh x$ is an odd function, just like $y = \tan x$. (We drew this in Section 5.A.(e).)

(3)   You should get

$$\tanh x = \frac{e^{-x} (e^x - e^{-x})}{e^{-x} (e^x + e^{-x})} = \frac{1 - e^{-2x}}{1 + e^{-2x}}.$$

The value of tanh x will become closer and closer to one as the value of x increases because e^{-2x} becomes extremely small when x takes large positive values.

Similarly, multiplying the top and bottom of the fraction by e^x shows that tanh x gets closer and closer to -1 when x takes large negative values, since e^{2x} then becomes extremely small.

(4) d/dx (tanh x) = sech^2 x, so the gradient of $y = \tanh x$ when $x = 0$ is 1, because sech$(0) = 1$. Also, since sech^2 x is positive, the gradient of $y = \tanh x$ is always positive.

(5)   Putting all this information together gives us the graph sketch shown in Figure 8.D.7.

The lines $y = 1$ and $y = -1$ are horizontal asymptotes for this graph.

I have also drawn on the graph a line showing how we could find the value of x when tanh $x = \frac{1}{2}$.

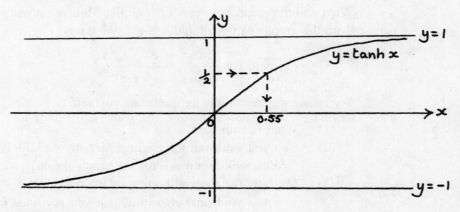

Figure 8.D.7

If you use your calculator to find $\tanh^{-1}(\frac{1}{2})$, you will get $x = 0.55$ to 2 d.p.

We can see from the shape of the graph that each value of $\tanh x$ can only come from one possible value of x, so therefore the function $y = \tanh x$ will have an inverse function. Now we'll find the rule that gives us this. We have

$$y = \tanh x = \frac{e^x - e^{-x}}{e^x + e^{-x}} \quad \text{so} \quad y(e^x + e^{-x}) = e^x - e^{-x}.$$

Multiplying all through by e^x gives $y(e^{2x} + 1) = e^{2x} - 1$, so

$$ye^{2x} + y = e^{2x} - 1 \quad \text{and} \quad y + 1 = e^{2x} - ye^{2x} = e^{2x}(1 - y).$$

Therefore

$$e^{2x} = \frac{1 + y}{1 - y}.$$

Taking logs both sides, we have

$$\ln(e^{2x}) = 2x = \ln\left(\frac{1 + y}{1 - y}\right) \quad \text{so} \quad x = \tfrac{1}{2} \ln\left(\frac{1 + y}{1 - y}\right).$$

We now have the rule to get back to the original x if we know y. Use it to check that, if you put $y = \frac{1}{2}$, you do get $x = 0.55$ to 2 d.p.

Interchanging x and y as before, so that we have this rule as a function of x, we get the inverse function of

$$\boxed{\tanh^{-1} x = \tfrac{1}{2} \ln\left(\frac{1 + x}{1 - x}\right).}$$

To give the log of a positive quantity, the possible values of x will have to lie between -1 and $+1$.

We can see that this is where the values of x must lie from looking at the graph sketch of $y = \tanh^{-1} x$ which I have drawn with $y = \tanh x$ in Figure 8.D.8.

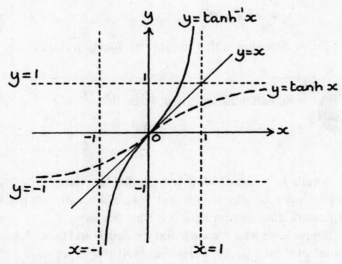

Figure 8.D.8

I have used the line of symmetry $y = x$ to draw this sketch. I have also used the answer to Question (4) which was that the gradient of $y = \tanh x$ when $x = 0$ is 1. This means that $y = x$ is a tangent to both $y = \tanh x$ and $y = \tanh^{-1} x$. It is a very interesting tangent because it crosses both of the curves, which sort of flex themselves when $x = 0$. The line $y = x$ does exactly the same thing with $y = \sinh x$ and $y = \sinh^{-1} x$ at the origin, as you'll see if you draw it in on Figure 8.D.4(a) and (b). We shall look at points of inflection like this in more detail in Section 8.E.(b).

You may find it helpful here to emphasise the separateness of the two curves by using two colours on them. Be careful to put the colour correctly on the two separate halves of each graph! (The tanh graph is a flattened S shape.)

We were able to see from the graph that $y = \tanh x$ must have an inverse function, but suppose we didn't know what the graph looked like? Can we still show that the inverse relation will be a function?

To do this, we have to show that it isn't possible to get the same value for $\tanh x$ from two different values for x, so that, when we go back the other way, there is only one possible answer.

In other words, we have to show that the only way that $\tanh a = \tanh b$ is for a and b to be themselves equal.

We put $\tanh a = \tanh b$ so

$$\frac{e^a - e^{-a}}{e^a + e^{-a}} = \frac{e^b - e^{-b}}{e^b + e^{-b}}$$

and see what happens. Try tidying this up for yourself, and see if you can show that a and b must be equal.

Multiplying by $(e^a + e^{-a})(e^b + e^{-b})$ to get rid of fractions, we get

$$(e^a - e^{-a})(e^b + e^{-b}) = (e^b - e^{-b})(e^a + e^{-a})$$

so

$$e^{(a+b)} - e^{(b-a)} + e^{(a-b)} - e^{-(a+b)} = e^{(a+b)} - e^{(a-b)} + e^{(b-a)} - e^{-(a+b)}$$

so

$$2e^{(a-b)} = 2e^{(b-a)} \quad \text{so} \quad a - b = b - a \quad \text{so} \quad 2a = 2b \quad \text{and} \quad a = b.$$

We've now shown that the inverse function does exist, without reference to the graph.

 Remember that it is *not* true that $e^a \times e^b = e^{ab}$. We must add the powers.

8.D.(h)   **What's in a name? Why 'hyperbolic' functions?**

The mystery of why $\sinh x$ and $\cosh x$ are called hyperbolic functions has not yet been explained. This section tells you why this is so.

Suppose we let $x = \cosh \theta$ and $y = \sinh \theta$ and then plot the points that we get for different values of θ on a graph. For example, if $\theta = 0$, we have $x = \cosh \theta = 1$ and $y = \sinh \theta = 0$, so one point on this graph will be (1,0).

Since $\cosh^2 \theta - \sinh^2 \theta = 1$, we know that the equation of this graph will be $x^2 - y^2 = 1$. This is the equation of the **hyperbola** which I show below in Figure 8.D.9.

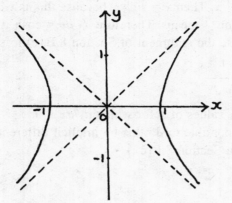

Figure 8.D.9

This graph may look a more familiar shape if you turn it through 45° anticlockwise. The two dashed lines make this resemblance easier to see.

Actually, only the right-hand side of it is given by $x = \cosh \theta$ and $y = \sinh \theta$. Can you see why this is? Can you think of a way that we could get the whole graph?

cosh θ can't be negative, and the points on the left-hand side of the graph have negative values for x.

We could get the whole graph by putting $x = \sec \theta$ and $y = \tan \theta$.

Since $\sec^2 \theta - \tan^2 \theta = 1$, we still have $x^2 - y^2 = 1$, and we have the left-hand side of the graph too, since sec θ can take negative values.

In a similar way, $x = \cos \theta$ and $y = \sin \theta$ are linked to the circle $x^2 + y^2 = 1$. Indeed, it was this circle which we used to define the sin and cos of angles greater than 90° in Section 5.A.(c).

The variable θ which we have used for this hyperbola and circle is called a **parameter**. We can get other curves of the same type by subtly adjusting how we use it. For example, $x = 2 \cosh \theta$ and $y = 3 \sinh \theta$ gives the hyperbola $(x/2)^2 - (y/3)^2 = 1$ and $x = 5 \cos \theta$ with $y = 5 \sin \theta$ gives $x^2 + y^2 = 25$, the circle with centre (0,0) and radius 5 units.

Unbalancing them to give $x = 4 \cos \theta$ and $y = 3 \sin \theta$, say, gives a squashed circle, or **ellipse**, with the equation $(x/4)^2 + (y/3)^2 = 1$. This is centred at the origin and cuts the axes at (4, 0), (0, 3), (−4, 0) and (0, −3).

There isn't space to go into this in more detail just now, but you will find that this use of parameters to describe particular curves is often of great practical use in extracting further information from relationships between physical quantities.

Finally, you may be thinking that the name 'hyperbolic' isn't the only strange thing about these functions. Why is there this curious link between them and the trig functions? I'll show you the reason for this in Section 10.C.(b).

8.D.(i)   Differentiating inverse trig and hyperbolic functions

This is something which students quite often find difficult, but if you have worked through the earlier parts of this section so that you are now happy with what these inverse functions do, you should find it quite straightforward. We'll look at two examples of differentiation, and then see how using the Chain Rule makes it possible to get lots of other similar results very easily.

EXAMPLE (1) How can we find dy/dx if $y = \sinh^{-1} x$?

We could set about doing this in two ways.

METHOD (1) Let $y = \sinh^{-1} x$. Then $x = \sinh y$ because this is what the inverse function of $\sinh^{-1}$ means. Therefore $dx/dy = \cosh y$.

Now we use the argument of Section 8.B.(e) to say

$$\frac{dx}{dy} = \frac{1}{dy/dx},$$

excluding any values of x for which $dy/dx = 0$.

(It is also possible to do this by implicit differentiation. I show you this method in Section 8.F.(c).)

Therefore

$$\frac{dy}{dx} = \frac{1}{\cosh y}.$$

But $\cosh^2 y = \sinh^2 y + 1$, so

$$\cosh^2 y = x^2 + 1 \quad \text{and} \quad \cosh y = \pm \sqrt{x^2 + 1}.$$

But we know that the gradient of $y = \sinh x$ is always positive. (How do we know this? What is d/dx ($\sinh x$)?)

It is $\cosh x$ and $\cosh x$ is always positive. Therefore

$$\frac{dy}{dx} = \frac{1}{\sqrt{x^2 + 1}}$$

and we have the result that

$$\boxed{\frac{d}{dx} (\sinh^{-1} x) = \frac{1}{\sqrt{x^2 + 1}}.}$$

METHOD (2) This uses the result which we found in Section 8.D.(e) that $\sinh^{-1} x = \ln (x + \sqrt{x^2 + 1})$.

Therefore we can say

$$\frac{d}{dx} (\sinh^{-1} x) = \frac{d}{dx} (\ln (x + \sqrt{x^2 + 1})) = \frac{1 + \frac{1}{2} (x^2 + 1)^{-1/2} (2x)}{x + \sqrt{x^2 + 1}}.$$

This doesn't look too good, but it is tidied up amazingly by multiplying the top and bottom by $\sqrt{x^2 + 1}$. We then get

$$\frac{\sqrt{x^2 + 1} + x}{\sqrt{x^2 + 1} \, (x + \sqrt{x^2 + 1})} = \frac{1}{\sqrt{x^2 + 1}} \ldots \text{neat!}$$

EXAMPLE (2) This time, we differentiate an inverse trig function.

We will find dy/dx if $y = \tan^{-1} x$ (or arctan x as it is also known).

Remember that $y = \tan^{-1} x$ means that y is the angle between $-\pi/2$ and $\pi/2$ whose tan is x. I explained this in Section 5.A.(i).

We start by saying that

$$x = \tan y \quad \text{so} \quad \frac{dx}{dy} = \sec^2 y.$$

Then we use the identity $\tan^2 y + 1 = \sec^2 y$ to get $\sec^2 y = x^2 + 1$, so

$$\frac{dx}{dy} = x^2 + 1 \quad \text{and} \quad \frac{dy}{dx} = \frac{1}{x^2 + 1}$$

giving us the result

$$\boxed{\frac{d}{dx} (\tan^{-1} x) = \frac{1}{x^2 + 1}.}$$

EXAMPLE (3) This example shows how we can apply the above result.

Suppose we need to find

$$\frac{d}{dx} (\tan^{-1} (2x + 3)).$$

We don't need to do all the previous working again because $2x + 3$ is itself a function of x. Therefore we can just use the Chain Rule, putting $X = 2x + 3$, and remembering that $dy/dx = (dy/dX)(dX/dx)$. (See Section 8.C.(a) if necessary.)

Here, we have

$$y = \tan^{-1} (2x + 3) = \tan^{-1} X \quad \text{and} \quad X = 2x + 3$$

so

$$\frac{dy}{dX} = \frac{1}{X^2 + 1} \quad \text{and} \quad \frac{dX}{dx} = 2.$$

Therefore

$$\frac{dy}{dx} = \frac{2}{X^2 + 1} = \frac{2}{(2x + 3)^2 + 1} = \frac{2}{4x^2 + 12x + 10}.$$

In general, we can say that if $y = \tan^{-1}$ (lump), and the lump is a function of x, then

$$\frac{d}{dx} (\tan^{-1} (\text{lump})) = \left(\frac{1}{(\text{lump})^2 + 1} \right) \times \frac{d}{dx} (\text{lump}).$$

If you think of it this way, you will probably be able to write the answers down straight away.

We get a particularly useful version of this if we put (lump) $= x/a$ where a is a constant. This gives us

$$\frac{d}{dx} \left(\tan^{-1} \frac{x}{a} \right) = \frac{1}{x^2/a^2 + 1} \times \frac{1}{a} = \frac{a^2}{x^2 + a^2} \times \frac{1}{a} = \frac{a}{x^2 + a^2}.$$

This result is very useful for finding some particular kinds of integral, as we shall see in Section 9.B.(d).

Exactly the same system can be used to differentiate inverse functions of other more complicated functions.

So, for example, if we have $\sinh^{-1}$ (lump), then

$$\frac{d}{dx}(\sinh^{-1}(\text{lump})) = \frac{1}{\sqrt{(\text{lump})^2 + 1}} \times \frac{d}{dx}(\text{lump}).$$

In particular, if (lump) $= x/a$, we have

$$\frac{d}{dx}\left(\sinh^{-1}\frac{x}{a}\right) = \frac{1}{\sqrt{x^2/a^2 + 1}} \times \frac{1}{a} = \frac{a}{\sqrt{x^2 + a^2}} \times \frac{1}{a} = \frac{1}{\sqrt{x^2 + a^2}}.$$

I have tidied up the first fraction by multiplying it top and bottom by a, remembering that a put *inside* a square root must be written a^2.

EXERCISE 8.D.3

(1) By choosing suitable values for a, and using the pair of results

$$\frac{d}{dx}(\tan^{-1}(x/a)) = \frac{a}{x^2 + a^2} \quad \text{and} \quad \frac{d}{dx}(\sinh^{-1}(x/a)) = \frac{1}{\sqrt{x^2 + a^2}},$$

differentiate the following with respect to x.

(a) $\tan^{-1}\left(\dfrac{x}{3}\right)$ (b) $\sinh^{-1}\left(\dfrac{x}{5}\right)$ (c) $\tan^{-1}\left(\dfrac{3x}{2}\right)$ (d) $\sinh^{-1}\left(\dfrac{2x}{3}\right)$

(2) Use the Chain Rule to differentiate the following with respect to x.

(a) $\tan^{-1}(5x)$ (b) $\sinh^{-1}(3x)$ (c) $\tan^{-1}(x+3)$ (d) $\tan^{-1}(3x+4)$

(e) $\tan^{-1}(1-x)$ (f) $\sinh^{-1}(2x+1)$ (g) $\sinh^{-1}(3-2x)$

(h) $\tan^{-1}\left(\dfrac{x+3}{4}\right)$ (i) $\tan^{-1}\left(\dfrac{2x+1}{3}\right)$ (j) $\tan^{-1}\left(\dfrac{3x+2}{4}\right)$

(3) Show that $\dfrac{d}{dx}(\tanh^{-1}x) = \dfrac{1}{1-x^2}$ using $\tanh^{-1}x = \frac{1}{2}\ln\left(\dfrac{1+x}{1-x}\right)$.

(4) Solve the equation $8 \sinh x = 3 \operatorname{sech} x$.

(5) Find all the possible solutions of the following equations.

(a) $2 \sinh^2 x - 5 \cosh x - 1 = 0$ (b) $3 \operatorname{sech}^2 x + 8 \tanh x - 7 = 0$

8.E   Some uses for differentiation

8.E.(a)   Finding the equations of tangents to particular curves

In Section 8.C.(e), we found the gradient of two of the tangents to the curve

$$y = (x + 3)/(x - 2)$$

by using the Quotient Rule to find dy/dx for this curve.

Since dy/dx tells us the steepness or gradient of a curve at any given point on it, it makes it possible for us to find the equation of the tangent to the curve at any point on it, provided that this is a point where the curve *has* a tangent, and none of the problems of Section 8.A.(f) exist.

Here are two examples of doing this.

EXAMPLE (1) Find the equations of the tangents to the curve $y = x^2 - 4x + 3$ at the two points (a) (5,8) and (b) (2, –1).

To find the gradients of the tangents we differentiate $y = x^2 - 4x + 3$ with respect to x giving $dy/dx = 2x - 4$.

> This gives us the rule to find the *gradient* of the tangent for any value of x. It is not the equation of the tangent.

(a) When $x = 5$, $dy/dx = 10 - 4 = 6$ so $m = 6$ for the tangent at (5,8).
     Using $y - y_1 = m(x - x_1)$ for the equation of the tangent, from Section 2.B.(f), we have $y - 8 = 6(x - 5)$, so $y = 6x - 22$ is the equation of tangent (a).
(b) When $x = 2$, $dy/dx = 0$.
     What is happening here? Try drawing your own sketch to show how this makes sense.

If $dy/dx = 0$, the tangent is horizontal. This tangent is at the lowest point of the curve $y = x^2 - 4x + 3$, and its equation is $y = -1$.

I show a sketch of the curve and these two tangents in Figure 8.E.1.

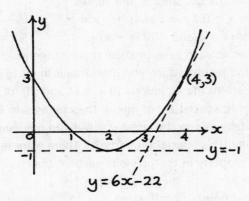

Figure 8.E.1

EXAMPLE (2) Find the equations of the tangents to the curve $y = \cos x$ when
(a) $x = \pi/2$, (b) $x = \pi/6$ and (c) $x = \pi$.

If $y = \cos x$ then $dy/dx = - \sin x$ so the gradient of tangent (a) is $- \sin \pi/2 = - 1$. It touches the curve $y = \cos x$ at the point $(\pi/2, 0)$ and its equation is $y = -1(x - \pi/2)$ or $y + x = \pi/2$.
The gradient of tangent (b) is $- \sin \pi/6 = -\frac{1}{2}$.
(We found the sin, cos and tan of $\pi/6$, $\pi/4$, and $\pi/3$ (that is, 30°, 45° and 60°), in Section 4.A.(g).)
Tangent (b) touches the curve $y = \cos x$ at $(\pi/6, \sqrt{3}/2)$ and its equation is $y - \sqrt{3}/2 = -\frac{1}{2}(x - \pi/6)$ or $2y + x = \pi/6 + \sqrt{3}$
or $y = -\frac{1}{2}x + (\pi/12 + \sqrt{3}/2)$.

This looks a little unfriendly, but it is not surprising that the equation of a tangent to a cos curve should involve numbers like π and $\sqrt{3}$. The value of $\pi/12 + \sqrt{3}/2$ is 1.13 to 2 d.p. and this agrees with the look of the y intercept of tangent (b) on the graph sketch which I have drawn below.

The gradient of tangent (c) is $-\sin\pi = 0$ so this tangent is horizontal. Its equation is $y = -1$.

All three tangents are shown here in Figure 8.E.2.

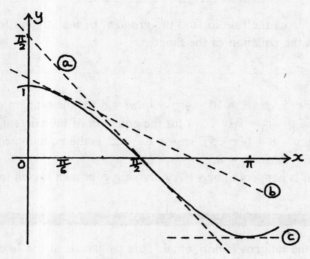

Figure 8.E.2

EXERCISE 8.E.1

Find the equations of the tangents to the curves
(1) $y = e^x$ at   (a) $x = 0$ (b) $x = 1$ and   (c) $x = 2$.
(2) $y = \tan x$ at   (a) $x = 0$ and   (b) $x = \pi/4$.

Draw sketches in each case to show these tangents.

Use one of the results which you have found in (1) to decide how many solutions there are to the equations (i) $x = e^x$ and (ii) $3x = e^x$.

(3) There is something special about one of the tangents in Example (2) above and one of the tangents to the curve $y = \tan x$ in question (2) in this exercise. Can you spot what this special property is? There were examples of tangents with this same property in the previous section, too.

8.E.(b)   Finding turning points and points of inflection

A **turning point** on a curve with the equation $y = f(x)$ is a point at which $dy/dx = 0$, or $f'(x) = 0$, writing the same thing in function notation. Turning points are also sometimes called stationary points, and the values of $f(x)$ where $f'(x) = 0$ are called **stationary values**.

From the examples which we have just looked at in the previous section we can see that it will be useful when sketching curves if we can find where the horizontal tangents are, that is the points where $dy/dx = 0$. Finding the answer to this will not only help us to draw graph sketches, but also to extract useful information about physical relationships. (For example, in Section 2.D.(g), the horizontal tangent is at the point of the curve corresponding to the highest point reached by the ball, so $ds/dt = 0$ at this point.)

Sometimes it is also helpful to know what the value of d^2y/dx^2 is for particular values of x. d^2y/dx^2 means $d/dx\,(dy/dx)$, so it tells us the rate of change of the rate of change with respect to x. We used it, but with a different letter, in Section 8.A.(e) when we found d^2x/dt^2 for $x = \cos t$.

To help you to understand the different possibilities, I have drawn sketches showing interesting points on the curves of some simple functions in Figure 8.E.3. You should fill in your own answers to the questions I have asked you in the table below the drawings.

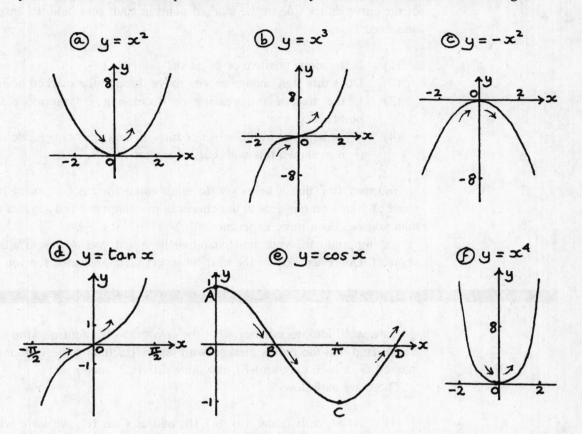

In all these sketches O marks both the origin and zero.

Figure 8.E.3

Function	$\dfrac{dy}{dx}$	Value of $\dfrac{dy}{dx}$	$\dfrac{dy^2}{dx^2}$	Is $\dfrac{dy^2}{dx^2}$ +, − or 0?
(a) $y = x^2$ at O	$2x$	0	2	+
(b) $y = x^3$ at O				
(c) $y = -x^2$ at O				
(d) $y = \tan x$ at O				

(e) $y = \cos x$
  (i) at A
  (ii) at B
  (iii) at C
  (iv) at D				

(f) $y = x^4$ at O				

Now check your answers to this table. These are given at the back of the book as Table 8.E.2 after the answers to Exercise 8.E.1.

Next, go back to the curves in Figure 8.E.3 and look at what is happening to the steepness of the curve either side of the marked point in each case, and try answering the following questions.

(1)   Is the slope positive or negative?

(2)   Does this sign change as you move through the marked point?

(3)   Is the steepness increasing or decreasing as you move through the marked point?

(4)   What happens to the sense of turn of the curve either side of the marked point? (I have shown this with curved arrows.)

You may find that it helps you to think about what is happening here if you sketch in some of your own tangents to the curves in my diagrams. (I'd suggest using pencil for this, then you can do it more experimentally.)

It's important for your understanding here that you *do* try to answer these questions yourself. Don't just skip to the next bit to get them answered for you.

Now, we'll look together at what the answers to the four questions above tell us.

We find that the points marked with letters (including the various points at the origin, marked *O* in each diagram) fit into three different categories.

These are as follows:

(1)   At *O* in diagrams (a) and (f), and at *C* in (e), we have what is called a **local minimum** ('local' because sometimes curves may dip down below this value somewhere else). At these points, the value of dy/dx is zero because the tangent to the curve is horizontal. As we pass through these points, the slope of the tangents changes from negative to positive as the value of *x* increases. The sense of turn remains anticlockwise through these points.

(2)   At *O* in diagram (c), and at *A* in (e) we have what is called a **local maximum**. Again, the value of dy/dx is zero at these points. As we pass through these points, the slope of the tangents changes from positive to negative as the value of *x* increases. The sense of turn remains clockwise through these points.

(1) and (2) give the result that

$$\frac{dy}{dx} = 0 \text{ at any local maximum or minimum.}$$

(3)   At *O* in diagrams (b) and (d), and at *B* and *D* in diagram (e), we have points where the curve flexes itself. These are called **points of inflection**. The tangent to each curve at these points crosses the curve there, and the sense of turn changes. At *O* in (b) and (d), and at *B* in (e), it changes from clockwise to anticlockwise, and at *D* in (e) it changes from anticlockwise to clockwise.

O in (b) is the only one of these points where we also have $dy/dx = 0$.

Either side of each of these points the slope of the tangents remains either positive or negative. In the first three cases, the slopes of the tangents first become flatter as we approach the point and then steeper again once we are through it. This means that the slope itself has a local minimum at the point concerned. In other words, $d/dx\,(dy/dx) = d^2y/dx^2 = 0$ at each of these points.

In the fourth case, of curve (e) at D, the slope becomes steeper as we approach D, and then less steep once we have passed D, so this slope has a local maximum at D. Again, $d^2y/dx^2 = 0$.

If you find d^2y/dx^2 for the other examples we have met of tangents crossing curves, you'll see that it is also zero at these points. (You could check for yourself with $y = \sin x$, $y = \sinh x$ and $y = \tanh x$, all at the origin.)

$$\frac{d^2y}{dx^2} = 0 \text{ at any point of inflection}$$

We have seen that $d^2y/dx^2 = 0$ at any point of inflection.

What will happen to the value of d^2y/dx^2 at a local maximum or minimum?

At each of the local maximum points from (c) and (e), the slope of the curve goes from positive to negative, so the change in the slope is negative.

In both cases, d^2y/dx^2 is negative at the maximum point.

At the two local minimum points of (a) and (e), the slope of the curve goes from negative to positive, so the change in the slope is positive.

In both cases, d^2y/dx^2 is positive at the minimum point.

The case of the local minimum in (f) works out a little differently. The slope of the curve goes from negative to positive, and its rate of change is positive *except* at the point O itself.

We have $d^2y/dx^2 = 12x^2 = 0$ at the point O, although it is positive either side of O. At O itself, the curve is very blunt because it has its four roots of $x = 0$ all bunched together here. This has the effect of making the rate of change of dy/dx at this point (that is, d^2y/dx^2) equal to zero. This effect, which will happen whenever a curve is blunt like this, makes the rules for testing for maximum and minimum points slightly more complicated, because it is only sometimes possible to use the sign of d^2y/dx^2 to test which we've got.

Here is a summary of the above results, so that we can use them to find out how particular curves will behave.

Finding and classifying turning points and points of inflection

For a point to be a local maximum, dy/dx must be equal to zero. Then use either

Test (1): the gradients of the tangents move through the point in the sequence + 0 −, so test the value of dy/dx either side of this point,

or

Test (2): if the value of d^2y/dx^2 is negative at this point, then it is a local maximum, but if $d^2y/dx^2 = 0$ then Test (1) must be used.

For a point to be a local minimum, dy/dx must be equal to zero. Then use either

Test (1): the gradients of the tangents move through the point in the sequence − 0 +, so test the value of dy/dx either side of this point,

or

Test (2): if the value of d^2y/dx^2 is positive at this point, then it is a local minimum, but if $d^2y/dx^2 = 0$ then Test (1) must be used.

For a point of inflection,

(1) the value of dy/dx does not change sign as it moves through the point (it may or may not be equal to zero at the point itself),

and

(2) the value of d^2y/dx^2 at the point *must* be equal to zero.

8.E.(c) **General rules for sketching curves**

The tests outlined in this previous section give us useful extra information which we can use for sketching graphs.

I have already listed informally the other questions which we need to answer in order to draw a graph sketch in Section 3.B.(i) where we sketched $y = (x + 3)/(x - 2)$. You should look back at how we built up this sketch before going on.

Now that we can include finding the turning points, I can give you a complete summary of the questions which you need to answer in order to sketch a curve.

For convenience, I will call this curve $y = f(x)$ but, of course, other letters can be used.

Questions to answer in order to draw a graph sketch

(1) Does the curve cut the y-axis? If so, where? (Try putting $x = 0$.)

(2) Does the curve cut the x-axis? If so where?
(This is the same as asking if the equation $f(x) = 0$ has any roots on the x-axis.)

(3) Are there any values of x which have to be excluded because they would mean trying to divide by zero?
If so, what are they? (Such values of x will give you **vertical asymptotes**. An asymptote is a line which the curve of the graph of the function becomes closer and closer to.)
What happens to the values of $f(x)$ for values of x just either side of the forbidden values?

(4) What happens to the values of $f(x)$ when x takes very large positive or negative values?
(If it gets closer and closer to some fixed limit then this will give you a **horizontal asymptote**.)

(5) Are there any turning points? (That is, are there any values of x for which $f'(x)$ or $dy/dx = 0$?) If so, what are they?
You will need to find the value of $f(x)$ (the **stationary value**) for each of these values of x.
Test each turning point to find whether it is a local maximum, local minimum or point of inflection. (The tests for this are at the end of the previous section.) You don't usually need to find points of inflection where $dy/dx \neq 0$ unless you are specifically asked to do so.

An example to show these tests in action

We'll draw a sketch of

$$y = f(x) = \frac{x - 5}{x^2 - 9},$$

so we go through answering each of the questions in the list above in turn.

(1)   Putting $x = 0$ gives $f(0) = \frac{5}{9}$ so the curve $y = f(x)$ cuts the y-axis at the point $(0, \frac{5}{9})$.

(2) $f(x) = 0$ if $x - 5 = 0$ so if $x = 5$.
      The curve $y = f(x)$ cuts the x-axis at $(5,0)$.

(3)   Any value of x which makes $x^2 - 9 = 0$ must be excluded.
      $x^2 - 9 = (x + 3)(x - 3)$ so we can't have $x = -3$ or $x = 3$.
      The lines $x = -3$ and $x = 3$ are vertical asympotes of $y = f(x)$.
      Testing with nearby values of x, using a calculator, gives:
      The value of $f(x)$ is large and negative if x is just less than -3.
      The value of $f(x)$ is large and positive if x is just greater than -3.
      The value of $f(x)$ is large and positive if x is just less than $+3$.
      The value of $f(x)$ is large and negative if x is just greater than $+3$.

(4)   The easiest way to see what will happen to $y = f(x) = (x-5)/(x^2-9)$ if x takes very large positive values, is to divide the top and bottom of the fraction by x^2. This gives us

$$y = \frac{x-5}{x^2-9} = \frac{1/x - 5/x^2}{1 - 9/x^2}.$$

Now, as x becomes very large, each of $1/x$, $-5/x^2$ and $-9/x^2$ becomes very small.

We can say that, as $x \to \infty$, each of $1/x$, $-5/x^2$, and $-9/x^2 \to 0$.

So we will have $y \to \frac{0}{1}$ or 0 as $x \to \infty$.

Exactly the same thing happens for large negative values of x, so the line $y = 0$ (which is the x-axis – be careful here!) is also an asymptote.

Check with some large values of x on your calculator that the value of y really is getting close to zero.

You could also look at what is happening entirely experimentally by using your calculator, but you might then be left with a sneaky feeling that perhaps the curve does some strange unforeseen wiggle which your calculator hasn't revealed. Remember that you can't ever prove what a curve will do by testing with numerical values, but you can certainly prove that it *won't* do something. It is always wise to check your ideas of what it does do.

A mistake which students quite often make when graph-sketching is to work out exact values for some very boring bit of the curve which is almost a straight line. Then they think that the whole thing is probably a straight line, so getting a total disaster. The method I have given you here shows you how to find all the interesting bits.

(5)   Differentiating $y = f(x)$, using the Quotient Rule, we get

$$\frac{dy}{dx} = f'(x) = \frac{(x^2-9)(1) - (x-5)(2x)}{(x^2-9)^2} = \frac{10x - x^2 - 9}{(x^2-9)^2}$$

so $\dfrac{dy}{dx} = 0$ if $10x - x^2 - 9 = 0$ or $x^2 - 10x + 9 = 0$.

Factorising gives $(x-1)(x-9) = 0$ so $x = 1$ or $x = 9$ for the stationary values.

(You could also find these by using the quadratic formula to solve the equation.) The two stationary values are $f(1) = \frac{1}{2}$ and $f(9) = \frac{4}{72} = \frac{1}{18}$, so the turning points are $(1, \frac{1}{2})$ and $(9, \frac{1}{18})$.

Remember that these turning points are points on the original curve, so that to find them you must substitute the two values of x which give them into the equation of the original curve.

Now we want to know whether there are local maximum or minimum points on the curve.

Finding d^2y/dx^2 is not a pleasant prospect here, so we look at the values of dy/dx or $f'(x)$ either side of $x = 1$ and $x = 9$.

Passing through $x = 1$, the sequence goes $-$ 0 $+$ giving a local minimum of $f(1) = \frac{1}{2}$. You can show this here by choosing, say, $x = 0$ and $x = 2$ and substituting these values into the expression which we have found for dy/dx. These particular values give $-\frac{1}{9}$ and $\frac{7}{25}$, confirming the sequence of $-$ 0 $+$.

Similarly, passing through $x = 9$, the sequence goes $+$ 0 $-$ giving a local maximum of $f(9) = \frac{1}{18}$.

Notice that the value of the local minimum is actually *greater* than the value of the local maximum for this curve.

We now have all the information we need to draw the graph sketch. I show this in Figure 8.E.4.

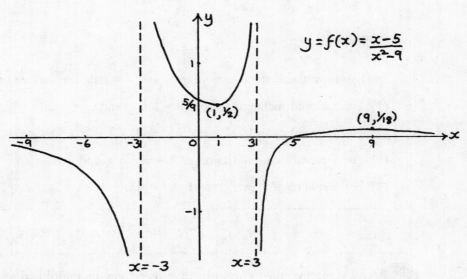

Figure 8.E.4

EXERCISE 8.E.2

Now try sketching the graphs of the following functions yourself.

Students often find graph-sketching difficult, but if you answer all the questions in my list for each curve, you should find that you can draw the sketches successfully. You will also understand *why* the curve behaves as it does, which won't be the case if you just use a graph-sketching calculator.

(a) $y = f(x) = \dfrac{x - 1}{x^2 - 4}$ (b) $y = g(x) = \dfrac{x}{1 + x^2}$ (c) $y = h(x) = x + 4/x$

(d) $y = p(x) = x - \dfrac{9}{x}$ (e) $y = f(x) = x^2 e^x$

8.E.(d) Some practical uses of turning points

Being able to find the turning points of a function can have much wider implications than just making it easier to sketch its graph. In particular, it gives us a method of answering many practical questions.

Since most of the examples we shall look at together in this section involve the volumes and surface areas of solid shapes, I am putting in a table here to give some of these.

A summary of volumes and surface areas of the commonest solids

The four solids are shown in Figure 8.E.5.

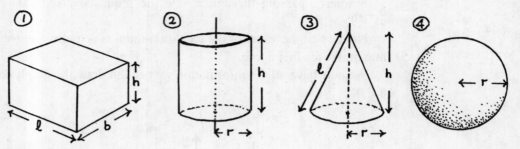

Figure 8.E.5

In each formula, V stands for volume and A stands for surface area.

(1) For a closed rectangular box, $V = lbh$ and $A = 2lb + 2bh + 2lh$.

(2) For a closed cylinder, $V = \pi r^2 h$ and $A = 2\pi r^2 + 2\pi rh$.

(3) For a cone, including its base, $V = \frac{1}{3}\pi r^2 h$ and $A = \pi rl + \pi r^2$.

(4) For a sphere, $V = \frac{4}{3}\pi r^3$ and $A = 4\pi r^2$

NOTE
A volume must always involve three lengths multiplied together. A surface area must always involve two lengths multiplied together. If you find that you have an equation for which this isn't true, *go back and recheck*! Something has gone wrong somewhere.

EXAMPLE (1) This is typical of the sort of problem which we can now solve. It comes in two parts.

(a) A manufacturer wishes to construct a metal can to hold a given volume of liquid. If the can is made entirely of the same thickness of metal, what is the best ratio of the height of the can to its radius so that the least amount of metal is used?

(b) To make the construction more rigid, it is decided that it will be necessary to use a double thickness of metal for the top and bottom of the can. In order to keep the cost of production to a minimum, what dimensions should the can now have? Give the answer again in the form of the best ratio of its height to its radius.

(a) We start by drawing a sketch of the can (which I have done in Figure 8.E.6, giving it a height of h and a radius of r).
Next, we label the other quantities we shall need to deal with.
Let the volume be V, the area be A, and the ratio of h/r be x.

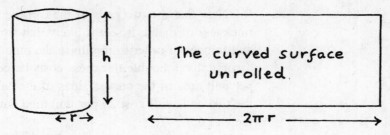

Figure 8.E.6

Since the can is being made to hold a given quantity of liquid, we know that V is a fixed quantity. We have $V = \pi r^2 h$, and $h/r = x$, so

$$V = \pi r^3 x \quad \text{and} \quad x = \frac{V}{\pi r^3}.$$

The surface area, A, is made up of the two circular ends of the can and its curved surface, which would unroll to give a rectangle.
This gives us $\quad A = 2\pi r^2 + 2\pi rh$.

At present, we only know how to differentiate functions with one variable, but the expression which we have for A involves the two variables, r and h.

However, we know that the fixed quantity $V = \pi r^2 h$, so $h = V/\pi r^2$.
Substituting this for h gives us

$$A = 2\pi r^2 + 2\pi r \left(\frac{V}{\pi r^2} \right) = 2\pi r^2 + \frac{2V}{r}.$$

We've now got A described entirely in terms of the one variable, r.
Since we want a minimum value of A, what should we do next?

We should find dA/dr, and look for values of r which make it equal to zero. We get:

$$\frac{dA}{dr} = 4\pi r - \frac{2V}{r^2} = 0 \quad \text{if} \quad \pi r^3 = \frac{V}{2}$$

so the ratio

$$\frac{h}{r} = x = \frac{V}{V/2} = 2.$$

(Remember when you differentiate that both π and V are constants.)
Now we check for certain that this gives a minimum value for A. We get

$$\frac{d^2A}{dr^2} = 4\pi + \frac{4V}{r^3}$$

which is positive since the value for r which we have found is positive. Therefore, we have found the ratio which gives a minimum value for A.

We have found that the surface area is smallest when the radius of the cylinder is half its height. This means that the vertical cross-section through the central axis will be a square.

(b) Now that the two ends of the can are to be made from a double thickness of metal, it seems likely that we should make the can taller and thinner in order to minimise the amount of metal we use. We will assume that a double thickness costs twice as much, and take the cost per unit area of the curved sides of the can to be c. Then the metal in the two ends will cost $2c$ per unit area, and we will call the total cost of the can C.

V and x will have the same equations as before but we will now have

$$C = 4\pi r^2 c + 2\pi rhc = \left(4\pi r^2 + \frac{2V}{r}\right)c$$

so

$$\frac{dC}{dr} = \left(8\pi r - \frac{2V}{r^2}\right)c = 0 \quad \text{if} \quad \pi r^3 = \frac{V}{4} \quad \text{so} \quad x = 4.$$

In the same way as before, this gives a minimum for the cost, so we have found that the height should now be four times the radius.

We can see how this pair of answers might work out numerically by taking the particular case of a half-litre can. This makes $V = 500$ cubic centimetres.

Then, in case (a) where $h = 2r$ we have $500 = 2\pi r^3$ so $r = 4.30$ cm to 2 d.p. and $h = 8.60$ cm to 2 d.p.

In case (b) where $h = 4r$ we have $500 = 4\pi r^3$ so $r = 3.41$ cm to 2 d.p. and $h = 13.66$ cm to 2 d.p.

EXAMPLE (2) What is the volume of the largest cylinder which can be placed inside a cone of fixed height H and radius R so that it just touches it inside, as I show in Figure 8.E.7(a)? Is it possible to fill in more than half the space inside this cone with such a cylinder?

We can see that the possible shape of the cylinder can vary between a sort of thin pencil to a flat biscuit. The largest possible size will occur somewhere between these two extremes.

I will call the height of the cylinder h and its radius r.

Then its volume V is given by $V = \pi r^2 h$ and we have to find the largest possible value of V (which will be in terms of R and H, the radius and height of the cone), as r and h vary.

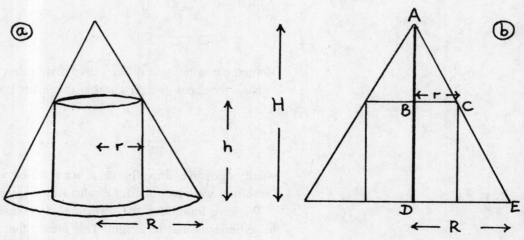

Figure 8.E.7

Since, at present, we can only differentiate functions with one variable, we must somehow use the physical relationship of the cone to the cylinder to find h in terms of r. To see how we can do this, we take a vertical cross-section along the joint axis of the cone and cylinder which gives us Figure 8.E.7(b).

We can now use the two similar triangles, *ABC* and *ADE*. These triangles nest into each other, so their sides are in the same proportion. Therefore

$$\frac{BC}{DE} = \frac{AB}{AD} \qquad \text{so} \qquad \frac{r}{R} = \frac{H-h}{H}$$

so $\quad rH = RH - Rh \quad$ and $\quad Rh = RH - rH = H(R - r)$.

Therefore

$$h = \frac{H(R-r)}{R}.$$

Substituting this for h in the equation $V = \pi r^2 h$ we get

$$V = \frac{\pi r^2 H(R-r)}{R} = \pi H r^2 - \frac{\pi H r^3}{R}.$$

We can now find dV/dr (remembering that π, H and R are all constants). We get

$$\frac{dV}{dr} = 2\pi H r - \frac{3\pi H r^2}{R} = \pi H r \left(2 - \frac{3r}{R}\right).$$

To find the maximum V, we put $dV/dr = 0$. Now

$$\pi H r \left(2 - \frac{3r}{R}\right) = 0 \quad \text{if} \quad r = 0 \quad \text{or} \quad \frac{3r}{R} = 2 \quad \text{so} \quad r = \frac{2R}{3}.$$

We can see physically that $r = 0$ gives us the minimum value of zero for V.

Also, d^2V/dr^2 is negative if $r = 2R/3$. (Check this for yourself.) Therefore, this value of r gives us the maximum volume.

How high will this cylinder be? We have

$$h = \frac{H(R-r)}{R} = \frac{H(R - 2R/3)}{R} = \frac{H}{3}$$

so it is one third of the height of the cone.

The volume of this cylinder is

$$\pi \left(\frac{2R}{3}\right)^2 \left(\frac{H}{3}\right) = \frac{4\pi R^2 H}{27}.$$

The volume of the cone is $\frac{1}{3}\pi R^2 H$ so the proportion of it which is filled by this largest possible cylinder is $(\frac{4}{27}) \div (\frac{1}{3}) = \frac{4}{9}$, that is, less than half of it.

EXERCISE 8.E.3

Try these for yourself.

(1) What is the maximum volume of a square-based open box made by cutting squares from the corners of a square piece of cardboard with sides 10 cm long, and then bending up the sides. I'm assuming here that the sides will then be taped together – you don't have to make allowances for overlap.

(2) What are the dimensions of the largest cylinder which can be placed inside a sphere of fixed radius R so that its two ends just touch the sphere? Is it possible to fill more than half of the interior of the sphere this way?

(3) What is the maximum distance from the origin of a particle moving on the x-axis so that its distance from O is given by the equation $x = 3 \cos t + 4 \sin t$?

Before rushing into differentiating here, have a think about how else you could write $3 \cos t + 4 \sin t$. (Look back at Section 5.D.(f) if necessary.)

8.E.(e)   **A clever use for tangents – the Newton–Raphson Rule**

This is an ingenious application of the properties of tangents which makes it possible to find closer and closer approximations to the roots of equations which are too difficult to solve exactly. (In the United States, the credit for this method is usually given entirely to Isaac Newton and it is called Newton's method.)

First, I'll explain graphically how it works.

Suppose you have some equation $f(x) = 0$ which you want to solve, so you want to find as accurately as possible the point where the curve $y = f(x)$ crosses the x-axis. It may, of course, do this more than once, but we will look at just one crossing point, where $x = a$, say.

In order to start the Newton–Raphson process, we need to have some idea of where a is. Suppose that by some ingenious method we have been able to find that $x = x_1$ is a value close to a. Figure 8.E.8(a) shows the curve of $f(x)$ near a and x_1.

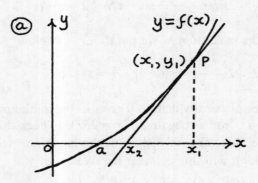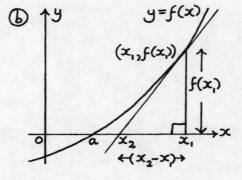

Figure 8.E.8

Then, if the curve really looks like my drawing, the tangent to the curve at $x = x_1$ will cut the x-axis at a point x_2 which is closer to the true root a than x_1 was.

How can we find out what x_2 is, from knowing what $y = f(x)$ and x_1 are?

The point P has coordinates (x_1, y_1), or $(x_1, f(x_1))$, so we do know some measurements.

From Figure 8.E.8(b) we can say that the gradient of the tangent at P is $f(x_1)/(x_1 - x_2)$. But the gradient of any tangent is also given by dy/dx or $f'(x)$ at the point concerned. This means that we know that the gradient of this particular tangent is $f'(x_1)$. (Using f' here instead of dy/dx is very handy as it makes it easier for us to talk about particular gradients.) We can now say that

$$f'(x_1) = \frac{f(x_1)}{x_1 - x_2}.$$

Next, we need to rearrange this to give us a rule for finding x_2. We get

$$(x_1 - x_2) f'(x_1) = f(x_1) \quad \text{so} \quad x_1 - x_2 = \frac{f(x_1)}{f'(x_1)}.$$

This gives us

The Newton–Raphson Rule

$$x_2 = x_1 - \frac{f(x_1)}{f'(x_1)}$$

(One of my students gave me a handy way of remembering which way round the last bit goes. She said '**d**ashed goes **d**own'.)

If x_1 is close to the root $x = a$, and if the curve is not too wiggly or behaving in other unexpected ways, then x_2 will be closer to a than x_1 was. (Sorting out these 'ifs' is what the subject of mathematical analysis does. It makes it possible to get results like this by analysing just what properties the curve must have near $x = a$ for the method to work. For example, we certainly won't want any of the complications described in Section 8.A.(f).)

Having found x_2, we can then repeat the process to get an even better approximation of x_3 to a, and so on, until we have as many decimal places of accuracy as we require.

Next, we will look at how this process works taking some particular examples.

EXAMPLE (1) For my first example, I will take an equation that we can solve exactly, so that you will be able to see how this process actually gives the right answer.

Suppose $f(x) = x^3 + x^2 - 9x - 9 = (x - 3)(x + 1)(x + 3)$ so $f'(x) = 3x^2 + 2x - 9$.

I show a picture of $f(x)$ in Figure 8.E.9.

We can see from the factorisation that one of the roots of $f(x) = 0$ is $x = 3$.

We'll take $x = 4$ as a starting value, and see if the Newton–Raphson process takes us towards the true root of $x = 3$. We have

$$x_2 = x_1 - \frac{f(x_1)}{f'(x_1)} = 4 - \frac{f(4)}{f'(4)} = 4 - \frac{35}{47} = 3.26 \text{ to 2 d.p.}$$

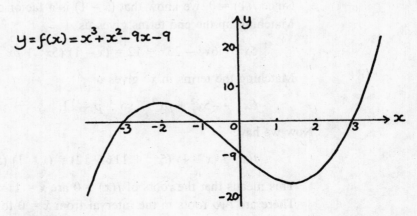

$$y = f(x) = x^3 + x^2 - 9x - 9$$

Figure 8.E.9

$$x_3 = 3.26 - \frac{f(3.26)}{f'(3.26)} = 3.024 \text{ to 3 d.p.} \qquad \text{Check this for yourself.}$$

$$x_4 = 3.024 - \frac{f(3.024)}{f'(3.024)} = 3.000 \text{ to 3 d.p.} \qquad \text{Check this one, too.}$$

In this particular example, the process is working beautifully, and you can see the successive answers homing in on $x = 3$.

If we hadn't known where the roots were, we could have used the changes in sign of $f(x)$ to show us where to look.

Working out some values gives us $f(4) = 35$, $f(2) = -15$, $f(0) = -9$, $f(-2) = 5$ and $f(-4) = -21$.

Looking at the picture of Figure 8.E.9, you can see the sign changing either side of each root, as the curve crosses the x-axis.

Have we made a brilliant discovery here?

Will we always be able to use this system to find an interval in which a root must lie? Try deciding for yourself whether the following two statements are true.

STATEMENT (1)   If $f(x) = 5x^3 + 6x^2 - 23x + 12$ then $f(0) = 12$ and $f(2) = 30$.

Therefore there is no root between $x = 0$ and $x = 2$.

STATEMENT (2)   If $f(x) = \dfrac{(x + 3)}{(x - 2)}$ then $f(1) = -4$ and $f(3) = 6$.

Therefore $f(x)$ has a root between $x = 1$ and $x = 3$.

If you don't agree with these statements, see if you can work out what is really happening.

Everything you need to be able to do this has already come in this book.

We can see what is really happening in the first case by using the methods of Section 2.E.(a).

We have $f(x) = 5x^3 + 6x^2 - 23x + 12$, and $f(1) = 0$, so immediately we know that statement (1) is false.

What is actually going on?

Since $f(1) = 0$, we know that $(x - 1)$ is a factor of $f(x)$.

Matching up the end terms gives us

$$5x^3 + 6x^2 - 23x + 12 = (x - 1)(5x^2 + px - 12).$$

Matching the terms in x^2 gives us

$$6x^2 = -5x^2 + px^2 \quad \text{so} \quad p = 11.$$

Now we have

$$f(x) = (x - 1)(5x^2 + 11x - 12) = (x - 1)(5x - 4)(x + 3).$$

This means that the roots of $f(x) = 0$ are $x = 1$, $x = \frac{4}{5}$ and $x = -3$.

There are *two* roots in the interval from $x = 0$ to $x = 2$. I show a sketch of $y = f(x)$ in Figure 8.E.10.

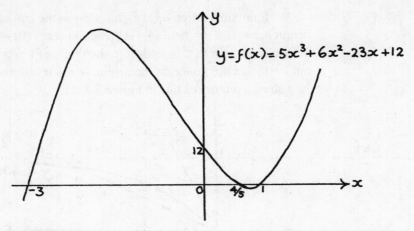

$$y = f(x) = 5x^3 + 6x^2 - 23x + 12$$

12

−3   0   4/5   1

Figure 8.E.10

We would only see the sign change for the roots $x = \frac{4}{5}$ and $x = 1$ by taking a value between them. Check for yourself that choosing such a value does make $f(x)$ come out negative.

Statement (2) is wrong for quite a different reason. We drew a picture of this function in Section 3.B.(i). The sign change here doesn't mean that $f(x)$ has crossed the x-axis between $x = 1$ and $x = 3$. The curve has a jump or discontinuity when $x = 2$ and gets to the other side of the x-axis this way.

The two examples above give us two useful rules to remember when looking for roots.

> **Rules for using a sign change when looking for roots**
>
> (1) If $f(x_1)$ and $f(x_2)$ have different signs, there must be at least one root between x_1 and x_2 provided that $f(x)$ is continuous from x_1 to x_2.
>
> (2) If $f(x)$ *is* continuous, then a sign-change tells us that there is an odd number of roots in the interval.

You can think of 'continuous' here as meaning that $f(x)$ can be drawn with a continuous straight line. The subtle mathematical non-pictorial meaning of this word is described in courses on mathematical analysis.

Obviously too, in order to be able to use the Newton–Raphson method, we must be able to differentiate $f(x)$. It mustn't have any of the problems which were described in Section 8.A.(f), in the part where we are working.

EXAMPLE (2) Next, we'll use the Newton–Raphson method to find all the roots of
(a) $\tanh x = 2x$ and (b) $\tanh x = \frac{1}{2}x$.
How many will there be? Will it be the same number for both (a) and (b)?
Try sketching what you think will happen, using Section 8.D.(g) if you need to.

We found in Section 8.D.(g) that $y = x$ is the tangent to $y = \tanh x$ at the origin because $d/dx\,(\tanh x) = \operatorname{sech}^2 x$ and $\operatorname{sech}^2(0) = 1$.

We can see from this, and from the shape of $y = \tanh x$, that $y = 2x$ will cut $y = \tanh x$ only once, at the origin, so $x = 0$ is the only solution of (a).

I show a picture of this in Figure 8.E.11.

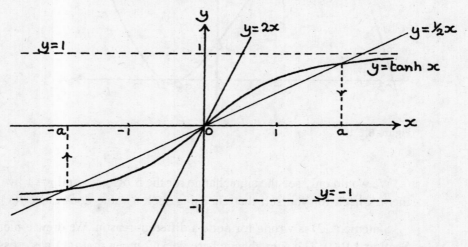

Figure 8.E.11

$y = \frac{1}{2}x$ will cut $y = \tanh x$ three times, once at the origin and also at two other points symmetrically placed either side of the origin because $y = \tanh x$ is odd. (Turn it upside down and it looks the same.) So we just have one solution to find.

We want the value of x on the right-hand side of the graph for which

$$\tanh x = \tfrac{1}{2} x \quad \text{so} \quad \tanh x - \tfrac{1}{2}x = 0.$$

We let $f(x) = \tanh x - \frac{1}{2}x$ and look for the solution here of $f(x) = 0$.

From the sketch we can see that if $x < a$ then $\tanh x > \frac{1}{2}x$. It looks as though a may be quite close to 2.

$f(2) = -0.036$ to 3 d.p. so the root is to the left of this.

$f(1.9) = 0.006$ to 3 d.p. The change in sign confirms that the root lies between 1.9 and 2, since $f(x)$ is continuous.

Since $f(1.9)$ is closer to zero, we'll start with $x_1 = 1.9$.

We have $f(x) = \tanh x - \frac{1}{2}x$ and $f'(x) = \operatorname{sech}^2 x - \frac{1}{2}$. This gives us

$$x_2 = 1.9 - \left(\frac{0.006237}{-0.414390}\right) = 1.915 \text{ to 3 d.p.}$$

$$x_3 = 1.915 - \left(\frac{3.354578 \times 10^{-6}}{-0.416813}\right) = 1.915 \text{ to 3 d.p.}$$

The three solutions of $\tanh x = \frac{1}{2}x$ are $x = -1.915$, $x = 0$ and $x = 1.915$, correct to 3 d.p.

EXAMPLE (3) Show, by drawing a sketch, that $\sin x = 3 - 2x$ has just one solution. Find this solution correct to 3 d.p.

See how far you can get with this one yourself before you look at my solution.

We want to solve $\sin x = 3 - 2x$ which is the same as $\sin x - 3 + 2x = 0$.

We let $f(x) = \sin x - 3 + 2x$ so $f'(x) = \cos x + 2$.

We can see from the sketch of Figure 8.E.12 that the root is less than $\frac{3}{2}$. It also looks as though it could be greater than 1.

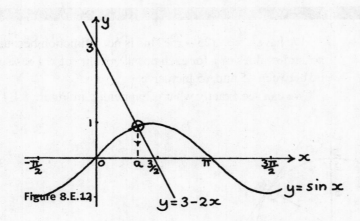

Figure 8.E.12

$f(1) = -0.159$ and $f(1.5) = 0.997$ so there is a root between $x = 1$ and $x = 1.5$ since $f(x)$ is continuous.

Since $f(1)$ is closer to zero, we'll start with $x_1 = 1$. This gives us

$$x_2 = x_1 - \left(\frac{-0.159}{2.540}\right) = 1.063.$$

$$x_3 = 1.063 - \left(\frac{-1.81829 \times 10^{-4}}{2.48625}\right) = 1.063 \text{ to 3 d.p.}$$

so the solution is $x = 1.063$ radians correct to 3 d.p.

EXERCISE 8.E.4

For each of the following, draw a sketch to help you decide where the roots of the following equations might lie, and then use the Newton–Raphson process to find these roots correct to 3 d.p.

(1) $2x^3 - 3x^2 + 6x + 1 = 0$ (2) $e^x = 3 - x$,   (3) (a) $\sinh x = \frac{1}{2}x$ (b) $\sinh x = 2x$

8.F   Implicit differentiation

8.F.(a)   How implicit differentiation works, using circles as examples

How could we find the rate of change of y with respect to x if we have a relation between them which does not give y described in terms of x? Let's look at two examples.

EXAMPLE (1)   Suppose we are given the equation $x^2 + y^2 = 25$.

This is the equation of the circle whose centre is at the origin and whose radius is five units. (See Section 4.C.(d) if necessary.)

The relationship here between x and y is called **implicit**, because we don't have it in the form of y given as some expression in x.

We can easily draw a sketch of this circle, and we can see how steep the curve looks at any point on it by sketching the tangent at that point. (Indeed, we can actually *find* this slope, using the property of the tangent being perpendicular to the radius, as we did in Section 4.C.(f).)

But how can we find dy/dx for this circle? Developing a technique to do this will make it possible for us to find dy/dx for other curves where we have no alternative method of finding the gradient.

One possibility would be to start by rearranging its equation so that we have $y^2 = 25 - x^2$.

What is y? Can you see a possible complication here?

We have $y = \pm \sqrt{25 - x^2}$. This is not a function because there are two possible values of y for each possible value of x. These possible values of x lie between -5 and $+5$ inclusive.

We can see exactly what is happening in Figure 8.F.1.

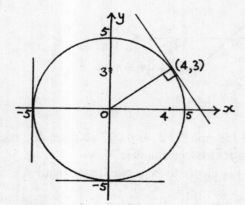

Figure 8.F.1

The equation $y = + \sqrt{25 - x^2}$ gives the top half of the circle.

The equation $y = - \sqrt{25 - x^2}$ gives the bottom half of the circle, and each of these are functions.

Differentiating these square roots would not be very pleasant. We therefore argue that it would seem reasonable to go through the equation $x^2 + y^2 = 25$ differentiating it term by term with respect to x in just the same way that we differentiated the equation $y = x^2 - 4x + 3$ term by term to give $dy/dx = 2x - 4$ in Section 8.E.(a).

(The equation $y = x^2 - 4x + 3$ gives y *explicitly* in terms of x.)

The problem that we have with this new equation is that we shall need to differentiate y^2 with respect to x.

We can do this by using the Chain Rule. We know that y^2 differentiated with respect to y is $2y$, and we then multiply this answer by dy/dx.

We are saying that

$$\frac{d}{dx}(y^2) = \left(\frac{d}{dy}(y^2)\right)\left(\frac{dy}{dx}\right) = 2y\,\frac{dy}{dx}.$$

Now, differentiating $x^2 + y^2 = 25$ term by term with respect to x gives us

$$2x + 2y\,\frac{dy}{dx} = 0 \quad \text{so} \quad 2y\,\frac{dy}{dx} = -2x \quad \text{and} \quad \frac{dy}{dx} = -\frac{x}{y}.$$

How does this result fit in with the particular examples shown on Figure 8.F.1?

At the point (4,3), $dy/dx = -\frac{4}{3}$.

This agrees with what we know the gradient of the tangent here must be, because the gradient of the radius to this point is $\frac{3}{4}$. The tangent is perpendicular to the radius so its gradient is $-\frac{4}{3}$. (This uses $m_1\,m_2 = -1$ from Section 2.B.(h).)

In fact, at any point on this circle with coordinates (x, y), the gradient of the radius is y/x and the gradient of the tangent is $-x/y$. We can see here that geometry and calculus both give us the same result.

This extends to special cases like the gradient at the point $(0, -5)$ which is zero, and the gradient at the point $(-5, 0)$ where dy/dx is undefined. Clearly from the diagram this must be so, because the tangent here is vertical.

EXAMPLE (2) Suppose this time that we want to find dy/dx for the implicit equation $x^2 - 6x + y^2 - 4y = 12$.

What kind of curve does this give?

This equation can be written in the form $(x-3)^2 + (y-2)^2 = 25$.

It describes the circle whose centre is at (3,2) and whose radius is 5 units.

We drew this particular circle in Section 4.C.(f) and found the gradients and equations of some of its tangents. This will help us now to see what is happening geometrically, and so to be able to make sense of some of the answers which we get by calculus.

To find dy/dx for this circle we again differentiate its equation term by term, remembering that y differentiated with repect to x is simply dy/dx. We get

$$2x - 6 + 2y\,\frac{dy}{dx} - 4\,\frac{dy}{dx} = 0.$$

Remember that differentiating a number gives zero, so $d/dx\,(25) = 0$. It *doesn't* change so its rate of change is zero.

Tidying up gives

$$\frac{dy}{dx}\,(2y - 4) = 6 - 2x \quad \text{so} \quad \frac{dy}{dx} = \frac{6 - 2x}{2y - 4} = \frac{3 - x}{y - 2}$$

dividing top and bottom of this fraction by 2. Always simplify when you can.

We can now see how this ties in with the gradients of the four tangents which we already found for this particular circle in Section 4.C.(f).

The points of contact of these tangents are (7,5), (−1,−1), (3,7) and (8,2).

Try using dy/dx yourself here to find the gradients of these four tangents. Use Figure 4.C.11 in Section 4.C.(f) to sort out what is happening if some of your results seem rather curious.

Substituting in these pairs of values for x and y in turn, we get

$$\frac{dy}{dx} \text{ at } (7,5) \text{ is } -\tfrac{4}{3} \text{ and } \frac{dy}{dx} \text{ at } (-1, -1) \text{ is also } -\tfrac{4}{3}.$$

(We can see that this is right on Figure 4.C.11. The gradients of the two radii to the points of contact are both $\tfrac{3}{4}$.)

dy/dx at $(3,7)$ is zero and the tangent there is horizontal.

When you find that the gradient of the tangent at the point $(8,2)$ is $dy/dx = -5/0$, don't be tempted to cross your fingers and say that this is zero as many students do! dy/dx at $(8,2)$ is undefined because the tangent is vertical.

Using Figure 4.C.11 we have seen geometrically that the answers which we have found by differentiating do make sense.

Also, from this same diagram, we can see that the gradient of the radius to the point (x,y) on this circle is $(y-2)/(x-3)$.

Therefore, using $m_1\, m_2 = -1$, the gradient of the tangent at this point is

$$-\frac{x-3}{y-2} = \frac{3-x}{y-2}$$

which is exactly what we get by differentiating.

EXERCISE 8.F.1

Find dy/dx for the circle whose equation is

$$x^2 + 16x + y^2 - 4y - 101 = 0.$$

Use this result to find the gradient of the tangents to this circle at the four points with coordinates $(4, -3)$, $(-3, 14)$, $(-8, 11)$ and $(-21, 2)$.

Draw a sketch of this circle showing these four tangents.

Check your results with the answers to Exercise 4.C.3 which find these same gradients without differentiating.

8.F.(b) **Using implicit differentiation with more complicated relationships**

What will we do if we have a curve whose equation has a term with x and y multiplied together? Let's look at an example.

EXAMPLE (1) Suppose we have the equation $2x^2 + xy - y^2 = 5$.

In order to differentiate xy with respect to x we use the Product Rule because we have the two variables x and y multiplied together. (See Section 8.C.(d) if necessary.)

This gives us

$$\frac{d}{dx}(xy) = y(1) + x\frac{dy}{dx} = y + x\frac{dy}{dx}.$$

So differentiating $2x^2 + xy - y^2 = 5$ with respect to x gives

$$4x + \left(y + x\frac{dy}{dx}\right) - 2y\frac{dy}{dx} = 0.$$

So
$$\frac{dy}{dx}(x - 2y) = -4x - y \quad \text{or} \quad \frac{dy}{dx} = \frac{-4x - y}{x - 2y} = \frac{4x + y}{2y - x},$$
multiplying top and bottom of the fraction by -1 to make it easier to handle.

At the point $(2,3)$ on this curve, (check that it is!), we have $dy/dx = \frac{11}{4}$ giving the slope of the tangent here, and therefore the gradient of the curve at this point.

We can find the equation of this tangent using $y - y_1 = m(x - x_1)$ from Section 2.B.(f). It is $y - 3 = \frac{11}{4}(x - 2)$ or $4y = 11x - 10$.

As students sometimes find this process slightly tricky, I shall give you another example which needs the use of the Product Rule.

EXAMPLE (2) Find dy/dx for the equation $2x^3 - 3x^2y + 5xy^2 + 2y^3 = 6$ and hence find the gradient of the tangent at the point $(1,1)$.

I think it is easier if you take the numbers outside as factors for the terms involving the Product Rule, particularly if they are negative, as it is easy to lose one of the minus signs otherwise.

I shall start by showing the equation split up in a working sort of way as follows:
$$2x^3 - 3([x^2]\,[y]) + 5\,([x]\,[y^2]) + 2y^3 = 6.$$
Now, differentiating all through with respect to x we get
$$6x^2 - 3\big([y]\,[2x] + [x^2]\,[dy/dx]\big) + 5\big([y^2]\,[1] + [x]\,[2y\,dy/dx]\big) + 6y^2\frac{dy}{dx} = 0.$$
Tidying up, we get
$$6x^2 - 6xy - 3x^2\frac{dy}{dx} + 5y^2 + 10xy\frac{dy}{dx} + 6y^2\frac{dy}{dx} = 0$$
so
$$\frac{dy}{dx}(6y^2 + 10xy - 3x^2) = 6xy - 5y^2 - 6x^2$$

$$\frac{dy}{dx} = \frac{6xy - 5y^2 - 6x^2}{6y^2 + 10xy - 3x^2}.$$

The slope of the tangent at the point $(1,1)$, i.e. the gradient of the curve at that point, is given by
$$\frac{dy}{dx} = \frac{6 - 5 - 6}{6 + 10 - 3} = -\frac{5}{13}.$$

Try this similar example for yourself now.

EXAMPLE (3) Find dy/dx for the curve given by the equation $3x^3 + 7x^2y - 3xy^2 + 2y^3 = 21$. Hence find the gradient of the tangent at the point $(1, 2)$ on this curve.

Make sure that you haven't started off your answer with '$dy/dx =$' because this is not at all what you mean.

What you are doing here is differentiating the whole expression with respect to x, so your answer should only start with '$dy/dx =$' if the original equation starts in the form '$y =$'.

Setting out the given equation in a working sort of way as I did in the last example gives

$$3x^3 + 7\,([x^2][y]) - 3\,([x][y^2]) + 2y^3 = 21.$$

(You don't have to do this step, but if you are at all unsure about keeping track of where you are then I think it will help you.)

Differentiating this expression term by term with respect to x gives

$$9x^2 + 7\left([y][2x] + [x^2]\left[\frac{dy}{dx}\right]\right) - 3\left([y^2][1] + [x]\left[2y\,\frac{dy}{dx}\right]\right) + 6y^2\,\frac{dy}{dx} = 0.$$

Again I have used square brackets so I can show you how the Product Rule is working on each bit. Tidying this up gives

$$9x^2 + 14xy + 7x^2\,\frac{dy}{dx} - 3y^2 - 6xy\,\frac{dy}{dx} + 6y^2\,\frac{dy}{dx} = 0$$

so $\quad \dfrac{dy}{dx}\,(7x^2 + 6y^2 - 6xy) = 3y^2 - 9x^2 - 14xy$

therefore $\quad \dfrac{dy}{dx} = \dfrac{3y^2 - 9x^2 - 14xy}{6y^2 + 7x^2 - 6xy}.$

At the point (1,2) on the curve, the slope of the tangent is $dy/dx = -25/19$.

8.F.(c)   **Differentiating inverse functions implicitly**

In Section 8.D.(i), we differentiated inverse trig and hyperbolic functions using $dy/dx = 1/(dx/dy)$. This seems reasonable if we take these as the limiting cases of $\delta y/\delta x$ and $1/(\delta x/\delta y)$ respectively. For this rule to work, we have to be sure that dy/dx is not equal to zero for any particular value of x which we might want to consider.

It is also possible to differentiate these inverse functions implicitly.

I'll show you how you can do this by taking the example of $y = f(x) = \sin^{-1}(x/a)$.

We know from Section 5.A.(h) that y must lie between $-\pi/2$ and $+\pi/2$ inclusive. The function itself looks like the sketch in Figure 8.F.2.

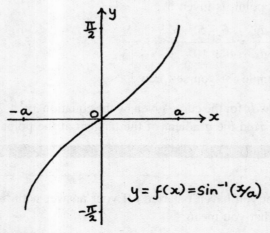

Figure 8.F.2

Notice that for this function we must have $-a \leq x \leq a$.

We have $y = f(x) = \sin^{-1} x/a$, so therefore $x/a = \sin y$ because this is an inverse function, and there is only the one possible value for y from a given value of x/a.

Differentiating $x/a = \sin y$ implicitly with respect to x gives

$$1/a = \cos y \, \frac{dy}{dx} \quad \text{so} \quad \frac{dy}{dx} = \frac{1}{a \cos y}.$$

But

$$\sin^2 y + \cos^2 y = 1 \quad \text{so} \quad \cos^2 y = 1 - \sin^2 y = 1 - \frac{x^2}{a^2} = \frac{a^2 - x^2}{a^2}.$$

We can see from Figure 8.F.2 that the gradient of $y = \sin^{-1} (x/a)$ is positive, and therefore we can say

$$\frac{dy}{dx} = \frac{1}{a \cos y} = \frac{1}{a} \sqrt{\frac{a^2}{a^2 - x^2}} = \frac{1}{\sqrt{a^2 - x^2}}.$$

(The $\sqrt{}$ means that we are taking the positive square root here.)

We know that for this function to work we must have $-a \leq x \leq a$.

Now, we must also exclude $x = a$ and $x = -a$. Why do we have to do this?

The answer is: because when $x = a$ or $x = -a$ the tangent to $y = \sin^{-1} (x/a)$ is vertical. The fraction $1/\sqrt{a^2 - x^2}$ is undefined. We've now seen what happens when $dy/dx = 1/0$, and why we must exclude this possibility.

This gives us the following result.

$$\boxed{\frac{d}{dx}\left(\sin^{-1} \left(\frac{x}{a} \right) \right) = \frac{1}{\sqrt{a^2 - x^2}} \quad \text{for } -a < x < a \quad \text{or} \quad |x| < a}$$

In Section 8.D.(f) we showed that $\cosh^{-1} x = \ln (x + \sqrt{x^2 - 1})$, and that, to have this inverse *function* for $y = \cosh x$, we needed to restrict the possible values of x by saying that $x \geq 0$, so that we were taking the cosh of a positive quantity.

We'll now differentiate $f(x) = \cosh^{-1} (x/a)$ implicitly. In Figure 8.F.3 I show a sketch of $y = \cosh (x/a)$ and $y = \cosh^{-1} (x/a)$. It is very similar to Figure 8.D.6 except that 1 is replaced by a.

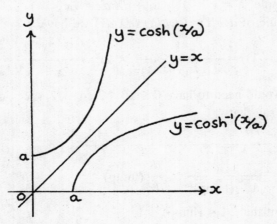

Figure 8.F.3

We can see from this sketch that, if $y = \cosh^{-1}(x/a)$, then we must have $x \geq a$.

Try finding $d/dx\ (\cosh^{-1}(x/a))$ for yourself by using implicit differentiation.

The method is very similar to what we just did for $y = \sin^{-1}(x/a)$.

We have $y = f(x) = \cosh^{-1}(x/a)$ so $x/a = \cosh y$.

Differentiating implicitly with respect to x gives

$$\frac{1}{a} = \sinh y \frac{dy}{dx} \quad \text{so} \quad \frac{dy}{dx} = \frac{1}{a \sinh y}.$$

Using $\cosh^2 y - \sinh^2 y = 1$ gives us

$$\sinh^2 y = \left(\frac{x}{a}\right)^2 - 1 = \frac{x^2 - a^2}{a^2}.$$

We can also see that the gradient of $y = \cosh^{-1}(x/a)$ is positive, so we can say that

$$\frac{dy}{dx} = \frac{1}{a}\sqrt{\frac{a^2}{x^2 - a^2}} = \frac{1}{\sqrt{x^2 - a^2}}.$$

Also, like last time, we can't have $x = a$ because this is where the tangent to $y = \cosh^{-1}(x/a)$ is vertical, and we would have the undefined fraction of $\frac{1}{0}$.

We now have the result that

$$\boxed{\frac{d}{dx}\left(\cosh^{-1}\left(\frac{x}{a}\right)\right) = \frac{1}{\sqrt{x^2 - a^2}} \quad \text{for } x > a.}$$

We can use the Chain Rule with these results to differentiate more fancy functions in exactly the same way that we used it in Section 8.D.(i) for fancy inverse functions of sinh and tanh.

I'll show you how to find $d/dx\ (\sin^{-1}(x^2 - 1))$ as an example.

We have just shown that if $y = \sin^{-1} X$ then

$$\frac{dy}{dX} = \frac{1}{\sqrt{1 - X^2}}.$$

In this particular example, $X = x^2 - 1$ so $dX/dx = 2x$.

Using the Chain Rule of $dy/dx = (dy/dX)(dX/dx)$ we have

$$\frac{d}{dx}(\sin^{-1}(x^2 - 1)) = \frac{1}{\sqrt{1 - (x^2 - 1)^2}} \times 2x = \frac{2x}{\sqrt{2x^2 - x^4}} = \frac{2}{\sqrt{2 - x^2}}.$$

For this to work, we would need to have $0 \leq x^2 < 2$ so $-\sqrt{2} < x < \sqrt{2}$.

In general, we can say that

$$\boxed{\frac{d}{dx}(\sin^{-1}(\text{lump})) = \frac{1}{\sqrt{1 - (\text{lump})^2}} \times \frac{d}{dx}(\text{lump})}$$

with the requirement that $-1 < (\text{lump}) < 1$.

Similarly, we have

$$\frac{d}{dx}(\cosh^{-1}(\text{lump})) = \frac{1}{\sqrt{(\text{lump})^2 - 1}} \times \frac{d}{dx}(\text{lump})$$

with the requirement that (lump) > 1.

You may be given the formula below as a rule for differentiating inverse functions.

If $y = f^{-1}(x)$ and $f^{-1}(x) = F(x)$ then $\dfrac{dy}{dx} = F'(x) = \dfrac{1}{f'(F(x))}$.

This has exactly the same meaning as what we have been doing above. The only difference is that it is written in function notation.

I think that you might need reminding just how this works.

- $f'(x)$ means $f(x)$ differentiated with respect to x.
- $f'(X)$ means $f'(X)$ differentiated with respect to X.
- $f'(f(x))$ means $f(f(x))$ differentiated with respect to $f(x)$.

In general, we can say that

$f'(\text{lump})$ means $f(\text{lump})$ differentiated with respect to (lump).

I can now show you where this formula comes from.
Suppose we have an inverse function $y = f^{-1}(x)$.
Then $x = f(y)$ because this is what an inverse function means.
Differentiating implicitly with respect to x, using the Chain Rule, gives

$$1 = \frac{d}{dy}(f(y))\frac{dy}{dx} = f'(y)\frac{dy}{dx} \quad \text{so} \quad \frac{dy}{dx} = \frac{1}{f'(y)}.$$

But $y = f^{-1}(x)$ and $f^{-1}(x)$ is also a function of x.
Suppose we call it $F(x)$. Then we can say

$$\frac{dy}{dx} = \frac{d}{dx}(F(x)) = F'(x) = \frac{1}{f'(F(x))}.$$

8.F.(d)   **Differentiating exponential functions like $x = 2^t$**

This particular function is the one which we used in Section 3.C.(a) to describe an example of cell growth. We said there that the rate of increase at any time t is equal to some constant, k, multiplied by the number of cells present at that time, but we couldn't then find the value of k.

It's now easy for us to do this. We have $x = 2^t$ so, taking natural logs both sides of this equation, $\ln x = \ln(2^t) = t \ln 2$ using the third rule of logs of Section 3.C.(d).

Differentiating $\ln x = t \ln 2$ implicitly with respect to t, we get

$$\frac{1}{x}\frac{dx}{dt} = \ln 2 \quad \text{so} \quad \frac{dx}{dt} = x \ln 2.$$

We now know that the value of k is $\ln 2$. (I said it would be this in Section 3.C.(e).)

We can also write this answer in terms of t if we want to. We have

$$\frac{dx}{dt} = x \ln 2 = (\ln 2)\, 2^t \quad \text{since } x = 2^t.$$

The $\ln 2$ is the scaling factor which gives us the difference between the rate of increase of $x = 2^t$ and $x = e^t$. If $x = e^t$, the scaling factor, k, is equal to 1.

The rate of increase at any given time is the *same* as the quantity of the substance present at that time.

Here is a second rather nastier-looking example. (Although it looks nasty, it is actually quite simple to do.)

If $y = x^{x^2 + 1}$, what is dy/dx?

Again we take natural logs both sides of the equation. This gives us

$$\ln y = \ln (x^{x^2 + 1}) = (x^2 + 1)\ln x.$$

Differentiating implicitly with respect to x, using the Product Rule, gives us

$$\frac{1}{y}\frac{dy}{dx} = 2x \ln x + (x^2 + 1)\left(\frac{1}{x}\right) = 2x \ln x + x + \frac{1}{x}.$$

Therefore

$$\frac{dy}{dx} = y\left(2x \ln x + x + \frac{1}{x}\right) = \left(2x \ln x + x + \frac{1}{x}\right)(x^{x^2 + 1}).$$

8.F.(e)   A practical application of implicit differentiation

I shall finish this section on implicit differentiation by giving you an example of a practical use for it.

The volume of metal in a hollow sphere remains constant. If the inner radius is increasing at the rate of $3\,\text{cm s}^{-1}$ find the rate of increase of the outer radius when the two radii are 2 cm and 4 cm respectively.

The volume of a sphere of radius r is $\frac{4}{3}\pi r^3$. I show a drawing of a cross-section through the centre of the hollow sphere in Figure 8.F.4.

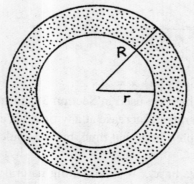

Figure 8.F.4

The volume of metal, V, is given by

$$V = \tfrac{4}{3} \pi R^3 - \tfrac{4}{3} \pi r^3.$$

V, R and r are functions of time, t, so differentiating implicitly with respect to t gives

$$\frac{dV}{dt} = 4\pi R^2 \frac{dR}{dt} - 4\pi r^2 \frac{dr}{dt}.$$

The volume V remains constant so $dV/dt = 0$.
Therefore we have

$$4\pi R^2 \frac{dR}{dt} = 4\pi r^2 \frac{dr}{dt}.$$

At the instant we are interested in, $R = 4$, $r = 2$ and $dr/dt = 3$. Therefore, $dR/dt = \tfrac{3}{4}$.
At this instant, the outer radius is increasing at a rate of $\tfrac{3}{4}$ cm s^{-1}.

In the first three questions, differentiate the given equation implicitly, rearranging in each case to find an expression for dy/dx. Then use this answer to find

(a) the gradient of each curve at the point whose coordinates are given in the question, and

(b) the equation of the tangent to the curve at this point.

(1) $x^2 y^2 = 25.$ The point is (1,5).

(2) $x^2 + 3xy - y^2 = 2.$ The point is (1,2).

(3) $1/y + 2/x = 1$
Do this one by differentiating the equation given above. Then do it a second time by differentiating the equation you get if you multiply all through by xy to get rid of the fractions.
In each case, use your expression for dy/dx to find the gradient of the curve at the point (4,2), and the equation of the tangent there.

(4) Use the Chain Rule to differentiate the following two functions with respect to x
(a) $\sin^{-1}(2x - 5)$ (b) $\cosh^{-1}(3x + 1)$
In both cases, say what restrictions you must put on x for the answer to work.

(5) Show that $\dfrac{d}{dx}(\tanh^{-1} x) = \dfrac{1}{1 - x^2}.$

(6) Differentiate the following three functions with respect to x
(a) $y = f(x) = x^x$ (b) $y = f(x) = 3x^{(x+2)}$ (c) $y = f(x) = (3x)^{(x+2)}$

8.G   Writing functions in an alternative form using series

We know already that we can write

$$1 + x + x^2 + x^3 + x^4 + \ldots \quad \text{as} \quad \frac{1}{1 - x} \quad \text{provided that } |x| < 1,$$

using the rule for the sum to infinity of a GP, from Section 6.C.(c).

We also found in Section 7.C.(a) that if we did a binomial expansion on $(1 - x)^{-1}$, keeping our fingers crossed that it would work with $n = -1$, we did in fact get the same series.

Might it be possible to find a way of writing other functions in the alternative form of series?

The answer to this question is a qualified 'yes'. When calculus is approached entirely through proofs and arguments based on taking mathematical limits, one of the most powerful results to come out of this is Taylor's Theorem. This gives a proof that such series expansions do indeed exist if certain conditions are met. (We saw some reasons why such a formal approach is necessary to lay mathematically firm foundations for calculus in Section 8.A.(f).) This rigorous approach is beyond my scope here because to do it properly takes a great deal of space and a dedicated book. My purpose is to give you a working knowledge that you can use in other areas together with an intuitive feel and understanding for what is happening, which will then make a pathway to lead you into whatever depth you later need to go. It is, however, possible for me to show you what some of the results of this approach will be, so that you can see why they are important.

Taylor's Theorem gives a way of approximating to functions by considering their rates of change, and the rates of their rates of change, and so on. In other words, if we have a function $x = f(t)$, then the theorem makes it possible (with certain qualifications concerning how this function behaves) to write the function in terms of dx/dt, d^2x/dt^2 and so on, or, in function notation, in terms of $f'(t)$, $f''(t)$ and so on. There is a very good description of how this works in Louis Lyons' book *All you wanted to know about mathematics but were afraid to ask* (Cambridge University Press 1995).

Because these series are so important we will look at some straightforward cases together now. We will only consider functions like e^t or $\sin t$ or $\cos t$ where we know that we can continue differentiating for ever if we want to. Also, we will only look at functions for which we know that there is no problem differentiating when $t = 0$.

Let us suppose that it *is* possible to write such a function as a series expansion so that we have

$$f(t) = a_0 + a_1 t + a_2 t^2 + a_3 t^3 + a_4 t^4 + \ldots \qquad \text{which I will call (1)}$$

a_0, a_1, etc. are coefficients which will depend upon the particular function $f(t)$ which we are considering.

If we put $t = 0$ in (1) above, we get $f(0) = a_0$ because everything else disappears.

Also, if the series of (1) is truly representing the function $f(t)$, we would expect that differentiating it term by term should give the same result as differentiating the function itself. (This isn't obvious – we have seen in Section 6.F that infinite series can behave in very odd ways indeed. It can in fact be shown that it *does* work for all the examples which we shall look at here.)

If we differentiate (1) with respect to t, we get

$$f'(t) = a_1 + 2a_2 t + 3a_3 t^2 + 4a_4 t^3 + 5a_5 t^4 + \ldots \qquad (2)$$

Putting $t = 0$ now gives us $f'(0) = a_1$ because all the further terms disappear. This system is beginning to look promising. Similarly

$$f''(t) = 2(1)a_2 + 3(2)a_3 t + 4(3)a_4 t^2 + 5(4)a_5 t^3 \ldots \qquad (3)$$

so $f''(0) = 2(1)a_2$. And

$$f'''(t) = 3(2)(1)a_3 + 4(3)(2)a_4 t + 5(4)(3)a_5 t^2 + \ldots \qquad (4)$$

so $f'''(0) = 3(2)(1)a_3$.

Writing all these dashes for the successive differentiations is beginning to be a little clumsy, so I will replace them from now on with a number which stands for the number of times that $f(t)$ has been differentiated.

Do the next differentiation yourself, so finding $f^4(0)$.

You should have $f^4(0) = 4(3)(2)(1)a_4$.

We can see that these are building up using factorials and we can write the results so far as

$$a_0 = f(0) \qquad a_1 = \frac{f^1(0)}{1!} \qquad a_2 = \frac{f^2(0)}{2!} \qquad a_3 = \frac{f^3(0)}{3!} \qquad a_4 = \frac{f^4(0)}{4!}$$

Since we are only considering functions here which can be differentiated as often as we like, we could say that $a_r = f^r(0)/r!$ where a_r is the rth coefficient. (We started the count with $r = 0$).

This gives us the following result.

Provided that certain conditions are met, it is possible to say

$$f(t) = f(0) + \frac{f^1(0)}{1!}t + \frac{f^2(0)}{2!}t^2 + \frac{f^3(0)}{3!}t^3 + \frac{f^4(0)}{4!}t^4 + \dots$$

 The little superscript numbers in the above expression refer to how many times $f(t)$ has been differentiated; they are *not* powers.

Series like this, which are a special case of Taylor series, are called Maclaurin series after the Scots mathematician Colin Maclaurin.

We can now use these results to write down some particular examples.

EXAMPLE (1) $f(t) = e^t$.

We have already found in Section 8.B.(a) that we can write e in the form of the infinite series

$$1 + \frac{1}{1!} + \frac{1}{2!} + \frac{1}{3!} + \dots$$

Will we get a similar series for $f(t) = e^t$ if we use the above process on it?

We know that $f(t) = e^t$ remains unchanged when it is differentiated with respect to t, so we have $f(t) = f'(t) = f''(t) = f^r(t) = e^t$ for all values of r. Also $e^0 = 1$, so we get

$$e^t = 1 + \frac{t}{1!} + \frac{t^2}{2!} + \frac{t^3}{3!} + \dots + \frac{t^r}{r!} + \dots$$

which agrees with our previous series if we put $t = 1$ so that we start with e.

This series can also be written in the Σ form which we described in Section 6.D.

We would have

$$e^t = \sum_{r=0}^{\infty} \frac{t^r}{r!}.$$

Notice that we are starting the count with $r = 0$.

EXAMPLE (2) $f(t) = \sin t$.

If $f(t) = \sin t$ then $f'(t) = \cos t$, $f''(t) = -\sin t$, $f'''(t) = -\cos t$ and $f''''(t) = \sin t$.

At this point, we have come back to the beginning of the cycle again.

Now, $\sin(0) = 0$ and $\cos(0) = 1$ so we have

$$a_0 = 0 \quad a_1 = \frac{1}{1!} \quad a_2 = 0 \quad a_3 = -\frac{1}{3!} \quad a_4 = 0 \quad \text{and so on.}$$

So

$$\sin t = \frac{t}{1!} - \frac{t^3}{3!} + \frac{t^5}{5!} - \frac{t^7}{7!} + \ldots + (-1)^r \frac{t^{2r+1}}{(2r+1)!} + \ldots$$

starting the count with $r = 0$.

Notice the flip of the signs given by the $(-1)^r$. If r is even, the term is positive, but if r is odd then the term is negative.

This series can be written in the Σ form as

$$\sin t = \sum_{r=0}^{\infty} (-1)^r \frac{t^{2r+1}}{(2r+1)!}.$$

We have already found experimentally and geometrically in Section 4.D.(e) that the first term of this series gives us a very good approximation for $\sin t$ if t is small.

For the same reason as then, it is essential that t is measured in radians for any trig series.

THINKING
POINT

How far do you need to go with summing this series for $\sin t$ in the case of $t = \pi/6$ to get a good approximation to the exact answer of $\frac{1}{2}$? (Let's say 'good' means to 4 d.p.) It works amazingly quickly. Try $t = \pi/2$ too.

For this process to work *at all*, we are assuming that adding the terms of the various series that we get will bring us closer and closer to some definite sum the further we go, that is, we are assuming that these series are convergent. (I describe the meaning of 'convergent' in Sections 6.C.(c) and (d), through the various possibilities of what can happen when we sum GPs.) If any of these series isn't convergent, we would find as we did there with the geometric series 2, 6, 18, . . . and its non-existent sum to infinity, that we were writing nonsense.

EXAMPLE (3) $f(t) = \cosh t$.

This time, the answers repeat in pairs when we differentiate.
We have $f'(t) = \sinh t$ and $f''(t) = \cosh t$ and so on. Also,

$$\cosh(0) = \frac{e^0 + e^{-0}}{2} = \frac{2}{2} = 1 \quad \text{and} \quad \sinh(0) = \frac{e^0 - e^{-0}}{2} = \frac{1-1}{2} = 0.$$

This gives us

$$\cosh t = 1 + \frac{t^2}{2!} + \frac{t^4}{4!} + \ldots + \frac{t^{2r}}{(2r)!} + \ldots$$

Writing this in Σ form gives us

$$\cosh t = \sum_{r=0}^{\infty} \frac{t^{2r}}{(2r)!}$$

again starting the sum with $r = 0$.

Now, suppose instead that we had wanted the series for $f(t) = \cosh(2t)$.

We could do this in two ways, of which the easiest is simply to replace the 't' in the series above by $2t$. This then gives us

$$\cosh 2t = 1 + \frac{4t^2}{2!} + \frac{16t^4}{4!} + \ldots + \frac{(2t)^{2r}}{(2r)!} + \ldots$$

Alternatively, we could have successively differentiated $f(t) = \cosh(2t)$ with respect to t to find the coefficients.

Each time that $f(t) = \cosh(2t)$ is differentiated with respect to t, it gets multiplied by 2 because of the Chain Rule, so the answer comes out exactly the same.

EXAMPLE (4) $f(t) = \ln(1 + t)$

(Why couldn't we find a series for $\ln t$?)

(Because $\ln 0$ is undefined and so we immediately run into an impossible situation.)

If $\quad f(t) = \ln(1 + t) \quad$ then $\quad f'(t) = \dfrac{1}{1+t}$.

To find $f''(t)$, it is easiest to write this as $(1 + t)^{-1}$. Then differentiating again, we get

$$f''(t) = -(1+t)^{-2} = -\frac{1}{(1+t)^2}.$$

Find $f^3(t), f^4(t)$ and $f^5(t)$ for yourself.

You should have the following answers.

$$f^3(t) = (-1)(-2)(1+t)^{-3} = \frac{2!}{(1+t)^3}$$

$$f^4(t) = (2)(-3)(1+t)^{-4} = -\frac{3!}{(1+t)^4}$$

$$f^5(t) = (2)(-3)(-4)(1+t)^{-5} = \frac{4!}{(1+t)^5}.$$

This gives us $f(0) = \ln 1 = 0, f^1(0) = 1, f^2(0) = -1, f^3(0) = 2!, f^4(0) = -3!$, and $f^5(0) = 4!$.

If this pattern continues, what will we have for $f^r(0)$?

We would get $(-1)^{r-1}(r-1)!$.

Putting all this information together gives us the series

$$\ln(1+t) = 0 + \frac{t}{1} - \frac{t^2}{2!} + \frac{2!}{3!}t^3 - \frac{3!}{4!}t^4 + \frac{4!}{5!}t^5 - \ldots$$

$$+ (-1)^{r-1}\frac{(r-1)!}{r!}t^r + \ldots$$

Cancelling as much as possible, we get the series

$$\ln(1+t) = t - \frac{t^2}{2} + \frac{t^3}{3} - \frac{t^4}{4} + \frac{t^5}{5} - \ldots + \frac{(-1)^{r-1}}{r}t^r + \ldots$$

We can write this in the Σ form as

$$\ln(1+t) = \sum_{r=1}^{\infty} \frac{(-1)^{r-1}}{r}t^r.$$

Notice that this time we are starting the count with $r = 1$.

Try putting $t = 1$, and see if you can find a good approximation to $\ln 2$ from summing the first few terms of the series above.

You have $\ln 2 = 1 - \frac{1}{2} + \frac{1}{3} - \frac{1}{4} + \frac{1}{5} - \frac{1}{6} + \ldots$

This is the series which we met briefly at the end of Section 6.F when we said that it is convergent, unlike the frog down the well, or harmonic series of $1 + \frac{1}{2} + \frac{1}{3} + \frac{1}{4} + \ldots$ which isn't.

As you feed in each successive term, you will see the sums flipping from one side to the other of the actual value of $\ln 2$. Unfortunately, although they *are* getting closer to $\ln 2$, this is happening extremely slowly.

A much faster way of finding $\ln 2$ by means of a series comes from using

$$f(t) = \ln\left(\frac{1+t}{1-t}\right) = \ln(1+t) - \ln(1-t).$$

We know that $\quad \ln(1+t) = t - \frac{1}{2}t^2 + \frac{1}{3}t^3 - \frac{1}{4}t^4 + \frac{1}{5}t^5 - \ldots$

Putting 't' $= -t$ gives $\quad \ln(1-t) = -t - \frac{1}{2}t^2 - \frac{1}{3}t^3 - \frac{1}{4}t^4 - \frac{1}{5}t^5 - \ldots$

So $\quad \ln(1+t) - \ln(1-t) = 2(t + \frac{1}{3}t^3 + \frac{1}{5}t^5 + \ldots)$

assuming that we can do this tricky move.

What value of t would you have to use to find $\ln 2$ from

$$\ln\left(\frac{1+t}{1-t}\right)?$$

$$\text{Putting } \frac{1+t}{1-t} = 2 \quad \text{gives} \quad 1+t = 2-2t \quad \text{so} \quad 3t = 1 \quad \text{and} \quad t = \tfrac{1}{3}.$$

Try substituting $t = \tfrac{1}{3}$ in the series above and see how you now get on with finding a value for ln 2. You will see that this series converges much more rapidly.

There is an important question that we need to ask here.

Do all of these series work for *any* value of t?

For example, we already know from Section 6.C.(c) that the GP of $1 + x + x^2 + x^3 + \ldots$ is only convergent if $|x| < 1$.

It seems quite possible that the series which we have been looking at here only converge to their functions for a restricted set of values for t.

In fact, it can be shown that the series for e^t, $\sin t$, $\cos t$, $\sinh t$, and $\cosh t$ are convergent for all values of t. However, the series for $\ln(1 + t)$ is only convergent if $-1 < t \leq 1$.

EXERCISE 8.G.1

Now try the following questions yourself.

(1) **Use the series expansion for e^t which we found in the first example to write down the series expansions of (a) e^{2t} and (b) e^{-t}.**

(2) **Use the series expansion which we found for $\sin t$ in the second example to write down the series expansions for (a) $\sin 2t$ and (b) $\sin(-t)$.**

(3) **Find the series expansion for $f(t) = \cos t$ and then use this to write down expansions for (a) $\cos 2t$ (b) $\cos(-t)$ and (c) $\cos(nt)$ where n is standing for a positive whole number.**

(4) **Find the series expansion for $f(t) = \sinh t$.**
   Now check that you've got the right answer by adding your series together and using $\cosh t + \sinh t = \tfrac{1}{2}(e^2 + e^{-t}) + \tfrac{1}{2}(e^t - e^{-t}) = e^t$ from Section 8.D.(a).

Now that we have a way of writing e^t, $\cosh t$, $\sinh t$, $\cos t$ and $\sin t$ in the form of series, the curious similarities which we have been finding in the behaviour of the trig functions $\cos t$ and $\sin t$ and the hyperbolic functions $\cosh t$ and $\sinh t$ no longer seem quite so surprising.

We know that $e^t = \cosh t + \sinh t$.

It looks as though, with a bit of cunning juggling, we ought to be able to link e^t, $\cos t$ and $\sin t$. Can you see any way of doing this? Maybe if we put 't' $= -t$? Or if we subtracted the series?

Whatever we do, we cannot quite make it work – yet. We shall find out in Section 10.C.(b) that these series do all slot together most beautifully.

9 Integration

In this chapter, we discover what kinds of information we can find out if we do the reverse process to differentiating. This includes finding areas and also finding the equations of functions if we know their rates of change.

The chapter is split up into the following sections.

9.A Doing the opposite of differentiating
(a) What could this tell us? (b) A physical interpretation of this process,
(c) Finding the area under a curve,
(d) What happens if the area we are finding is below the horizontal axis?
(e) What happens if we change the order of the limits? (f) What is $\int(1/x)dx$?

9.B Techniques of integration
(a) Making use of what we already know, (b) Integration by substitution,
(c) A selection of trig integrals with some hyperbolic cousins,
(d) Integrals which use inverse trig and hyperbolic functions,
(e) Using partial fractions in integration, (f) Integration by parts,
(g) Finding rules for doing integrals like $I_n = \int \sin^n x \, dx$,
(h) Using the $t = \tan(x/2)$ substitution

9.C Solving some more differential equations
(a) Solving equations where we can split up the variables,
(b) Putting flesh on the bones – some practical uses for differential equations,
(c) A forwards look at some other kinds of differential equation, including ones which describe SHM,

---

9.A   Doing the opposite of differentiating

9.A.(a)   What could this tell us?

We know from the last chapter that, when we differentiate a function $y = f(x)$, we find a rule which gives us the slope or gradient of $f(x)$ at any point on its curve. So, for example, if we have $y = f(x) = x^2$ then $dy/dx = f'(x) = 2x$, and the gradients of this curve when $x = 0, 1$ and 3 are given by $dy/dx = 0, 2$ and 6 respectively. We can use this rule of $dy/dx = 2x$ to find its gradient for any given value of x.

Now suppose that we start from the other end, and we are told that we have a function for which $dy/dx = f'(x) = 2x$. Since we know the slope of $f(x)$ for any given value of x, we must know what this curve looks like, so we must surely be able to sketch it. Try doing this for yourself.

It is true that we know what this curve looks like but we can't sketch it because we don't know where it is. Any curve shaped in the same way as $y = x^2$, but shifted up or down, will have a slope described by $dy/dx = 2x$ so there are infinitely many possibilities.

Can you think of an extra piece of information which we could be given, so that we would know which particular curve we were talking about?

If we are told a particular point which lies on the curve then there is only one possible answer. As the curve of $y = x^2$ slides up and down, it only goes through this point in one possible position. For example, if we know that this particular $f(x)$ passes through the point (1,3), then its equation must be $y = x^2 + 2$.

Figure 9.A.1 below shows four possible curves for which $dy/dx = 2x$, including the curve $y = x^2 + 2$. The other three are $y = x^2 + 5$, $y = x^2$ and $y = x^2 - 3$.

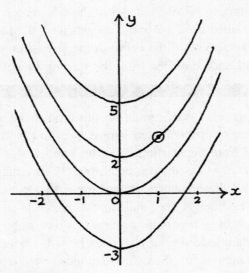

Figure 9.A.1

All the curves for which $dy/dx = f'(x) = 2x$ can be described by the equation $f(x) = x^2 + C$ where C is the number giving the shift up from the x-axis of that particular curve. When you differentiate $f(x)$, the C disappears because $dC/dx = 0$.

The results of these forwards and backwards processes have special names.

$f'(x)$ is called the **derivative** of $f(x)$.

Any $f(x)$ that we get by working backwards from $f'(x)$ is called an **antiderivative** of $f(x)$.

The difference is that there is only one derivative but infinitely many antiderivatives.

The process of finding the antiderivative is called **antidifferentiation**.

EXERCISE 9.A.1

If I tell you what $f'(x)$ is for three functions, and also give you one point which each curve goes through, see if you can find the equation of the curve and sketch it.
(1) $f'(x) = e^x$ and $y = f(x)$ passes through the point (0,2).
(2) $g'(x) = \sinh x$ and $y = g(x)$ passes through the point (0,0).
(3) $h'(x) = 3x^2$ and $y = h(x)$ passes through the point (1,2).

9.A.(b)   **A physical interpretation of this process**

Suppose that there is a race which takes place on a straight track. (The track is straight so that I can make everything happen in the same direction. This will mean that velocity will be the same as speed, since velocity is speed with its direction given also.)

9.A Doing the opposite of differentiating   369

At the beginning of this particular race, all the competitors start running with the same constant acceleration of $2\,\mathrm{m\,s^{-2}}$ (that is, 2 metres per second per second).

Velocity is the rate of change of distance with respect to time so, for travel in a straight line, we can say that $v = dx/dt$. (v stands for the velocity after a time t, and x is the distance travelled from the starting line.)

Since acceleration is the rate of change of velocity with respect to time, we can also say that $dv/dt = 2$. (We also talked about acceleration in Section 8.A.(e).)

We take $t = 0$ when the starting pistol fires. All of the competitors then start running, so after a time t we have $v = 2t$ since $dv/dt = 2$ and $v = 0$ when $t = 0$. (This physical knowledge is the equivalent of our known point on the curve in the last section.)

Now, what will be the distance travelled from the starting line after t seconds?

Again we do the reverse process to differentiation, this time on the equation $v = dx/dt = 2t$. This gives us $x = t^2$ (since t^2 differentiated with respect to t is $2t$), so the distance that these runners have travelled from the starting line after t seconds is t^2 metres.

Or is it? Knowing that $dx/dt = 2t$, can you be absolutely sure that $x = t^2$?

Suppose that, although all these runners started off with identical accelerations, after some further amount of time some get tired and slow down, perhaps because they are very young. If this is known in advance, the runners might be given staggered starts. If a runner begins at the starting line, then indeed after t seconds he will have run t^2 metres. But if a runner was given a start of 3 metres, say, then after t seconds he will actually be $t^2 + 3$ metres from the starting line.

If another runner is given a start of C metres, she will be $t^2 + C$ metres from the starting line after a time t. The physical meaning of the C here is the start which the runner has been given. A runner who is so good that he is placed behind the starting line will have a negative C.

Similar processes will be involved in any physical situation involving distance, velocity and time.

Here is a second example.

If a particle is moving in a straight line so that $v = 2t + e^{2t}$ find (a) the acceleration, a, and (b) the distance x travelled after a time t, given that $x = \frac{1}{2}$ when $t = 0$.

What are the particular values of a and x when $t = 2$?

(a)   We have $a = dv/dt = 2 + 2e^{2t}$.

(b)   Since $v = dx/dt = 2t + e^{2t}$, we have $x = t^2 + \frac{1}{2}e^{2t} + C$. (Check this by differentiating back!)

We know that $x = \frac{1}{2}$ when $t = 0$ so we know $\frac{1}{2} = 0 + \frac{1}{2} + C$, giving $C = 0$.
When $t = 2$, $a = 2 + 2e^4 = 111.2$ to 1 d.p. and $x = 4 + \frac{1}{2}e^4 = 31.3$ to 1 d.p.

EXERCISE 9.A.2

Try the following questions yourself. Questions (1)–(4) have the same meanings for x, v and a as in the example above.

For questions (5) and (6) it is helpful to know that, since $a = dv/dt$ and $v = dx/dt$, we can also write a as $d/dt\,(dx/dt) = d^2x/dt^2$.

(1) If $x = 5t^3 + 3t^2$ find (a) $v = dx/dt$ and (b) $a = dv/dt$.
(2) If $v = 3t^2 + 2t$ and $x = 5$ when $t = 0$, find (a) x and (b) $a = dv/dt$.
(3) If $a = dv/dt = 2t + 1$ and $v = 5$ when $t = 0$, and $x = 0$ when $t = 0$, find (a)v and (b)x.
(4) If $v = 4e^{2t}$ and $x = 1$ when $t = 0$, find (a) x and (b) a.
(5) If $v = -\sin t$ and $x = 1$ when $t = 0$, find (a) x and (b) a.
(6) If $v = \cos(t + \pi/6)$ and $x = \frac{1}{2}$ when $t = 0$, find (a) x and (b) a.

9.A.(c)   Finding the area under a curve

We know that differentiation has the graphical meaning of telling us the steepness or gradient of a curve at any given point on it, provided none of the problems described in Section 8.A.(f) is present.

This means that we can draw a second graph using the gradients of the first graph. Here is a physical example of this.

Suppose we have a particle moving outwards from the origin along the positive x-axis so that its distance x from O is given by $x = t^3 - t^2 + 2t$. From this equation, we can see that $x = 0$ when $t = 0$. This particle is at the origin when we start measuring the time.

The velocity v of the particle is given by $v = dx/dt$, so we have $v = 3t^2 - 2t + 2$. From this equation we see that $v = 2$ when $t = 0$.

We can show both the change in distance over time and the change in velocity over time graphically, and I have done this in Figure 9.A.2(a) and (b).

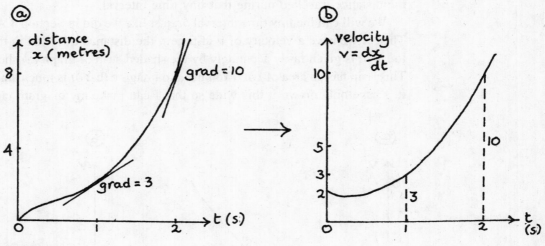

Figure 9.A.2

These two graphs include the pairs of values

t	0	1	2
x	0	2	8

on (a)   and

t	0	1	2
v	2	3	10

on (b)

The actual path of the particle looks like Figure 9.A.3.

Figure 9.A.3

The gradients of the tangents to graph (a) in Figure 9.A.2 for each particular instant in time give us the height of the curve of graph (b) at that same instant in time. This is because these gradients tell us the rate of change of x with respect to t at these points, so they tell us the velocity of the particle at those times. We have got a direct geometrical link going from graph (a) to graph (b).

9.A Doing the opposite of differentiating

Now comes what is probably the most important question in this chapter. Is there a geometrical way of looking at graph (b) which relates it back to graph (a)?

Is there a way of showing the distance travelled by using the graph of velocity against time?

If the velocity was constant, say $3 \, \text{m s}^{-1}$, then the distance travelled in 2 seconds would be $3 \times 2 = 6$ metres. But the curve of graph (b) shows a velocity which is continually changing, so we know that the distance travelled will depend upon the particular time interval that we choose. Suppose we consider the time interval from $t = 1$ to $t = 2$. How could we find the distance travelled in this time interval using this continually changing velocity? (We know what it should be from graph (a), but how could we find it using only graph (b)?)

We could say that if we took a very tiny time interval indeed then the velocity would hardly have changed during that time interval. Therefore, multiplying the velocity at the start of that tiny time interval by the length of the interval would give us a very good estimate of the distance travelled during that tiny time interval.

We will call the tiny time interval δt, just like we did in Section 8.A.(a) in the last chapter. Then if we have a velocity of v at time t, the distance travelled in the time interval from t to $t + \delta t$ is given fairly accurately by the shaded strip of graph (a) shown in Figure 9.A.4(a). This strip has an area of $v \delta t$. (You need to imagine that δt is much smaller than I have drawn it. I have only drawn it this wide so that I can make my diagram clearer.)

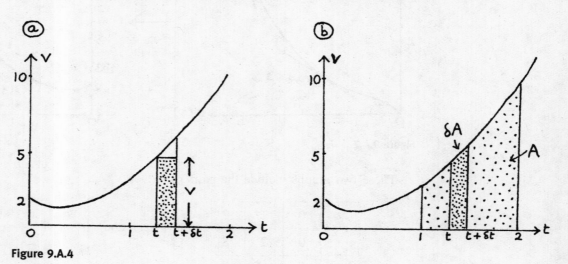

Figure 9.A.4

This strip is a rectangle so it does not follow the shape of the curve at its top but, if we make the tiny interval small enough, we shall be very close to the area under the curve for this small time interval.

I'll take A to be the total area under the graph over the time interval from $t = 1$ to $t = 2$, and I'll let δA be the exact area under the curve over the time interval δt. I show A and δA in Figure 9.A.4(b).

A is made up of the sum of all the little strips like the one I've shown, each of them over a time interval δt. Therefore, if we use the sign '$\approx$' which means 'is approximately equal to', we can say

$$A = \sum_{t=1}^{t=2} \delta A \approx \sum_{t=1}^{t=2} v \, \delta t.$$

(The particular v depends on the particular strip you are looking at.)

Now, the smaller we make the time interval δt, the closer this sum gets to the exact area of A. We call this sum the **integral** of v from $t = 1$ to $t = 2$ and write it as

$$\int_1^2 v\, dt = \int_1^2 (3t^2 - 2t + 2)\, dt.$$

The sign $\int$ comes from an elongated S for 'sum'.

The two values of t, which here show the time interval we are considering, are called the **limits** of the integral.

This integral is now giving us the distance travelled from $t = 1$ to $t = 2$, but what does it actually mean?

From Figure 9.A.4(a) and (b), we can say that $\delta A \approx v\delta t$ so $\delta A/\delta t \approx v$.

But, from the last chapter, if we let $\delta t \to 0$, we will have $dA/dt = v$.

Therefore we have the amazing result that A, the area under the curve, must be what we get if we do the reverse process to differentiation on v.

Integration and differentiation are the reverse processes of each other.

This is called the **Fundamental Theorem of the Calculus** and is crucially important to all mathematicians, scientists and engineers.

Surely now we have the link which takes us back from graph (b) to graph (a) in Figure 9.A.2, since we know that doing the reverse process to differentiation on v gives us x. We know that antidifferentiation gives us lots of possible answers, so how do we use the values of $t = 1$ and $t = 2$ at the ends of our particular time interval to get both the particular A of graph (b) and the particular distance travelled from $t = 1$ to $t = 2$ of graph (a)?

Doing the reverse process to differentiation gives us

$$\int_1^2 (3t^2 - 2t + 2)\, dt = t^3 - t^2 + 2t + C \quad \text{from } t = 1 \text{ to } t = 2.$$

We know that our particular x is actually $t^3 - t^2 + 2t$, but the process of antidifferentiation won't tell us this, so I have included a constant, C, here.

If we put $t = 2$, we get $8 + C$.

If we put $t = 1$, we get $2 + C$.

The distance travelled from $t = 1$ to $t = 2$ is given by the difference between these two answers, which is $8 - 2 = 6$. This agrees with what we get by feeding in $t = 1$ and $t = 2$ directly to x, and then subtracting to find the distance travelled in this time interval.

Notice that the C has cancelled out in the working we did from the integral. We can see why this happens by thinking about why the C is there in the first place. Its physical meaning is that graph (b) in Figure 9.A.2 could equally fit a second particle which starts C units from the origin when $t = 0$, but which moves with exactly the same velocity as the first particle. (This particle is doing exactly the same thing as a runner given a lead of C from the starting line in Section 9.A.(a).) The velocity/time graphs for both particles are identical, but the distance/time graph for the second particle would be pushed up C units from the t-axis. Both particles travel identical distances in any given time interval. In particular, they would each travel 6 units in the interval from $t = 1$ to $t = 2$.

We have now found the geometrical link which takes us back from graph (b) to graph (a). I show this happening in Figure 9.A.5.

9.A Doing the opposite of differentiating 373

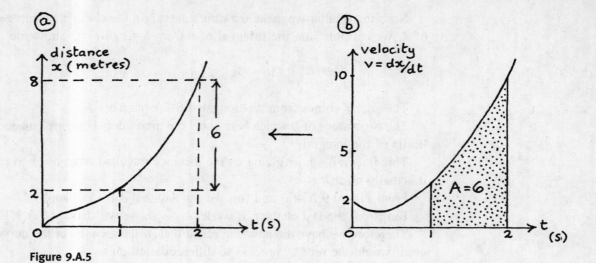

Figure 9.A.5

We have the extraordinary result that the area under a curve made up of the sum of tiny strips, each depending on the equation of the curve, can be found from a calculation just involving the values at the two end points.

As you go further with learning about integration in later parts of your course, you will see this same kind of simplification happening again, so that an area surrounded by a closed curve can be given in terms of its boundary curve, and a volume can be given in terms of the surface enclosing it. This is your first step along this powerful path of simplification.

Any other distance travelled by the particle of our example can be found just by changing the limits on the integral.

For example, the distance travelled from $t = 0$ to $t = 2$ is given by

$$\int_0^2 (3t^2 - 2t + 2)dt = \left[t^3 - t^2 + 2t\right]_0^2 = 8 - 0 = 8.$$

Notice that I have also written in the limits on the square brackets above. Sometimes these are written either end of a single vertical line. The working would then look like this:

$$\int_0^2 (3t^2 - 2t + 2)dt = t^3 - t^2 + 2t\Big|_0^2 = 8 - 0 = 8.$$

The opposite process to differentiation which we have done here is called **integrating**.

An integral which has given limits is called a **definite integral**.

$\int_a^b f(t)dt$ tells us the area under the curve $x = f(t)$ between the two limits $t = a$ and $t = b$.

To find what this area is, first integrate $f(t)$ to give $F(t)$, say. Then use the limits to find $F(a)$ and $F(b)$. The area is given by $F(a) - F(b)$.

An integral with no limits is called an **indefinite integral**. Integrating this gives us $\int f(t)\, dt = F(t) + C$ where $F'(t) = f(t)$. Integrals can also be written using other letters, most usually x.

You can think of $F(t)$ as either giving you a rule to find any particular area if you put in given limits, or as describing the curve whose rate of change with respect to t is $f(t)$, so $F'(t) = f(t)$. This description needs to include the constant C because we don't know the amount of vertical shift of $F(t)$ away from the horizontal axis. Just as with the curves in Section 9.A.(a), we can only find the value of C if we are given more information. C is called the **constant of integration**. It doesn't affect any area which we find because it cancels out when the limits are put in.

> If you are finding a solution to a physical problem by using indefinite integrals, it is extremely important to include all the constants of integration, so that you have allowed for all the possible solutions. Often, you will have extra information which will make it possible to find the values of these, but leaving them out will make all your subsequent working wrong.

We have just looked at one physical example here, so that we can actually see what is happening, but it can be shown by the methods of mathematical analysis that it is always possible to find areas under curves in this way provided none of the complications of Section 8.A.(f) is present, and that we do indeed get closer and closer to a particular A as we make the strips narrower and narrower and then add them all together.

EXERCISE 9.A.3

Try finding the following integrals yourself. Because integration is the reverse process to differentiation, you can check any answers you are doubtful about by differentiating them back before putting in your limits.

Draw sketches to show the areas you are finding for questions (3), (4) (5) and (6).

(1) $\int_1^3 (2t^2 + 3t + 1)dt$ (2) $\int_0^4 (2t^3 - t)dt$ (3) $\int_0^{\pi/6} \sin x \, dx$

(4) $\int_0^{\pi/6} 2 \sin x \, dx$ (5) $\int_0^{\pi/6} \sin 2x \, dx$ (6) $\int_0^1 \cosh x \, dx$

(7) $\int_0^{\pi/4} 3 \cos 2u \, du$ (8) $\int_1^2 e^{2x} \, dx$ (9) $\int_0^1 e^{3x+1} \, dx$

Now try finding the following indefinite integrals. Remember to include a constant of integration in your answers.

(10) $\int (t^2 + 2t) \, dt$ (11) $\int \sin 3x \, dx$ (12) $\int \cos (2x + \pi/2) \, dx$

(13) $\int \sinh 2u \, du$ (14) $\int e^{5x} \, dx$ (15) $\int e^{2x+3} \, dx$

The integrals in the next three questions each represent the distance x travelled by a particle moving in a straight line. Find x in each case, using the given information to find the value of each of the constants of integration.

(16) $\int (8t^3 + t)dt$ given that $x = 2$ when $t = 0$

(17) $\int \sin 2t \, dt$ given that $x = 1$ when $t = 0$

(18) $\int (t + e^{2t})dt$ given that $x = 0$ when $t = 0$

We shall look at what happens in the special case when the acceleration is constant in question (2) of Exercise 9.C.2 at the end of Section 9.C.(b).

To get the maximum worth from the distance/time and velocity/time graphs of Figure 9.A.2(a) and (b), we can link them back to Section 8.E.(b) on turning points and points of inflection. I show these two graphs again in Figure 9.A.6. I have also added the acceleration/time graph and the sketch of the actual path of the particle.

See if you can describe physically what is happening to this particle around the instant in time when $t = \frac{1}{3}$. Think about what happens to its acceleration, its velocity and how the distance it has travelled changes. See how this fits in with the point of inflection and the local minimum you can see on the two graphs. The answer to this thinking point is given at the back of the book after the answers to Exercise 9.A.3.

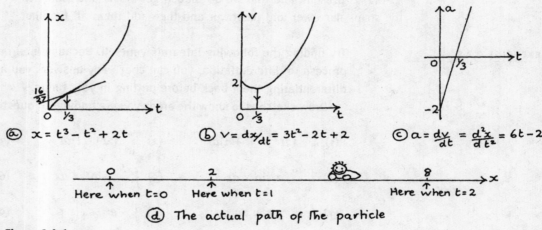

(a) $x = t^3 - t^2 + 2t$ (b) $v = \frac{dx}{dt} = 3t^2 - 2t + 2$ (c) $a = \frac{dv}{dt} = \frac{d^2x}{dt^2} = 6t - 2$

Here when $t=0$ Here when $t=1$ Here when $t=2$

(d) The actual path of the particle

Figure 9.A.6

9.A.(d)   What happens if the area we are finding is below the horizontal axis?

As an example of this, suppose we have $v = f(t) = 8t^3 - 24t^2 + 16t$. This gives a graph like the one I show in Figure 9.A.7.

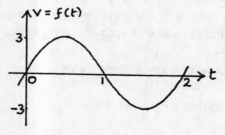

Figure 9.A.7

Integrating, we have

$$\int_0^1 (8t^3 - 24t^2 + 16t)dt = \left[2t^4 - 8t^3 + 8t^2\right]_0^1 = 2$$

and

$$\int_1^2 (8t^3 - 24t^2 + 16t)dt = \left[2t^4 - 8t^3 + 8t^2\right]_1^2 = -2. \qquad \text{(Check this!)}$$

The second area, which is below the horizontal axis, has worked out as negative. This also means that

$$\int_0^2 (8t^3 - 24t^2 + 16t)dt = \left[2t^4 - 8t^3 + 8t^2\right]_0^2 = 0.$$

What kind of physical meaning could we give to this?

The graph could represent the velocity/time graph of a particle which moves 2 units away from the origin along the positive x-axis in the time interval from $t = 0$ to $t = 1$ and then 2 units back again in the time interval from $t = 1$ to $t = 2$. This means that effectively it hasn't moved from its starting point over the interval from $t = 0$ to $t = 2$, and the minus sign of the second integral shows that this part of its journey was in the backwards direction. If it is claiming travel expenses, it will have to work out the distance travelled in two separate bits, taking the absolute value of each. Doing this gives a total of $2 + |-2| = 4$ units of distance travelled.

Any area below the horizontal axis will integrate as negative.

9.A.(e)   What happens if we change the order of the limits?

Sometimes it is possible to get an integral in which the lower limit is a larger number than the upper limit, so that in a sense you are looking at the area backwards. You can probably see what will happen; the integral will change its sign.

Using Figure 9.A.6 above as an example, we would have

$$\int_1^0 (8t^3 - 24t^2 + 16t)dt = \left[2t^4 - 8t^3 + 8t^2\right]_1^0 = -2.$$

You could give a physical meaning to what is happening here by saying that, looking backwards over the first time interval from $t = 0$ to $t = 1$, the particle was 2 units nearer to the origin at the beginning of this time interval.

Having the limits swapped in this way is something which actually happens. You will find when you do integration by substitution that sometimes the altered limits do have the larger number as the lower limit. You can see from the above example that it would then be possible to swap them round to the usual position, provided that you also change the sign of the integral.

In general, we can say

$$\int_a^b f(t)dt = -\int_b^a f(t)dt.$$

As in all these rules, you can replace t by x, or any other letter of your choice.

9.A.(f)   **What is $\int (1/x)\ dx$?**

Since

$$\frac{d}{dx}(x^n) = nx^{n-1},$$

we know that

$$\int x^n\ dx = \frac{x^{n+1}}{n+1} + C.$$

But this rule will not work if $n = -1$ because we will get a zero underneath the fraction and therefore it will be undefined.

So what is $\int x^{-1}\ dx$ or $\int (1/x)\ dx$?

From Section 8.B.(e), we know that $d/dx\ (\ln x) = 1/x$. Therefore, reversing the process, we have $\int (1/x)\ dx = \ln x + C$.

We can see what is happening graphically by looking at the particular example shown in the two drawings of Figure 9.A.8(a) and (b). Figure 9.A.8(a) shows ln 3 on the graph of $y = \ln x$ and Figure 9.A.8(b) shows ln 3 as an area on the graph of $y = 1/x$.

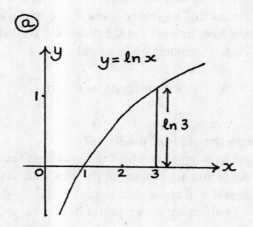
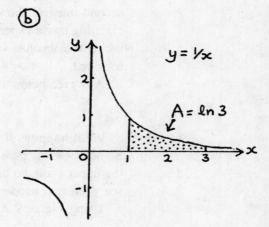

Figure 9.A.8

We have

$$\int_1^3 \frac{1}{x}\ dx = \left[\ln x\right]_1^3 = \ln 3 - \ln 1 = \ln 3 \text{ since } \ln 1 = 0.$$

In a similar way, we could say

$$\int_1^X \frac{1}{x}\ dx = \left[\ln x\right]_1^X = \ln X - \ln 1 = \ln X$$

for any value of X greater than 1. It is possible to *define* the log function from the function $f(x) = 1/x$ in this way. If this is done, e then comes from the inverse function e^x of this log function, rather than through considering rates of growth as we did in Section 8.B.(a). We extend this rule to cover values of X between one and zero by reversing the limits as we did in the previous section. For example,

$$\ln \tfrac{1}{2} = \int_1^{1/2} \frac{1}{x}\ dx = -\int_{1/2}^1 \frac{1}{x}\ dx.$$

From Figure 9.A.8(b), you will see that we can't find any area under the curve of $f(x) = 1/x$ which involves crossing through the value $x = 0$ because the graph does an infinite jump there. (It has what is called a discontinuity.)

However, we *can* find an area which is completely on the left of the origin, like the one shown as area B in Figure 9.A.9. Can we then give this area a meaning in terms of logs?

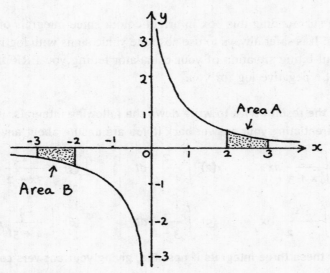

Figure 9.A.9

Area B is given by

$$\int_{-3}^{-2} \frac{1}{x} \, dx = \left[\ln x \right]_{-3}^{-2}.$$

We feel a bit tempted to say that area $B = \ln(-2) - \ln(-3)$ except that we know that we can't because these logs don't exist. (You can see in Figure 9.A.8(a) that they just aren't there.) But, from the symmetry of the graph, we know that

$$\int_{-3}^{-2} \frac{1}{x} \, dx = -\int_{2}^{3} \frac{1}{x} \, dx = -\left[\ln x \right]_{2}^{3} = -(\ln 3 - \ln 2).$$

This means that we can sidestep the problem of trying to find the undefined $\ln(-3)$ and $\ln(-2)$ by saying

$$\text{area } B = \ln |-2| - \ln |-3| = \ln 2 - \ln 3 = -\text{ area } A.$$

(Remember that $|x|$ means 'take the positive value of the x inside the $|\ |$'.)

Notice that this result also fits in with the second rule of logs.

$$\text{area } A = \ln 3 - \ln 2 = \ln \left(\tfrac{3}{2} \right) \quad \text{and} \quad \text{area } B = \ln 2 - \ln 3 = \ln \left(\tfrac{2}{3} \right).$$

Checking on your calculator will show you that $\ln \left(\tfrac{3}{2} \right) = -\ln \left(\tfrac{2}{3} \right)$.

We know from Chapter 8 that

$$\frac{d}{dx} (\ln (ax + b)) = \frac{a}{ax + b}.$$

From this, and the working above, we get the reverse result that

$$\int \frac{dx}{ax + b} = \frac{1}{a} \ln |ax + b| + C.$$

It is worth surrounding this box in bright colour since integrals of fractions like this are very common. It is safer always to use absolute value signs with log integrals. This prevents that horrid nail-biting situation of your calculator telling you ERROR when you try to get it to work out a negative log for you.

EXERCISE 9.A.4

Use the result above to write down the following integrals. (Check by differentiating your answer back if you are unsure about any of your numbers.)

(1) $\int \dfrac{1}{x + 3}\, dx$ (2) $\int \dfrac{2}{t - 4}\, dt$ (3) $\int \dfrac{1}{3x + 2}\, dx$

(4) $\int \dfrac{1}{5 - 2x}\, dx$ (5) $\int \dfrac{4}{3 - t}\, dt$ (6) $\int \dfrac{-2}{4 + 5t}\, dt$

Find these three integrals if possible, giving your answers correct to 3 d.p.

(7) $\int_{-1}^{2} \dfrac{1}{5 - 2x}\, dx$ (8) $\int_{1}^{3} \dfrac{1}{3x - 4}\, dx$ (9) $\int_{-4}^{-2} \dfrac{1}{3 + 2x}\, dx$

You might be interested to notice here that the right-hand side of $f(x) = 1/x$ is the smoothed-out continuous version of the frog-down-the-well or harmonic series of $1 + \frac{1}{2} + \frac{1}{3} + \frac{1}{4} + \dots$ of Section 6.F. You can see that the curve of $y = 1/x$ could be drawn through the tops of the separate jumps which I showed laid out side by side there in Figure 6.F.1.

We showed in that section that, although the individual terms of this series become smaller and smaller the further we go, the sum of the series is infinitely large. If we stack the terms one on top of the other, we can reach any height we want to. The frog always escapes from the well eventually, however deep down he is to start with.

In a similar way, it can be shown that the area under the graph from 1 to t becomes slowly but inexorably larger and larger, beyond any value we care to think of, as t increases.

This might seem surprising since the strip giving the increase in area is becoming so thin, but the log function of $\ln X = \int_1^X (1/x)\, dt$ can be made as large as we please by choosing a sufficiently large value of X, in just the same way that the sum of the harmonic series can be made as large as we please if we take a sufficiently large number of terms.

You can also see that the log function will continue to increase if you look at Figure 9.A.8(a).

Its gradient, given by $d/dx\, (\ln x) = 1/x$, continues to be positive, although it becomes very small indeed as the value of x increases.

9.B   Techniques of integration

In this section we look at various techniques which will help us to find more complicated integrals. When we have done this, we shall be able to apply this knowledge to solve some more physical problems. This will then form part of the last section of this chapter.

9.B.(a) Making use of what we already know

When finding integrals the most important rule to remember is that integration and differentiation are reverse processes of each other. (In the last section, we looked at examples which showed this fundamental rule of calculus happening.) To check if you have integrated correctly, you can always differentiate your answer to see if you get what was inside the integral sign. (You would need to do this check before you put in any limits.) See how you get on doing the following exercise.

EXERCISE 9.B.1

This exercise is in three parts. Part (A) consists of quickie questions on differentiation similar to the ones in Section 8.C.(a).

(A) Differentiate each of the following with respect to the letter used to describe the function.

(1) $f(t) = 4t^2$ (2) $f(x) = 3x^{3/2}$ (3) $f(t) = e^{2t}$ (4) $f(x) = e^{5x+2}$

(5) $f(t) = e^{-3t}$ (6) $f(t) = \ln 3t$ (7) $f(x) = \ln (3 + x)$ (8) $f(x) = \ln (2 - x)$

(9) $f(t) = \ln (3 - 5t)$ (10) $f(t) = \sin 3t$ (11) $f(x) = \cos (x + \pi/3)$

(12) $f(x) = \sin (2x + \pi/6)$ (13) $f(t) = \cosh (2t + 5)$ (14) $f(x) = \sinh (8 - 5x)$

(15) $f(x) = \sin^3 x$ (16) $f(x) = \tan x$

Check your answers to these before going on to (B). I think you will probably have found them very easy. (If you did have any problems here, it is important to go back to Section 8.C before you go any further with integration.)

Part (B) has quickie questions for you to take back the other way by antidifferentiating or integrating. They are close enough to the questions in (A) that you should have a good idea of what to do with them. Remember that you need to add in a constant to these answers to show that you don't know the shift of your curve from the horizontal axis. The need for this is explained in Section 9.A.(a).

(B) (1) $f'(x) = 2x^3$ (2) $f'(t) = t^{1/2}$ (3) $f'(t) = e^{3t}$ (4) $f'(x) = e^{2x+3}$

(5) $f'(t) = e^{-2t}$ (6) $f'(t) = e^{3-4t}$ (7) $f'(x) = \dfrac{1}{x + 2}$ (8) $f'(x) = \dfrac{1}{3x + 2}$

(9) $f'(x) = \dfrac{1}{1 - x}$ (10) $f'(t) = \sin 5t$ (11) $f'(x) = \cos (x + \pi/6)$

(12) $f'(t) = \sin (2t + \pi/4)$ (13) $f'(x) = \cosh 3x$ (14) $f'(x) = \sinh (2x + 3)$

(15) $f'(x) = 4 \sin^3 x \cos x$ (16) $f'(x) = \sec^2 2x$

Check your answers to (B) before going on to (C).

(C) This has some straightforward integrals with limits (that is, *definite* integrals), for you to work out. If you are not sure what to do here, you need to go back to the first section of this chapter.

(1) $\displaystyle\int_0^2 (u^2 + 3u)\, du$ (2) $\displaystyle\int_1^3 (5t^2 + 1/t)\, dt$ (3) $\displaystyle\int_0^{\pi/8} \cos 4t\, dt$

(4) $\displaystyle\int_0^{\pi/6} \sin 2x\, dx$ (5) $\displaystyle\int_0^{\pi/2} \sin (2x + \pi/3)\, dx$ (6) $\displaystyle\int_{-1}^1 e^{5t}\, dt$

(7) $\displaystyle\int_{-1}^0 e^{2t+3}\, dt$ (8) $\displaystyle\int_{-1}^1 \cosh 2x\, dx$ (9) $\displaystyle\int_0^{\pi/4} \sec^2 x\, dx$

(10) $\displaystyle\int_0^3 \dfrac{du}{2u + 3}$ (11) $\displaystyle\int_0^1 \dfrac{1}{x + 2} + \dfrac{4}{2x + 5}\,dx$ (12) $\displaystyle\int_{-1}^1 \dfrac{1}{2 - x} - \dfrac{1}{2x + 3}\,dx$

Before going on to the next part of this section, we'll collect together the general rules we have been using above.

Some rules for integration

(1) If $f'(x) = x^n$ then $f(x) = \dfrac{x^{n+1}}{n+1} + C$ if $n \neq -1$.

(2) (a) If $f'(x) = \dfrac{1}{x}$ then $f(x) = \ln|x| + C$.

(b) If $f'(x) = \dfrac{1}{ax+b}$ then $f(x) = \dfrac{1}{a} \ln|ax+b| + C$.

(3) If $f'(x) = e^{ax}$ then $f(x) = \dfrac{1}{a} e^{ax} + C$.

(4) If $f'(x) = \sin(ax+b)$ then $f(x) = \dfrac{-1}{a} \cos(ax+b) + C$.

(5) If $f'(x) = \cos(ax+b)$ then $f(x) = \dfrac{1}{a} \sin(ax+b) + C$.

(6) If $f'(x) = \sinh(ax+b)$ then $f(x) = \dfrac{1}{a} \cosh(ax+b) + C$.

(7) If $f'(x) = \cosh(ax+b)$ then $f(x) = \dfrac{1}{a} \sinh(ax+b) + C$.

(8) If $f'(x) = \sec^2 x$ then $f(x) = \tan x + C$.

THINKING POINT

See if you can work out what the answers to these three similar looking integrals are. Everything you need to be able to do them has come in previous chapters.

(1) $\displaystyle\int \dfrac{x}{1+x^2}\, dx$ (2) $\displaystyle\int \dfrac{1}{1+x^2}\, dx$ (3) $\dfrac{1}{1-x^2}\, dx$

9.B.(b)   **Integration by substitution**

This is just the Chain Rule in reverse.

In fact, you have already been using it when you wrote down answers like

$$\int \cos 4t\, dt = \tfrac{1}{4} \sin 4t + C \quad \text{and} \quad \int e^{2t+3}\, dt = \tfrac{1}{2} e^{2t+3} + C.$$

Putting in the $\tfrac{1}{4}$ and the $\tfrac{1}{2}$ in these two answers means that you are allowing for the fact that $4t$ and $2t + 3$ are themselves functions of t.

Just as the Chain Rule makes it possible to differentiate complicated functions of functions, so the method of substitution makes it possible for us to integrate more complicated functions of functions.

The easiest way of explaining how this works is to look at some particular examples.

EXAMPLE (1) $I = \int 2x \, (x^2 + 3)^7 \, dx$

One way to do this would be to multiply out the bracket using the Binomial Theorem, and then integrate term by term, but this would be very lengthy and tedious.

There is something much neater and quicker which we can do.

If we let $t = x^2 + 3$, we shall then have

$$I = \int 2x \, t^7 \, dx.$$

(I have used the letter t here because I want to save u and v for later purposes.)

This new integral looks a lot nicer in many ways, but it has a peculiar mixture of variables which means that it can't be related back to any particular curve.

 You can't find an integral which has mixed variables in it.

So how can we get round this problem here? In particular, we need some way of relating the dx which we have to the dt which we would like to have.

We do this by using our chosen relation between x and t of $t = x^2 + 3$. From this, we can say that $dt/dx = 2x$. So therefore we can replace the $2x \, dx$ by dt.

We now have

$$I = \int t^7 \, dt = \frac{t^8}{8} + C = \frac{(x^2 + 3)^8}{8} + C,$$

replacing the original variable of x in the final answer. You should always do this unless the integral has limits. The next example will show you what to do in this case.

Notice how fortunate we were in the above example to have exactly $2x \, dx$! This was not by chance – I wanted an easy example to start with. When you have done a number of integrals like this, you get used to looking for this kind of helpful situation. When it happens, you will often find that you can work out what the answer must be in your head, but it's wise to check back by differentiating; it is horribly easy to get the arithmetic slightly wrong otherwise.

EXAMPLE (2) $I = \int_0^1 \dfrac{3x}{\sqrt{x^2 + 1}} \, dx$

We can use a very similar process here, to get rid of the problem of the complicated square root underneath.

If we put $t = x^2 + 1$ then $dt/dx = 2x$. Therefore we can replace $x \, dx$ by $\frac{1}{2} \, dt$. (Notice that in this example too we have an x exactly where we need it.)

This then gives us

$$I = \int_{x=0}^{x=1} \frac{\frac{3}{2} \, dt}{\sqrt{t}} = \int_{x=0}^{x=1} \tfrac{3}{2} t^{-1/2} \, dt.$$

We must now change the limits of the integral so that they fit the new variable, t.

If $x = 0$ then $t = 1$, and if $x = 1$ then $t = 2$, using $t = x^2 + 1$.

So now we have

$$I = \int_1^2 \tfrac{3}{2} t^{-1/2}\, dt = \left[3t^{1/2} \right]_1^2 = 3\sqrt{2} - 3.$$

(You can replace t by $x^2 + 1$ and then use the original limits if you prefer, but doing this is generally more messy and time-consuming.)

EXAMPLE (3) $\displaystyle I = \int \frac{x}{(x+5)^9}\, dx.$

If we let $t = x + 5$, we have $dt/dx = 1$ so we can replace dx by dt. But this will leave us with an x sitting on the top. What can we do with it so that the whole integral is in terms of t?

We can say $t = x + 5$ so $x = t - 5$. Doing this gives us

$$I = \int \frac{(t-5)}{t^9}\, dt = \int t^{-8} - 5t^{-9}\, dt$$

$$= -\frac{t^{-7}}{7} + \frac{5t^{-8}}{8} + C$$

$$= \frac{-1}{7(x+5)^7} + \frac{5}{8(x+5)^8} + C$$

which will tidy up as

$$\frac{-8(x+5) + 35}{56(x+5)^8} + C = \frac{-8x - 5}{56(x+5)^8} + C.$$

From the examples above, we can make these general rules.

Integration by substitution

(1) Replace the nastiest bit by something simple, so that you are working with a new function of your own choosing.

(2) Find the relation between the old dx and the new dt (or whatever letters you are using). If this turns out to be rather complicated, you may have to rethink step (1).

(3) Make sure that you rewrite your new integral so that it is completely in terms of your new variable. (No stray left-behind xs!)

(4) Change the limits to fit your new variable if it's a definite integral.

(5) Integrate!

(6) Replace the old variable if this was an indefinite integral.

The common mistake that students make when they write, for example, $\int \sin^3 x\, dx = \tfrac{1}{4} \sin^4 x + C$ shows that their minds *are* working on the right track. They just haven't gone quite far enough.

What they are finding is the equivalent of $\int (\text{lump})^3 \, d(\text{lump}) = \frac{1}{4}(\text{lump})^4 + C$ when what they've actually got is $\int (\text{lump})^3 \, dx$.

EXERCISE 9.B.2

Try these three for yourself now.

(1) $\displaystyle\int \frac{3x}{(2x^2 - 3)^4} \, dx$ (2) $\displaystyle\int_0^2 \frac{6x^2}{\sqrt{2x^3 + 9}} \, dx$ (3) $\displaystyle\int_4^5 x(x-4)^9 \, dx$

EXAMPLE (4) $I = \displaystyle\int \frac{2x + 6}{x^2 + 6x + 25} \, dx$

This integral looks quite difficult but it is actually very simple to do. Can you spot why this is?

If we differentiate the bottom, we get the top.
This means that we can write down the answer straight away.

It must be $\ln |x^2 + 6x + 25| + C$. Check that this answer works by differentiating it back.

It is also possible to find this answer by substituting $u = x^2 + 6x + 25$. This gives

$$du/dx = 2x + 6 \quad \text{so} \quad I = \int \frac{du}{u} = \ln|u| + C = \ln |x^2 + 6x + 25| + C$$

but the first method is much simpler once you know that it is worth looking out for.

Sometimes we need to do a bit of juggling with the numbers, so for example,

$$\int \frac{4x + 12}{x^2 + 6x + 25} \, dx = 2 \int \frac{2x + 6}{x^2 + 6x + 25} \, dx = 2 \ln |x^2 + 6x + 25| + C$$

but $\displaystyle\int \frac{2x + 7}{x^2 + 6x + 25} \, dx$ is not yet possible for us to do.

We shall find out how to do this one in Example (1) of Section 9.B.(d).
Meanwhile, we have this useful rule:

$$\boxed{\quad \text{if } I = \int \frac{f'(x)}{f(x)} \, dx \quad \text{then} \quad I = \ln|f(x)| + C. \quad}$$

We can now also do the first of the three integrals in the thinking point at the end of Section 9.B.(a).
This was

$$\int \frac{x}{1 + x^2} \, dx$$

and the answer is

$$\tfrac{1}{2} \ln |1 + x^2| + C.$$

9.B Techniques of integration

EXAMPLE (5) A special property of $\int \left(\dfrac{1}{x} \right) dx.$

Suppose we have $\displaystyle\int_1^6 \dfrac{1}{x}\, dx - \int_1^2 \dfrac{1}{x}\, dx.$

We can see from Figure 9.B.1 that this leaves us with $\int_2^6 (1/x) dx.$

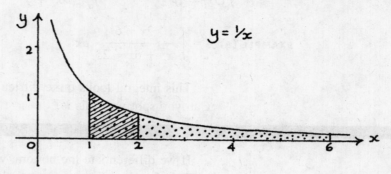

Figure 9.B.1

But from Section 9.A.(f) we know that

$$\int_1^6 \dfrac{1}{x}\, dx = \ln 6 \quad \text{and} \quad \int_1^2 \dfrac{1}{x}\, dx = \ln 2.$$

Also,

$$\ln 6 - \ln 2 = \ln \left(\dfrac{6}{2} \right) = \ln 3 = \int_1^3 \dfrac{1}{x}\, dx.$$

This means that we have found the two different answers of $\int_2^6 (1/x) dx$ and $\int_1^3 (1/x) dx$ for the spotty area. What has gone wrong? The answer is that nothing has. If we let $x = 2t$ in $\int_2^6 (1/x)\, dx$, we get

$$\int_1^3 \left(\dfrac{1}{2t} \times 2 \right) dt = \int_1^3 \dfrac{1}{t}\, dt = \ln 3.$$

The letter used here makes no difference to the final answer because it disappears when we put in the limits. We have discovered that this area can be pushed around without changing its size, like toothpaste in a tube. I've shown this in Figure 9.B.2.

Multiplying or dividing the limits by any constant number will make no difference to the resulting area because the effects cancel each other out.

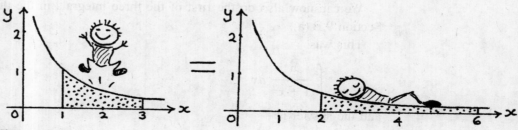

Figure 9.B.2

Integration

Now try these for yourself. You may find that you can do some of these questions in your head, just as it not necessary to write down all the details every time you use the Chain Rule. It is excellent if you can do this, but you should always check by differentiating back that you have found the right answer.

Always include the *dx* or *du* or whatever when you write down an integral. If you don't do this you will make mistakes when you substitute a new variable.

(1) $\int \dfrac{2x + 3}{x^2 + 3x + 9}\, dx$ 　　(2) $\int \dfrac{6x + 2}{3x^2 + 2x + 7}\, dx$ 　　(3) $\int \dfrac{4x + 8}{x^2 + 4x + 7}\, dx$

(4) $\int \dfrac{x\, dx}{(3x^2 + 4)^3}$ 　　(5) $\int \dfrac{x^2}{\sqrt{x^3 - 1}}\, dx$ 　　(6) $\int \dfrac{x}{\sqrt{1 - 4x^2}}\, dx$

(7) $\int x\sqrt{1 - x^2}\, dx$ 　　(8) $\int \dfrac{1 + 2x}{\sqrt{x + x^2}}\, dx$ 　　(9) $\int \dfrac{dx}{(2x - 7)^2}$

9.B.(c)　A selection of trig integrals with some hyperbolic cousins

Integrals of trig functions are often involved in solutions to physical problems, so this section deals with some of the ones which you may meet.

We start with a very easy one.

EXAMPLE (1) 　$I = \int \cos x \sin^6 x\, dx$

This integral is similar to the ones in the previous section. The presence of the cos *x* means that you can probably write the answer straight down. Try doing this.

In case you couldn't, here's how the integration works by substitution. We put $t = \sin x$ so $dt/dx = \cos x$. We then get

$$I = \int t^6\, dt = \tfrac{1}{7} t^7 + C = \tfrac{1}{7} \sin^7 x + C.$$

EXAMPLE (2) 　$I = \int \tan x\, dx = \int \dfrac{\sin x}{\cos x}\, dx$

We can now find the integral of tan *x* using the same method as in Example (1).

If we let $t = \cos x$ 　then we have 　$dt/dx = -\sin x$, so

$$I = -\int \dfrac{dt}{t} = -\ln|t| + C = -\ln|\cos x| + C.$$

Because $\sec x = \dfrac{1}{\cos x}$, we can also say

$$\ln|\sec x| = \ln(1) - \ln|\cos x| = -\ln|\cos x|.$$

This gives the two forms of

$$\boxed{\int \tan x\, dx = \ln|\sec x| + C = -\ln|\cos x| + C.}$$

This is another result worth surrounding in bright colour.

EXAMPLE (3) $I = \int_0^{\pi/3} \dfrac{2\sin x}{(1 + \cos x)^3}\, dx$

This one is very easy, because $d/dx(1 + \cos x) = -\sin x$, so we can find it by putting $t = 1 + \cos x$. Try finishing the working off for yourself.

We can replace $\sin x\, dx$ by $-dt$.
Also, if $x = \pi/3$ then $t = \frac{3}{2}$, and if $x = 0$ then $t = 2$. This gives

$$I = \int_2^{3/2} -\frac{2}{t^3}\, dt = \int_{3/2}^2 \frac{2}{t^3}\, dt = \left[\frac{1}{t^2}\right]_{3/2}^2 = \frac{1}{4} - \frac{4}{9} = -\frac{7}{36}.$$

This is an example of an integral where the limits get swapped over in order of size when we do a substitution. We then use the result of Section 9.A.(e) to write them in the usual order.

EXAMPLE (4) Finding $\int \sin^2 x\, dx$ and $\int \sin^3 2x\, dx$.

I have chosen these two to show you, but the methods we will use on them will carry over with only slight adjustments if we replace sin by cos or sinh or cosh.

Both $I_1 = \int \sin^2 x\, dx$ and $I_2 = \int \sin^3 2x\, dx$ are quite straightforward.

For $I_1 = \int \sin^2 x\, dx$, we use the identity $\cos 2x = 1 - 2\sin^2 x$. We showed this in Section 5.D.(d). Using it gives us $\sin^2 x = \frac{1}{2}(1 - \cos 2x)$.
Therefore $I_1 = \frac{1}{2}\int 1 - \cos 2x\, dx = \frac{1}{2}x - \frac{1}{4}\sin 2x + C$.

To find $I_2 = \int \sin^3 2x\, dx$, we cunningly rewrite it to make it into two easy integrals. We say

$$I_2 = \int \sin 2x(1 - \cos^2 2x)\, dx = \int \sin 2x - \sin 2x \cos^2 2x\, dx$$
$$= -\frac{1}{2}\cos 2x + \frac{1}{6}\cos^3 2x + C.$$

Differentiate this back to see that it is right.

It isn't possible for me to put in all the trig integrals I want to show you here, because some of them depend on future work, so I'll give you a list now of where to find these.

Where to look for other trig and hyperbolic integrals
- In Section 9.B.(g) we shall discover how we can get rules to find integrals like $\int \sin^n x\, dx$ where n is a positive whole number.
- In Section 9.B.(f) we shall discover how to find integrals involving multiplications like $\int \sin 2x\, e^{3x}\, dx$, and also what to do with integrals like $\int \sin 4x \cos 2x\, dx$.
- In Section 9.B.(d) we shall find out how we can use the inverse trig and hyperbolic functions to give us answers to another collection of integrals.
- In Section 9.B.(h) we shall discover how to find integrals like

$$\int \frac{1}{1 + \cos x}\, dx$$

which is surprisingly tricky, unlike

$$\int \frac{\sin x}{1 + \cos x}\, dx$$

which you can do in your head (yes?)

It is $-\ln (1 + \cos x) + C.$

EXERCISE 9.B.4

Try these for yourself now. As usual, try to do the easy ones in your head, but differentiate back to check the arithmetic.

(1) $\int \sin x \cos^5 x \, dx$ (2) $\int \sin 2x \cos^3 2x \, dx$ (3) $\int \cos (x + \pi/4) \sin^2 (x + \pi/4) \, dx$

(4) $\displaystyle\int \frac{\sin 3x}{\cos^4 3x}\, dx$ (5) $\displaystyle\int \frac{\cos x}{(1 + \sin x)^3}\, dx$ (6) $\displaystyle\int \cot x \, dx = \int \frac{\cos x}{\sin x}\, dx$

(7) $\displaystyle\int \cos^2 x \, dx$ (8) $\displaystyle\int \sin^2 3x \, dx$ (9) $\displaystyle\int \cos^3 x \, dx$

(10) $\displaystyle\int \cos^3 5x \, dx$ (11) $\displaystyle\int \cosh^2 x \, dx$ (12) $\displaystyle\int \sinh (x + 1) \cosh^4 (x + 1) \, dx$

For the next two questions, if you put $t = \sin x$ in (13) and $t = \cosh x$ in (14), you will turn them into something you can do.

(13) $\displaystyle\int_0^{\pi/6} \frac{3 \cos x}{\sin^2 x + 5 \sin x + 6}\, dx$ (14) $\displaystyle\int \frac{7 \sinh x}{2 \cosh^2 x + 5 \cosh x - 3}\, dx$

(15) $\displaystyle\int_0^1 \frac{dx}{\sqrt{4 - x^2}}$ (Put $x = 2 \sin t$.)   (16) $\displaystyle\int_0^{1/2} \sqrt{(1 - x^2)}\, dx$ (Put $x = \sin t$.)

(17) $\displaystyle\int_0^{\pi/2} \frac{\cos x}{(4 + \sin x)^2}\, dx$ (18) $\displaystyle\int_1^{\sqrt{2}} \frac{1}{x^2 \sqrt{4 - x^2}}\, dx$ (Put $x = 2 \cos t$.)

9.B.(d)   Integrals which use inverse trig and hyperbolic functions

In Section 8.D.(i) and Section 8.F.(c) we found some results which will be very useful to us now for finding certain kinds of integral. I have collected the most important of them together for you in the box below.

Four useful rules about integrals

(a) $\displaystyle\int \frac{dx}{\sqrt{a^2 - x^2}} = \sin^{-1}\left(\frac{x}{a}\right)$ $(|x| < a)$ (b) $\displaystyle\int \frac{dx}{x^2 + a^2} = \frac{1}{a} \tan^{-1}\left(\frac{x}{a}\right)$

(c) $\displaystyle\int \frac{dx}{\sqrt{x^2 + a^2}} = \sinh^{-1}\left(\frac{x}{a}\right)$ (d) $\displaystyle\int \frac{dx}{\sqrt{x^2 - a^2}} = \cosh^{-1}\left(\frac{x}{a}\right)$ $(x > a)$

NOTE (1)   We take the value of a to be positive. If you put $a = 1$, you will get the simplest case of the integral.

NOTE (2)   I have not included a constant of integration for any of these integrals both for reasons of space and because you may be using these results to find integrals with given limits.

NOTE (3)   Notice the restrictions on the values of x in (a) and (d). We must have these to avoid problems like wanting the square roots of negative numbers, trying to find angles whose sin is greater than one, or trying to find values of x whose cosh is less than 1.

Just as when we did the differentiation in Sections 8.D.(i) and 8.F.(c), we can match up the as in the box above to fit particular circumstances. For example, we now have the answer to the second part of the thinking point at the end of Section 9.B.(a).
Putting $a = 1$ in (b) gives

$$\int \frac{dx}{x^2 + 1} = \tan^{-1} x.$$

The rest of this thinking point is answered in the next section.

EXERCISE 9.B.5   Try the following three yourself, by picking out the right integral and choosing the value of a which will make it work.

$$(1) \int \frac{3 \, dx}{\sqrt{x^2 + 25}} \qquad (2) \int \frac{2 \, dx}{\sqrt{9 - x^2}} \qquad (3) \int_0^\pi \frac{dx}{x^2 + 16}$$

A SPECIAL WARNING EXAMPLE   What happens if we have $I = \int \dfrac{du}{\sqrt{4u^2 - 9}}$?

Can you write down the answer for this one? Be careful! There is something about this integral which makes it not entirely straightforward.

There are two possible approaches to doing this one.
The first is to use substitution, putting $u = \frac{3}{2} \sin \theta$, to simplify the square-rooted expression.
This gives us $du/d\theta = \frac{3}{2} \cos \theta$, and

$$I = \frac{3}{2} \int \frac{\cos \theta}{\sqrt{9 \sin^2 \theta - 9}} \, d\theta = \frac{3}{2} \int \frac{\cos \theta}{3 \cos \theta} \, d\theta = \frac{1}{2} \theta + C.$$

Now $\frac{2}{3} u = \sin \theta$ so $\theta = \sin^{-1} \left(\frac{2}{3} u \right)$. This gives us $I = \frac{1}{2} \sin^{-1} \left(\frac{2}{3} u \right) + C$.
Did you include the $\frac{1}{2}$ in your answer? This is what makes this integral not quite straightforward. (Excellent, if you got it!)
The other way of approaching this integral is to use Rule (a) above.
In this case, there are again two things which you can do.
The first is to convert the integral into the same form as (a) by rearranging it so it has u^2 instead of $4u^2$ underneath.

This is probably the simplest approach and we do it as follows:

$$I = \int \frac{du}{\sqrt{4u^2 - 9}} = \int \frac{du}{2\sqrt{u^2 - \frac{9}{4}}} = \frac{1}{2} \int \frac{du}{\sqrt{u^2 - \frac{9}{4}}}.$$

Now $a^2 = \frac{9}{4}$ so $a = \frac{3}{2}$ (since a is positive). This gives $I = \frac{1}{2} \sin^{-1}\left(\frac{2}{3}u\right) + C$.

Alternatively, you can put $x = 2u$ and $a = 3$. If you do this, you have to remember that putting $x = 2u$ will give $dx/du = 2$ and so du must be replaced by $\frac{1}{2} dx$.

Doing this gives the same answer yet again of $I = \frac{1}{2} \sin^{-1}\left(\frac{2}{3}u\right) + C$.

Notice, too, that I've made the integral easier to see by taking the constant multiplying number outside it. We can do this because it is just a scaling factor for the area if the integral has limits. (You can see this happening in questions (3) and (4) of exercise 9.A.3.)

 Never make an integral look nicer by taking a variable like an x outside. It may look better but it means something completely different.

Three examples which use the technique of completing the square

EXAMPLE (1) In Section 9.B.(b), we briefly met the integral $I = \int \frac{2x + 7}{x^2 + 6x + 25} dx$.

We could not do this then, but we did find that

$$\int \frac{2x + 6}{x^2 + 6x + 25} dx = \ln|x^2 + 6x + 25| + C.$$

We therefore split I to make use of this, writing

$$I = I_1 + I_2 = \int \frac{2x + 6}{x^2 + 6x + 25} dx + \int \frac{1}{x^2 + 6x + 25} dx.$$

We then make it possible to use the inverse tan rule on I_2 by cunningly completing the square underneath.

We say

$$x^2 + 6x + 25 = (x + 3)^2 - 9 + 25 = (x + 3)^2 + 16.$$

So we have

$$I_2 = \int \frac{dx}{(x + 3)^2 + 4^2} = \frac{1}{4} \tan^{-1}\left(\frac{x + 3}{4}\right) + C$$

using Rule (b) for the inverse tan from the last section with 'x' $= x + 3$ and 'a' $= 4$.

Therefore

$$I = \ln|x^2 + 6x + 25| + \frac{1}{4} \tan^{-1}\left(\frac{x + 3}{4}\right) + C.$$

EXAMPLE (2) Sometimes you need to do some juggling with the numbers to make the rules fit the particular circumstances. For example, suppose you have

$$\int \frac{5t + 3}{t^2 + 10t + 29} \, dt.$$

Differentiating the bottom with respect to t gives $2t + 10$, so we rearrange the top to take advantage of a log integral getting rid of the t for us. We can do this by saying

$$I = \frac{1}{2} \int \frac{10t + 6}{t^2 + 10t + 29} \, dt = \frac{1}{2} \int \frac{10t + 50 - 44}{t^2 + 10t + 29} \, dt$$

$$= \frac{5}{2} \int \frac{2t + 10}{t^2 + 10t + 29} \, dt - \int \frac{22}{t^2 + 10t + 29} \, dt$$

$$= \frac{5}{2} \ln |t^2 + 10t + 29| - 22 \int \frac{dt}{(t + 5)^2 + 4}.$$

Now, we do the second integral using Rule (b), and letting $x = t + 5$ and $a = 2$. This gives us

$$I = \frac{5}{2} \ln |t^2 + 10t + 29| - 11 \tan^{-1} \left(\frac{t + 5}{2} \right) + C.$$

EXAMPLE (3) Sometimes we have to be careful about applying Rule (b) when we have completed the square, just as in the special warning example which we had earlier. Suppose we have

$$I = \int \frac{3dt}{4t^2 + 12t + 25} = \int \frac{3dt}{(2t + 3)^2 + 16}.$$

We put $a = 4$ and $x = 2t + 3$. Doing this substitution gives $dx/dt = 2$ so we must replace dt by $\frac{1}{2} \, dx$ here. I now becomes

$$\int \frac{\frac{3}{2}}{x^2 + 16} \, dx = \frac{3}{2} \times \frac{1}{4} \tan^{-1} \left(\frac{x}{4} \right) + C = \frac{3}{8} \tan^{-1} \left(\frac{2t + 3}{4} \right) + C.$$

If you do this replacement step in your head, you must remember to include the extra $\frac{1}{2}$ which comes from replacing $2t + 3$ by x.

Alternatively, you can say

$$\int \frac{3dt}{4t^2 + 12t + 25} = \frac{1}{4} \int \frac{3dt}{t^2 + 3t + \frac{25}{4}}$$

$$= \frac{3}{4} \int \frac{dt}{(t + \frac{3}{2})^2 + 4} \quad \text{since } -\frac{9}{4} + \frac{25}{4} = 4.$$

Now, using Rule (b) with $x = t + \frac{3}{2}$ and $a = 2$ we get

$$I = \frac{3}{4} \times \frac{1}{2} \times \tan^{-1} \left(\frac{2t + 3}{4} \right) = \frac{3}{8} \tan^{-1} \left(\frac{2t + 3}{4} \right).$$

This method involves more fractions but means you can use the $\tan^{-1}$ rule directly, without having to make adjustments between the dt and the dx.

Now try finding the following integrals yourself.

(1) $\displaystyle\int \frac{dt}{9 + t^2}$ (2) $\displaystyle\int \frac{dx}{x^2 + 6x + 10}$ (3) $\displaystyle\int \frac{2x + 5}{x^2 + 4x + 13}\, dx$

(4) $\displaystyle\int \frac{4t + 3}{t^2 + 8t + 17}\, dt$ (5) $\displaystyle\int \frac{dx}{4x^2 + 9}$ (6) $\displaystyle\int \frac{3dx}{9x^2 + 16}$

(7) $\displaystyle\int \frac{3u + 2}{u^2 + 6u + 13}\, du$ (8) $\displaystyle\int \frac{3du}{9u^2 + 24u + 17}$ (9) $\displaystyle\int \frac{3du}{2u^2 + 2u + 5}$

9.B.(e)   Using partial fractions in integration

Being able to write single complicated fractions in the form of two or more much simpler fractions makes integration much easier. I have described how we can do this in Section 6.E. If you are at all unsure about how to find partial fractions, you should go back there now and check it out before going on.

Here are three examples showing their use in integration.

EXAMPLE (1) $\displaystyle\int \frac{10}{x(x - 1)(x + 4)}\, dx$

In question (3) of Exercise 6.E.1 we found that

$$\frac{10}{x(x - 1)(x + 4)} = \frac{-\frac{5}{2}}{x} + \frac{2}{x - 1} + \frac{\frac{1}{2}}{x + 4}$$

so $\displaystyle I = \int \left(\frac{-\frac{5}{2}}{x} + \frac{2}{x - 1} + \frac{\frac{1}{2}}{x + 4} \right) dx$

$$= -\tfrac{5}{2} \ln |x| + 2 \ln |x - 1| + \tfrac{1}{2} \ln |x + 4| + C.$$

HELPFUL HINT

> It is much easier to do the integration if you keep the number fractions on the top as I have done, and think of them as scaling factors multiplying each answer. You can see how this works in the answers to questions (3) and (4) of Exercise 9.A.3.

EXAMPLE (2) This will give you the answer to (3) of the thinking point at the end of Section 9.B.(a). I asked you there if you could find

$$\int \frac{1}{1 - x^2}\, dx.$$

You might know now what to do even if you didn't then. If so, try it now before looking at what I have done.

We can use partial fractions to say that

$$\int \frac{1}{1 - x^2}\, dx = \int \left(\frac{\frac{1}{2}}{1 + x} + \frac{\frac{1}{2}}{1 - x} \right) dx = \tfrac{1}{2} \ln |1 + x| - \tfrac{1}{2} \ln |1 - x| + C.$$

You might also have realised that you could use the result that

$$\frac{d}{dx}(\tanh^{-1} x) = \frac{1}{1 - x^2} \quad \text{(this was question (3) of Exercise 8.D.3)}$$

so you could say that $I = \tanh^{-1} x + C$.

Does this mean that we've got two different answers?

No, because if we use the laws of logs on the first answer, we get

$$I = \ln \sqrt{\frac{1 + x}{1 - x}} + C$$

and we know that

$$\tanh^{-1} x = \tfrac{1}{2} \ln\left(\frac{1 + x}{1 - x}\right) = \ln \sqrt{\frac{1 + x}{1 - x}}$$

from Section 8.D.(g).

When we combine the logs together to give a single log in this way, it is often convenient to include the constant term in this log too. We can do this by letting $C = \ln A$. This then gives us the answer that

$$I = \ln \left(A \sqrt{\frac{1 + x}{1 - x}}\right).$$

We shall find that this kind of combination of an answer into a single log will be particularly useful when we solve some differential equations in Section 9.C.

EXAMPLE (3) What will happen if the partial fractions which we find involve the special cases of Section 6.E? Will it still be easy to integrate the answer? The following integral has examples of two of these special cases in its second and third fractions.

$$I = \int \left(\frac{2}{x - 1} + \frac{1}{(x - 1^2)} + \frac{2x + 3}{x^2 + 4}\right) dx.$$

Try writing down the answer to it yourself.

Three separate integrals make up the answer here. The first two are

$$I_1 = \int \frac{2}{x - 1}\, dx = 2 \ln |x - 1| \quad \text{and} \quad I_2 = \int (x - 1)^{-2}\, dx = -\frac{1}{x - 1}.$$

 I_2 isn't a log!

I_3 is a little bit tricky, but we know how to do it from Section 9.B.(d). We have

$$I_3 = \int \frac{2x + 3}{x^2 + 4}\, dx = \int \frac{2x}{x^2 + 4}\, dx + \int \frac{3}{x^2 + 4}\, dx$$

$$= \ln |x^2 + 4| + \tfrac{3}{2} \tan^{-1} (x/2).$$

This gives a complete answer of

$$I = \ln [(x - 1)^2 (x^2 + 4)] - \frac{1}{x - 1} + \tfrac{3}{2} \tan^{-1} \left(\frac{x}{2} \right) + C.$$

We know from looking at it that the inside of the log above must be positive.

EXAMPLE (4) Suppose we are given the three similar-looking integrals below.

$$I_1 = \int \frac{2x}{x^2 + 5x + 6} \, dx \qquad I_2 = \int \frac{2x + 3}{x^2 + 3x + 4} \, dx \qquad I_3 = \int \frac{dx}{x^2 + 8x + 25}$$

Try finding each of these for yourself before looking at what I have done with them.

$$I_1 = \int \frac{6}{x + 3} - \frac{4}{x + 2} \, dx = 6 \ln |x + 3| - 4 \ln |x + 2| + C.$$

To find I_2, we notice that $x^2 + 3x + 4$ will not factorise. Indeed, it *has* no real factors. But $d/dx \, (x^2 + 3x + 4) = 2x + 3$. So $I_2 = \ln |x^2 + 3x + 4| + C$.

Now for I_3. We can't factorise $x^2 + 8x + 25$, and there is no handy $2x + 8$ on the top, so it isn't a straightforward log. This is an integral which uses the methods of the last section.

Putting 'x' $= x + 4$ and 'a' $= 3$ and using Rule (b) from there, we get

$$I_3 = \int \frac{dx}{(x + 4)^2 + 9} = \tfrac{1}{3} \tan^{-1} \left(\frac{x + 4}{3} \right) + C.$$

EXERCISE 9.B.7

Try the following for yourself now.

(1) $\int \dfrac{4}{(x + 2)(x + 3)} \, dx$ (2) $\int \dfrac{5}{(x - 2)(x + 3)^2} \, dx$ (3) $\int \dfrac{x + 1}{x - 1} \, dx$

(4) $\int \dfrac{4}{y(y^2 + 1)} \, dy$ (5) $\int \dfrac{3t + 1}{(2t - 1)(t + 2)^2} \, dt$ (6) $\int \dfrac{10y}{(y - 1)(y^2 + 9)} \, dy$

(7) $\int \dfrac{10x}{(x - 1)(x^2 - 9)} \, dx$ (8) $\int \dfrac{t^2 + 1}{t^2 - 1} \, dt$ (9) $\int \dfrac{x^2 + 1}{(x + 2)(x + 4)} \, dx$

(10) $\int \dfrac{1}{y^3 - y^2} \, dy$ (11) $\int \dfrac{u^2 - 1}{u^2(2u + 1)} \, du$ (12) $\int \dfrac{e^x}{e^{2x} + 5e^x + 6} \, dx$

9.B.(f)   **Integration by parts**

Suppose that you need to find $\int x \cos 3x \, dx$.

If instead you had wanted to find $\int (x + \cos 3x) \, dx$, you would just have integrated each bit in turn, giving you the answer of $\frac{1}{2} x^2 + \frac{1}{3} \sin 3x + C$. Is it true that

$$I = \int x \cos 3x \, dx = (\tfrac{1}{2} x^2)(\tfrac{1}{3} \sin 3x) + C = \tfrac{1}{6} x^2 \sin 3x + C?$$

Try differentiating this back and see what you get.

At this point you should have decided that this method won't work.

We have to use the Product Rule to differentiate $\frac{1}{6}x^2 \sin 3x$, and doing this gives us

$$\frac{d}{dx}(\tfrac{1}{6}x^2 \sin 3x) = \tfrac{1}{3}x \sin 3x + \tfrac{1}{2}x^2 \cos 3x.$$

(If you are at all unsure about this last step, you need to go back to Section 8.C.(d) urgently before going on any further with this section.)

To do integrals which are made up of two functions multiplied together, like the one above, we use the Product Rule turned backwards. We used the Chain Rule turned backwards, in a very similar way, to let us do integration by substitution.

In Section 8.C.(d), we wrote down the Product Rule in the form

$$\frac{d}{dx}(uv) = v\,\frac{du}{dx} + u\,\frac{dv}{dx}$$

where u and v are both themselves functions of x.

Integrating both sides with respect to x, and remembering that integration is the reverse process to differentiation, gives us

$$uv = \int v\,\frac{du}{dx}\,dx + \int u\,\frac{dv}{dx}\,dx.$$

We then rearrange this so that we get

The rule for integration by parts

$$\int u\,\frac{dv}{dx}\,dx = uv - \int v\,\frac{du}{dx}\,dx.$$

This is also sometimes written in the form

$$\int uv'\,dx = uv - \int vu'\,dx.$$

 As we have just seen, it isn't true that you can find $\int uv\,dx$ by integrating each of u and v and then multiplying these answers together (though students quite often do this).

I will now show you how the above rule works by taking some examples.

EXAMPLE (1) We'll start with the correct version of $I = \int x \cos 3x\,dx$.

Here, we can easily differentiate and integrate both x and $\cos 3x$, but differentiating x looks promising since it gives us something simpler. Integrating x would gives us a more complicated integral to find, involving $\frac{1}{2}x^2$. Therefore we let

$$u = x \quad \text{so} \quad \frac{du}{dx} = 1 \quad \text{and} \quad \frac{dv}{dx} = \cos 3x \quad \text{so} \quad v = \tfrac{1}{3}\sin 3x.$$

You should always write down these bits of working because it helps you not to make arithmetical slips, and it makes it easier to check back if something should go wrong. You don't need to worry about putting in a constant when you integrate dv/dx; all we will need is a constant of integration in the final answer if the given integral has no limits, and so we are not finding a definite area.

Next, we use this working to substitute into the rule above. Doing this gives us

$$I = \int x \cos 3x \, dx = \tfrac{1}{3} x \sin 3x - \int \tfrac{1}{3} \sin 3x \, dx$$

and we see that we've now got a nice easy integral to do, so giving us the final answer of

$$I = \tfrac{1}{3} x \sin 3x + \tfrac{1}{9} \cos 3x + C.$$

Differentiate this answer back for yourself to check that it really does work.

If we had started with $I = \int x^2 \cos 3x \, dx$, we would have done exactly the same sort of thing, with $u = x^2$ and $dv/dx = \cos 3x$, and then used the method of integration by parts twice.

EXAMPLE (2) *A tricky pair.* Suppose we have $I_1 = \int xe^x \, dx$ and $I_2 = \int xe^{x^2} \, dx$.

Have a go at doing each of these yourself.

We can use integration by parts for I_1 by putting $u = x$ so $du/dx = 1$ and $dv/dx = e^x$ so $v = e^x$. Doing this gives us

$$I_1 = xe^x - \int e^x \, dx = xe^x - e^x + C.$$

When we come to try I_2, we can't use integration by parts because we can't integrate e^{x^2}. (If you thought that you could, try differentiating back!) If, on the other hand, we make $x = dv/dx$, the integral we have to find gets worse instead of better.

The answer here is to do the substitution $x^2 = t$ so that $dt/dx = 2x$ and $x \, dx$ can be replaced by $\tfrac{1}{2} dt$. Doing this gives us

$$I_2 = \tfrac{1}{2} \int e^t \, dt = \tfrac{1}{2} e^t + C = \tfrac{1}{2} e^{x^2} + C.$$

You might perhaps have spotted that this answer will work just by looking at the integral. This example shows us that not every integral which is made up of two functions multiplied together will respond to being attacked by the method of integration by parts.

EXAMPLE (3) $I = \int x^3 \ln x \, dx$

Integrating $\ln x$ is tricky (we shall see how to do it in the next example), but differentiating it is easy. Also, although getting a higher power of x often makes an integral worse, in this particular case it won't cause us a problem because we also have a log.

We let

$$u = \ln x \quad \text{so} \quad \frac{du}{dx} = \frac{1}{x} \quad \text{and} \quad \frac{dv}{dx} = x^3 \quad \text{so} \quad v = \frac{x^4}{4}.$$

This gives us

$$I = \tfrac{1}{4}x^4 \ln x - \int \left(\tfrac{1}{4}\, x^4 \times \frac{1}{x}\right) dx$$

(I have used $\times$ to show the multiplication inside the integral.)

$$= \tfrac{1}{4}\, x^4 \ln x - \tfrac{1}{4} \int x^3 \; dx = \tfrac{1}{4}\, x^4 \ln x - \tfrac{1}{16}\, x^4 + C.$$

EXAMPLE (4) Now we can find $I = \int \ln x \; dx$.

To do this, we use a very useful trick which is to think of $\ln x$ as being $1 \times \ln x$.

Then $I = \int 1 \times \ln x \; dx$, and we put $\quad u = \ln x$
so $\quad du/dx = 1/x \quad$ and $\quad dv/dx = 1 \quad$ so $\quad v = x.$
This then gives us

$$I = \int \ln x \; dx = x \ln x - \int x \times \frac{1}{x} \; dx = x \ln x - x + C.$$

EXAMPLE (5) This is another example which uses the same cunning trick.

If we want

$$\int_0^{1/2} \tan^{-1}(2x) \; dx$$

then we write it as

$$\int_0^{1/2} 1 \times \tan^{-1}(2x) \; dx.$$

Now we let $u = \tan^{-1}(2x)$ so

$$\frac{du}{dx} = \frac{2}{1 + 4x^2} \quad \text{and} \quad \frac{dv}{dx} = 1 \quad \text{so} \quad v = x.$$

(This step uses the work of Section 8.D.(i).)
This gives

$$I = \left[x \tan^{-1}(2x) \right]_0^{1/2} - \int_0^{1/2} \frac{2x}{1 + 4x^2} \; dx$$

$$= \tfrac{1}{2}(\pi/4) - \left[\ln |1 + 4x^2| \right]_0^{1/2} = \pi/8 - \tfrac{1}{4} \ln 2.$$

All the inverse trig and hyperbolic functions can be integrated by using this method.

EXAMPLE (6) Now, suppose we want $\int 5^x dx$.

Students often find this integral rather awkward to do. The cunning trick which we used above won't work because it actually makes things worse, so what can we do? We can think that 5^x behaves very much (but not quite) like e^x, so we look what happens if we differentiate it instead.

If $y = 5^x$ then $\ln y = \ln (5^x) = x \ln 5$ so $(1/y)dy/dx = \ln 5$ differentiating implicitly, like we did in Section 8.F.(d).

Therefore, if $y = 5^x$, we have

$$\frac{dy}{dx} = y\ln 5 = (\ln 5)5^x.$$

Therefore

$$\int 5^x\, dx = \left(\frac{1}{\ln 5}\right)5^x + C$$

since integration is the reverse process to differentiation.

EXAMPLE (7) $\int e^{3u} \sin 2u\, du$

The safest first step here is to replace u by x so that you don't get into a horrid confusion of us which mean two different things. Since this integral has no limits, we shall have to replace x by the given variable u as our last step. If an integral does have limits, so that it is finding a particular area, you can rewrite it using any letter of your choice – the answer will come out exactly the same. For this reason, the letter that we use is called a **dummy variable**. This worked in exactly the same way when we used Σ in Section 6.D.(a).

We start, then, by rewriting I as

$$\int e^{3x} \sin 2x\, dx.$$

Both of e^{3x} and $\sin 2x$ are easy to differentiate and integrate, and it doesn't matter which way round you choose your u and dv/dx here.

I will put $u = e^{3x}$ so $du/dx = 3e^{3x}$ and $dv/dx = \sin 2x$ so $v = -\frac{1}{2} \cos 2x$. This then gives me

$$I = -\tfrac{1}{2} e^{3x} \cos 2x + \int \tfrac{3}{2}e^{3x} \cos 2x\, dx.$$

At this stage, you may think that things are no better, since we have a very similar integral to find to the one which we started with. However, if we repeat the process of integration by parts on this new integral, we will find that everything will work out very nicely.

 It is very important at this stage to stick to the same sort of choice that we made at the beginning. If we don't do this, the whole thing unravels like knitting and we finish up with $I = I$ which, though true, is not very helpful.

This means that I must put $u = e^{3x}$ so $du/dx = 3e^{3x}$ and $dv/dx = \cos 2x$ so $v = \frac{1}{2} \sin 2x$. This then gives me

$$\int e^{3x} \cos 2x\, dx = \tfrac{1}{2} e^{3x} \sin 2x - \int \tfrac{3}{2} e^{3x} \sin 2x\, dx.$$

Substituting this back in the original equation gives

$$I = -\tfrac{1}{2} e^{3x} \cos 2x + \tfrac{3}{2} \left(\tfrac{1}{2} e^{3x} \sin 2x - \tfrac{3}{2} \int e^{3x} \sin 2x\, dx \right).$$

Again you may think that we are not getting anywhere at all as we now have the same integral back again that we started with. But, if we call this I, you will see that we are in a good position after all.

We have

$$I = -\tfrac{1}{2} e^{3x} \cos 2x + \tfrac{3}{4} e^{3x} \sin 2x - \tfrac{9}{4} I$$

and rearranging gives us

$$\tfrac{13}{4} I = \tfrac{1}{4} e^{3x} (-2 \cos 2x + 3 \sin 2x).$$

Finally, we must replace x by the original u. Doing this gives us

$$I = \tfrac{1}{13} e^{3u} (3 \sin 2u - 2 \cos 2u) + C$$

putting in a constant of integration.

EXAMPLE (8) $\displaystyle\int \sin 4x \cos 2x \, dx$

We can do this in the same way as the last example by using integration by parts, but it is very much quicker (and safer from arithmetical slips) to use the rule

$$\sin A \cos B = \tfrac{1}{2} [\sin (A + B) + \sin (A - B)]$$

from Section 5.D.(h).

Here, $A = 4x$ and $B = 2x$ so we get $\tfrac{1}{2} (\sin 6x + \sin 2x)$ giving

$$I = \tfrac{1}{2} \int \sin 6x + \sin 2x \, dx = \tfrac{1}{2}(-\tfrac{1}{6} \cos 6x - \tfrac{1}{2} \cos 2x) + C$$
$$= -\tfrac{1}{12} \cos 6x - \tfrac{1}{4} \cos 2x + C.$$

EXERCISE 9.B.8

Students quite often make slips with the little bits of differentiating and integrating which come in the working steps for integration by parts, so this first question starts you off with some extra practice on this.

(1) For each of the following, write down either du/dx (or du/dt if the function is in terms of t), or v as required.

(a) $u = e^{5x}$ (b) $u = \sin 4t$ (c) $dv/dx = e^{4x}$ (d) $dv/dt = e^{-t}$

(e) $u = \ln (2x)$ (f) $dv/dt = \sinh 2t$ (g) $dv/dx = \sin 3x$ (h) $u = \cos 2x$

Check your answers to this question before going on to integrate the following.

(2) $\displaystyle\int 2t \sin 3t \, dt$ (3) $\displaystyle\int x e^x \, dx$ (4) $\displaystyle\int x^2 \sin x \, dx$

(5) (a) $\displaystyle\int x^2 \ln x \, dx$ (b) $\displaystyle\int \tfrac{1}{x} \ln x \, dx$ (c) $\displaystyle\int x^2 (\ln x)^2 \, dx$

(6) $\displaystyle\int_0^1 x^2 \tan^{-1} x \, dx$ (7) $\displaystyle\int_0^\pi e^{-x} \sin x \, dx$ (8) $\displaystyle\int e^{-t} \cosh 2t \, dt$

(9) $\displaystyle\int \sin 4x \sin 3x \, dx$ (10) $\displaystyle\int \cos 3t \cos 6t \, dt$

9.B.(g)   Finding rules for doing integrals like $I_n = \int \sin^n x \, dx$

What we want to do here is to find a pattern which will take us down from I_n to I_{n-1} or possibly to I_{n-2}. (Which of these we get will depend on the particular integral we are considering.) If we then find I_0 and I_1, we shall be able to find I_n for any positive whole number value of n.

I will show you how this actually works by taking some examples.

EXAMPLE (1) If $I_n = \int_0^{\pi/2} \sin^n x \, dx$ then answer the following.

(a) Find I_0 and I_1.
(b) Find a rule which links I_n to I_{n-2}.
(c) Use (a) and (b) to find I_3, I_4, I_5 and I_6.

(a) Since $(\sin x)^0 = 1$, we have

$$I_0 = \int_0^{\pi/2} dx = \left[x \right]_0^{\pi/2} = \pi/2$$

$$I_1 = \int_0^{\pi/2} \sin x \, dx = \left[-\cos x \right]_0^{\pi/2} = 1.$$

(b) Now we use integration by parts on I_n because this will make it possible to find a pattern which will take us down from the starting number of n. This means that we are using the work of the previous section.

First, we have to decide on the best way to split the integral into two parts, remembering that we must have one part that we can integrate easily.

Here, the most promising idea seems to be to write

$$I_n = \int_0^{\pi/2} \sin^n x \, dx = \int_0^{\pi/2} \sin x \times \sin^{n-1} x \, dx.$$

(It would not be a good idea to choose $\int_0^{\pi/2} 1 \times \sin^n x \, dx$ because integrating the 1 would give us an x which would just be an embarrasment.)

Now we let $dv/dx = \sin x$ so $v = -\cos x$ and $u = \sin^{n-1} x$ so

$$\frac{du}{dx} = (n-1) \sin^{n-2} x (\cos x).$$

Using the rule for integration by parts gives us

$$I_n = \left[-\cos x \sin^{n-1} x \right]_0^{\pi/2} - \int_0^{\pi/2} (-\cos x)((n-1) \sin^{n-2} x (\cos x)) \, dx.$$

Now,

$$\sin 0 = 0 \quad \text{and} \quad \cos \pi/2 = 0 \quad \text{so} \quad \left[-\cos x \sin^{n-1} x \right]_0^{\pi/2} = 0.$$

This gives us

$$I_n = (n-1) \int_0^{\pi/2} \cos^2 x \sin^{n-2} x \, dx.$$

We are interested in sines here, so we use the identity $\cos^2 x = 1 - \sin^2 x$. This gives us

$$I_n = (n-1) \left[\int_0^{\pi/2} \sin^{n-2} x \, dx - \int_0^{\pi/2} \sin^n x \, dx \right].$$

But this is the same as

$$I_n = (n-1) I_{n-2} - (n-1) I_n.$$

Therefore

$$I_n + (n-1)I_n = (n-1) I_{n-2} \quad \text{so} \quad I_n = \left(\frac{n-1}{n} \right) I_{n-2}.$$

We now have the rule which gives us the pattern of how the integrals run down from n. A rule like this is called a **reduction formula**. Notice that this particular rule is going down in double jumps from n to $n - 2$. This means that we need to know both I_0 and I_1 to be able to use it for all values of n.

(c) Using the rule we have just found and the answers from (a), we can say that

$$I_2 = \tfrac{1}{2} I_0 = \pi/4 \quad \text{and} \quad I_4 = \tfrac{3}{4} I_2 = \tfrac{3}{4} \times \pi/4 = 3\pi/16 \quad \text{and}$$
$$I_3 = \tfrac{2}{3} I_1 = \tfrac{2}{3} \quad \text{and} \quad I_5 = \tfrac{4}{5} I_3 = \tfrac{4}{5} \times \tfrac{2}{3} = \tfrac{8}{15}.$$

EXAMPLE (2) If $I_n = \displaystyle\int_1^e \ln^n x \, dx$, answer the following.

   (a) Find I_0.
   (b) Find a rule to write I_n in terms of I_{n-1}.
   (c) Use this rule to find I_1, I_2 and I_3.

(a) $I_0 = \displaystyle\int_1^e dx = \left[x\right]_1^e = e - 1.$

(b) This time it looks best to split the integral as

$$I_n = \int_1^e 1 \times \ln^n x \, dx$$

since getting an x from integrating the 1 will be no problem. If we do this, we are working in a very similar way to Example (4) in Section 9.B.(f). We will have

$$\frac{dv}{dx} = 1 \quad \text{so} \quad v = x \quad \text{and} \quad u = \ln^n x$$

so $\dfrac{du}{dx} = (n \ln^{n-1} x)\left(\dfrac{1}{x}\right) = \dfrac{n}{x} \ln^{n-1}x.$

Therefore

$$I_n = \left[x \ln^n x\right]_1^e - \int_1^e x \times \frac{n}{x} \ln^{n-1} x \, dx.$$

Now $\ln e = 1$ and $\ln 1 = 0$, so we have

$$I_n = e - n\, I_{n-1}.$$

This gives us the pattern of how the integrals run down from n. Notice that it is just single jumps this time, so we only need to know the value of I_0 to use it.

(c) We now find I_1, I_2 and I_3.

$$I_1 = e - I_0 = e - (e - 1) = 1$$
$$I_2 = e - 2I_1 = e - 2$$
$$I_3 = e - 3I_2 = e - 3(e - 2) = 6 - 2e$$

EXAMPLE (3) Find a reduction formula for

$$I_n = \int_0^{\pi/4} \tan^n x \, dx \text{ for } n \geq 2$$

and use it to evaluate I_2, I_3 and I_4.

This time we have to be a bit ingenious with our choice of split. Using $\int_0^{\pi/4} 1 \times \tan^n x \, dx$ will not work because it will give us an awkward x to deal with. (We want an answer involving tans.)

Using $\int_0^{\pi/4} \tan x \times \tan^{n-1} x \, dx$ doesn't look very good either since $\int \tan x \, dx = -\ln(\cos x)$ which will make things much more complicated.

So, can we do anything with

$$\int_0^{\pi/4} \tan^2 x \times \tan^{n-2} x \, dx?$$

$\int \tan^2 x \, dx$ is not something which we can immediately do, but we can rewrite this in a way which makes it much easier to handle by using the identity $\tan^2 x + 1 = \sec^2 x$.

Now we have

$$I_n = \int_0^{\pi/4} \tan^2 x \times \tan^{n-2} x \, dx$$

$$= \int_0^{\pi/4} (\sec^2 x - 1) \tan^{n-2} x \, dx$$

so

$$I_n = \int_0^{\pi/4} \sec^2 x \times \tan^{n-2} x \, dx - I_{n-2}.$$

The neatest method of going on from here is not to use integration by parts at all, but to use the substitution $t = \tan x$. This then gives us $dt/dx = \sec^2 x$ so we can replace $\sec^2 x \, dx$ by dt.

Also, if $x = \pi/4$ then $t = 1$ and if $x = 0$ then $t = 0$. This then gives us

$$\int_0^{\pi/4} \sec^2 x \times \tan^{n-2} x \, dx = \int_0^1 t^{n-2} \, dt$$

$$= \left[\frac{t^{n-1}}{n-1} \right]_0^1 = \frac{1}{n-1}$$

so $\quad I_n = \dfrac{1}{n-1} - I_{n-2}.$

We can now see why we had to say $n \geq 2$ at the beginning. We can't have $n = 1$.

(c) To answer this, we shall first need to find both I_0 and I_1 because again we have a rule which is going in double jumps.

$$I_0 = \int_0^{\pi/4} dx = \left[x \right]_0^{\pi/4} = \pi/4$$

$$I_1 = \int_0^{\pi/4} \tan x \, dx = \int_0^{\pi/4} \frac{\sin x}{\cos x} \, dx$$

$$= \left[-\ln \cos x \right]_0^{\pi/4} = -\ln\left(1/\sqrt{2}\right) = \ln \sqrt{2}.$$

Also, $I_2 = 1 - I_0 = 1 - \pi/4 \quad$ and $\quad I_3 = \frac{1}{2} - I_1 = \frac{1}{2} - \ln \sqrt{2}.$

and $\quad I_4 = \frac{1}{3} - I_2 = \frac{1}{3} - (1 - \pi/4) = \pi/4) = \pi/4 - \frac{2}{3}.$

(1) (a) Find a rule which relates $I_n = \int_0^{\pi/2} \cos^n x \, dx$ to I_{n-2}.

(b) Find I_0 and I_1 and then use these and the rule you have found to evaluate I_2, I_3 and I_4.

(c) Compare these answers with the corresponding answers for example (1). What do you notice? Draw sketches of I_1 and I_2 in each case to see how this happens.

(2) The similarities in the answers to question (1) and Example (1) apply to many sin and cos integrals over particular equal intervals. Such integrals are very important for finding Fourier series so I am asking you to find some of them here, since you have now met all the necessary methods. (Fourier series are used to describe various wave functions in terms of sin and cos functions. You were able to get an experimental idea of how this might be possible in Section 5.c.(c).)

Find the answers for the following four integrals if n and m are non-zero whole numbers.

(a) $\int_{-\pi}^{\pi} \sin nx \, dx$ (b) $\int_{-\pi}^{\pi} \cos nx \, dx$ (c) $\int_{-\pi}^{\pi} \sin^2 nx \, dx$

(d) $\int_{-\pi}^{\pi} \sin nx \cos mx \, dx$ if   (i) $m \neq n$ and   (ii) $m = n$.

Draw sketches showing the areas involved for the case when $n = m = 1$.

(3) If $I_n = \int \cosh^n x \, dx$, show that $n I_n = \cosh^{n-1} x \sinh x + (n-1) I_{n-2}$.

(4) If $I_n = \int \sec^n x \, dx$, show that, for $n \geq 2$, $(n-1) I_n = \sec^{n-2} x \tan x + (n-2) I_{n-2}$. Use this result to find I_6, and hence to show that $\int_0^{\pi/4} \sec^6 x \, dx = \frac{28}{15}$.

9.B.(h)   Using the $t = \tan (x/2)$ substitution

It is surprisingly hard to tell how difficult it will be to find an integral just by looking at it, for example, we have seen already that $\int \sin^2 x \cos x \, dx$ is easier to do than $\int \sin^2 x \, dx$.

Similarly, $I_1 = \int (1/\cos x) \, dx$ is considerably trickier to solve than $I_2 = \int (1/\cos^2 x) \, dx$. Can you recognise I_2, even though it's slightly disguised?

$$I_2 = \int \sec^2 x \, dx = \tan x + C \quad \text{because} \quad \frac{d}{dx} (\tan x) = \sec^2 x.$$

But how could we find I_1? You just might know what it is, because we found in question (5) of Exercise 8.C.5 that

$$\frac{d}{dx} (\ln (\sec x + \tan x)) = \sec x.$$

However, this is not the kind of answer which springs immediately to the memory.

How can we *show* that this is the correct answer, starting from I_1?

In order to do this, we let $t = \tan(x/2)$. To show you how nicely this improbable substitution works, I will start with a simpler example and then come back to I_1. My simpler example is

$$I = \int \frac{dx}{1 + \sin x + \cos x}.$$

First, we have to see how we can replace $\sin x$, $\cos x$ and dx in terms of t.

We already know that

$$\tan 2A = \frac{2 \tan A}{1 - \tan^2 A} \text{ from Section 5.D.(d).}$$

From this we can say

$$\tan x = \frac{2t}{1 - t^2}.$$

Also, we can say
$$\sin x = 2 \sin (x/2) \cos (x/2) = 2 \tan (x/2) \cos^2 (x/2)$$

$$= \frac{2 \tan (x/2)}{\sec^2 (x/2)} = \frac{2 \tan (x/2)}{1 + \tan^2 (x/2)}$$

so

$$\sin x = \frac{2t}{1 + t^2}$$

using the two identities $\sin 2A = 2 \sin A \cos A$ and $1 + \tan^2 A = \sec^2 A$.

Since $\tan x = \dfrac{\sin x}{\cos x}$, we also have

$$\cos x = \frac{1 - t^2}{1 + t^2}.$$

Finally, if $t = \tan (x/2)$ then

$$\frac{dt}{dx} = \tfrac{1}{2} \sec^2 (x/2) = \tfrac{1}{2} (1 + t^2)$$

again using the identity $1 + \tan^2 A = \sec^2 A$.

At this point, your courage may be beginning to fail, but we now have all the information we need to do any integral which needs a t substitution. I am putting it in a box for you.

What you need when you use the substitution $t = \tan (x/2)$

$$\sin x = \frac{2t}{1 + t^2}, \quad \cos x = \frac{1 - t^2}{1 + t^2}, \quad \tan x = \frac{2t}{1 - t^2} \quad \text{and} \quad dx = \frac{2dt}{1 + t^2}.$$

Now we see how it works, using

$$I = \int \frac{1}{1 + \sin x + \cos x} \, dx.$$

Substituting into this, using the information from the box above, gives

$$I = \int \left(\frac{1}{1 + \left(\frac{2t}{1 + t^2}\right) + \left(\frac{1 - t^2}{1 + t^2}\right)} \right) \left(\frac{2 \, dt}{1 + t^2} \right).$$

This looks far worse than what we started with, but multiplying the whole of the bottom of the first fraction by the $(1 + t^2)$ tidies it up amazingly to give

$$\int \frac{dt}{1 + t} = \ln |1 + t| + C = \ln |1 + \tan (x/2)| + C.$$

EXAMPLE (2) Now we return to $I_1 = \int \dfrac{1}{\cos x}\, dx$.

Substituting in a similar way to the previous example gives us

$$\int \frac{1}{\cos x}\, dx = \int \frac{1}{\frac{1-t^2}{1+t^2}} \left(\frac{2\, dt}{1+t^2} \right) = \int \frac{2\, dt}{1-t^2}.$$

The integral has now turned into something we know how to do.

We can use the method of partial fractions to give us

$$I_1 = \int \frac{dt}{1-t} + \int \frac{dt}{1+t} = -\ln|1-t| + \ln|1+t| + C = \ln \left| \frac{1+t}{1-t} \right| + C.$$

Putting back $\tan(x/2)$ for t gives us

$$I_1 = \ln \left| \frac{1 + \tan(x/2)}{1 - \tan(x/2)} \right| + C = \ln \left| \frac{\cos(x/2) + \sin(x/2)}{\cos(x/2) - \sin(x/2)} \right| + C.$$

We are not quite there yet! We want to go back from the half-angles to an answer in terms of the whole angle, x.

We do it like this:

$$\frac{\cos(x/2) + \sin(x/2)}{\cos(x/2) - \sin(x/2)} = \frac{(\cos(x/2) + \sin(x/2))(\cos(x/2) + \sin(x/2))}{(\cos(x/2) - \sin(x/2))(\cos(x/2) + \sin(x/2))}$$

$$= \frac{\cos^2(x/2) + 2\sin(x/2)\,(\cos(x/2) + \sin^2(x/2)}{\cos^2(x/2) - \sin^2(x/2)}$$

$$= \frac{1 + \sin x}{\cos x} = \sec x + \tan x$$

using the identities $\sin 2A = 2\sin A \cos A$ and $\cos 2A = \cos^2 A - \sin^2 A$.

This finally gives us

$$\int \frac{1}{\cos x}\, dx = \int \sec x\, dx = \ln|\sec x + \tan x| + C.$$

Because this result is useful but rather hideous to prove, it is often given on formula sheets, and I have put it in a box for you for this reason. It is by far the most complicated example of this type that you are likely to meet because of the rearrangement at the end. The three in the exercise below are easier.

EXERCISE 9.B.10

Try these for yourself now.

(1) $\displaystyle\int_0^{\pi/2} \frac{dx}{2 + 3\sin x + 2\cos x}$ (2) $\displaystyle\int_0^{\pi/2} \frac{dx}{1 + \cos x}$ (3) $\displaystyle\int_0^{2\pi/2} \frac{dx}{2 + \cos x + 2\sin x}$

Show that (1) is 0.305 to 3 d.p., (2) is 1 and (3) is ln ($\sqrt{3}$).

Differential equations are equations which include terms like dy/dx and d^2y/dx^2. We have already solved some simple examples in Exercise 9.A.1 and Exercise 9.A.2. All of these could be integrated straight away because they were given in terms of the variable with respect to which the differentiation had been done. (An example of this is $v = 3t^2 + 2$. Since $v = dx/dt$, we have $x = t^3 + 2t + C$.)

This section shows you how to deal with some examples which have mixed variables.

9.C.(a)   **Solving equations where we can split up the variables**

In all the cases which we will solve here, it will be possible to split up the variables so that we finish up with two integrals, each entirely in terms of its own variable. Here is an example of such a differential equation.

If $dy/dx = \dfrac{3-x}{y-2}$,

find the equation of the curve which has this gradient, given that $y = 5$ when $x = 7$.

We split up the diffential equation so that all the ys are together and all the xs are together, and then write both sides as integrals. Doing this gives us

$$\int (y-2)\, dy = \int (3-x)\, dx \quad \text{so} \quad \frac{y^2}{2} - 2y = 3x - \frac{x^2}{2} + C.$$

(We only need one constant of integration because this takes care of the combined constants on both sides.)

Now we have

$$x^2 + y^2 - 4y - 6x = 2C$$

and putting $y = 5$ and $x = 7$ gives $2C = 12$.

We now have

$$x^2 - 6x + y^2 - 4y = 12 \quad \text{or} \quad (x-3)^2 + (y-2)^2 = 25.$$

This is the same circle which we used as an example for implicit differentiation in Section 8.F.(a), so we know that we have found the right answer. In fact, we have just done the reverse process here to the one that we did there.

Students sometimes find the splitting up in this kind of equation a little bit difficult, so we will get some more practice in this before moving on to some physical examples.

EXAMPLE (1) $\tan x\, dy/dx = 1/y \sec^2 x$

Separating the variables and writing both sides of the equation as integrals, we get

$$\int y\, dy = \int \frac{\sec^2 x}{\tan x}\, dx.$$

Try doing the integration for each of these sides yourself. (The RHS is easier than it looks because $d/dx\,(\tan x) = \sec^2 x$.)

The LHS $= y^2/2$, and the RHS $= \ln |\tan x|$ so we have $\frac{1}{2} y^2 = \ln |\tan x| + C$.

Also, $y = 0$ when $x = \pi/4$, so $0 = 0 + C$ giving $C = 0$, and we finish up with the solution $y^2 = 2 \ln |\tan x|$.

EXAMPLE (2) $(x + 1)(y + 1) = xy \, dy/dx$

Separating the variables, and writing both sides of the equation as integrals, we get

$$\int \left(\frac{x + 1}{x} \right) dx = \int \left(\frac{y}{y + 1} \right) dy.$$

Both these fractions are top-heavy, so we tidy up and rewrite as

$$\int (1 + 1/x) \, dx = \int \frac{y + 1 - 1}{y + 1} \, dy = \int \left(1 - \frac{1}{y + 1} \right) dy$$

giving

$$x + \ln |x| = y - \ln |y + 1| + C \quad \text{so} \quad \ln |x(y + 1)| = y - x + C.$$

We can also write this as

$$x(y + 1) = e^{y - x + c} = e^c \, e^{y - x}.$$

If we now write the constant of e^c as A, we get $\quad x(y + 1) = Ae^{y - x}$.
We could also have got this answer by letting $C = \ln A$.

EXAMPLE (3) If $dy/dx = (2y + 3)(y + 2)$
(a) find x in terms of y, if $x = \ln (3/2)$ when $y = 0$,
(b) from this, find y in terms of x.

Separating the variables, and writing both sides of the equation as integrals, we get

$$\int dx = \int \frac{dy}{(2y + 3)(y + 2)} = \int \left(\frac{2}{2y + 3} - \frac{1}{y + 2} \right) dx$$

writing the second integral in partial fractions.
Integrating both sides gives

$$x = \ln |2y + 3| - \ln |y + 2| + C.$$

This can be tidied up better if we let $C = \ln A$. Then we have

$$x = \ln \left| A \left(\frac{2y + 3}{y + 2} \right) \right|.$$

Putting $x = \ln(\frac{3}{2})$ and $y = 0$ gives $A = 1$, so we have

$$x = \ln \left| \frac{2y + 3}{y + 2} \right| \quad \text{and this is the answer to (a).}$$

To find the answer to (b), we start by doing the reverse process to taking logs on both sides of the answer above. This is sometimes called antilogging. Doing this gives us

$$e^x = \left(\frac{2y + 3)}{y + 2} \right)$$

so $\quad ye^x + 2e^x = 2y + 3$

so $\quad y(e^x - 2) = 3 - 2e^x \quad$ so $\quad y = \dfrac{3 - 2e^x}{e^x - 2}.$

This is the answer to (b).

Try finding the general solutions to the following differential equations yourself now. Each of these must include a constant of integration. In each case, I then give you further information, as in Example (3) above, so that you can find the individual solution which fits that particular case. You can see the difference between these two kinds of solution in Section 9.A.(a).

Remember that you must do your rearranging so that each integral only involves one variable. (It is possible to do this for each of the questions given here.) Remember, too, that the dx or dy or whatever must be on the top of any fraction inside its integral. (I have sometimes seen desperate attempts at integration when it has been on the bottom.)

(1) $y(dy/dx) = \cos x$ with $y = 1$ when $x = \pi/6$
(2) $dx/dt = 3x$ with $x = 1$ when $t = 0$
(3) $x^2 (dy/dx) = y^2$ with $y = \frac{1}{2}$ when $x = 1$
(4) $y^2 (dy/dx) = x^2$ with $y = 2$ when $x = 1$
(5) $2xy(dy/dx) = y^2 + 1$ with $y = 0$ when $x = 1$
(6) $(x - 5) (dy/dx) = 2y$ with $y = 2$ when $x = 4$
(7) $\cot y (dy/dx) = x$ with $y = \pi/2$ when $x = \sqrt{2}$
(8) $rd\theta/dr = \cos^2 \theta$ with $r = 2$ when $\theta = 0$
(9) $e^{x+y} (dy/dx) = x$ with $y = 0$ when $x = 0$
(10) $dy/dx = (1 + y^2)(1 + x^2)$ with $y = 0$ when $x = 0$

9.C.(b)   **Putting flesh on the bones – some practical uses for differential equations**
There are many situations where we know the rate of change of a physical quantity but we need information about the quantity itself, as when we found velocity from acceleration, or distance travelled from a known but changing velocity. In this section, we will use the techniques of the earlier part of this chapter to work through some practical examples of differential equations together.

Radioactive decay
I mentioned this in Section 3.C.(f) as an example of a physical process which is described using e. We can now see how this happens.

The decay takes place because unstable nuclei of a radioactive element (called radionuclides), emit particles so that they are then converted into nuclei of another element; so that, for example, a certain type of uranium gets changed into lead. These particles don't all get emitted simultaneously; the physical law which governs this process says that, if there are N radionuclides present after a time t, then the rate of decay at this time is directly proportional to N.

Try writing down an equation for yourself which says this.

We can say that the rate of change of N with t is dN/dt.

It is directly proportional to N, so it is equal to kN where k is a constant. (See Section 3.A.(a) if necessary.)

The rate of change is negative because N is decreasing so, if we take k to be positive, we have $dN/dt = -k N$.

Physicists usually use, λ, the Greek letter l, called lambda, instead of k for this constant, so I shall do this too. λ is called the **disintegration constant**, and depends on the actual radioactive substance concerned.

Suppose that the number of radionuclides present when $t = 0$ is N_0. We know that both N and N_0 are positive because they represent physical quantities. We now have

$$dN/dt = -\lambda N \quad \text{and} \quad N = N_0 \text{ when } t = 0.$$

Try solving this equation yourself, giving both t in terms of N and N in terms of t.

We have

$$dN/dt = -\lambda N \quad \text{so} \quad \int \frac{dN}{N} = -\int \lambda dt \quad \text{so} \quad \ln N = -\lambda t + C.$$

Since $N = N_0$ when $t = 0$, we know that $C = \ln N_0$.

You can see here that, if we had left out the constant of integration, any later working would all be wrong. Also, because both N and N_0 are positive, we don't have to bother writing $\ln |N|$ and $\ln |N_0|$.

Now we have

$$\ln N = -\lambda t + \ln N_0 \quad \text{so} \quad \ln \left(\frac{N}{N_0} \right) = -\lambda t$$

using the second rule for logs from Section 3.C.(d).

Doing the reverse process to taking logs on this equation, we can say

$$\frac{N}{N_0} = e^{-\lambda t} \quad \text{so} \quad N = N_0 e^{-\lambda t}.$$

It is often very useful to know the time taken for half the radionuclides to have decayed. This will depend on the particular value of λ for the substance concerned. We have

$$\ln (N/N_0) = -\lambda t.$$

If $N = \frac{1}{2} N_0$ we have $\ln (\frac{1}{2}) = -\lambda t$, but

$$\ln (\tfrac{1}{2}) = \ln (1) - \ln (2) = -\ln (2)$$

so

$$-\ln 2 = -\lambda t \quad \text{so} \quad t = (\ln 2)/\lambda.$$

This special value of t is called τ, the Greek letter t, pronounced to rhyme with Ow.

It is called the **half-life** of the substance. We now know that

the half-life of the substance is given by $\tau = (\ln 2)/\lambda$.

Half-lives vary enormously. For example, ^{40}K, a radioactive form of potassium, has a half-life of 1.25×10^9 years. It decays into a form of argon, and the age of rocks can be calculated from the ratio of the amounts of these two substances in them. On the other hand, there is a radioactive form of iodine, ^{128}I, which has a half-life of 25 minutes. It is used medically as a tracer to measure the rate at which iodine is absorbed by the thyroid gland.

EXAMPLE (1) *A calculation based on radioactive decay.* There is a radioactive form of carbon which is produced at a constant rate in the upper atmosphere, and which then mixes with the ordinary carbon which is present in the air as carbon dioxide. All living plants and animals need air to keep them alive, so they are taking in a certain amount of this radioactive carbon all the while. Once they die, they no longer take in fresh supplies, and the radioactive carbon already present in them will decay. It has a half-life of 5730 years.

A human skeleton is dug up in a field. If the bones of this skeleton have 90% of the radioactive carbon which would be present in living bones, how long is it since the owner of the skeleton died?

Have a go at working this out for yourself.

We have

$$\tau = 5730 = (\ln 2)/\lambda \quad \text{so} \quad \lambda = \frac{\ln 2}{5730} \quad \text{and also} \quad N = 90\% \, N_0 = \tfrac{9}{10} N_0.$$

Using $\ln (N/N_0) = -\lambda t$ gives

$$\ln (0.9) = -\frac{\ln 2}{5730} t \quad \text{so} \quad t = 870 \text{ years to 2 s.f.}$$

Two chemical examples

EXAMPLE (1) *Using a first-order rate law.* If the rate of a chemical reaction can be described by the equation $-dc/dt = kc$ where k is a constant and c is the concentration of a solution in moles per litre, then this is called a first-order rate equation. (The rate of change of the concentration depends on the concentration at that time.)

If c_0 is the concentration when $t = 0$, show for yourself that $\ln (c_0/c) = kt$.

This uses exactly the same maths as the previous example although it is describing a different physical situation. We have

$$\int \frac{dc}{c} = -\int k \, dt \quad \text{so} \quad \ln c = -kt + C.$$

Now $c = c_0$ when $t = 0$, so $C = \ln c_0$, and

$$\ln (c_0/c) = kt. \tag{1}$$

9.C Solving some differential equations

If you have experimental pairs of results for c and t, you can see from this equation that plotting $\ln (c_0/c)$ against t will give you a straight line through the origin with a gradient of k. (I have described how this works in more detail in Section 3.D.)

If you are working using logs to base 10, you can use the rule from Section 8.B.(d) that $\ln a = (\ln 10)(\log a)$.

This gives us

$$\ln \left(\frac{c_0}{c} \right) = (\ln 10) \log \left(\frac{c_0}{c} \right) = 2.303 \log \left(\frac{c_0}{c} \right).$$

(Remember that 'log' means $\log_{10}$.)

This means that we can write equation (1) above as

$$2.303 \log \left(\frac{c_0}{c} \right) = kt \quad \text{or} \quad \log c = \frac{-kt}{2.303} + \log c_0$$

using the second law of logs.

From this, we can see that plotting $\log c$ against t will also give a straight line, with a gradient of $-k/2.303$, and an intercept on the vertical axis of $\log c_0$, which means that we can find k and c_0 for that particular reaction.

EXAMPLE (2) *Plotting results which are connected by Arrhenius' equation*. This is a mathematically similar equation which you might meet in chemistry or theory of materials courses. We do the same kind of thing with it to get a straight line graph.

Arrhenius' equation says that

$$k = Ae^{-E_a/RT} \quad \text{or} \quad k = A \exp \left(\frac{-E_a}{RT} \right)$$

using the alternative way of writing powers of e.

A is a constant called the frequency factor or Arrhenius constant, E_a is the activation energy, R is the gas constant, and T is the absolute temperature. (The meanings of these letters won't affect the maths of how we get the straight line, but of course they will be very important to you if you are a chemist or a materials engineer.)

We have

$$k/A = \exp \left(-\frac{E_a}{RT} \right) \quad \text{so} \quad \ln (k/A) = -\frac{E_a}{RT}$$

$$\text{or} \quad \ln k = \ln A - \frac{E_a}{RT}.$$

Comparing this with $y = mx + c$, as I showed you how to do in Section 3.D, you will see that, if you plot $\ln k$ against $1/T$, you will get a straight line which has a gradient of $-E_a/R$, and which cuts the vertical axis at $\ln A$. From these measurements, you can find both A and E_a.

As in the last example, you can work with logs to base 10 if you prefer. Your working equation then becomes

$$\log k = \log A - \frac{E_a}{2.303RT}.$$

Newton's Law of Cooling

This law states that the rate of cooling of a previously heated body is proportional to the *difference* between its temperature T after time t, and the temperature T_c of the surrounding environment. Try answering the following questions about this law.

(1) When will the heated body cool fastest?

(1) It cools fastest when it is first brought into the cooler environment. 'Come and eat it while it's hot!' makes good physical sense.

(2) Write down an equation for the rate of change of temperature with time.

(2) $dT/dt = -k(T - T_c)$ where k is a positive constant.

(3) If the temperature of the body is T_h when $t = 0$, solve this equation. (I've used h for 'hot' and c for 'cold' here.)

(3) Separating the variables, we have

$$\int \frac{dT}{T - T_c} = -k \int dt \quad \text{so} \quad \ln |T - T_c| = -kt + C.$$

Since $T = T_h$ when $t = 0$, we have $C = \ln |T_h - T_c|$. Also, while the object is cooling, we have $T_h > T > T_c$, so we can say

$$\ln(T - T_c) = -kt + \ln(T_h - T_c). \tag{1}$$

Using the second law of logs on equation (1) gives us

$$\ln \left(\frac{T - T_c}{T_h - T_c} \right) = -kt. \tag{2}$$

Antilogging, we can write this as

$$\frac{T - T_c}{T_h - T_c} = e^{-kt}. \tag{3}$$

Which one of these equations we find easiest to work with will depend on what we are looking for.

(4) We'll assume that Newton's Law of Cooling will apply to a cup of coffee. (This is a rather complicated physical object, so experimental verification of this law might be sensible in this case. The maths will only tell you the consequences of a physical law; it won't tell you how well it fits the actual physical circumstances.)

Suppose that the temperature of the coffee is 90 °C when it is brought into a room at 25 °C, and that after ten minutes it cools down to 60 °C.

Working in minutes and °C, find the value of k to 2 s.f.

(4) Using equation (2), we have

$$\ln \left(\frac{35}{65} \right) = -10k \quad \text{so} \quad k = 0.062 \text{ to 2 s.f.}$$

(5)   (a) Find its temperature after 5 minutes to the nearest degree.

       (b) Find its temperature after an hour. What do you notice? What happens after two hours?

       (c) How long will it take to cool down to 37 °C? (This is about blood heat.)

---

(5)   (a) When $t = 5$, we have $T - 25 = 65e^{-0.31}$ from equation (3), so $T = 73$. After 5 minutes, it has cooled to about 73 °C.

       (b) When $t = 60$, we have $T - 25 = 65e^{-3.72}$ so $T = 27$ to 2 s.f.

           The coffee is now very close to the temperature of the room. After two hours, it will be equal to the room temperature to 2 s.f. We can see mathematically that it gets closer and closer to 25 because $e^{-0.062t}$ gets closer and closer to zero as the value of t gets larger.

       (c) Using equation (2), when $T = 37$ we have $\ln\left(\frac{12}{65}\right) = -0.062t$ so $t = 27$. It takes about 27 minutes to cool to blood heat.

A leaking cone

An upside-down cone of height H and radius R is filled with water. I show a drawing of it in Figure 9.C.1.

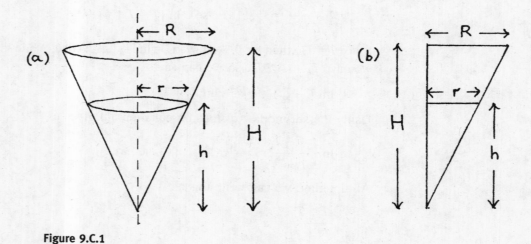

Figure 9.C.1

Suddenly, the tip of the cone springs a leak, and the water begins to run out. The rate at which the volume of water in the cone decreases is proportional to its depth.

We'll now work together to find out the rule which describes how the depth of the water in the cone changes.

If the volume of water left in the cone is V after a time of t, and its depth is h, write down an equation for dv/dt.

We will have $dv/dt = -kh$ where k is a positive constant.

At first sight, this equation looks as if it will work in the same kind of way as our previous examples so that it will give us a log. This is not so, however, because here we are talking about a rate of change of volume given in terms of depth. We have an equation with three variables, V, h and t. We also know that $V = \frac{1}{3}\pi r^2 h$.

If we are using the techniques described in this book, we can only move forwards now if we can find a way to describe V in terms of just the one variable h. (There was a very similar situation in Example (2) of Section 8.E.(d).)

Use Figure 9.C.1(b) above to find r in terms of R, H and h.

From this, what is V in terms of R, H and h?

Using what you now have, what is dV/dh?

We have $r/h = R/H$, because we have similar triangles, so $r = Rh/H$

Therefore,

$$V = \tfrac{1}{3}\pi r^2\, h = \tfrac{1}{3}\pi \left(\frac{R}{H}\right)^2 h^3 \quad \text{so} \quad \frac{dV}{dh} = \pi \left(\frac{R}{H}\right)^2 h^2.$$

We can now use the Chain Rule to say that $dV/dt = (dV/dh)\,(dh/dt)$. Doing this, what is dh/dt?

We have

$$\frac{dV}{dt} = -kh = \pi \left(\frac{R}{H}\right)^2 h^2 \frac{dh}{dt} \quad \text{so} \quad \frac{dh}{dt} = -\frac{1}{h}\left(\frac{kH^2}{\pi R^2}\right).$$

The quantity in the bracket is itself a constant because it is made up of constant quantities.

Now separate the variables so that you can integrate.

See what you get when you do this, remembering that $h = H$ when $t = 0$.

You should have

$$\int h \, dh = -\left(\frac{kH^2}{\pi R^2}\right) \int dt \quad \text{so} \quad \tfrac{1}{2} h^2 = -\left(\frac{kH^2}{\pi R^2}\right) t + C$$

and, since $h = H$ when $t = 0$, we have $C = \tfrac{1}{2} H^2$. This gives us

$$\left(\frac{kH^2}{\pi R^2}\right) t = \tfrac{1}{2} (H^2 - h^2). \tag{1}$$

If the time taken for the cone to empty completely is T, show that $k = \dfrac{\pi R^2}{2T}$.

What does equation (1) now become if you substitute this instead of k?

$$t = T \quad \text{when} \quad h = 0 \quad \text{so} \quad \left(\frac{kH^2}{\pi R^2}\right) T = \tfrac{1}{2} H^2 \quad \text{so} \quad k = \frac{\pi R^2}{2T}.$$

Replacing k by this in equation (1) gives

$$\left(\frac{\pi R^2}{2T}\right)\left(\frac{H^2}{\pi R^2}\right) t = \frac{H^2 t}{2T} = \tfrac{1}{2} (H^2 - h^2). \tag{2}$$

9.C Solving some differential equations

Now answer the following two questions, using equation (2).

(a)   How long does it take for the water to get to a depth of $\frac{1}{2} H$?

(b)   How deep is the water after half the time which it takes for all the water to leak out of the cone?

(a)   When $h = \frac{1}{2} H$, we have $\quad \dfrac{H^2 t}{2T} = \frac{1}{2} \left(\frac{3}{4} H^2 \right) \quad$ so $\quad t = \frac{3}{4} T.$

(b)   When $t = \frac{1}{2} T$, we have $\quad \dfrac{H^2}{4} = \frac{1}{2} (H^2 - h^2) \quad$ so $\quad h^2 = \frac{1}{2} H^2 \quad$ so $\quad h = \dfrac{H}{\sqrt{2}}.$

EXERCISE 9.C.2

(1) Here are two examples of second-order rate equations from chemistry.

    (a) $- dc/dt = kc^2$ (b) $dx/dt = k (a - x)(b - x)$

(A second-order rate equation is one in which the rate of the reaction is proportional to the square of the concentration of one of the reagents, or to the product of the concentrations of two species of reagents.)

For (a), show that, if c stands for the varying concentration of the reagent, and $c = c_0$ when $t = 0$, then $1/c - 1/c_0 = kt$.

For (b), show that, if $a - x$ and $b - x$ are the concentrations of reagents A and B after time t, with a and b being the concentrations of A and B when $t = 0$, then

$$\frac{1}{a - b} \ln \left(\frac{b(a - x)}{a(b - x)} \right) = kt.$$

(2) A particle moves in a straight line so that its distance from a fixed point after a time t is s. Its velocity after this time is v. If $s = 0$ and $v = u$ when $t = 0$, and the particle has a constant acceleration a, show that the motion of the particle can be described by the following four equations.

    (a) $v = u + at$ (b) $s = ut + \frac{1}{2} at^2$

    (c) $s = \left(\dfrac{u + v}{2} \right) t$ (d) $v^2 = u^2 + 2as.$

(3) A particle is falling under gravity but with a drag force acting on it resisting its downwards motion. This drag force is proportional to its velocity, so that its acceleration at time t is given by $g - kv$.

If the particle starts from rest when $t = 0$, show that the effect of the drag force is to stop it falling faster and faster, and that its velocity gets closer and closer to the constant value of g/k. This is called its **terminal velocity**.

(4) A second particle which is falling from rest under gravity also has a drag force acting on it, but this time the force is proportional to the square of its velocity. If its velocity is v after a time t, and its acceleration is given by $dv/dt = g - kv^2$, show that x, the distance travelled after time t, is given by the equation

$$x = \frac{1}{2k} \ln \left(\frac{g}{g - kv^2} \right).$$

To show this, you will need to use the Chain Rule in the form

$$\frac{dv}{dt} = \frac{dv}{dx}\frac{dx}{dt} = v\frac{dv}{dx} \quad \text{so} \quad v\frac{dv}{dx} = g - kv^2.$$

Now rearrange the answer you have already got which gives x in terms of v so that it gives v in terms of x instead. Then use this rearranged form to show that, as x increases, v will become closer and closer to the constant value of $\sqrt{g/k}$.

9.C.(c) **A forwards look at some other kinds of differential equation, including ones which describe SHM**

We have been able to solve all the differential equations which we have looked at so far in this section by separating the variables into two integrals, one for each variable. This is the only kind of differential equation which I show you how to do in detail in this book, but we can look at a few special cases of other types here. This is because we have already met them working the other way round, and also because they are so important physically that I think it will help you to have met some examples informally when you come to study their solutions in your courses.

We'll go back first to Section 8.A.(e) where we showed that if $x = \cos t$ then $d^2x/dt^2 = -x$. As we said there, this is an example of a differential equation which describes SHM. This means that it describes the motion of a particle on a straight line whose acceleration is always towards a fixed point, and proportional to the distance of the particle from this point. SHM is of enormous importance to physicists and engineers, because it can be used to describe the effect of combinations of small vibrations.

In Section 8.A.(e), we couldn't yet differentiate more complicated sin and cos functions.

Now that this is no problem, find dx/dt and d^2x/dt^2 for the following functions and write down an equation relating x and d^2x/dt^2 in each case.

(a) $x = 3 \cos t$ (b) $x = \cos 2t$ (c) $x = \cos (t + \pi/6)$

(d) $x = 2 \cos (t - \pi/6)$ (e) $x = 5 \sin (3t + \pi/6)$ (f) $x = A \cos (\omega t + \varepsilon)$

You should find that each of the given equations for x is a solution of the SHM differential equation of $d^2x/dt^2 = -k^2x$ where k is some positive number.

Your working should go as follows:

(a) $dx/dt = -3 \sin t$ and $d^2x/dt^2 = -3 \cos t$ so $d^2x/dt^2 = -x$.

(b) $dx/dt = -2 \sin 2t$ and $d^2x/dt^2 = -4 \cos 2t$ so $d^2x/dt^2 = -4x$.

(c) $dx/dt = -\sin (t + \pi/6)$ and $d^2x/dt^2 = -\cos (t + \pi/6)$ so $d^2x/dt^2 = -x$.

(d) $dx/dt = -2 \sin (t - \pi/6)$ and $d^2x/dt^2 = -2 \cos (t - \pi/6)$ so $d^2x/dt^2 = -x$.

(e) $dx/dt = 15 \cos (3t + \pi/6)$ and $d^2x/dt^2 = -45 \sin (3t + \pi/6)$ so $d^2x/dt^2 = -9x$.

(f) $dx/dt = -A\omega \sin (\omega t + \varepsilon)$ and $d^2x/dt^2 = -A\omega^2 \cos (\omega t + \varepsilon)$ so $d^2x/dt^2 = -\omega^2 x$.

These examples all come from Section 5.C.(b). If you go back there, you will find drawings of (a), (b) and (c), and you can see the distance x marked on them. This will help you to picture how this distance will change with time, and what it is when $t = 0$.

Examples (d) and (e) are questions (7) and (8) from Exercise 5.C.2. Again, you can use your drawings to help you to picture what is happening physically here.

Example (f) above gives you a general form of the SHM equation, using the letters which you will commonly find in scientific and engineering applications. A and ω are explained in Sections 5.C.(a) and (b), and ε, which gives the starting position when $t = 0$, is the Greek letter e which is called epsilon. You can see from Sections 5.D.(f) and (g) exactly how the ε works, and also that $A \cos(\omega t + \varepsilon)$ can be written in the form of two separate sin and cos functions.

It would also be possible to use the form $A \sin(\omega t + \varepsilon)$ if you wish. As an example of this, we showed in Section 5.D.(g) that $3 \sin 2t + \cos 2t$ can also be written as $\sqrt{10} \sin(2t + \alpha)$, with $\alpha = 0.32$ radians to 2 d.p.

You can see a sketch of the graph of $x = \sqrt{10} \sin(2t + \alpha)$ in Figure 5.D.9(b), and the sketch showing the motion of P on its circle giving the vertical distance x in Figure 5.D.9(a). x could represent the displacement from O of a particle moving in SHM as t changes. In this example, $A = \sqrt{10}$, $\omega = 2$ and $\varepsilon = \alpha = 0.32$.

If there are other forces acting as well as a force which on its own would produce SHM, more complicated equations are needed to describe the motion. Questions (9) and (10) in Exercise 8.C.4 show you examples of such equations, and they also give you some idea of how we could set about solving them by using powers of e found from the roots of quadratic equations.

Have another look at these two questions now and then see if you can solve the differential equation $d^2x/dt^2 - 5dx/dt + 6 = 0$ by putting $x = e^{kt}$.

Finding d^2x/dt^2 and dx/dt and substituting these into the equation, you should find that $e^{kt}(k^2 - 5k + 6) = 0$ so $k^2 - 5k + 6 = 0$, since e^{kt} is never equal to zero.

This means that $k = 2$ or $k = 3$.

Making use of both these possibilities, and writing the solution in the most general way, gives $x = Ae^{2t} + Be^{3t}$ where A and B are two constants of integration.

Now, if we also know in this particular case that $x = 0$ when $t = 0$, and that $dx/dt = 2$ when $t = 0$, see if you can find out what the values of A and B are.

Putting $x = 0$ and $t = 0$ in $x = Ae^{2t} + Be^{3t}$ gives $0 = A + B$.

Now, $dx/dt = 2Ae^{2t} + 3Be^{3t}$, so if $dx/dt = 2$ when $t = 0$ we also know that $2 = 2A + 3B$.

Solving these two equations, we find that $B = 2$ and $A = -2$. Therefore the solution of the differential equation is $x = 2e^{3t} - 2e^{2t}$.

It seems a little curious that this answer appears to come out so differently to the solutions of the SHM equations from the beginning of this section, since these also involved $d^2 x/dt^2$ and x. You would expect that the method of putting $x = e^{kt}$ which we have just used should also work for these SHM equations, and yet they all had solutions for x which involved a sin or a cos instead of powers of e. What actually happens if you put $x = e^{kt}$ in the equation $d^2x/dt^2 = -9x$, for example?

We shall find out in Section 10.C.(b) that sin, cos and e are not as unrelated as they might seem. I plan to describe the methods of solution of differential equations like the ones above in much more detail in another book leading on from this one, but I shall show you how to solve just one differential equation describing SHM in Section 10.C.(i) because its solution connects together so many of the ideas in this book.

Physically, differential equations of this type are very important. Two examples of situations which can be described mathematically by such equations are the possible fate of a suspension bridge if a platoon of soldiers marches across it in step, and the combined effect of the suspension of a car, its dampers and the bumps in the road on its vertical motion as it travels along.

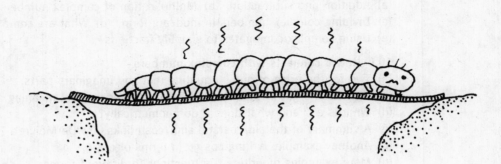

10 Complex numbers

In this final chapter we powerfully extend the possibilities of what we can do with our number system. This extension leads us to simpler ways of finding mathematical rules. It also has many very important physical applications.

The chapter is divided into the following sections.

10.A A new sort of number
(a) Finding the missing roots, (b) Finding roots for *all* quadratic equations,
(c) Modulus and argument (or mod and arg for short)

10.B Doing arithmetic with complex numbers
(a) Addition and subtraction, (b) Multiplication of complex numbers,
(c) Dividing complex numbers in mod/arg form, (d) What are complex conjugates?
(e) Using complex conjugates to simplify fractions

10.C How *e* connects with complex numbers
(a) Two for the price of one – equating real and imaginary parts,
(b) How does *e* get involved? (c) What is the geometrical meaning of $z = e^{j\theta}$?
(d) What is $e^{-j\theta}$ and what does it do geometrically?
(e) A summary of the sin/cos and sinh/cosh links, (f) De Moivre's Theorem,
(g) Another example: writing $\cos 5\theta$ in terms of $\cos \theta$,
(h) More examples of writing trig functions in different forms,
(i) Solving a differential equation which describes SHM,
(j) A first look at how we can use complex numbers to describe electric circuits

10.D Using complex numbers to solve more equations
(a) Finding the *n* roots of $z^n = a + bj$,
(b) Solving quadratic equations with complex coefficients,
(c) Solving cubic and quartic equations with complex roots

10.E Finding where *z* can be if it must fit particular rules
(a) Some simple examples of paths or regions where *z* must lie,
(b) What do we do if *z* has been shifted? (c) Using algebra to find where *z* can be,
(d) Another example involving a relationship between *w* and *z*

10.A A new sort of number

10.A.(a) Finding the missing roots

So far, we have been able to show all of the answers or roots of the equations which we have been able to solve as points on the horizontal axis of a graph. If we write the equation as $f(x) = 0$, then its solutions are where the curve of $y = f(x)$ cuts the *x*-axis, so that $y = 0$.

These solutions or roots have included different kinds of numbers. For example:

(a) $2x^2 - x - 6 = 0$ has the two roots $x = 2$ or $x = -\frac{3}{2}$ since
      $2x^2 - x - 6 = (x - 2)(2x + 3)$.

(b) $x^2 - 4x - 1 = 0$ has the two roots $x = 2 \pm \sqrt{5}$, using the quadratic formula from
      Section 2.D.(d).

(c) $2x^3 - x^2 - 5x - 2 = 0$ has the three roots $x = -\frac{1}{2}$, $x = 2$ and $x = -1$
      since $2x^3 - x^2 - 5x - 2 = (2x + 1)(x - 2)(x + 1)$.

(d) $4x^4 - 5x^2 + 1 = 0$ has the four roots of $x = 1$, $x = -1$, $x = \frac{1}{2}$ and $x = -\frac{1}{2}$
      since $4x^4 - 5x^2 + 1 = (x - 1)(x + 1)(2x - 1)(2x + 1)$.

In order to have all these solutions, we have had to extend the system of counting numbers 1, 2, 3, 4, . . . to include negative numbers, fractions, and numbers like $\sqrt{5}$ which can't be written as fractions. I described all these different kinds of numbers in Section 1.E. Together, they make up what are known as the **real numbers**. We can think of the horizontal axis of a graph as a number line which contains all the solutions to equations like (a), (b), (c) and (d) above.

But we have also found that sometimes equations which *look* as though they ought to have solutions, such as $x^2 - 2x + 2 = 0$, have no solution given by any number which we can find on the x-axis. It appears that sometimes the number of solutions tallies with the highest power of x, as in the four examples of (a), (b), (c) and (d) above, but sometimes this doesn't seem to work. This peculiar situation is shown particularly clearly by what seem at first sight to be the simplest possible equations of this kind.

We know that the equation $x^2 = 1$ has the two solutions or roots of $x = +1$ or $x = -1$.

However, the equation $x^3 = 1$, unlike (c) above with its three solutions, only has the one solution of $x = 1$.

The equation $x^4 = 1$ again has the two solutions of $x = +1$ or $x = -1$.

With $x^5 = 1$, we are back to just the one solution of $x = 1$.

As we take higher powers of x, this pattern will continue, with just one solution of $x = +1$ for odd powers, and the two solutions of $x = +1$ or $x = -1$ for even powers.

This is somehow not very satisfying; would it not feel more correct if $x^2 = 1$ had two roots, $x^3 = 1$ had three roots, $x^4 = 1$ had four roots and so on? But where would they be? How could we widen our number system so that we would have these extra roots or solutions?

Suppose we take the horizontal axis out of the graph paper, and lay it out separately as a number line which contains all the roots of the equations which we can so far solve, including equations (a), (b), (c) and (d) above. This number line, which I have drawn in Figure 10.A.1, shows all the *real* numbers. (The arrows are there to show that this line can be infinitely extended in either direction.)

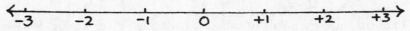

Figure 10.A.1

Now imagine that we are looking down on this number line and seeing all these roots. We are also seeing the various roots of $x = +1$ and $x = -1$ for the equations $x^n = 1$. (We see both +1 and −1 if n is even, and just +1 if n is odd.)

If we take the particular case of $x^4 = 1$, we have $x = +1$ and $x = -1$ as two of its roots. If we want this equation to have four roots, where could we think of the other two roots as being?

Suppose we think that up to now we have just seen the possible answers through a slit which allows us to see the number line of Figure 10.A.1. Sometimes this has meant that we could see *all* the possible solutions to an equation and sometimes it has meant that there are solutions which are somehow off to the side so that they are hidden from us. If so, it seems reasonable that the four roots of $x^4 = 1$ should be symmetrically placed. This would give us the four roots (a) (b) (c) and (d) which I have shown in Figure 10.A.2(a).

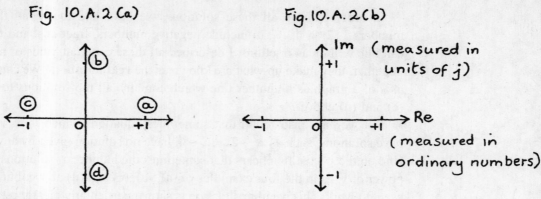

Figure 10.A.2

It seems a good idea that the two roots on the vertical axis should also be placed one unit away from O. But what are they?

They would have to be two different numbers, each of which multiplied by itself four times would give the answer of $+1$.

We know that $(-1)^2 = +1$, so if we can somehow think of the vertical axis as showing units of $\sqrt{-1}$, as the horizontal axis shows units of $\sqrt{+1}$, we shall be able to have the four roots which we would like.

Each root is one unit away from the origin, so root (b) would be $\sqrt{-1}$ and root (d) would be $-\sqrt{-1}$.

If we now let ourselves say that $\sqrt{-1} \times \sqrt{-1} = -1$, the two extra roots will work in the following way.

For (b), we shall have

$$(\sqrt{-1})^4 = (\sqrt{-1}) \times (\sqrt{-1}) \times (\sqrt{-1}) \times (\sqrt{-1}) = (-1) \times (-1) = +1.$$

For (d), we shall have

$$(-\sqrt{-1})^4 = (-\sqrt{-1}) \times (-\sqrt{-1}) \times (-\sqrt{-1}) \times (-\sqrt{-1}) = +1$$

like (b), since each pair of minuses multiplied gives a plus.

To emphasise that we have invented a new number here, which needs a separate direction of its own to show it, we shall call $\sqrt{-1}$ by the letter j.

We are defining a new number j such that $j^2 = -1$ so $j = \sqrt{-1}$.

We can now write the four roots of $x^4 = 1$ as $x = +1$, $x = -1$, $x = j$ and $x = -j$ with each of these roots being one unit away from the origin. I show them in Figure 10.A.2(b).

The horizontal axis shows the real numbers (just like any ordinary x-axis) so it is called the **real axis** and labelled **Re** instead of x.

The numbers shown on the vertical axis, which are measured in units of j, are called **imaginary numbers.** (This curious name is for historical reasons.) We therefore call the vertical axis the **imaginary axis** and label it **Im** instead of y.

Figure 10.A.2(b) is an example of what is called an **Argand diagram.** It is named after the Swiss mathematician who first thought of showing complex numbers in this way.

NOTE
Mathematicians often use i rather than j for $\sqrt{-1}$. However, physicists and engineers usually use j, because imaginary numbers have important physical applications in the study of electric circuits where the letter i is often used for current.

THINKING POINT

Can you draw a sketch showing where you think the eight roots of $x^8 = 1$ might be? We shall come back to this at the end of Section 10.B.(b).

10.A.(b)   Finding roots for *all* quadratic equations

Next we shall find out whether having this new number j will make it possible for us to find solutions for all quadratic equations.

Suppose we take as an example the equation $x^2 - 8x + 25 = 0$.

It is usual to use z instead of x if we are extending the possibilities for roots by using these new numbers, so I shall rewrite this equation as $z^2 - 8z + 25 = 0$.

Now we see what happens if we use the quadratic formula. Using z, this formula says

$$\text{If} \quad az^2 + bz + c = 0 \quad \text{then} \quad z = \frac{-b \pm \sqrt{b^2 - 4ac}}{2a}. \qquad \text{(Section 2.D.(d))}$$

Here, we get

$$z = \frac{+8 \pm \sqrt{64 - 100}}{2} = \frac{+8 \pm \sqrt{-36}}{2} = \frac{+8 \pm 6\sqrt{-1}}{2}$$

so

$$z = +4 \pm 3\sqrt{-1} \quad \text{giving} \quad z_1 = 4 + 3j \quad \text{and} \quad z_2 = 4 - 3j$$

using our new number j for $\sqrt{-1}$, and calling the roots z_1 and z_2.

(Try checking for yourself whether you think each of these two roots *do* fit the equation by substituting each of them back into the equation in turn and seeing what happens.)

Notice that the two roots which we now have for this equation are each made up of two parts, the $+4$ which is a real number, and the $3j$ which is then either added or subtracted. We can show these two roots on the same sort of diagram that we used in Figure 10.A.2(b), with the real parts lying along the horizontal or real axis, and the imaginary parts lying along the vertical or imaginary axis. I have drawn this particular pair of roots in Figure 10.A.3.

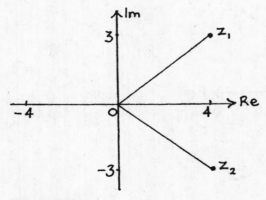

Figure 10.A.3

Notice that they have the property that they are symmetrically placed either side of the real axis. We shall look at the special properties of pairs like this in Section 10.B.(c).

I shall now use the number $4 - 3j$ shown in Figure 10.A.3 to give you some definitions.

A number like $4 - 3j$ is called a **complex number**.

4 is called its **real part** and -3 is called its **imaginary part**.

The number 4 gives the measurement along the real axis.
The number -3 gives the measurement along the imaginary axis.

If $z = 4 - 3j$ then we say Re $(z) = 4$ and Im $(z) = -3$.

(Notice that the imaginary part tells us how many units of j we have; it does not actually include the j.)

EXERCISE 10.A.1

Solve the following equations writing your answers in the same way that we used in the example above. (This is called writing them in the form $a \pm bj$.)

Show each pair of roots on a separate Argand diagram. Save these answers and sketches as you will need them for the next exercise.

(If you feel shaky about $\sqrt{36}$ being the same as 6, or $\sqrt{12}$ being the same as $2\sqrt{3}$, you should read through Section 1.F.(c) and then do Exercise 1.F.3 before continuing.)

(1) $z^2 - 2z + 2 = 0$ (2) $z^2 - 4z + 13 = 0$ (3) $z^2 + 4z + 5 = 0$ (4) $z^2 + 2z + 6 = 0$

10.A.(c)   Modulus and argument (or mod and arg for short)

In the last section, we found that the two roots of the equation $z^2 - 8z + 25 = 0$ are $z_1 = 4 + 3j$ and $z_2 = 4 - 3j$ and I showed them on the Argand diagram of Figure 10.A.3. In Figure 10.A.4, I show an alternative way of describing the positions of z_1 and z_2 on an Argand diagram. Instead of using the pairs of coordinates given by their real and imaginary parts, we can use two other measurements.

The first of these is their lengths which I have labelled $|z_1|$ and $|z_2|$ on the diagram.

The length of a complex number is called its **modulus** or **mod**, and it is written as mod z or $|z|$.

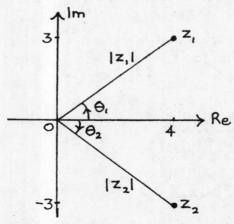

Figure 10.A.4

From Figure 10.A.4 we see that $|z_1| = \sqrt{4^2 + 3^2} = 5$ by Pythagoras' Theorem.
Also $|z_2| = |z_1|$, since $|z_2| = \sqrt{4^2 + (-3)^2} = 5$.
For any complex number $z = x + yj$ we have the following definition.

> The **modulus** or length of the complex number $z = x + yj$ is given by $|z| = \sqrt{x^2 + y^2}$.
>
> Because $|z|$ is a length, it is always positive.

If we want to talk about more than one modulus, we call them moduli and not moduluses. (This is because 'modulus' is a Latin word.)

To draw a complex number z on an Argand diagram it will not be enough just to know its length. We can see this in the example above where both z_1 and z_2 have the same length. In order to describe them fully, we also need to know the direction in which we should draw them. This direction can be described by using the angles turned through from the positive real axis to get to each of them. I have called these two angles θ_1 and θ_2 on Figure 10.A.4. In this particular example, these two angles are equal in size, but are turning in opposite directions from the positive real axis.

> The angle turned through from the positive real axis to give a complex number, z, is called its **argument**.
> It is written as arg z.
> An anticlockwise turn gives a positive angle.
> A clockwise turn gives a negative angle.

Here,
$$\arg z_1 = \theta_1 = \tan^{-1}\left(\tfrac{3}{4}\right) = 0.644 \text{ radians to 3 d.p.}$$

and $$\arg z_2 = \theta_2 = -\tan^{-1}\left(\tfrac{3}{4}\right) = -0.644 \text{ radians to 3 d.p.}$$

Writing this pair of angles in this way shows their symmetry very nicely. It is true that we could also get to the same position of z_2 by turning the other way. This would give an angle $\theta_2 = 2\pi - \tan^{-1}\left(\tfrac{3}{4}\right) = 5.640$ radians to 3 d.p. but this hides the symmetry of θ_1 and θ_2, coming from the symmetrical pair of $4 \pm 3j$.

For this reason we give this further definition.

> The **principal value of the argument** of a complex number z lies between $-\pi$ and π, so that $-\pi < \theta \leq \pi$.

Therefore, the values I have given above for θ_1 and θ_2 are the principal values for the arguments of z_1 and z_2.

It is rather easy to make mistakes when finding the arguments of complex numbers if you use a calculator to find $\tan^{-1}\theta$. The following example shows why this is.

We will find the mod and arg of the pair of solutions to the equation $z^2 + 2z + 4 = 0$. Using the formula gives

$$z = \frac{-2 \pm \sqrt{-12}}{2} = \frac{-2 \pm 2\sqrt{-3}}{2} = -1 \pm \sqrt{3}j.$$

I show the pair of roots $z_1 = -1 + \sqrt{3}j$ and $z_2 = -1 - \sqrt{3}j$ in Figure 10.A.5.

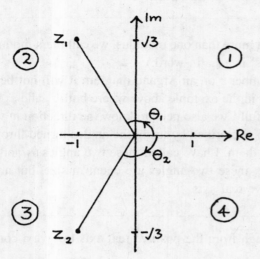

Figure 10.A.5

We have

$$|z_1| = \sqrt{(-1)^2 + (\sqrt{3})^2} = \sqrt{4} = 2 \quad \text{and} \quad |z_2| = \sqrt{(-1)^2 + (-\sqrt{3})^2} = 2.$$

We can also see from the diagram that

$$\arg z_1 = \theta_1 = \pi - \tan^{-1}(\sqrt{3}) = \pi - \pi/3 = 2\pi/3,$$

and

$$\arg z_2 = \theta_2 = -\pi + \tan^{-1}(\sqrt{3}) = -\pi + \pi/3 = -2\pi/3.$$

But if you had used $\tan^{-1}(\sqrt{3}/-1) = \tan^{-1}(-\sqrt{3})$ to find θ_1 you would have got $-\pi/3$ which is an angle in the fourth quadrant, and so not the right place at all. To remind you where each quadrant is, I have labelled them in the diagram above as (1), (2), (3) and (4).

Similarly, if you had said $\theta_2 = \tan^{-1}(-\sqrt{3}/-1) = \tan^{-1}\sqrt{3}$, you would have got the angle of $\pi/3$ in the first quadrant which is again the wrong place. The problem is that the function $\tan^{-1}$ is only defined for the range from $-\pi/2$ to $+\pi/2$. (We saw why this is in Section 5.A.(i).)

In this particular example, you could of course find θ_2 from its symmetry with θ_1.

You should always draw a little sketch of what is happening when finding arguments so that you don't get them in the wrong place.

(To help you to check your answers, remember that there are roughly 3 radians in a half-turn of π, and one radian is roughly 60°.)

A shorthand version which is often used to write a complex number z in mod/arg form, is to write z as $[r, \theta]$ where $r = |z|$ and $\theta = \arg z$. In the example above, we would have $z_1 = [2, 2\pi/3]$ and $z_2 = [2, -2\pi/3]$.

Another shorthand version, which is often used by engineers, is to write $r \angle \theta$ instead of $[r, \theta]$.

> **NOTE**
>
> r and θ can be used instead of x and y in *any* graph. If the graph is not an Argand diagram, then θ is measured between 0 and 2π. r and θ are called **polar coordinates** and x and y are called **Cartesian coordinates**. This name is after the French mathematician and philosopher, Réné Descartes, who is also famous for having said 'Cogito ergo sum' or 'I think, therefore I am'.

What is the connection between $z = x + yj$ and $z = [r, \theta]$?
I show this in Figure 10.A.6.

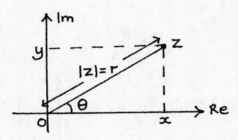

Figure 10.A.6

From this diagram, we see the following pair of results.

> $x = |z| \cos \theta = r \cos \theta$ and $y = |z| \sin \theta = r \sin \theta$.
>
> So $z = x + yj = r (\cos \theta + j \sin \theta)$.

Looking back, you will see that these relationships are true for all the diagrams wherever z is. The size of θ automatically takes care of the signs of x and y. We are making use here of the definitions for the sin and the cos of angles greater than $\pi/2$. These came from the turn of a unit length about the origin (in Section 5.A.(c)), so they are very much related to what we are doing now.

EXERCISE 10.A.2

In each of the following questions, give the arguments in radians correct to 2 d.p. unless they are an exact fraction of π.
(1) Find the modulus and argument of each of the following complex numbers.
  (a) 3   (b) −2   (c) 2j   (d) −5j   (e) 12j
  (f) −5 − 12j   (g) 7 − 24j   (h) −7 + 24j   (i) − $\sqrt{3}$ + j   (j) −1 − j
(2) Figure 10.A.2(b) in Section 10.A.(a) shows the four roots of $z^4 = 1$.
  What is the modulus and argument of each of these?
(3) Find the modulus and argument for each of the pairs of roots of the quadratic equations which you solved in Exercise 10.A.1. Use your sketches of these roots to make sure that your arguments are in the right place.

When we do arithmetic with these numbers, we assume that each separate part of them, the real and the imaginary, behaves within itself according to the usual rules for numbers. The only extra property is that these two tracks of numbers are not forever running beside each other with no communication; every time we get a j^2 we get a cross-over from the imaginary to the real. In fact, any even power of j will do this. This makes their possibilities much more interesting.

10.B.(a)   Addition and subtraction

Suppose we have two complex numbers $z_1 = 3 + j$ and $z_2 = 2 + 5j$.

Since each number is made up of two separate independent measurements, we add the numbers by adding each of these pairs, giving us $(3 + j) + (2 + 5j) = 5 + 6j$.

The order of addition will not matter because $(2 + 5j) + (3 + j) = 5 + 6j$ also.

We can show this addition on an Argand diagram if we allow ourselves to shift the second number. (See Figure 10.B.1(a).)

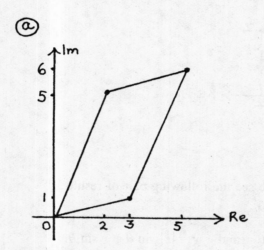
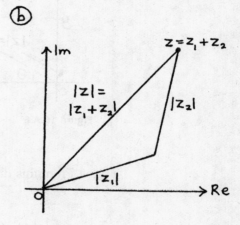

Figure 10.B.1

The two separate displacements of $(2 + 5j)$ and $(3 + j)$ add together to give the single displacement of $(5 + 6j)$. We see that the final result is the same whichever order we do the addition in – we just get there by a different route. These two different routes put together make a parallelogram.

> To add two complex numbers together, we add the real parts and the imaginary parts separately.
> For example, $(3 - 2j) + (5 + 4j) = 8 + 2j$.

Subtracting complex numbers works equally easily.

> To subtract two complex numbers we subtract the real parts and the imaginary parts separately.
> For example, $(4 + 3j) - (1 - 2j) = 3 + 5j$.

If we have any three complex numbers z, z_1 and z_2 so that $z = z_1 + z_2$, and we draw them on an Argand diagram, we can see a useful relationship between their lengths. I show this in Figure 10.B.1(b) above.

Since the length of the third side of a triangle must be shorter than the lengths of the other two sides added together (unless the triangle is squashed completely flat, when side (3) = side (1) + side (2)), we have the following result.

If $z = z_1 + z_2$ then $|z| \leq |z_1| + |z_2|$, that is, $|z_1 + z_2| \leq |z_1| + |z_2|$.

We could show the addition of any quantity of complex numbers on an Argand diagram. The resulting final displacement will give the sum of the complex numbers.

I show an example of this in Figure 10.B.2 where the sum of the complex numbers is zero. We have $(4 + j) + (2 + 3j) + (-3 + 2j) + (-3 - 6j) = (0 + 0j)$.

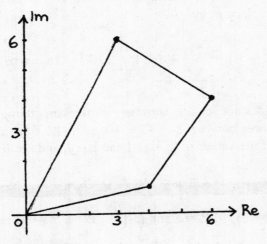

Figure 10.B.2

This is also the way in which the addition of **vectors** works. Vectors are quantities which have both magnitude and direction. Complex numbers have direction built into them because of their structure of two separate parts written in a particular order.

EXERCISE 10.B.1

If $z_1 = 3 + 5j$ and $z_2 = -2 + 2j$ and $z_3 = -7 - 2j$ find the following.

(1) $z_1 + z_2$ (2) $z_1 + z_2 + z_3$ (3) $z_2 - z_3$ (4) $z_3 - z_1$

10.B.(b) Multiplication of complex numbers

Multiplying by an ordinary number or scalar

If a complex number is multiplied by an ordinary number, the effect is to change its size by the scale given by this number.

For example,

$$3(2 + j) = 6 + 3j \quad \text{and} \quad \tfrac{1}{2}(4 + 2j) = 2 + j.$$

I show these in Figure 10.B.3. The numbers 3 and $\tfrac{1}{2}$ have no direction themselves. They simply enlarge or shrink the two complex numbers.

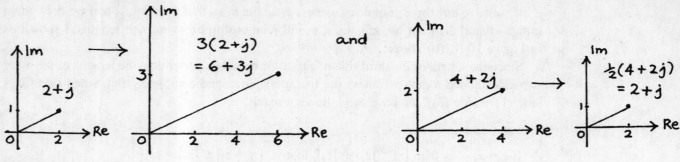

Figure 10.B.3

Multiplying two complex numbers together

Again we use the ordinary rules of arithmetic on the separate parts of the two numbers.

> To multiply two complex numbers together we multiply out the brackets in the usual way except that we replace j^2 by -1.
> For example,
>
> $$(3 + 4j) \times (12 + 5j) = 36 + 48j + 15j + 20j^2 = 16 + 63j.$$

The above example does not seem to show any particular pattern, but there are two very nice links between the three numbers $z_1 = (3 + 4j)$, $z_2 = (12 + 5j)$ and $z_1 z_2 = 16 + 63j$.

Work out for yourself the values of $|z_1|$, $|z_2|$ and $|z_1 z_2|$, and see if you can spot the first link.

You should have $|z_1| = \sqrt{3^2 + 4^2} = 5$ and $|z_2| = \sqrt{12^2 + 5^2} = 13$ and $|z_1 z_2| = \sqrt{16^2 + 63^2} = 65$ so $|z_1| \, |z_2| = |z_1 \, z_2|$.

Now work out for yourself $\arg z_1$, $\arg z_2$ and $\arg(z_1 z_2)$ and again see if you can spot a link between them.

You should have $\arg z_1 = \tan^{-1}(\frac{4}{3}) = 0.927$ rad to 3 d.p. and $\arg z_2 = \tan^{-1}(\frac{5}{12}) = 0.395$ rad to 3 d.p. and $\arg(z_1 \, z_2) = \tan^{-1}(\frac{63}{16}) = 1.322$ rad to 3 d.p.

It looks as though $\arg z_1 + \arg z_2 = \arg(z_1 z_2)$.

We can see that this is exactly so by using the rule

$$\tan(A + B) = \frac{\tan A + \tan B}{1 - \tan A \tan B} \quad \text{from Section 5.D.(c).}$$

Putting $A = \arg z_1$ and $B = \arg z_2$ we get $\tan A = \frac{4}{3}$ and $\tan B = \frac{5}{12}$. So

$$\tan(\arg z_1 + \arg z_2) = \frac{\frac{4}{3} + \frac{5}{12}}{1 - (\frac{4}{3})(\frac{5}{12})} = \frac{48 + 15}{36 - 20} = \frac{63}{16} = \tan(\arg(z_1 \, z_2))$$

therefore

$$\arg(z_1 z_2) = \arg z_1 + \arg z_2.$$

Multiplying complex numbers in mod/arg form

Is the result which we have just found above a special coincidence because of our particular choice of z_1 and z_2 or is there a reason why they should behave like this?

We show that there is an underlying reason for this result by taking any two complex numbers z_1 and z_2 and writing them in modulus/argument form.

To do this, we let $|z_1| = r_1$ and arg $z_1 = \theta_1$ so $z_1 = r_1 (\cos\theta_1 + j \sin\theta_1)$, and $|z_2| = r_2$ and arg $z_2 = \theta_2$ so $z_2 = r_2(\cos \theta_1 + j \sin \theta_2)$.

Then

$$z_1z_2 = r_1r_2 (\cos \theta_1 \cos \theta_2 + j^2 \sin \theta_1 \sin \theta_2 + j \sin \theta_1 \cos \theta_2 + j \cos \theta_1 \sin \theta_2).$$

$$= (\cos \theta_1 \cos \theta_2 - \sin \theta_1 \sin \theta_2) + j (\sin \theta_1 \cos \theta_2 + \cos \theta_1 \sin \theta_2).$$

We can replace the real and imaginary parts of the inside of this bracket using the double angle rules of Section 5.D.(b). This gives us

$$z_1z_2 = r_1r_2 (\cos (\theta_1 + \cos \theta_2) + j \sin(\theta_1 + \theta_2)).$$

We now have a very neat way of working out the result of multiplying two complex numbers together.

> To multiply two complex numbers together we *multiply* their moduli and *add* their arguments.

As an example, we'll take the two roots $z_1 = 4 + 3j$ and $z_2 = 4 - 3j$ of the quadratic equation $z^2 - 8z + 25 = 0$. Multiplying them together, we get

$$z_1 z_2 = (4 + 3j) (4 - 3j) = 16 + 12j - 12j - 9j^2 = 25.$$

We also know from Section 10.A.(c) that $|z_1| = |z_2| = 5$ and arg $z_1 = 0.644$ radians while arg $z_2 = -0.644$ radians.

Multiplying the two moduli gives us 25, and adding the two arguments, taking account of their signs, gives us zero, so both ways of multiplying give us the same answer which is the real number of 25.

Notice that this answer agrees with the rule of Section 2.D.(e) which says that the product of the roots of a quadratic equation is given by c/a. (Here, $c = 25$ and $a = 1$.)

The sum of the roots also agrees with the rule given there, since $(4 + 3j) + (4 - 3j) = 8 = -b/a$. (Here, $b = -8$.)

These two rules continue to work for quadratic equations with complex roots for exactly the same reason which we found they worked in Section 2.D.(e). The square root of $b^2 - 4ac$, which is where potential problems can arise, cancels out when the roots are added or multiplied.

The geometrical effect of multiplying complex numbers

It is now very easy for us to see geometrically what happens to complex numbers when they are multiplied together.

In particular, if we return to the equation $z^4 = 1$ in Section 10.A.(a), and its solutions shown in Figure 10.A.2, we see that multiplying j by itself turns it through 90° or $\pi/2$. Multiplying this result by j turns it through another right angle.

A final multiplication by j makes the last turn through 90° to finish at +1. We see geometrically that $j^4 = +1$.

The next root of $z = -1$ turns through $180°$ when it is multiplied by itself, which brings it to $+1$. Repeating this process takes it through a full circle again to $+1$.

Think for yourself what happens when you multiply $-j$ by itself and how it is that repeating this makes you end up at $+1$ for $(-j)^4$.

The results of all the above multiplications keep us on the unit circle because each of these roots is of unit length, but multiplying *any* complex number by j will turn it through a right angle. (This is because $|j| = 1$ and $\arg j = \pi/2$.)

For example, $j(4 + 2j) = 4j + 2j^2 = -2 + 4j$. I show this pair in Figure 10.B.4(a).

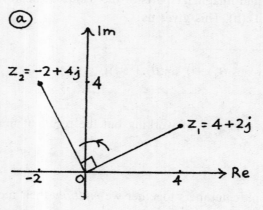

Figure 10.B.4

You should know now whether you got the right answer to the thinking point at the end of Section 10.A.(a) which asked you where the eight roots of $z^8 = 1$ are.

They all lie on the unit circle with angles of $\pi/4$ between them.

Each of them has a modulus of 1 and there is one eighth of a full turn between each pair of roots so their arguments are 0, $\pi/4$, $\pi/2$, $3\pi/4$, π, $-\pi/4$, $-\pi/2$ and $-3\pi/4$.

If you haven't yet made your own drawing of these, draw them in now on the unit circle shown above in Figure 10.B.4(b) in a different colour to make them show up. Think how each of them will turn as it is multiplied by itself.

EXERCISE 10.B.2

(1) If $z_1 = 3 + 2j$, $z_2 = -2 + j$ and $z_3 = -3 - 5j$ find the following.
    (a) $3z_1$ (b) $2z_1 + 3z_2$ (c) $2z_1 + z_2 + z_3$ (d) $3z_2 - 4z_3$

(2) (a) If $w = z^2$ find w for the following values of z:
      (i) $z = 2 + j$ (ii) $z = 4 + j$ (iii) $z = 4 + 3j$.
      In each case, also find $|z|$ and $|w|$ and show w on an Argand diagram.

    (b) If $z = x + jy$ and $w = u + jv$, find $|z|$, $|w|$, u and v each in terms of x and y. If x and y are whole numbers, which of $|z|$, $|w|$, u and v must also be whole numbers?

    (c) Each of the Argand diagrams from (a) will show you a right-angled triangle with sides whose lengths are whole numbers since, in each case, $|w|$, u and v are all whole numbers.

      Sets of whole numbers which give the sides of right-angled triangles are called Pythagorean triples because they fit Pythagoras' Theorem.

      Two examples are 3, 4, 5 and 5, 12, 13 because $5^2 = 4^2 + 3^2$ and $13^2 = 12^2 + 5^2$.

      It is possible to get some very strange shapes of triangle with sides given by Pythagorean triples.

Try finding $w = (12 + 5j)^2$. Then find $|w|$ and draw a sketch to show w on an Argand diagram.

Now use your result for w to find $W = w^2 = (12 + 5j)^4$. Find $|W|$ and then show W on a new Argand diagram.

You could also try finding some new triangles like this for yourself. You can check if your working is correct by seeing if the lengths of the sides do fit Pythagoras' Theorem.

(3) Check that the pairs of roots which you found for the four equations in Exercise 10.A.1 *do* fit the equations by substituting them back in. Then find the sum and the product for each of these pairs of roots and check that they *do* come to '$-b/a$' and 'c/a' respectively in each case.

(4) One of the roots of the equation $z^3 = 1$ is $z = 1$.
   (a) Draw a little sketch showing all three roots of this equation.
   (b) Find these three roots giving them both in mod/arg form and in $a + bj$ form.
   (c) These three roots are known as 1, ω and ω^2, as we move round anticlockwise. Check that $\omega \times \omega$ gives ω^2 and that $\omega^2 \times \omega^2$ gives ω.

THINKING POINT

Can you work out where the four roots of $z^4 = -1$ will be, drawing them on a little sketch and giving each of them both in mod/arg form and in $a + bj$ form?

10.B.(c)   Dividing complex numbers in mod/arg form

We can get the rules for this by thinking of dividing as a rearranged multiplying.

Just as with ordinary numbers, we can say $(z_1/z_2)\, z_2 = z_1$ so, from the rules for multiplying, we get the following pair of results.

$$\left|\frac{z_1}{z_2}\right| |z_2| = |z_1| \quad \text{so} \quad \left|\frac{z_1}{z_2}\right| = \frac{|z_1|}{|z_2|}$$

and

$$\arg(z_1/z_2) + \arg(z_2) = \arg(z_1) \quad \text{so} \quad \arg(z_1/z_2) = \arg(z_1) - \arg(z_2).$$

To divide two complex numbers we *divide* their moduli and *subtract* their arguments.

Notice that, just as with ordinary numbers, the order matters here, unlike when we multiply. z_1/z_2 is not the same as z_2/z_1, so you must be careful when you work out the new modulus and argument.

As an example of using these rules, we'll find $(1 + j) \div (1 - j)$. I have shown these two complex numbers on the Argand diagram in Figure 10.B.5.

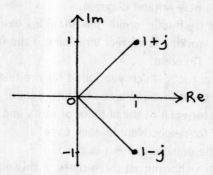

Figure 10.B.5

Using this, we can easily write the two complex numbers involved in this division in mod/arg form, that is, in the form $z = [r, \theta]$ where $|z| = r$ and $\arg z = \theta$.

We have

$$\frac{1 + j}{1 - j} = \frac{[\sqrt{2}, \pi/4]}{[\sqrt{2}, -\pi/4]} = \left[\frac{\sqrt{2}}{\sqrt{2}}, \frac{\pi}{4} - \left(-\frac{\pi}{4} \right) \right] = \left[1, \frac{\pi}{2} \right] = j.$$

10.B.(d) What are complex conjugates?

In Section 10.A.(b) we found that the roots of the equation $z^2 - 8z + 25 = 0$ are $z_1 = 4 + 3j$ and $z_2 = 4 - 3j$. We saw in the Argand diagram of Figure 10.A.3 that they formed a symmetrical pair either side of the real axis.

> Pairs of numbers which can be written in the form $a + bj$ and $a - bj$ are called **complex conjugates.**
> If $z = a + bj$ then its conjugate of $a - bj$ is written as $\bar{z}$.
> The conjugate of $\quad a - bj \quad$ is $\quad a + bj$.

'Conjugates' means a sort of married couple – in fact the word stems from this original meaning.

> You will see that since $|z| = \sqrt{a^2 + b^2}$ and $|\bar{z}| = \sqrt{a^2 + (-b)^2}$ it must be true that $|z| = |\bar{z}|$.
>   Also,
>
> $$z\bar{z} = (a + bj)(a - bj) = a^2 + b^2 = (|z|)^2 = |z^2|$$
>
> because
>
> $$|z^2| = |(a + bj)^2| = |(a^2 - b^2) + 2abj| = \sqrt{(a^2 - b^2)^2 + 4a^2b^2} = a^2 + b^2.$$

You can also see that this must be so from the way in which they lie on the Argand diagram. $\bar{z}$ is the reflection of z in the real axis, so it must be the same length.

I have shown five complex numbers in Figure 10.B.6. Draw in their conjugates for yourself, labelling each one so that you can check if you have put them in the right places.

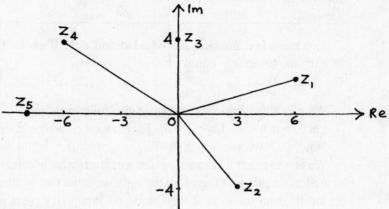

Figure 10.B.6

10.B.(e)   Using complex conjugates to simplify fractions

Complex conjugates are important in many areas. One of their uses is in simplifying fractions which have complex numbers underneath. Suppose we have the fraction

$$\frac{3 + j}{5 - 2j}.$$

There is a very neat way of simplifying fractions like this, in which we convert the bottom of the fraction (its denominator) into a real number.

Can you see how we can do this? (We have already used this method for simplifying fractions with other square roots underneath in Section 1.F.(d). The only difference here is that now we are dealing with j, the special square root of -1.)

The trick is to multiply the bottom of the fraction by its conjugate. (Of course, in this case we must also multiply the top of the fraction by the same thing to keep its whole value unchanged.) This gives us

$$\frac{(3 + j)}{(5 - 2j)} \frac{(5 + 2j)}{(5 + 2j)} = \frac{15 + 2j^2 + 5j + 6j}{25 - 4j^2 - 10j + 10j} = \frac{13 + 11j}{29}.$$

This is now in the very much more convenient form of the single complex number $\frac{1}{29}(13 + 11j)$. What we have done here is, in effect, a division.

We have found that $(3 + j) \div (5 - 2j) = \frac{1}{29}(13 + 11j)$.

Try simplifying these yourself.

(1) $\dfrac{3}{2 + j}$ (2) $\dfrac{4}{3 - 2j}$ (3) $\dfrac{1 + j}{1 - j}$ (4) $\dfrac{1 - 2j}{5 + 3j}$

(5) $\dfrac{2 - j}{(3 + j)(3 - 2j)}$ (Multiply out the bottom first.)

(6) $\dfrac{3}{2 - j} + \dfrac{2}{3 + 2j}$ (Simplify each separate fraction first, and then add.)

How *e* connects with complex numbers

Two for the price of one – equating real and imaginary parts

Because of their structure, complex numbers have a special property which is of enormous importance.

> Two complex numbers are equal if and only if each of their real and imaginary parts is separately equal.

We can show that this is true in the following way.

Let $z_1 = a + bj$ and $z_2 = c + dj$. Then, if $a = c$ and $b = d$, it is certainly true that $z_1 = z_2$.

Now we have to show that if $z_1 = z_2$ then $a = c$ and $b = d$.

We can see that this must be true geometrically because complex numbers have direction as well as length, and therefore the only way that two of them can be equal, and so lie exactly on top of each other, is if their real and imaginary parts are separately equal.

We can also see this rather nicely in the following way, using algebra.

If $z_1 = z_2$ then $a + bj = c + dj$ so

$$a - c = dj - bj = j(d - b).$$

Squaring both sides of this equation gives

$$(a - c)^2 = j^2(d - b)^2$$

so

$$(a - c)^2 = -(d - b)^2.$$

Therefore

$$(a - c)^2 + (d - b)^2 = 0.$$

But remember that a, b, c, and d are all real numbers. Because of this, we can say that $(a - c)^2 \geq 0$ and $(d - b)^2 \geq 0$.

Therefore, the only way for it to be possible that $(a - c)^2 + (d - b)^2 = 0$ is that each of $(a - c)$ and $(d - b)$ are equal to zero. Therefore $a = c$ and $d = b$.

This property of complex numbers is of huge importance in their application to physical situations, because it means that any equation involving complex numbers is actually made up of two separate equations. We are, in a sense, getting two for the price of one.

EXAMPLE (1)   To see an example of this in action, we will solve the equation $z^2 = 5 + 12j$.

Let $z = a + bj$. (We know the solution must be complex because its square is a complex number.)

We have $(a + bj)^2 = 5 + 12j$ so

$$a^2 - b^2 + 2abj = 5 + 12j.$$

This can only be true if both the real parts and the imaginary parts are separately equal.

Equating the real parts gives

$$a^2 - b^2 = 5 \qquad (1)$$

Equating the imaginary parts gives

$$2ab = 12 \qquad (2)$$

From (2) we have $a = 12/2b = 6/b$. Substituting this in equation (1) gives

$$\frac{36}{b^2} - b^2 = 5 \quad \text{so} \quad 36 - b^4 = 5b^2 \quad \text{so} \quad b^4 + 5b^2 - 36 = 0.$$

This is much easier to solve than it looks at first, being one of those quadratic equations which are masquerading as something much nastier which we met in Example (4) in Section 2.E.(d). Factorising this equation gives $(b^2 - 4)(b^2 + 9) = 0$, (or you could put $b^2 = y$, say, and then use the formula). This gives us $b^2 = 4$ or $b^2 = -9$.

Because b is a real number, we discard $b^2 = -9$ as it will only give us imaginary solutions. This leaves us with the two solutions of $b = +2$ or $b = -2$.

The corresponding answers for a are $a = 6/2 = 3$ or $a = 6/-2 = -3$.

So the two solutions of the equation $z^2 = 5 + 12j$ are $z_1 = 3 + 2j$ and $z_2 = -3 - 2j$. Check for yourself that they really work!

I show these two solutions in Figure 10.C.1(a).

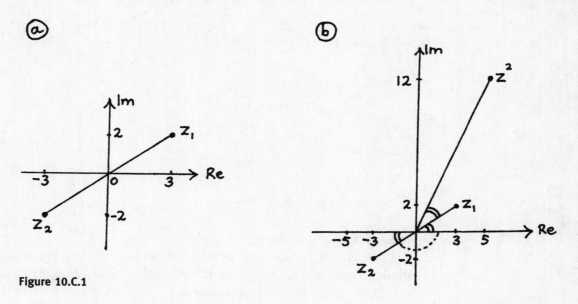

Figure 10.C.1

Geometrically, we can see straight away that z_1 is a solution of $z^2 = 5 + 12j$. This is because $|z^2| = 13$ and $|z_1| = \sqrt{13}$, and also

$$\arg z_1 = \tan^{-1}(\tfrac{2}{3}) \quad \text{and} \quad \arg(z^2) = \tan^{-1}(\tfrac{12}{5}) \quad \text{so} \quad \arg z_1 = \tfrac{1}{2}(\arg z^2).$$

(Check this numerically on your calculator.)

We know from the rules for complex multiplication given in Section 10.B.(b) that $|z_1| = |z_2| = \sqrt{|z|}$. We also know that $2 \arg z_1 = 2 \arg z_2 = \arg(z^2)$.

In Figure 10.B.1.(b), you can see how these angles actually work. Doubling both $\arg z_1$ and $\arg z_2$ brings you round to the direction of z^2. (Since $\arg z_2$ is negative, doubling it will mean that you are moving in a clockwise direction about the origin.)

The two square roots of a complex number will always be in the form $\pm(x + jy)$, so together they will always make a straight line on their Argand diagram.

We shall need to be able to find these square roots when we solve quadratic equations with complex coefficients in Section 10.D.(b).

(1) Find the square roots of each of the following complex numbers in $a + bj$ form and show each of these pairs of roots on an Argand diagram.

    (a) $3 + 4j$ (b) $15 + 8j$ (c) $5 - 12j$

(2) Find the modulus and argument of each of the given complex numbers and sketch each of them on its own Argand diagram. Then use this to help you to find both of the square roots of each of the given numbers, showing them also on the Argand diagrams, and giving your answers in mod/arg form, that is, in the form $z = [r, \theta]$ where $r = |z|$ and $\theta = \arg z$.

    (a) $4j$ (b) $1 + \sqrt{3}j$ (c) $-\sqrt{2} + \sqrt{2}j$ (d) $-1 - \sqrt{3}j$

10.C.(b)   **How does e get involved?**

At last we are able to solve the mystery of the resemblance between the trigonometrical functions $\cos t$ and $\sin t$ and the hyperbolic functions $\cosh t$ and $\sinh t$.

To do this, we will need to use the series which we found in Section 8.G. I list below the ones which we shall need.

$$e^t = 1 + \frac{t}{1!} + \frac{t^2}{2!} + \frac{t^3}{3!} + \frac{t^4}{4!} + \frac{t^5}{5!} + \frac{t^6}{6!} + \dots \tag{1}$$

$$\cos t = 1 - \frac{t^2}{2!} + \frac{t^4}{4!} - \frac{t^6}{6!} + \dots \tag{2}$$

$$\sin t = \frac{t}{1!} - \frac{t^3}{3!} + \frac{t^5}{5!} - \frac{t^7}{7!} + \dots \tag{3}$$

$$\cosh t = 1 + \frac{t^2}{2!} + \frac{t^4}{4!} + \frac{t^6}{6!} + \dots \tag{4}$$

$$\sinh t = \frac{t}{1!} + \frac{t^3}{3!} + \frac{t^5}{5!} \dots \tag{5}$$

We know that $e^t = \cosh t + \sinh t$.

At the end of Section 8.G, we were left with the feeling that it should also be possible to link e^t, $\cos t$ and $\sin t$ together, but we couldn't quite do it then. Now we have the extra possibilities given by complex numbers.

What happens if we put $t = j\theta$ in series (1)? We shall get

$$e^{j\theta} = 1 + \frac{j\theta}{1!} + \frac{j^2\theta^2}{2!} + \frac{j^3\theta^3}{3!} + \frac{j^4\theta^4}{4!} + \frac{j^5\theta^5}{5!} + \frac{j^6\theta^6}{6!} + \dots$$

so $$e^{j\theta} = 1 + \frac{j\theta}{1!} - \frac{\theta^2}{2!} - \frac{j\theta^3}{3!} + \frac{\theta^4}{4!} + \frac{j\theta^5}{5!} - \frac{\theta^6}{6!} + \dots \tag{6}$$

Now answer these questions yourself.

(1)   Pick out the real parts in series (6) above to give a new series. What do you get?

(2)   Make a new series with just the imaginary parts of the series above. Can you see what doing this will give you?

(3)   What linking relationship have you now found?

This is what you should have.

The real parts of series (6) give the series for $\cos \theta$.

The imaginary parts of series (6) give the series for $\sin \theta$, so if we include the js too we have the series for $j \sin \theta$.

(Note: we are assuming here that it is all right to play around with these series in this way. In these particular cases, mathematicians have shown that it *is* all right but this is by no means a general rule. Infinite series have to be treated with great caution, as we saw for ourselves in Section 6.F.)

Putting together what we now have, we finally get our link, which is

$$e^{j\theta} = \cos \theta + j \sin \theta.$$

This is an amazing and beautiful result, and is due to the German mathematician Euler after whom e is named.

10.C.(c)   What is the geometrical meaning of $z = e^{j\theta}$?

Since $z = e^{j\theta} = \cos \theta + j \sin \theta$ we can see from Figure 10.C.2(a) that z must lie on a circle whose centre is at the origin and whose radius is one unit. This circle is known as the **unit circle** about the origin.

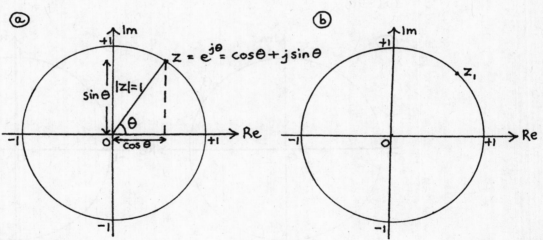

Figure 10.C.2

For any value of θ, the point representing z will be somewhere on this unit circle. Everywhere on this circle, $|z| = 1$ and arg $z = \theta$.

To get the feel of what is happening, we'll look at some particular possibilities for z as θ varies.

As a first example, if $\theta_1 = \pi/4$, we get the point $z_1 = \cos(\pi/4) + j \sin(\pi/4) = (1/\sqrt{2}) + (1/\sqrt{2})j$. I have marked z_1 on Figure 10.C.2(b).

What points do you get if you take (a) $\theta_2 = 0$ (b) $\theta_3 = \pi$ (c) $\theta_4 = \pi/2$ (d) $\theta_5 = -\pi/2$ (e) $\theta_6 = 2\pi/3$? Mark each of the corresponding z values on Figure 10.C.2(b).

You should have the following points.

(a) $z_2 = e^0 = 1$, (b) $z_3 = e^{j\pi} = -1$ because $e^{j\pi} = \cos \pi + j\sin \pi$.

The result $e^{j\pi} = -1$ is also due to Euler and is known as Euler's Formula. It links three extraordinary numbers, e, π and j, to give the simple answer of -1.

Similarly; (c) $z_4 = e^{j\pi/2} = j$, (d) $z_5 = e^{-j\pi/2} = -j$, and (e) $z_6 = e^{2j\pi/3} = \cos 2\pi/3 + j \sin 2\pi/3 = -1/2 + (\sqrt{3}/2)j$.

As θ increases, $z = e^{j\theta}$ is turning anticlockwise round the unit circle.

10.C.(d)   **What is $e^{-j\theta}$ and what does it do geometrically?**

We know that $e^{j\theta} = \cos \theta + j\sin \theta$. If we put $-\theta$ instead of θ, we will get

$$e^{-j\theta} = \cos(-\theta) + j \sin(-\theta).$$

From Section 5.A.(c) we know that $\cos(-\theta) = \cos \theta$ and $\sin(-\theta) = -\sin \theta$ using the turn of P on its unit circle. This gives us

$$\boxed{e^{-j\theta} = \cos \theta - j \sin \theta.}$$

What will we get if we plot $z = e^{-j\theta}$ on an Argand diagram?

If we start with $\theta_1 = \pi/4$ as before, and let $z_1^* = e^{-j\pi/4}$, using the * to distinguish it from the previous z_1, we shall have $z_1^* = \cos \pi/4 - j \sin \pi/4 = 1/\sqrt{2} - (1/\sqrt{2})j$.

I have shown the turn to θ_1 in Figure 10.C.3(a), and the position of z_1^* in Figure 10.C.3(b).

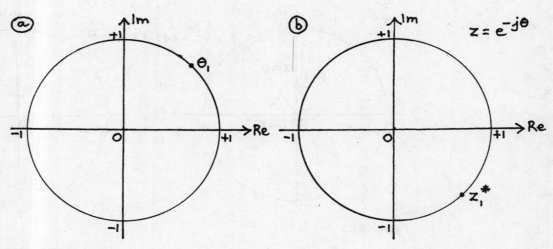

Figure 10.C.3

Now, using the same values of θ which you used in Section 10.C.(c), try finding the corresponding values for z^* yourself. Mark each of your angles on Figure 10.C.3(a) and each of your values for z^* on Figure 10.C.3(b).

You should have $z_2^* = 1$, $z_3^* = -1$, $z_4^* = -j$, $z_5^* = j$ and $z_6^* = -1/2 - (\sqrt{3}/2)j$.

As θ increases, $z = e^{-j\theta}$ moves *clockwise* round the unit circle about the origin. This means that each z^* is the reflection of its corresponding z in the real axis, so each $z^* = \bar{z}$ for that particular z.

> For any given value of θ, $e^{j\theta}$ and $e^{-j\theta}$ are conjugates of each other.

If $z = e^{j\theta}$ then $\bar{z} = e^{-j\theta}$. $|z| = |\bar{z}| = 1$ so $(e^{j\theta})(e^{-j\theta}) = 1$.

This exactly slots with the result given by multiplying $z_1 = e^{j\theta}$ and $z_2 = e^{-j\theta}$ and using the first rule of powers.

If $z_1 = e^{j\theta}$ and $z_2 = e^{-j\theta}$ then

$$z_1 z_2 = e^{j\theta} e^{-j\theta} = e^{(j\theta - j\theta)} = e^0 = 1.$$

10.C.(e)   A summary of the sin/cos and sinh/cosh links

We now have the following results:

> $e^{j\theta} = \cos\theta + j\sin\theta,$
>
> $e^{-j\theta} = \cos\theta - j\sin\theta.$

From this,

$$e^{j\theta} + e^{-j\theta} = 2\cos\theta \quad \text{giving} \quad \cos\theta = \tfrac{1}{2}(e^{j\theta} + e^{-j\theta})$$

and $e^{j\theta} - e^{-j\theta} = 2j\sin\theta \quad \text{giving} \quad j\sin\theta = \tfrac{1}{2}(e^{j\theta} - e^{-j\theta}).$

If we compare these results with $\cosh x = \tfrac{1}{2}(e^x + e^{-x})$ and $\sinh x = \tfrac{1}{2}(e^x - e^{-x})$ and put $x = j\theta$, we see that

> $\cosh(j\theta) = \cos\theta$
>
> and
>
> $\sinh(j\theta) = j\sin\theta.$

The similar behaviour of the sin/cos and sinh/cosh pairs is no longer mysterious. These functions are all intimately linked together and these links explain the differences between the rules for trigonometric and hyperbolic functions. Every time we square a sin or sinh as part of our working, we get the cross-over from the imaginary to the real from putting $j^2 = -1$. This is the reason for **Osborn's Rule** which says that the formulas for the trigonometrical and the hyperbolic functions are the same except that we must change the sign whenever the working involves multiplying two sins or two sinhs together. (We met this rule in Section 8.D.(d).)

Since it was the series expansions which showed us the links between the pairs of cos and sin, and cosh and sinh, we can begin to see that it might be rather good to actually *define* these functions by using these series. This may seem alarmingly peculiar when you first meet it – after all, it is a very long way away from those original right-angled triangles. However, it does have many advantages, not least of which is that it is then possible to define what we mean by e^z, $\sin z$, $\cos z$, $\sinh z$ and $\cosh z$ when z is complex. It is also possible to show that, if we do define sin and cos by means of series expansions, all the familiar properties are true if we consider only the cases for which z is real.

10.C.(f)   **De Moivre's Theorem**

We know from Section 10.C.(b) that $e^{j\theta} = \cos\theta + j\sin\theta$.

If we put $n\theta$ instead of θ, we get $e^{jn\theta} = \cos n\theta + j\sin n\theta$.

But it can be shown that we can say $e^{jn\theta} = (e^{j\theta})^n = (\cos\theta + j\sin\theta)^n$.

This gives us the following result which is known by the name of the French mathematician Abraham De Moivre.

De Moivre's Theorem

$(\cos\theta + j\sin\theta)^n = (\cos n\theta + j\sin n\theta)$

Here is one example of how we can use this.

Putting $n = 2$ gives us

$$(\cos\theta + j\sin\theta)^2 = \cos 2\theta + j\sin 2\theta$$

so $\cos^2\theta - \sin^2\theta + 2j\sin\theta\cos\theta = \cos 2\theta + j\sin 2\theta.$

Now we use the fact that this equation is really *two* equations – we have two for the price of one, as we saw in Section 10.C.(a).

Equating the real parts gives us

$$\cos 2\theta = \cos^2\theta - \sin^2\theta$$

Equating the imaginary parts gives us

$$\sin 2\theta = 2\sin\theta\cos\theta.$$

We have been able to show both the double angle rules from Section 5.D.(d) very simply. Notice the neat way in which using complex numbers gives us both rules from the same piece of working!

10.C.(g)   **Another example: writing cos 5θ in terms of cos θ**

De Moivre's Theorem makes it possible to find rules for multiple angles with a fraction of the work that would have been necessary up to now. As an example of this, I will show you how to write $\cos 5\theta$ in terms of $\cos\theta$ and $\sin 5\theta$ in terms of $\sin\theta$, two results which examiners seem to be rather fond of.

Since we want rules for $\cos 5\theta$ and $\sin 5\theta$, we put $n = 5$ in De Moivre's Theorem. This gives us

$$(\cos\theta + j\sin\theta)^5 = (\cos 5\theta + j\sin 5\theta).$$

To multiply out the left-hand side, we shall have to use a binomial expansion and this means that we shall need to know the right binomial coefficients to use.

   Don't leave these coefficients out!

Since $n = 5$, and 5 is quite a small number, much the easiest way to find the binomial coefficients is to use Pascal's Triangle. We wrote this down in Section 7.A.(a).

The line of coefficients given by the triangle when $n = 5$ is

$$1 \qquad 5 \qquad 10 \qquad 10 \qquad 5 \qquad 1$$

Because we shall be using Pascal's Triangle a lot in this chapter, here is a brief reminder of how it works. Each new line is made from adding the two numbers nearest in the line above, unless it is at the end of the line when the single number closest to it is used. So, for example, if $n = 6$, the *next* line of the triangle would be

$$1 \qquad 6 \qquad 15 \qquad 20 \qquad 15 \qquad 6 \qquad 1$$

Remember that the triangle starts with $\quad 1 \qquad 1 \quad$ when $n = 1$.

Putting in the coefficients in our example, we have

$$(\cos \theta + j \sin \theta)^5 = (\cos \theta)^5 + 5\,(\cos \theta)^4\,(j \sin \theta) + 10\,(\cos \theta)^3\,(j \sin \theta)^2$$
$$+ 10\,(\cos \theta)^2\,(j \sin \theta)^3 + 5\,(\cos \theta)\,(j \sin \theta)^4 + (j \sin \theta)^5.$$

I have written this working out in detail because mistakes often creep in here. In particular, notice that it is the *whole* of $(j \sin \theta)$ which is being raised to the different powers.

Now we replace j^2 by -1, j^3 by $-j$ etc. and tidy up generally.

Doing this gives us a right-hand side of

$$\cos^5 \theta + 5j \cos^4 \theta \sin \theta - 10 \cos^3 \theta \sin^2 \theta - 10j \cos^2 \theta \sin^3 \theta + 5 \cos \theta \sin^4 \theta$$
$$+ j \sin^5 \theta.$$

We also know from De Moivre's Theorem that $(\cos \theta + j \sin \theta)^5 = \cos 5\theta + j \sin 5\theta$.

Now we equate the real and imaginary parts of this equation, so getting double worth from our labours so far.

Equating the real parts gives us

$$\cos 5\theta = \cos^5 \theta - 10 \cos^3 \theta \sin^2 \theta + 5 \cos \theta \sin^4 \theta \qquad\qquad (1)$$

and equating the imaginary parts gives us

$$\sin 5\theta = 5 \cos^4 \theta \sin \theta - 10 \cos^2 \theta \sin^3 \theta + \sin^5 \theta. \qquad\qquad (2)$$

Finally, if we want the RHS of equation (1) to be entirely in terms of $\cos \theta$ and the RHS of equation (2) to be entirely in terms of $\sin \theta$, a proviso which examiners frequently make, what should we use?

We use the identity $\sin^2 \theta + \cos^2 \theta = 1$ to do the necessary adjustments. You may be wondering how this identity is going to help us with $\sin^4 \theta$. Can you see how we can use it?

We can say that

$$\sin^4 \theta = (\sin^2 \theta)^2 = (1 - \cos^2 \theta)^2 = 1 + \cos^4 \theta - 2 \cos^2 \theta.$$

This gives us

$$\cos 5\theta = \cos^5 \theta - 10 \cos^3 \theta\,(1 - \cos^2 \theta) + 5 \cos \theta\,(1 + \cos^4 \theta - 2 \cos^2 \theta).$$

Tidying this up, we have

$$\cos 5\theta = 16 \cos^5 \theta - 20 \cos^3 \theta + 5 \cos \theta.$$

Now do the very similar process for yourself of equating the imaginary parts to show that

$$\sin 5\theta = 16 \sin^5 \theta - 20 \sin^3 \theta + 5 \sin \theta.$$

EXERCISE 10.C.2

Try the following questions yourself.
(1) Show, using De Moivre's Theorem, that
    (a) $\cos 3\theta = 4 \cos^3 \theta - 3 \cos \theta$ and   (b) $\sin 3\theta = 3 \sin \theta - 4 \sin^3 \theta$.
(2) Show that $\cos 7\theta = 64 \cos^7 \theta - 112 \cos^5 \theta + 56 \cos^3 \theta - 7 \cos \theta$.

10.C.(h)   More examples of writing trig functions in different forms

There is a rather similar problem (also popular with examiners) which we can now solve. This is basically a question of going back the other way.

How, for example, could we write $\cos^7 \theta$ in terms of cosines of multiples of θ?

To do this, we use the results of Section 10.C.(e) that

$$\cos \theta = \tfrac{1}{2} (e^{j\theta} + e^{-j\theta}) \quad \text{and} \quad j \sin \theta = \tfrac{1}{2} (e^{j\theta} - e^{-j\theta}).$$

If we replace θ by $n\theta$ we get

$$\cos n\theta = \tfrac{1}{2} (e^{nj\theta} + e^{-nj\theta}) \quad \text{and} \quad j \sin n\theta = \tfrac{1}{2} (e^{nj\theta} - e^{-nj\theta}).$$

To save some writing, it is handy to let $e^{j\theta} = z$ at this point, so that

$$e^{nj\theta} = (e^{j\theta})^n = z^n \quad \text{and} \quad e^{-j\theta} = \tfrac{1}{e^{j\theta}} = 1/z.$$

We can then rewrite the above pair of results as follows:

If $e^{j\theta} = z$,   then $\cos \theta = \tfrac{1}{2} (z + 1/z)$ and $j \sin \theta = \tfrac{1}{2} (z - 1/z)$.

Also, $\cos n\theta = \tfrac{1}{2} (z^n + 1/z^n)$ and $j \sin n\theta = \tfrac{1}{2} (z^n - 1/z^n)$.

So how will we now set about writing $\cos^7 \theta$ in terms of cosines of multiples of θ? We can say

$$\cos^7 \theta = [\tfrac{1}{2} (z + 1/z)]^7 = (\tfrac{1}{2})^7 (z + 1/z)^7.$$

Next, we use Pascal's Triangle to get the binomial coefficients for the expansion of $(z + 1/z)^7$. The line for $n = 7$ comes immediately from the line for $n = 6$ which we wrote down in the previous section. This was

    1   6   15   20   15   6   1

So, when $n = 7$ we get

    1   7   21   35   35   21   7   1

Using these coefficients gives us

$$(\tfrac{1}{2})^7 (z + 1/z)^7 = (\tfrac{1}{2})^7 [z^7 + 7(z)^6 (1/z) + 21(z)^5 (1/z)^2 + 35(z)^4 (1/z)^3$$
$$+ 35 (z)^3 (1/z)^4 + 21 (z)^2 (1/z)^5 + 7(z)(1/z)^6 + (1/z)^7].$$

This tidies up very neatly as

$$(\tfrac{1}{2})^7 \, (z + 1/z)^7 = (\tfrac{1}{2})^7 \, [(z^7 + 1/z^7) + 7(z^5 + 1/z^5) + 21(z^3 + 1/z^3) + 35(z + 1/z)].$$

Now all that remains to be done is to use $\cos n\theta = \tfrac{1}{2} \, (z^n + 1/z^n)$ here. This gives us

$$\cos^7 \theta = (\tfrac{1}{2})^7 \, (2 \cos 7\theta + 14 \cos 5\theta + 42 \cos 3\theta + 70 \cos \theta).$$

$$= (\tfrac{1}{2})^6 \, (\cos 7\theta + 7 \cos 5\theta + 21 \cos 3\theta + 35 \cos \theta).$$

If you want $\sin^7 \theta$ in terms of multiples of $\sin \theta$, you just start with

$$(j \sin \theta)^7 = (\tfrac{1}{2})^7 \, (z - 1/z)^7,$$

and then work in a very similar way. You will find that the js will all cancel out.

You can also get slightly quirky extra results by differentiating a previous result.

For example, if we differentiate the expression which we have just found for $\cos^7 \theta$ with respect to θ, we get

$$-7 \cos^6 \theta \sin \theta = (\tfrac{1}{2})^6 \, (-7 \sin 7\theta - 35 \sin 5\theta - 63 \sin 3\theta - 35 \sin \theta)$$

so $\qquad \cos^6 \theta \sin \theta = (\tfrac{1}{2})^6 \, (\sin 7\theta + 5 \sin 5\theta + 9 \sin 3\theta + 5 \sin \theta).$

(Examiners rather like this one, too.)

EXERCISE 10.C.3

(1) Show that $\cos^5 \theta = (\tfrac{1}{2})^4 \, (\cos 5\theta + 5 \cos 3\theta + 10 \cos \theta)$.

(2) By rewriting $(j \sin \theta)^5$ in a different form, show that

$$\sin^5 \theta = (\tfrac{1}{2})^4 \, (\sin 5\theta - 5 \sin 3\theta + 10 \sin \theta).$$

10.C.(i) Solving a differential equation which describes SHM

We can now see that the method for solving differential equations described in Section 9.C.(c) and questions (9) and (10) of Exercise 8.C.4 will also work for the equation $d^2x/dt^2 = -x$. (Newton's dot notation is quite often used for this kind of equation. We write d^2x/dt^2 as $\ddot{x}$ and dx/dt as $\dot{x}$. The differential equation then becomes $\ddot{x} = -x$.)

We will find the solution for the particular case when we also know that $x = 3$ and $dx/dt = 4$ when $t = 0$. We have

$$\frac{d^2x}{dt^2} = -x$$

and we try putting $x = e^{kt}$. This gives us

$$\frac{dx}{dt} = ke^{kt} \quad \text{and} \quad \frac{d^2x}{dt^2} = k^2 \, e^{kt}.$$

If $x = e^{kt}$ is a solution of the given equation then we must have $k^2 e^{kt} = -e^{kt}$ or $(k^2 + 1) \, e^{kt} = 0$.

Since e^{kt} is never equal to zero, we would have to have $k^2 = -1$ giving us $k = j$ or $k = -j$.

Making the solution as general as possible gives us $\quad x = Ae^{jt} + Be^{-jt} \quad$ where A and B are constants.

Now we use Sections 10.C.(b) and (d) to say that

$$e^{jt} = \cos t + j \sin t \quad \text{and} \quad e^{-jt} = \cos t - j \sin t.$$

This gives us

$$x = A (\cos t + j \sin t) + B (\cos t - j \sin t)$$

so $\qquad x = (A + B) \cos t + (A - B) j \sin t.$

If we let $C = A + B$ and $D = j (A - B)$, we have

$$x = C \cos t + D \sin t.$$

We are told that $x = 3$ when $t = 0$ so $3 = C$.
Also, we know that $dx/dt = 4$ when $t = 0$.
Now,

$$\frac{dx}{dt} = - C \sin t + D \cos t \qquad \text{so} \quad 4 = D.$$

This gives us the solution

$$x = 3 \cos t + 4 \sin t.$$

Examples (a), (c) and (d) in Section 9.C.(c) also all give solutions for x from the differential equation $d^2x/dt^2 = -x$. Each of these solutions is different because the values of x and dx/dt when $t = 0$ are different in each case.

I showed in Section 5.D.(f) that $3 \cos t + 4 \sin t$ can also be written as $5 \sin (t + \alpha)$ with $\alpha = \tan^{-1} \frac{3}{4}$. You can see on Figure 5.D.3(a) how x changes with t.

This solution fits the general solution of an equation like this of $x = A \sin (\omega t + \varepsilon)$ if we put $A = 5$, $\omega = 1$ and $\varepsilon = \tan^{-1} \frac{3}{4}$.

Now try solving the equation $d^2x/dt^2 = -9x$ for yourself, given that we also know that $x = 0$ and $dx/dt = 6$ when $t = 0$.

This time, putting $x = e^{kt}$ gives us $\quad k^2 + 9 = 0 \quad$ so $\quad k = \pm 3j.$
We now have the solution $\quad x = Ae^{3jt} + Be^{-3jt}.$
Using

$$e^{3jt} = \cos 3t + j \sin 3t \quad \text{and} \quad e^{-3jt} = \cos 3t - j \sin 3t$$

gives us

$$x = (A + B) \cos 3t + j (A - B) \sin 3t = C \cos 3t + D \sin 3t$$

with $\qquad C = A + B \quad$ and $\quad D = j (A - B).$

Also, $x = 0$ when $t = 0$ so $0 = C$, and $dx/dt = 3D \cos 3t = 6$ when $t = 0$ so $D = 2$.

This gives us the solution that $x = 2 \sin 3t$. I showed what this particular x looks like in the first example at the end of Section 5.C.(b).

Comparing this solution with $x = A \sin (\omega t + \varepsilon)$, we have $A = 2$, $\omega = 3$ and $\varepsilon = 0$.

It is also possible to write the general SHM solution in the form $x = A \cos (\omega t + \varepsilon)$. In this particular case, this would give us $x = 2 \cos (3t - \pi/2)$.

10.C.(j)   **A first look at how we can use complex numbers to describe electric circuits**
In Section 5.D.(g) we said that the combined effect of particular components in electric circuits, such as inductors, capacitors or resistors, can have the effect of making the current and voltage be out of phase with each other if there is an alternating electromotive force

(e.m.f.). This phase difference will continue to be present even when the circuit has settled down after being switched on. (Initially there will also be transient responses, but these will die away leaving the circuit in a steady state.)

Suppose that the current in a single branch of a circuit network is given by $I = \sqrt{3} \cos t - \sin t$ and that the frequency is constant throughout the circuit.

At a junction, we know from Kirchhoff's junction rule that what goes in must come out, or that the net current flowing into a junction must be equal to the net current flowing out of it. We are supposing that the other currents involved at this junction have the same frequency, but they could certainly have different amplitudes and phase constants. Therefore we would be faced with having to combine a string of terms like $2 \cos (t + \pi/6)$, $3 \cos (t - \pi/4)$ etc.

Adding these in the form of trig functions is unpleasantly complicated, but we can use the properties of complex numbers to make things much simpler.

Here's how it would work in the case above.

From Section 10.C.(b), we know that $e^{j\theta} = \cos \theta + j \sin \theta$.

We have also seen there that, as θ increases, $e^{j\theta}$ moves round the unit circle.

If we call the variable t instead of θ, and let t represent time, then $z = e^{jt}$ will have moved right round the unit circle once in 2π seconds.

How can we now use this to help us describe the particular current which we started with, $I = \sqrt{3} \cos t - \sin t$?

Using the methods of Section 5.D.(g) we can rewrite $\sqrt{3} \cos t - \sin t$ in the form $2 \cos (t + \pi/6)$. (This is, in fact, the answer to the first question of Exercise 5.D.1 at the end of that section.)

As we saw there, we can then use the motion of P on its circle to help us to see how the wave function of the current behaves. I show this again below in Figure 10.C.4.

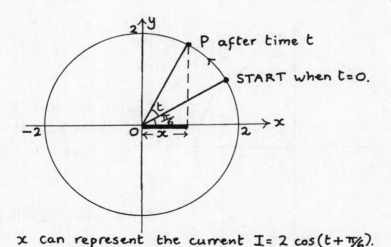

x can represent the current $I = 2 \cos (t + \pi/6)$.

Figure 10.C.4

Now here is the step which links all the work which we were doing there with our present work on complex numbers, and which makes it possible for us to avoid a lot of tedious trig.

We can think of the current $I = 2 \cos (t + \pi/6)$ as being the real part of the complex number $2e^{j(t + \pi/6)}$, or $2 \exp[j(t + \pi/6)]$ as this is sometimes written to avoid cramped-up complicated powers.

We can also use the first rule of powers to say that

$$2 \exp[j(t + \pi/6)] = 2 \exp(jt) \exp(j\pi/6) \quad \text{or} \quad (2e^{j\pi/6})(e^{jt}).$$

(We know from the rule for multiplying two complex numbers that we must multiply their moduli and add their arguments.)

Now, $2e^{j\pi/6}$ has a modulus of 2 (the radius of the circle), and an argument of $\pi/6$. It tells us the starting position of P and is called a **phasor**.

e^{jt} has a modulus of 1 and an argument of t, and this argument tells us how far P has turned after a time of t seconds.

Multiplying the two together has the effect of using the e^{jt} to drive P round the circle of radius two units. P starts from $2e^{j\,\pi/6}$, and one full cycle takes 2π seconds.

One benefit of this is that, if we have all the currents at a junction represented by the real parts of complex numbers, we will be able to add these currents by adding the complex numbers. We know that the real parts will remain separate from the imaginary parts so, at any stage of the working, the real part will still represent current. There are other advantages too, which you will discover if you are studying the theory of what is happening in these circuits as part of your other courses.

If the currents are given as sin functions, it is equally possible to represent them by the imaginary parts of complex numbers. The only difference would be that they would be being represented on the vertical or imaginary axis instead of the horizontal or real axis.

Also, everything would work equally well with a different common frequency for a circuit. So, for example, if we had started with the particular current of $I = 4 \cos(3t + \pi/8)$, then we would use the real part of the complex number $4 \exp[j\,(3t + \pi/8)]$ to represent this. The starting point of P would then be represented by the complex number $4e^{j\,\pi/8}$, and the e^{3jt} would then drive P round the circle as time passes. A full cycle would take $2\pi/3$ seconds. I show this in Figure 10.C.5 below.

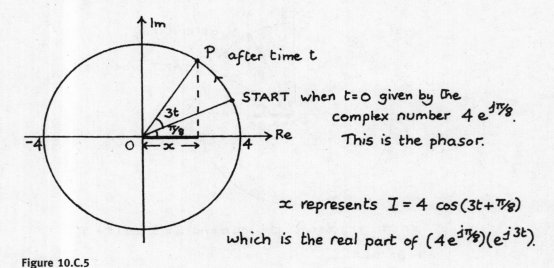

Figure 10.C.5

10.D   Using complex numbers to solve more equations

10.D.(a)   Finding the *n* roots of $z^n = a + bj$

We have already found solutions to some particular examples of this kind of equation, such as $z^4 = 1$ at the beginning of the chapter, and $z^3 = 1$ in question (3) of Exercise 10.B.2. We did this by thinking how the turns should go to make the answers fit in with the rules for multiplying complex numbers.

In this section, we look for some general rules which will help us to solve any equation of this kind. We do this by using the relationship $e^{j\theta} = \cos\theta + j\sin\theta$ which makes it possible for us to work out the answers for the n roots of the equation $z^n = a + bj$ in a very simple way. I shall use the thinking point from Exercise 10.B.2 at the end of Section 10.B.(c) to explain how this works. I asked you there if you could find the four roots of $z^4 = -1$, so you will also be able to see now if your answers were right.

We start by noticing that $|z^4| = 1$, because $z^4 = -1$.

We also know that $\arg(z^4) = \pi$.

I show a sketch of z^4 marked on the the unit circle in Figure 10.D.1(a).

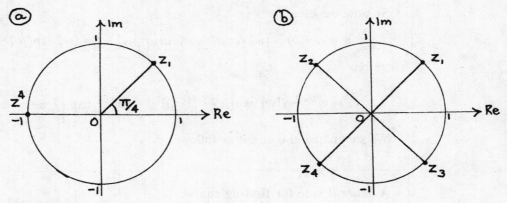

Figure 10.D.1

Geometrically, we can see that z_1 is a root since $|z_1| = 1$ and $\arg(z_1^4) = 4\arg z_1 = \pi$. The other three roots must also all have a modulus of 1 since, when multiplied by themselves, they give an answer with a modulus of 1. Because of this, all the roots will lie on the unit circle. They will be given by successive quarter turns from z_1, since four times any quarter turn makes a full turn. I have shown the four roots on Figure 10.D.1(b).

If we take z_2, for example, we will have $\arg(z_2) = \pi/4 + \pi/2$ giving $\arg(z_2^4) = 4(\pi/4 + \pi/2) = \pi + 2\pi$ using the rule for multiplying complex numbers from Section 10.B.(b). The extra full turn makes no difference to the point where we finish up, so $\arg(z_2^4) = \pi$.

Now we see how all this fits in with using $e^{j\theta} = \cos\theta + j\sin\theta$.

We have

$$z^4 = -1 = \cos(\pi) + j\sin(\pi) = e^{j\pi}$$

so $\quad z_1 = (e^{j\pi})^{1/4} = e^{j\pi/4} = \cos(\pi/4) + j\sin(\pi/4)$.

We can get *all* the roots by writing the arg of z^4 so that it includes the possibility of any whole number of extra full turns.

If we say $\arg(z^4) = \pi + 2k\pi$, where k is any whole number, then

$$z^4 = e^{j(\pi + 2k\pi)} \quad \text{so} \quad z = (e^{j(\pi + 2k\pi)})^{1/4} = e^{j(\pi + 2k\pi)/4}.$$

This power of e is rather cramped and difficult to read. Often such powers are written using exp instead of e so, for example, $e^{j\theta}$ would be written as $\exp(j\theta)$. Here we would have

$$z = \exp\left[j\frac{(\pi + 2k\pi)}{4}\right].$$

Putting $k = 0$, 1, -1, and -2 gives the four distinct roots of $\exp(j\pi/4)$, $\exp[j(3\pi/4)]$, $\exp[j(-\pi/4)]$, and $\exp[j(-3\pi/4)]$. These are the four roots z_1, z_2, z_3 and z_4 which I have shown in Figure 10.D.1(b) above. There is one quarter of a full turn between each root and the next one, and any other value of k will give you one of the roots which you already have.

In exactly the same way, we can make a general rule for solving any equation of the form $z^n = a + bj$.

Let $z^n = w$, so $|z^n| = |w|$ and $\arg(z^n) = \arg w = \theta$ say.

Then, just as we did in the previous example, we can say that $\arg z^n = \theta + 2k\pi$ where k is any integer (that is, whole number), since adding any number of full turns leaves the argument unchanged.

So now we have

$$w = a + bj = |w|\,(\cos\theta + j\,\sin\theta) = |w|\,e^{j\theta} = |w|\,\exp[j(\theta + 2k\pi)]$$

Therefore

$$z = w^{1/n} = (|w|\,\exp[j(\theta + 2k\pi)])^{1/n} = |w|^{1/n}\,\exp\left[j\,\frac{(\theta + 2k\pi)}{n}\right].$$

We summarise this result as follows:

A general rule for finding roots

If $z^n = w$ then $z = |w|^{1/n}\,\exp\left[j\,\frac{(\theta + 2k\pi)}{n}\right]$

where the n distinct roots are given by the n different integer values of k which make the arguments lie in the range $-\pi < \arg z \leq \pi$.

HELPFUL HINT   You should always draw a sketch of the roots on an Argand diagram so that you choose the right values for k.

EXAMPLE (1)   We will now find the eight distinct roots of $z^8 = 8 + 8\sqrt{3}j$ using the above result.

I let $z^8 = w$, and start by drawing the sketch of w shown in Figure 10.D.2(a). We have

$$|w| = \sqrt{8^2 + (8\sqrt{3})^2} = \sqrt{256} = 16$$

and

$$\arg w = \tan^{-1}\left(\frac{8\sqrt{3}}{8}\right) = \tan^{-1}\sqrt{3} = \frac{\pi}{3}.$$

Using the symmetry of the eight roots, I now draw the sketch of where they will be in Figure 10.D.2(b). I am using $\arg z_1 = \frac{1}{8}\arg w$.

Since we can add any number of full turns to $\arg w$ without altering it, we can say $\arg w = 2k\pi + \pi/3$ where k is any whole number.

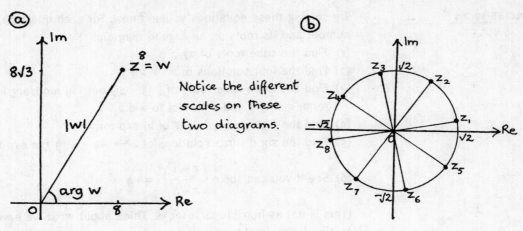

Figure 10.D.2

Therefore

$$w = 16[\cos(2k\pi + \pi/3) + j\sin(2k\pi + \pi/3)]$$

$$= 16 \exp [j(2k\pi + \pi/3)].$$

So, letting $z^8 = w$, we can say

$$z = w^{1/8} = (16)^{1/8} \exp \left[j \left(\frac{2k\pi + \pi/3}{8} \right) \right]$$

$$= \sqrt{2} \exp \left[j \left(\frac{6k\pi + \pi}{24} \right) \right].$$

From the sketch, we can see that we should put $k = 0, 1, 2, 3, -1, -2, -3$ and -4.

Equally, you could find the solutions to $z^8 = 8 + 8\sqrt{3}j$ by just using the geometry of your sketch.

You know that $|z^8| = 16$ and arg $z^8 = \pi/3$ so therefore z_1 (the first root you come to when you turn anticlockwise from the positive real axis) is given by $|z_1| = 16^{1/4} = \sqrt{2}$ and arg $z_1 = \frac{1}{8} \times \pi/3 = \pi/24$.

All of the roots have this same modulus of $\sqrt{2}$.

Also, each of the eight roots is one eighth of a full turn (that is, $2\pi/8$ or $\pi/4$) from the next one. Therefore, by looking at the diagram, it is easy to see what the correct argument for each one should be.

If you want your answers in the $a + bj$ form, you simply use the relation of $e^{j\theta} = \cos\theta + j\sin\theta$ to do the conversion.

For example, if you want z_3, all you need to do is to put $k = 2$. This gives

$$z_3 = \sqrt{2} \exp \left[j \left(\frac{12\pi + \pi}{24} \right) \right] = \sqrt{2} \exp j \left(\frac{13\pi}{24} \right)$$

$$= \sqrt{2} \cos \left(\frac{13\pi}{24} \right) + j\sin \left(\frac{13\pi}{24} \right) = -0.185 + 1.414j \text{ to 3 d.p.}$$

(Notice that these numbers look reasonable from the sketch.)

Try solving these equations yourself now. For each question, sketch the starting number and its roots on an Argand diagram.

(1) Find the cube roots of $27j$.

(2) Find the four solutions of $z^4 = 1 + j$.

(3) Find the three cube roots of $\frac{1}{5}(3 - 4j)$ both in mod/arg form, and in $a + bj$ form, giving a and b correct to 2 d.p.

(4) Find the fifth roots of $-4 + 4j$ in exp form.

(5) Find the six distinct solutions of $z^6 = -4 - 4\sqrt{3}\, j$ in exp form.

(6) See if you can solve $\left(\dfrac{z-1}{z+1}\right)^4 = -4$.

(This is not as horrible as it looks. Think about what we have already found in this section before rushing into binomial expansions.)

10.D.(b)   Solving quadratic equations with complex coefficients

Suppose you have an equation like $z^2 + (2 + j)z - (3 + j) = 0$ to solve.

Students are sometimes alarmed by the presence of complex coefficients. There is no cause for concern, however. The usual quadratic equation formula can be used with an equation like this.

Here, we have $a = 1$, $b = 2 + j$ and $c = -(3 + j)$ so substituting in the formula gives

$$z = \frac{-(2 + j) \pm \sqrt{(2 + j)^2 + 4(3 + j)}}{2} = \frac{-(2 + j) \pm \sqrt{15 + 8j}}{2}.$$

Check that you also get $15 + 8j$ when you tidy up under the square root sign.

Now we have to find the two square roots of $15 + 8j$.

We use the method of Section 10.C.(a) to do this, so we let $w^2 = 15 + 8j$ where $w = u + jv$. (I have not used the letter z since it is already being used in this example.) We have

$$(u + jv)^2 = u^2 - v^2 + 2juv = 15 + 8j.$$

Equating the real parts gives

$$u^2 - v^2 = 15.$$

Equating the imaginary parts gives

$$2uv = 8 \quad \text{so} \quad v = \frac{4}{u}.$$

Therefore

$$u^2 - (4/u^2) = 15 \quad \text{so} \quad u^4 - 15u^2 + 16 = 0 \quad \text{so} \quad (u^2 - 16)(u^2 + 1) = 0.$$

u is real so the only two possible solutions are $u = +4$ or -4 giving $v = +1$ or -1.

The square roots of $15 + 8j$ are $\pm(4 + j)$. (Notice that this method of solution automatically gives us the $\pm$ which is included in the quadratic equation formula, and that together these roots would make a straight line on an Argand diagram.)

Now, substituting them back in the quadratic equation formula, we find that the two possible solutions for z are given by

$$z = \frac{-(2 + j) \pm (4 + j)}{2} = 1 \quad \text{or} \quad -3 - j.$$

Notice that these two roots do *not* form a conjugate pair because the coefficients of the original equation were not all real, but they *do* fit the rules from Section 2.D.(e) that the sum of the roots is $-b/a$, and the product of the roots is c/a. (Check this for yourself.)

EXERCISE 10.D.2

Try solving the following quadratic equations yourself. (You may find that your answers to Exercise 10.C.1 are helpful here.)

(1) $z^2 - (3 - 2j)z - (1 + 3j) = 0$ (2) $z^2 + (3 - 4j)z - (22 + 6j) = 0$

(3) $z^2 + (2 + j)z - (3 + j) = 0$ (4) $z^2 + z - (\frac{1}{2} + j) = 0$ (5) $z^2 - z - 1 + 3j = 0$

(6) $z^2 - 2(1 + j)z + 5(1 - 2j) = 0$ (7) $z^4 - (1 + j)z^2 + j = 0$

10.D.(c)   Solving cubic and quartic equations with complex roots

How we do this is very similar in many ways to what we did in Section 2.E.(a) in order to solve cubic and quartic equations with real roots. You might find it helpful to go back there before going on with this section.

The following two examples will show you these similarities, and also how we use the extra information if we are given one complex root.

EXAMPLE (1)   We will solve $f(z) = z^3 - 5z^2 + 9z - 5 = 0$.

We can easily find one root here. (Can you spot it?)

Putting $z = 1$ gives $f(z) = 0$ so $z = 1$ is a root, and this means that $(z - 1)$ is a factor. Matching up what we know gives us

$$f(z) = z^3 - 5z^2 + 9z - 5 = (z - 1)(z^2 + pz + 5)$$

where p stands for the number which we don't yet know.

Looking at the terms in z^2, we get

$$-5z^2 = -z^2 + pz^2 \quad \text{so} \quad p = -4.$$

(Check for yourself that this is right by looking at the terms with z.)

The other two roots of $f(z) = 0$ will come from solving $z^2 - 4z + 5 = 0$. They are

$$z = \frac{4 \pm \sqrt{16 - 20}}{2} = \frac{4 \pm 2j}{2} = 2 \pm j.$$

The three roots of $z^3 - 5z^2 + 9z - 5 = 0$ are $z = 1$, and $z = 2 \pm j$.

EXAMPLE (2)   Now we'll solve $z^4 + az^3 + bz^2 - 5z - 26 = 0$ given that $z = 2 + 3j$ is one root of the equation, and a and b are real. We'll also find the values of a and b since examiners often like being given this information.

Since the coefficients of this equation are all real numbers, any complex roots of it must come in conjugate pairs, because otherwise multiplying out the factors would give us stray unwanted j's.

Therefore what else must be a root?

There must be a second root of $z = 2 - 3j$.

Now, in just the same way as we said in the last example that if $z = 1$ is a root of $f(z) = z^3 - 5z^2 + 9z - 5 = 0$ then $(z - 1)$ is a factor of $f(z)$, here we can say that both $(z - (2 + 3j))$ and $(z - (2 - 3j))$ are factors.

These look rather alarmingly complicated, but if we multiply them together we shall get something very much nicer. We shall have

$$(z - (2 + 3j))\,(z - (2 - 3j)) = z^2 - (2 + 3j)z - (2 - 3j)z$$
$$+ (2 + 3j)\,(2 - 3j) = z^2 - 4z + 13.$$

The imaginary part has disappeared *because* the roots are conjugates, so making it possible for the original equation to have only real coefficients.

Now we can say that

$$z^4 + az^3 + bz^2 - 5z - 26 = (z^2 - 4z + 13)\,(z^2 + pz - 2)$$

matching up the first and last terms in the second bracket, since we know that when we multiply the two brackets together they must give us z^4 and -26. Again, p is standing for the number which we don't yet know.

Just as in the first example, we can now match up the terms for the various powers of z, since the right-hand side is just another way of writing the left-hand side.

We have already matched the z^4 and number terms.

Matching the terms in z, we get

$$-5z = 13pz + 8z \quad \text{so} \quad p = -1.$$

Now we have

$$z^4 + az^3 + bz^2 - 5z - 26 = (z^2 - 4z + 13)(z^2 - z - 2)$$

so we'll match up the terms in z^3 to find a, and in z^2 to find b. Notice here that we have to get the arithmetic right first time; we can't use these equations as checks because we need them to find new information.

Matching up the terms with z^3 gives us

$$az^3 = -4z^3 - z^3 \quad \text{so} \quad a = -5.$$

Matching up the terms in z^2 gives us

$$bz^2 = 13z^2 - 2z^2 + 4z^2 \quad \text{so} \quad b = 15.$$

(We have to be careful with this one because there are three ways in which we can make z^2 on the right-hand side.)

We now have

$$z^4 - 5z^3 + 15z^2 - 5z - 26 = (z^2 - 4z + 13)(z^2 - z - 2) = 0.$$

To find the other two roots, we now just have to solve the equation $z^2 - z - 2 = 0$. Now $z^2 - z - 2 = (z - 2)(z + 1)$ so the two solutions are $z = 2$ and $z = -1$. In this particular example, they have both turned out to be real.

It has been shown by mathematicians that every equation like the one that we started with here will have the same number of roots as the highest power of z, if we both allow complex roots, and count roots such as $x = 1$ from $(x - 1)^2 = 0$ as a double root. (Here, the curve is just sitting on the x-axis instead of cutting it in two places.) Also, if all the coefficients of an equation are real numbers, then the roots must come in conjugate pairs. How does this apply to the two roots of $z = 2$ and $z = -1$ which we have just found?

Each of these is its own conjugate because they are real numbers.

EXAMPLE (3) Now we will try one together, before I give you an exercise on these.

Given that a and b are real, and that $z = 2 + j$ is a root of the equation $z^4 + az^3 - 9z^2 + bz + 30 = 0$, find the values of a and b and the other three roots.

(1) What else must be a root?

Since a and b are real, the conjugate of $2 + j$ which is $2 - j$ must also be a root.

(2) From these two roots, what two factors must you have?

We have the two factors

$$(z - (2 + j)) \quad \text{and} \quad (z - (2 - j)).$$

(3) When you multiply these two together, what single nice factor do you get?

You get

$$z^2 - (2 + j)z - (2 - j)z + (2 + j)(2 - j) = z^2 - 4z + 5.$$

(4) Use all this information to write $z^4 + az^3 - 9z^2 + bz + 30$ as two factors multiplied together.

You should have

$$z^4 + az^3 - 9z^2 + bz + 30 = (z^2 - 4z + 5)(z^2 + pz + 6)$$

where p is standing for the number which we still have to find.

(5) Now, match up the terms in z^2 to find p.

You should have

$$-9z^2 = 5z^2 + 6z^2 - 4pz^2 \quad \text{so} \quad p = 5.$$

(Notice the three ways of getting z^2 on the RHS again!)
We now have
$$z^4 + az^3 - 9z^2 + bz + 30 = (z^2 - 4z + 5)(z^2 + 5z + 6).$$

(6) Now match up the terms in z^3 and z to find a and b.

You should have

$$az^3 = -4z^3 + 5z^3 \quad \text{so} \quad a = 1,$$

$$bz = -24z + 25z \quad \text{so} \quad b = 1 \text{ also.}$$

(7) Finally, what are the other two roots of the given equation?

We have

$$z^2 + 5z + 6 = 0 \text{ if } (z + 3)(z + 2) = 0$$

so the other two roots are $z = -3$ and $z = -2$.

It's worth noticing that if you were only asked for the roots then you could have left out step (6) altogether. You don't need to know the values of a and b to solve the equation.

It is also possible to answer the original question by substituting $z = 2 + j$ in the original equation, since we know that it fits it. Then we would equate the real and imaginary parts to find a and b, and also find the first factor with z^2 as we did above. Then you either use long division to find the other factor, or match up the terms as we did above. This method involves using binomial expansions to work out $(2+j)^4$ and $(2+j)^3$, always a fruitful source of arithmetical mistakes. I think that you will find that the method I have shown you is easier.

EXERCISE 10.D.3

Now have a go at solving these equations yourself.
(1) Given that a and b are real numbers, and that $z = 1 - 2j$ is a root of the equation $z^4 - 3z^3 + az^2 + bz - 30 = 0$, find a and b and the other three roots.
(2) Given that a and b are real numbers, and that $z = j$ is a root of the equation $z^4 + az^3 + bz^2 - 4z + 13 = 0$, find the values of a and b and the other three roots.
(3) Given that a and b are real numbers, and that $z = 1 - j$ is a root of the equation $z^4 + az^3 + bz^2 + 1 = 0$, find a and b and the other three roots.

10.E   Finding where z can be if it must fit particular rules

In Chapter 3, when we were working with functions, we often found that we had to restrict the choice of possible values for x in order to make the functions work as we wanted. One example of this is given by $f(x) = \sqrt{4 - x}$. To make $f(x)$ real, we have to have $x \leq 4$. This means that we have to restrict the possible values of x to just one part of the number line which makes up the horizontal axis.

The same kind of thing can happen with applications of complex numbers. To make things work in the way that we want, we may often find that we have to restrict the possible values of z. Since z is made up of both an x and y component from its real and imaginary parts, this restriction may lead to the exclusion of any area or region of the complex plane because it may affect the possible values of both x and y. (The complex plane is the flat surface shown on an Argand diagram.)

Physical quantities which have both magnitude and direction, and which are acting in a flat surface, can often be represented very conveniently by complex numbers. Examples of such applications are two-dimensional problems involving lines of electric or magnetic force, or streamlines in fluid flow. If the physical quantities you are considering need three dimensions to describe them, complex numbers will no longer be any use. You would probably then use vectors, which I plan to include in a second book.

The following section is designed to give you more practice in using complex numbers and seeing how particular restrictions could be described.

10.E.(a)   Some simple examples of paths or regions where z must lie

If complex numbers are described by saying that they obey some rule, we can use this information to show the possible points in the complex plane where these numbers may lie.

EXAMPLE (1) Suppose we are told that $|z| = 3$.

This means that the distance of the point z from the origin is always 3 units, so therefore z can lie anywhere on the circle shown in the Argand diagram in Figure 10.E.1.

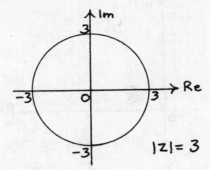

Figure 10.E.1

We could also write the equation of this circle using xs and ys as we did in Section 4.C.(d) by letting $z = x + jy$ and using the relationship $|z| = \sqrt{x^2 + y^2}$. Since $|z| = 3$, we then have $x^2 + y^2 = 3^2 = 9$ which is the equation of the circle with centre (0,0) and radius 3 units. All the possible positions of z lie on the path given by this circle. Such a path is sometimes called the **locus** of z.

There is a third way of describing this particular circle. We know from Section 10.C.(c) that $z = e^{j\theta}$ gives the unit circle about the origin as θ varies. Therefore $z = 3e^{j\theta}$ gives the unit circle enlarged by a factor of 3, and this is our circle.

EXAMPLE (2) Suppose we are told that $|z| \leq 2$.

We want the distance of z from the origin to be less than or equal to 2 units, so it must lie either on or inside the circle shown in Figure 10.E.2 below. This time the possible positions of z make a *region* in the plane, rather than a path given by a line as in Example (1). (You may find it helpful to use your own colour on these diagrams to highlight the different possible positions of z.)

This region can be described as either $|z| \leq 2$ or as $x^2 + y^2 \leq 2^2$ which is the same as $x^2 + y^2 \leq 4$.

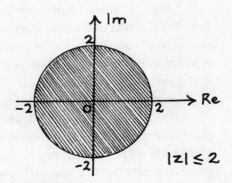

Figure 10.E.2

EXAMPLE (3) Suppose we are told that arg $z = \pi/3$.

The argument of these complex numbers is fixed but their modulus can take any value. This means that z can lie anywhere on the straight line shown below in Figure 10.E.3.

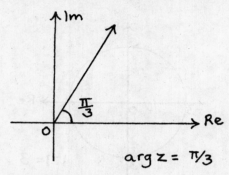

arg z = $\pi/3$

Figure 10.E.3

The arrow on the end of the line which represents the possible values of z shows that it can be extended indefinitely. Such a line is sometimes called a half-line because it can only be infinitely extended in one direction. Extending it in the opposite direction would include points for which arg $z = -2\pi/3$, and this doesn't fit the given condition.

EXERCISE 10.E.1

Use separate Argand diagrams to show the possible positions for z for each of the following.

(1) $|z| = 4$ (2) $|z| \leq 1$ (3) arg $z = -\pi/6$ (4) arg $z = 0$ (5) $|z| > 2$

(When the boundary of a region is not included, show it with a dashed line.)

10.E.(b)   **What do we do if z has been shifted?**

Each of the examples which we have looked at so far has been related directly to the origin. What will happen to the path if the information that we are given concerns a complex number which has been shifted away from there?

EXAMPLE (1) Suppose we are told that $|z - 5| = 2$. Where is the path which describes the possible positions of z now?

If we let $z - 5 = w$ then we know that the path for w is the circle about the origin with a radius of 2 units. But what do we have to do to z to get w?

We have to subtract 5 from z which means that we are taking 5 away from the real part of z since 5 itself is real. Therefore we are shifting z by 5 units to the left to get w. So the circle giving the path for z must be 5 units to the *right* of the origin. I show both these circles on the pair of Argand diagrams in Figure 10.E.4.

We can check that the path of z is in the right place by putting $z = 7$. Then $|7 - 5|$ is indeed equal to 2, and we see from the diagram that $z = 7$ *is* a point on the path of z. The equation of this circle in terms of x and y is given by $(x - 5)^2 + y^2 = 2^2$ or $x^2 - 10x + y^2 + 21 = 0$ because its centre is at $(5, 0)$ and its radius is 2 units. (If you need help with this, you should go to Section 4.C.(d).)

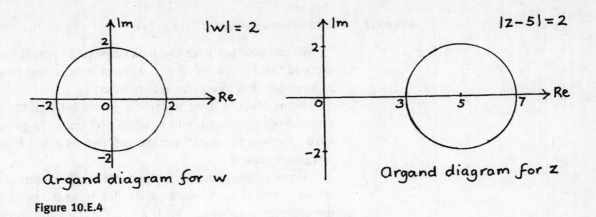

Figure 10.E.4

EXAMPLE (2) What would the path of z be if we are told that $|z - 3j| = 1$?

Try drawing a sketch for yourself of where you think it would be.
Check your sketch by using $z = 4j$ as a test point, and write down
the equation of the path of z in terms of x and y.

If we put $w = z - 3j$ then the path of w is the circle about the origin
with a radius of one unit since $|w| = 1$.

To get w from z, we have to subtract 3 from the imaginary part of z.

Therefore the path of z is the circle whose radius is one unit and
whose centre is at the point (0,3). That is, the centre of the circle is at
the point which represents the complex number 3j. I show the w circle
and the z circle in the pair of Argand diagrams in Figure 10.E.5.

The equation of this circle can also be written as $x^2 + (y - 3)^2 = 1^2$
or $x^2 + y^2 - 6y + 8 = 0$.

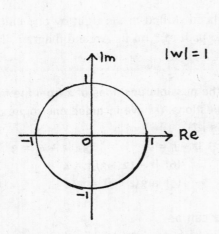

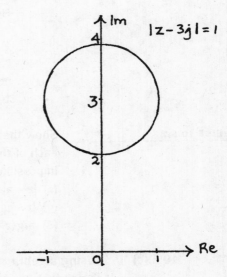

Argand diagram for w

Argand diagram for z

Figure 10.E.5

EXAMPLE (3) The general case of $|z - (a + bj)| = k$.

We can now see what the path of possible points for z will be if we are told that $|z - (a + bj)| = k$, where $a + bj$ is any given complex number and k is any given real number.

This is because if we put $w = z - (a + bj)$ then the path of w is the circle about the origin with a radius of k units. To get to w from z, we have to subtract a units from the real part of z and b units from the imaginary part of z.

Therefore the path of z is the circle whose radius is k units and whose centre is at the point (a, b), that is, at the point which represents the complex number $a + bj$. If $z = x + jy$ then the equation of this circle can be written as $(x - a)^2 + (y - b)^2 = k^2$.

EXAMPLE (4) Suppose we are told that arg $(z + 1) = \pi/3$.

This time, we want to find the path of z from knowing what its argument is after it has been shifted.

If we let $z + 1 = w$ then we can immediately sketch the position of w on an Argand diagram. I have done this in Figure 10.E.6(a).

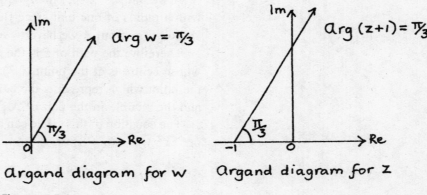

Figure 10.E.6

Since z has been shifted to the right by one unit to give w, we can now draw in the path of z on its Argand diagram. I show this in Figure 10.E.6(b).

EXERCISE 10.E.2

Show the paths giving the possible positions of z from the information given in each of the following questions. (I have included one rogue question which is impossible. Which one is it?)

(1) $|z - 3| = 1$ (2) $|z - j| = 2$ (3) $|z + 4| = 3$ (4) $|z + 2j| = 3$
(5) $|z - 1| = -2$ (6) $|z - (2 + 3j)| = 4$ (7) $|z + 3 - 2j| = 3$
(8) $\arg(z - 2) = \pi/6$ (9) $\arg(z + j) = \pi/4$ (10) $\arg(z + 3) = -\pi/3$

10.E.(c) **Using algebra to find where z can be**

It is possible to use algebra to find where z must be in the complex plane if it has to fit certain conditions. If the geometry of what is happening gets more complicated then it is often easier to take this approach.

EXAMPLE (1) I will start by showing you how we could have used algebra to solve the first example that we looked at in Section 10.E.(b), so that you can see how we get the same answer.

We were told that $|z - 5| = 2$ and we had to find the path which z could take in order to fit this condition.

We start by putting $z = x + jy$ which gives us $|x + jy - 5| = 2$.

Next, we tidy up the $x + jy - 5$ to show the real and imaginary parts separately and clearly. Doing this gives us

$$|(x - 5) + jy| = 2.$$

Therefore we can say that

$$\sqrt{(x - 5)^2 + y^2} = 2 \qquad (1)$$

using the rule that $|a + bj| = \sqrt{a^2 + b^2}$ from Section 10.A.(c).

Don't be tempted to think this should be $\sqrt{(x - 5)^2 - y^2} = 2$.

Remember that the j is telling you that the y is measured vertically. We then use Pythagoras' Theorem on the right-angled triangle shown in the sketch in Figure 10.E.7 to get $|w|$ where $w = (x - 5) + jy$.

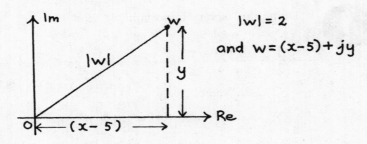

Argand diagram for w

Figure 10.E.7

Squaring both sides of equation (1) above gives us $(x - 5)^2 + y^2 = 2^2 = 4$.

This is the equation of the circle with centre $(5,0)$ and radius 2, which agrees exactly with what we obtained at the beginning of Section 10.E.(b) for the path of z by using the geometry of the diagram.

EXAMPLE (2) My second example involves a more complicated situation where it is definitely easier to use algebra.

We will find where z can be in the complex plane if

$$\left| \frac{z + 2}{z + j} \right| \leq 2.$$

From the rule for dividing complex numbers, we know that

$$\left|\frac{z_1}{z_2}\right| = \frac{|z_1|}{|z_2|}$$

so here we have

$$\left|\frac{z+2}{z+j}\right| = \frac{|z+2|}{|z+j|}$$

which means that

$$\frac{|z+2|}{|z+j|} \le 2.$$

This gives us

$$|z+2| \le 2\,|z+j|.$$

It is possible to do this last step because we know that $|z+j|$ is positive since it is a length. Therefore, multiplying both sides of the inequality by it does not change the $\le$ sign to a $\ge$ sign. (If k is a positive number, and $a \le b$, then $ka \le kb$ also.)

Next, we put $z = x + jy$ and then carefully tidy up to show the real and imaginary parts separately and clearly.

 Doing this is very important!

We get

$$|x + jy + 2| \le 2\,|x + jy + j|$$

so $\qquad |(x + 2) + jy| \le 2\,|x + j(y + 1)|$

so $\qquad \sqrt{(x + 2)^2 + y^2} \le 2\,\sqrt{x^2 + (y + 1)^2}.$

Since both sides are positive, the inequality remains true in the same sense if we square them both. (If they weren't both positive this would not necessarily be so. For example, $-3 < 2$ but $(-3)^2 > (2)^2$.)

Squaring both sides gives us

$$(x + 2)^2 + y^2 \le 4(x^2 + (y + 1)^2).$$

 Don't forget to square the 2!

Now we have

$$x^2 + 4x + 4 + y^2 \le 4x^2 + 4y^2 + 8y + 4$$

so $\qquad 0 \le 3x^2 - 4x + 3y^2 + 8y \quad$ or $\quad 3x^2 - 4x + 3y^2 + 8y \ge 0.$

We then use the method of completing the squares to get the centre and radius of the boundary circle. (See Sections 2.D.(b) and 4.C.(d) if necessary.) It is easier to complete the squares here if we divide all through by 3 to get x^2 and y^2.

Doing this gives

$$x^2 - \frac{4}{3}x + y^2 + \frac{8}{3}y \geq 0$$

so $\quad (x - \frac{2}{3})^2 - \frac{4}{9} + (y + \frac{4}{3})^2 - \frac{16}{9} \geq 0 \quad$ or $\quad (x - \frac{2}{3})^2 + (y + \frac{4}{3})^2 \geq \frac{20}{9}.$

The boundary of the region where z can lie is the circle whose centre is $(\frac{2}{3}, -\frac{4}{3})$ and whose radius is $\sqrt{20/9} = (2\sqrt{5})/3$. (If you have any trouble with this step, you should go back to Section 1.F.(c) for help.)

Since the form of the inequality tells us that the distance of the point representing z from the centre of this circle is greater than or equal to its radius, the region where z may lie is the circumference of the circle and everywhere outside it.

I show this in Figure 10.E.8. The centre of the circle is given by the complex number $\frac{2}{3} - \frac{4}{3}j$.

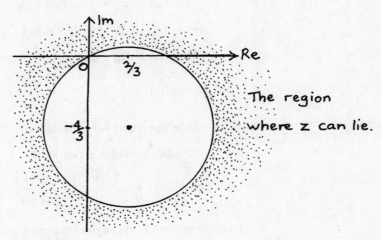

Figure 10.E.8

EXAMPLE (3) Find the region in the complex plane in which z can lie if

$$\text{Im}\left(\frac{z}{z + 4j}\right) \geq 2.$$

First, we put $z = x + jy$, giving us

$$\text{Im}\left(\frac{x + jy}{x + j(y + 4)}\right) \geq 2.$$

Next, we tidy up the fraction

$$\frac{x + jy}{x + j(y + 4)}$$

so that we can easily see what its imaginary part is.

We do this tidying up by multiplying its top and bottom by the conjugate of the bottom which gives

$$\frac{(x+jy)}{(x+j(y+4))}\frac{(x-j(y+4))}{(x-j(y+4))} = \frac{x^2 + y(y+4) + j(xy - xy - 4x)}{x^2 + (y+4)^2}$$

so

$$\text{Im}\left(\frac{z}{z+4j}\right) = \frac{-4x}{x^2 + (y+4)^2}.$$

Now we put

$$\frac{-4x}{x^2 + (y+4)^2} \geq 2$$

to see where z can lie. This gives us

$$-4x \geq 2x^2 + 2(y+4)^2.$$

(We can do the above multiplication because we know that $x^2 + (y+4)^2$ is positive since x and y are both real.)

Dividing by 2 gives us

$$-2x \geq x^2 + (y+4)^2 \quad \text{so} \quad x^2 + 2x + (y+4)^2 \leq 0$$

so

$$(x+1)^2 + (y+4)^2 \leq 1.$$

This means that z can lie anywhere on or inside the circle whose centre is at the point $-1 - 4j$ and whose radius is one unit. Sketch this for yourself.

EXERCISE 10.E.3

Now try these questions for yourself.

(1) Find the path on which z can lie if

  (a) $|z - 1| = |z + 1|$,   (b) $\left|\dfrac{z+1}{2z-j}\right| = 1$.

(2) Find and sketch the regions where z can lie in the complex plane for each of the following given conditions.

  (a) $|z| \leq |z - j|$ (b) $|z| \leq \sqrt{2}\,|z - j|$ (c) $2|z| \leq |2z - j|$

(3) Solve the equation $\text{Re}\left(\dfrac{z}{z+1}\right) = \dfrac{3}{2} + jz$.

10.E.(d)   **Another example involving a relationship between w and z**

In the examples of Section 10.E.(b) we saw how we could write rules given for a variable number z so that they were in terms of a new variable number w instead. I shall finish Section 10.E by looking at an example where we are given a relationship between z and w which isn't just a simple shift. We shall then use this to find out how making z obey a special rule will affect w.

We will start with the relationship

$$w = \frac{z+j}{z-j}$$

with $w = u + jv$ and $z = x + jy$.

We'll now answer some questions about how this relationship works.

(1)   Can you see a value of z which we must exclude?

---

(1)   We can't have $z = j$ since this would give us a zero on the bottom of the fraction.

(2)   Each value of z will give a corresponding value of w and vice versa.
       What will w be if (a) $z = 1 + j$,   (b) $z = 3 + j$,   (c) $z = 3 - j$?
       Try answering this yourself, giving your answers in as simple a form as you can

---

(2)   You should get

(a) $w = 1 + 2j$ (b) $w = \dfrac{3 + 2j}{3} = 1 + \tfrac{2}{3}j$ (c) $w = \dfrac{3}{3 - 2j}$.

Simplifying (c) by multiplying top and bottom by the conjugate $3 + 2j$ gives

$$w = \left(\frac{3}{3 - 2j}\right)\left(\frac{3 + 2j}{3 + 2j}\right) = \frac{9 + 6j}{9 + 4} = \tfrac{9}{13} + \tfrac{6}{13}j$$

(3)   What are the real and imaginary parts of w in terms of the real and imaginary parts of z?
       Putting $w = u + jv$ and $z = x + jy$ gives

$$w = u + jv = \frac{(x + jy) + j}{(x + jy) - j} = \frac{x + j(y + 1)}{x + j(y - 1)}.$$

Now we simplify this fraction by multiplying its top and bottom by the complex conjugate of the bottom. Try doing this for yourself.

---

(3)   You should have

$$u + jv = \left(\frac{x + j(y + 1)}{x + j(y - 1)}\right)\left(\frac{x - j(y - 1)}{x - j(y - 1)}\right) = \frac{x^2 + y^2 - 1 + 2jx}{x^2 + (y - 1)^2}.$$

(4)   Now see if you can write down what u and v are in terms of x and y.

---

(4)   Equating the real and imaginary parts from above should give you

$$u = \frac{x^2 + y^2 - 1}{x^2 + (y - 1)^2} \quad \text{and} \quad v = \frac{2x}{x^2 + (y - 1)^2}.$$

(5)   If we now make the special condition that the real and imaginary parts of w are equal, so that w lies on the straight line $u = v$, what path will z lie on?
       If $u = v$ it must be true that $x^2 + y^2 - 1 = 2x$ since the bottoms of the fractions are the same.
       This gives us $x^2 - 2x + y^2 - 1 = 0$ so $(x - 1)^2 + y^2 = 2$.
       Therefore z lies on a circle whose centre is at $(1,0)$ and whose radius is $\sqrt{2}$.

---

(5)   The easiest way to show what is happening geometrically is to draw two separate Argand diagrams for each of w and z.

Doing this gives us Figure 10.E.9

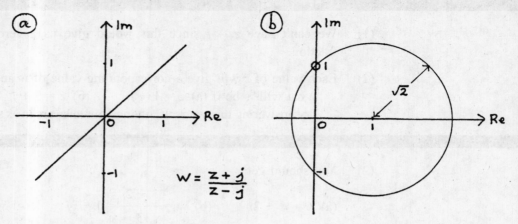

Argand diagram for $w = u + jv$ Argand diagram showing
with $u = v$ the corresponding path for

Figure 10.E.9

The little circle on the imaginary axis at $+1$ in (b) is to show that $z = j$ isn't allowed.

(6)   We will finish off this example by finding out how a few points will transfer from one Argand diagram to the other. (Notice that none of the points which we found in (1) will help us here as u and v are not equal for any of them.)

We will look first at how two points will move from (b) to (a).

What is w when (i) $z_1 = -j$ and (ii) $z_2 = 2 + j$?

---

(6)   (i)   If $z_1 = -j$ then $w_1 = 0$.
       (ii)   If $z_2 = 2 + j$ then

$$w_2 = \frac{2 + 2j}{2} = 1 + j.$$

Mark these two pairs of points on the Argand diagrams. (You can see that $z_2 = 2 + j$ really is on the circle of (b) if you put $x = 2$ and $y = 1$ in its equation.)

Now we'll go the other way. It will be easier to do the working out if we start by finding out how we can write z in terms of w. We have

$$w = \frac{z + j}{z - j} \quad \text{so} \quad wz - wj = z + j$$

so $z(w - 1) = j(w + 1)$ so $z = \dfrac{j(w + 1)}{w - 1}$

We only want points for which $u = v$.

What is z if (iii) $w_3 = \frac{1}{2} + \frac{1}{2}j$ and (iv) $w_4 = -1 - j$?

(iii)　If $w_3 = \frac{1}{2} + \frac{1}{2}j$ we get $z_3 = \dfrac{j\left(\frac{3}{2} + \frac{1}{2}j\right)}{-\frac{1}{2} + \frac{1}{2}j} = \dfrac{3j - 1}{j - 1}$.

Tidying up by multiplying top and bottom by $j + 1$, we get

$$\left(\frac{3j - 1}{j - 1}\right)\left(\frac{j + 1}{j + 1}\right) = \frac{-4 + 2j}{-2} = 2 - j = z_3.$$

(iv)　$w_4 = -1 - j$　gives　$z_4 = \dfrac{j(-j)}{-2 - j} = \left(\dfrac{-1}{2 + j}\right)\left(\dfrac{2 - j}{2 - j}\right) = -\dfrac{2}{5} + \dfrac{1}{5}j.$

Mark these two pairs of points on the two Argand diagrams of Figure 10.E.9. as well.

One important application of relationships like this is in the description of electric circuits by using complex numbers. Working as we have done above could make it possible to find out what happens to, say, the output impedance of a circuit from a given variation in the input impedance.

EXERCISE 10.E.4

Try this example for yourself now.

If $w = u + jv$ and $z = x + jy$ and w and z are related by the equation

$$w = \frac{z + j}{z - 1},$$

answer the following questions.

(1) What value of z must we exclude?

(2) What is w if (a) $z = 2$ and (b) $z = 1 - j$?

(3) Find the real and imaginary parts of w in terms of the real and imaginary parts of z.

(4) If the real and imaginary parts of w are equal, find the path on which z must lie.

(5) Show the paths of z and w on two separate Argand diagrams.

(6) Find w if　(a) $z_1 = -j$　(b) $z_2 = 2 - j$　(c) $z_3 = 1 - 2j$　(d) $z_4 = \frac{2}{5} - \frac{1}{5}j$
(e) $z_5 = \frac{1}{5} - \frac{2}{5}j$　(f) $z_6 = \frac{8}{5} - \frac{1}{5}j$.
Mark each of these pairs of points on your two Argand diagrams.

(7) If $u = -v$ what is the path on which z must lie?

(8) If $v = 0$, so that w lies on the real axis, where must z lie?

(9) Sketch two diagrams showing the three paths of w from (4), (7) and (8) on the first diagram and the corresponding three paths of z from (4), (7) and (8) on the second diagram.

(10) The three paths of w cut each other at the origin.
The three corresponding paths of z cut each other at $z = -j$.
Compare the angles between the paths at these two points on your two sketches.
What do you find?

Answers to the exercises

Chapter 1
Exercise 1.A.1

(1) (a) $8a + b + 3c$ ($1b$ is written as just b.)   (b) $5ab + 5a + 3b$ (since $ab = ba$.)
(c) $5p + 5pq + 8q$ (d) $2x + 5y + 3xy$

(2) (a) $2^3 = 8$ (b) $5a^2 = 5 \times 2^2 = 20$ (c) $(5a)^2 = 10^2 = 100$ (d) $1^2 = 1$
(e) $8 + 3 = 11$

(3) (a) $6xy$ (b) $15x^3y$ (c) $6a + 9b$ (d) $6a^2 + 10ab$ (e) $6p^3 + 4p^2q + 2pq^2$
(f) $6x^3 + 4x^3y + 2x^2y^2$

Exercise 1.A.2

(1) (a) $ab + cd = 2 \times 3 + 4 \times 5 = 6 + 20 = 26$ (b) $ab^2e = 2 \times 3^2 \times 0 = 0!$
(c) $ab^2d = 2 \times 3^2 \times 5 = 2 \times 9 \times 5 = 90$ (d) $(abd)^2 = (2 \times 3 \times 5)^2 = 30^2 = 900$
(e) $a(b + cd) = 2(3 + 4 \times 5) = 2(3 + 20) = 46$
(f) $ab^2d + c^3 = 2 \times 3^2 \times 5 + 4^3 = 90 + 64 = 154$
(g) $ab + d - c = 2 \times 3 + 5 - 4 = 6 + 5 - 4 = 7$
(h) $a(b + d) - c = 2(3 + 5) - 4 = 2 \times 8 - 4 = 16 - 4 = 12$

(2) (a) $3x(2x + 3y) + 4y(x + 7y) = 6x^2 + 9xy + 4xy + 28y^2 = 6x^2 + 13xy + 28y^2$
(b) $5p^2 (2p + 3q) + q^2 (3p + 5q) + pq(p + 2q) = 10p^3 + 15p^2q + 3pq^2 + 5q^3 + p^2q +$
$2pq^2 = 10p^3 + 16p^2q + 5pq^2 + 5q^3$

(3) (a) $a^2 = 3^2 = 3 \times 3 = 9$ (b) $3b^2 = 3 \times 4^2 = 3 \times 16 = 48$ (c) $(3b)^2 = (12)^2 = 144$

Notice this last pair! Students quite often confuse (b) and (c). In (b), only the b is squared.

(d) $c^2 = 1^2 = 1 \times 1 = 1$ (not 2!)   (e) $ab + c = 3 \times 4 + 1 = 12 + 1 = 13$
(f) $bd - ac = 4 \times 5 - 3 \times 1 = 20 - 3 = 17$ (g) $b(d - ac) = 4(5 - 3 \times 1) = 4(2) = 8$
(h) $d^2 - b^2 = 5^2 - 4^2 = 25 - 16 = 9$ (i) $(d - b) (d + b) = (5 - 4) (5 + 4) = 9$
(j) $d^2 + b^2 = 5^2 + 4^2 = 25 + 16 = 41$ (k) $(d + b)(d + b) = (5 + 4)(5 + 4) = 9 \times 9 = 81$
(l) $a^2 + c^2d = 3^2 \times 4 + 1^2 \times 5 = 9 \times 4 + 1 \times 5 = 36 + 5 = 41$
(m) $5e(a^2 - 3b^2) = 0$ since it is all multiplied by 0
(n) $a^b + d^a = 3^4 + 5^3 = 81 + 125 = 206$

(4) (a) $3a(2b + 3c) + 2a(b + 5c) = 6ab + 9ac + 2ab + 10ac = 8ab + 19ac$
(b) $2xy(3x^2 + 2xy + y^2) = 6x^3y + 4x^2y^2 + 2xy^3$
(c) $5p(2p + 3q) + 2q(3p + q) = 10p^2 + 15pq + 6pq + 2q^2 = 10p^2 + 21pq + 2q^2$
(d) $2c^2(3c + 2d) + 5d^2 (2c + d) = 6c^3 + 4c^2d + 10cd^2 + 5d^3$

Exercise 1.A.3

(1) $2x - (x - 2y) + 5y = 2x - x + 2y + 5y = x + 7y$
(2) $4(3a - 2b) - 6(2a - b) = 12a - 8b - 12a + 6b = -2b$
(3) $6(2c + d) - 2(3c - d) + 5 = 12c + 6d - 6c + 2d + 5 = 6c + 8d + 5$

(4) $6a - 2(3a - 5b) - (a + 4b) = 6a - 6a + 10b - a - 4b = -a + 6b$

(5) $3x(2x - 3y + 2z) - 4x(2x + 5y - 3z) = 6x^2 - 9xy + 6xz - 8x^2 - 20xy + 12xz = -2x^2 - 29xy$
       $+ 18xz$

(6) $2xy(3x - 4y) - 5xy(2x - y) = 6x^2y - 8xy^2 - 10x^2y + 5xy^2 = -4x^2y - 3xy^2$

(7) $2a^2(3a - 2ab) - 5ab(2a^2 - 4ab) = 6a^3 - 4a^3b - 10a^3b + 20a^2b^2 = 6a^3 - 14a^3b + 20a^2b^2$

(8) $-3p - p - q + 2pq - 6q = -4p - 7q + 2pq$

Exercise 1.A.4

(1) $5(a + 2b)$ (2) $a(3a + 2b)$ (3) $3a(a - 2b)$ (4) $x(5y + 8z)$

(5) $5x(y - 2z)$ (6) $ab(a + 3b)$ (7) $2pq(2q - 3p)$ (8) $x^2y^2(3y + 5x)$

(9) $2pq(2p + q - 3pq)$ (10) $a^2b^2(2b + 3a - 6)$

Exercise 1.B.1

(1) $(x + 2)(x + 3) = x^2 + 2x + 3x + 6 = x^2 + 5x + 6$

(2) $(a + 3)(a - 4) = a^2 + 3a - 4a - 12 = a^2 - a - 12$

(3) $(x - 2)(x - 3) = x^2 - 2x - 3x + 6 = x^2 - 5x + 6$

(4) $(p + 3)(2p + 1) = 2p^2 + 6p + p + 3 = 2p^2 + 7p + 3$

(5) $(3x - 2)(3x + 2) = 9x^2 - 6x + 6x - 4 = 9x^2 - 4$

(6) $(2x - 3y)(x + 2y) = 2x^2 - 3xy + 4xy - 6y^2 = 2x^2 + xy - 6y^2$

(7) $(3a - 2b)(2a - 5b) = 6a^2 - 4ab - 15ab + 10b^2 = 6a^2 - 19ab + 10b^2$

(8) $(3x + 4y)^2 = (3x + 4y)(3x + 4y) = 9x^2 + 12xy + 12xy + 16y^2 = 9x^2 + 24xy + 16y^2$

(9) $(3x - 4y)^2 = (3x - 4y)(3x - 4y) = 9x^2 - 12xy - 12xy + 16y^2 = 9x^2 - 24xy + 16y^2$

(10) $(3x + 4y)(3x - 4y) = 9x^2 + 12xy - 12xy - 16y^2 = 9x^2 - 16y^2$

(11) $(2p^2 + 3pq)(5p + 3q) = 10p^3 + 15p^2q + 6p^2q + 9pq^2 = 10p^3 + 21p^2q + 9pq^2$

(12) $(2ab - b^2)(a^2 - 3ab) = 2a^3b - a^2b^2 - 6a^2b^2 + 3ab^3 = 2a^3b - 7a^2b^2 + 3ab^3$

(13) $(a + b)(a^2 - ab + b^2) = a^3 - a^2b + ab^2 + a^2b - ab^2 + b^3 = a^3 + b^3$

(14) $(a - b)(a^2 + ab + b^2) = a^3 + a^2b + ab^2 - a^2b - ab^2 - b^3 = a^3 - b^3$
       These last two often have useful applications so I have separated them out.

(15)   The answer to (b) comes out as 1 less than the answer to (a) each time.
       Using n we have the starting number plus 1 is $n + 1$ and the starting number minus 1 is
       $n - 1$.
       These two, multiplied together, give $(n + 1)(n - 1)$ which is $n^2 - 1$.
       This is 1 less than the starting number squared, and therefore the answer to (b) will
       always be 1 less than the answer to (a).

Exercise 1.B.2

(1) $(x + 7)(x + 1)$ (2) $(p + 5)(p + 1)$ (3) $(x + 6)(x + 1)$ (4) $(x + 2)(x + 3)$

(5) $(y + 3)(y + 3)$ or $(y + 3)^2$ (6) $(x + 2)(x + 4)$ (7) $(a + 2)(a + 5)$

(8) $(x + 4)(x + 5)$ (9) $(x + 4)(x + 9)$

Exercise 1.B.3

(1) $(3x + 5)(x + 1)$ (2) $(2y + 1)(y + 7)$ (3) $(3a + 2)(a + 3)$

(4) $(3x + 1)(x + 6)$ (5) $(5p + 3)(p + 4)$ (6) $(5x + 6)(x + 2)$

Exercise 1.B.4

(1) $(x - 3)(x - 8)$ (2) $(y - 3)(y - 6)$ (3) $(x - 2)(x - 9)$ (4) $(p - 3)(p + 8)$

(5) $(x - 2)(x + 6)$ (6) $(2q + 1)(q - 3)$ (7) $(3x + 2)(x - 4)$ (8) $(2a - 5)(a + 1)$

(9) $(2x + 3)(x - 4)$ (10) $(3b - 2)(b - 6)$ (11) $(3x - 5y)(3x + 5y)$

(12) $(4x^2 - 9y^2)(4x^2 + 9y^2) = (2x - 3y)(2x + 3y)(4x^2 + 9y^2)$
       The $4x^2 + 9y^2$ won't factorise any further.

Exercise 1.C.1

(1) $\dfrac{3}{4}$ (2) $\dfrac{1}{5}$ (3) $\dfrac{5}{19}$ (4) $\dfrac{3}{8}$ (5) $\dfrac{5}{8}$ (6) $\dfrac{b}{c}$ (7) $\dfrac{3y}{2}$

(8) $4p$ (9) $\dfrac{2a}{b}$ (10) $\dfrac{3x}{2y}$ (11) $\dfrac{6p}{5q}$ (12) $\dfrac{5a}{b^2}$

Exercise 1.C.2

(1) (a) They are all equivalent except for $\frac{4}{9}$ and $\frac{2}{6}$.

     (b) They are all equivalent.

     (c) The middle one is the odd one out.

     (d) The first two are equivalent because the z cancels, but it isn't possible to cancel the p in the last one.

(2) (a) $\dfrac{2(x + 3y)}{2(3x - 4y)} = \dfrac{x + 3y}{3x - 4y}$ (b) $\dfrac{3(2a - 3b)}{2(2a - 3b)} = \dfrac{3}{2}$ (c) $\dfrac{p(x - q)}{p(p - x)} = \dfrac{x - q}{p - x}$

     (d) No simplification is possible.   (e) $\dfrac{x(2y + 5z)}{6x} = \dfrac{2y + 5z}{6}$ (f) $\dfrac{2z(2x + 3y)}{2x + 3y} = 2z$

     (g) No simplification is possible.   (h) $\dfrac{(x - y)(x + y)}{(x + y)(x + y)} = \dfrac{x - y}{x + y}$

     (i) $\dfrac{(x + 2)(x + 3)}{(x + 2)(x - 1)} = \dfrac{x + 3}{x - 1}$

Exercise 1.C.3

(1) $\dfrac{3}{4} + \dfrac{2}{7} = \dfrac{3 \times 7}{4 \times 7} + \dfrac{2 \times 4}{7 \times 4} = \dfrac{21}{28} + \dfrac{8}{28} = \dfrac{29}{28}$

(2) $\dfrac{2}{3} + \dfrac{4}{5} = \dfrac{2 \times 5}{3 \times 5} + \dfrac{4 \times 3}{5 \times 3} = \dfrac{10}{15} + \dfrac{12}{15} = \dfrac{22}{15}$

(3) $\dfrac{1}{2} + \dfrac{2}{3} + \dfrac{4}{5} = \dfrac{1 \times (3 \times 5)}{2 \times (3 \times 5)} + \dfrac{2 \times (2 \times 5)}{3 \times (2 \times 5)} + \dfrac{4 \times (2 \times 3)}{5 \times (2 \times 3)} = \dfrac{15}{30} + \dfrac{20}{30} + \dfrac{24}{30} = \dfrac{59}{30}$

Exercise 1.C.4

(1) $\dfrac{2}{9} + \dfrac{7}{15} = \dfrac{10}{45} + \dfrac{21}{45} = \dfrac{31}{45}$

(Notice, there was a common factor of 3 on the bottom.)

(2) $\dfrac{5}{6} + \dfrac{3}{8} = \dfrac{5 \times 4}{6 \times 4} + \dfrac{3 \times 3}{8 \times 3} = \dfrac{20}{24} + \dfrac{9}{24} = \dfrac{29}{24}$

(3) $\dfrac{1}{3} + \dfrac{3}{4} + \dfrac{5}{6} = \dfrac{1 \times 4}{3 \times 4} + \dfrac{3 \times 3}{4 \times 3} + \dfrac{5 \times 2}{6 \times 2} = \dfrac{4}{12} + \dfrac{9}{12} + \dfrac{10}{12} = \dfrac{23}{12}$

(4) There is a common factor of $(2x - y)$ on the bottom. So you should have

$$\dfrac{3x}{y(2x - y)} + \dfrac{5y}{x(2x - y)} = \dfrac{3x^2}{xy(2x - y)} + \dfrac{5y^2}{xy(2x - y)} = \dfrac{3x^2 + 5y^2}{xy(2x - y)}.$$

(5)   There is a common factor of x on the bottom, so we can say

$$\frac{2}{x(3x+1)} + \frac{5}{x(2x-1)} = \frac{2(2x-1)}{x(3x+1)(2x-1)} + \frac{5(3x+1)}{x(2x-1)(3x+1)}$$

$$= \frac{19x+3}{x(2x-1)(3x+1)}.$$

(6)   There is a hidden common factor of $(x+y)$ so we get

$$\frac{4}{x^2-y^2} + \frac{3}{(x+y)^2} = \frac{4}{(x-y)(x+y)} + \frac{3}{(x+y)(x+y)}$$

$$= \frac{4(x+y)}{(x-y)(x+y)(x+y)} + \frac{3(x-y)}{(x+y)(x+y)(x-y)}$$

$$= \frac{4x+4y+3x-3y}{(x-y)(x+y)^2}$$

$$= \frac{7x+y}{(x-y)(x+y)^2}$$

or you can write this as

$$\frac{7x+y}{(x^2-y^2)(x+y)}.$$

Exercise 1.C.5

I shall put in all the brackets from the start in these answers.

(1) $\dfrac{(3x-5)}{10} + \dfrac{(2x-3)}{15} = \dfrac{3(3x-5)}{3\times 10} + \dfrac{2(2x-3)}{2\times 15}$

$$= \frac{9x-15+4x-6}{30} = \frac{13x-21}{30}$$

(2) $\dfrac{(3a+5b)}{4} - \dfrac{(a-3b)}{2} = \dfrac{(3a+5b)}{4} - \dfrac{2(a-3b)}{2\times 2}$

$$= \frac{3a+5b-2a+6b}{4} = \frac{a+11b}{4}$$

(3) $\dfrac{(3m-5n)}{6} - \dfrac{(3m-7n)}{2} = \dfrac{(3m-5n)}{6} - \dfrac{3(3m-7n)}{3\times 2}$

$$= \frac{3m-5n-9m+21n}{6} = \frac{-6m+16n}{6}$$

$$= \frac{2(-3m+8n)}{6} = \frac{-3m+8n}{3}$$

(4) $\dfrac{2b}{a(2a+b)} + \dfrac{3a}{b(2a+b)} = \dfrac{2b^2}{ab(2a+b)} + \dfrac{3a^2}{ab(2a+b)}$

$$= \frac{3a^2+2b^2}{ab(2a+b)}$$

(5) $\dfrac{2a}{(a+b)(3a+b)} + \dfrac{3b}{(a-b)(3a+b)} = \dfrac{2a(a-b)}{(a+b)(3a+b)(a-b)} + \dfrac{3b(a+b)}{(a-b)(3a+b)(a+b)}$

$$= \dfrac{2a^2 - 2ab + 3ab + 3b^2}{(a+b)(3a+b)(a-b)} = \dfrac{2a^2 + ab + 3b^2}{(a-b)(3a+b)(a+b)}$$

(6) $\dfrac{5}{x^2 - y^2} - \dfrac{2}{x(x+y)} = \dfrac{5}{(x-y)(x+y)} - \dfrac{2}{x(x+y)}$

(using the difference of two squares! Did you spot it?)

$$= \dfrac{5x}{x(x-y)(x+y)} - \dfrac{2(x-y)}{x(x-y)(x+y)} = \dfrac{5x - 2x + 2y}{x(x-y)(x+y)}$$

$$= \dfrac{3x + 2y}{x(x-y)(x+y)} \text{ or, equally nice, } \dfrac{3x + 2y}{x(x^2 - y^2)}$$

Exercise 1.C.6

(1) The x underneath is a common factor, so

$$\dfrac{2}{x(2x - 3y)} - \dfrac{3}{2x(x + 4y)} = \dfrac{4(x + 4y)}{2x(2x - 3y)(x + 4y)} - \dfrac{3(2x - 3y)}{2x(x + 4y)(2x - 3y)}$$

(Notice, only the x is common, so we multiplied the first fraction top and bottom by $2(x + 4y)$.)

$$= \dfrac{4(x + 4y) - 3(2x - 3y)}{2x(2x - 3y)(x + 4y)} = \dfrac{25y - 2x}{2x(2x - 3y)(x + 4y)}.$$

When the two fractions are combined, the top has two chunks which are subtracted. This has been tidied up by multiplying out the brackets, and then putting together as much as possible. The bottom, with the brackets multiplied, is already in a neat factorised form, and so we generally leave it this way.

(2) First, we put in the brackets and get

$$\dfrac{(2x - 1)}{3} - \dfrac{(x - 7)}{5} = \dfrac{5(2x - 1)}{5 \times 3} - \dfrac{3(x - 7)}{3 \times 5}$$

$$= \dfrac{10x - 5 - 3x + 21}{15} = \dfrac{7x + 16}{15}.$$

(3) (a) $\dfrac{3a^2}{2b} \times \dfrac{ab}{6c} = \dfrac{3a^3b}{12bc} = \dfrac{a^3}{4c}$, dividing top and bottom by $3b$.

(b) $\dfrac{2a}{3b} \div \dfrac{b^2}{9a^2} = \dfrac{2a}{3b} \times \dfrac{9a^2}{b^2} = \dfrac{18a^3}{3b^3} = \dfrac{6a^3}{b^3}$

(c) $\dfrac{3x}{y^2 z} \div \dfrac{2x^2}{5yz^2} = \dfrac{3x}{y^2 z} \times \dfrac{5yz^2}{2x^2} = \dfrac{15xyz^2}{2x^2 y^2 z} = \dfrac{15z}{2xy}$

dividing top and bottom by xyz.

(4) (a) $\dfrac{3x^2 (2x + 3y)}{2y(x - y)} \times \dfrac{y^2 (x - y)}{x(x + 3y)} = \dfrac{3x^2 y^2 (2x + 3y) (x - y)}{2xy(x - y)(x + 3y)} = \dfrac{3xy(2x + 3y)}{2(x + 3y)}$

It is quicker to do the cancelling before the multiplying (which you can show by crossing through the same factors above and below, if you like).

Then you can leave out the second step of the above working.

But you must remember that only factors of the whole of the top and bottom can be cancelled.

(b) $\dfrac{5pq(p+q)}{(3p+2q)} \times \dfrac{(3p+2q)}{q^2\,(5p-q)} = \dfrac{5p(p+q)}{q(5p-q)}$

Don't be tempted to cancel the 5s!

(c) $\dfrac{(a^2-b^2)^4}{(a^2+b^2)} \times \dfrac{(a^4-b^4)}{(a+b)^4} = \dfrac{((a-b)\,(a+b))^4}{(a^2+b^2)} \times \dfrac{(a^2-b^2)\,(a^2+b^2)}{(a+b)^4}$

$$= (a-b)^4\,(a^2-b^2) \quad \text{or} \quad (a-b)^5\,(a+b) \text{ if you like}$$

Exercise 1.D.1

(1) $\;3^{-1} = \dfrac{1}{3}$ (2) $\;16^{1/2} = 4 \text{ or } -4$ (3) $\;9^{3/2} = (9^{1/2})^3 = (\pm 3)^3 = \pm 27$

(4) $\;27^{-1/3} = \dfrac{1}{27^{1/3}} = \dfrac{1}{3}$ (5) $\;4^0 = 1$ (6) $\;7^1 = 7$ (7) $\;7^{-2} = \dfrac{1}{7^2} = \dfrac{1}{49}$

(8) $\;4^{-1/2} = \dfrac{1}{4^{1/2}} = \dfrac{1}{\pm 2} = \pm \dfrac{1}{2}$ (9) $\;32^{1/5} = 2$

(10) $\;16^{-3/4} = \dfrac{1}{16^{3/4}} = \dfrac{1}{(16^{1/4})^3} = \dfrac{1}{(\pm 2)^3} = \dfrac{1}{\pm 8} = \pm \dfrac{1}{8}$ (11) $\;25^{3/2} = (25^{1/2})^3 = (\pm 5)^3 = \pm 125$

(12) $\;49^{-1/2} = \dfrac{1}{49^{1/2}} = \dfrac{1}{\pm 7} = \pm \dfrac{1}{7}$

Exercise 1.F.1

(1) $10111 = 1 + 1(2^1) + 1(2^2) + 0(2^3) + 1(2^4) = 1 + 2 + 4 + 16 = 23$

(2) $1111 = 1 + 1(2^1) + 1(2^2) + 1(2^3) = 1 + 2 + 4 + 8 = 15$

(3) $111011 = 1 + 1(2^1) + 0(2^2) + 1(2^3) + 1(2^4) + 1(2^5) = 1 + 2 + 8 + 16 + 32 = 59$

Exercise 1.F.2

(1)	2	72	Rem		(2)	2	2431	Rem		(3)	2	3251	Rem
	2	36	0			2	1215	1			2	1625	1
	2	18	0			2	607	1			2	812	1
	2	9	0			2	303	1			2	406	0
	2	4	1 ↑			2	151	1			2	203	0
	2	2	0			2	75	1			2	101	1
	2	1	0			2	37	1 ↑			2	50	1 ↑
		0	1			2	18	1			2	25	0
						2	9	0			2	12	1
						2	4	1			2	6	0
						2	2	0			2	3	0
						2	1	0			2	1	1
							0	1				0	1

$72_{10} = 1001000_2$

$2431_{10} = 100101111111_2$ $3251_{10} = 110010110011_2$

Exercise 1.F.3

(1) $\sqrt{28} = \sqrt{2 \times 2 \times 7} = \sqrt{2^2 \times 7} = 2\sqrt{7}$

(2) $\sqrt{45} = \sqrt{3 \times 3 \times 5} = \sqrt{3^2 \times 5} = 3\sqrt{5}$

(3) $\sqrt{50} = \sqrt{2 \times 5 \times 5} = \sqrt{2 \times 5^2} = 5\sqrt{2}$

(4) $\sqrt{44} = \sqrt{2 \times 2 \times 11} = \sqrt{2^2 \times 11} = 2\sqrt{11}$

(5) $\sqrt{63} = \sqrt{3 \times 3 \times 7} = \sqrt{3^2 \times 7} = 3\sqrt{7}$

(6) $\sqrt{40} = \sqrt{2 \times 2 \times 2 \times 5} = \sqrt{2^2 \times 2 \times 5} = 2\sqrt{10}$

Exercise 1.F.4

(1) $\dfrac{5}{3 + \sqrt{2}} = \dfrac{5(3 - \sqrt{2})}{(3 + \sqrt{2})(3 - \sqrt{2})} = \dfrac{15 - 5\sqrt{2}}{9 - 2} = \dfrac{15 - 5\sqrt{2}}{7}$

(2) $\dfrac{3 - \sqrt{5}}{3 + \sqrt{5}} = \dfrac{(3 - \sqrt{5})(3 - \sqrt{5})}{(3 + \sqrt{5})(3 - \sqrt{5})} = \dfrac{14 - 6\sqrt{5}}{4} = \dfrac{7 - 3\sqrt{5}}{2}$

(3) $\dfrac{3 - 2\sqrt{3}}{5 + 3\sqrt{2}} = \dfrac{(3 - 2\sqrt{3})(5 - 3\sqrt{2})}{(5 + 3\sqrt{2})(5 - 3\sqrt{2})} = \dfrac{15 + 6\sqrt{6} - 10\sqrt{3} - 9\sqrt{2}}{25 - 18} = \dfrac{15 + 6\sqrt{6} - 10\sqrt{3} - 9\sqrt{2}}{7}$

During this working we have used the fact that $\sqrt{x} \times \sqrt{y} = \sqrt{xy}$. An example of this is $\sqrt{4} \times \sqrt{9} = \sqrt{36}$ which is the same as $2 \times 3 = 6$.

 $\sqrt{x + y}$ is *not* the same as $\sqrt{x} + \sqrt{y}$. For example, $\sqrt{9 + 16} = \sqrt{25} = 5$ but $\sqrt{9} + \sqrt{16} = 3 + 4 = 7$.

Chapter 2
Exercise 2.A.1

(1) $x + 8 = 5$ so $x = 5 - 8 = -3$.   (2) $5y = 40$ so $y = \frac{40}{5} = 8$.

(3) $2y = 7$ so $y = \frac{7}{2}$.

(4) $7 + 2x = 5 - x$ so $2x + x = 5 - 7$ so $x = -\frac{2}{3}$.

(5) $4 + 2b = 5b + 9$ so $4 - 9 = 5b - 2b$ so $-5 = 3b$ and $b = -\frac{5}{3}$.

(6) $3(x - 3) = 6$ so $3x - 9 = 6$ so $3x = 15$ and $x = 5$.

(7) $3(y - 2) = 2(y - 1)$ so $3y - 6 = 2y - 2$ so $3y = -2 + 6$ so $y = 4$.

(8) $2(3a - 1) = 3(4a + 3)$ so $6a - 2 = 12a + 9$ so $-2 - 9 = 12a - 6a$ so $-11 = 6a$ and $a = -\frac{11}{6}$.

(9) $3x - 1 = 2(2x - 1) + 3$ so $3x - 1 = 4x - 2 + 3$ so $-1 - 1 = 4x - 3x$ so $x = -2$.

(10) $2(p + 2) = 6p - 3(p - 4)$ so $2p + 4 = 6p - 3p + 12$ so $4 - 12 = 3p - 2p$ and $p = -8$.

Exercise 2.A.2

(1) $\dfrac{5x}{3} = 2$ so $5x = 3 \times 2 = 6$ and $x = \frac{6}{5}$.

(2) $5 + x = \dfrac{2x}{3}$ so, multiplying by 3, $15 + 3x = 2x$ and $x = -15$.

(3) Multiplying both sides of $\dfrac{x}{3} - \dfrac{x}{4} = 1$ by $3 \times 4 = 12$ gives

$$12\left(\dfrac{x}{3} - \dfrac{x}{4}\right) = 12 \times 1 = 12 \quad \text{so} \quad 4x - 3x = 12 \quad \text{and} \quad x = 12.$$

(4) Start by putting in the two brackets, so you have $\dfrac{y}{3} - \dfrac{(3y-7)}{5} = \dfrac{(y-2)}{6}$.

Multiply by 30 to get rid of fractions (notice you don't need to use 90).
Then $10y - 6(3y-7) = 5(y-2)$ so $10y - 18y + 42 = 5y - 10$ so $52 = 13y$ and $y = 4$.

(5) First, put in brackets to give $\dfrac{(3m-5)}{4} - \dfrac{(9-2m)}{3} = 0.$

Then multiply by 12 to give $3(3m-5) - 4(9-2m) = 0$ (not $= 12$!) so $9m - 15 - 36 + 8m = 0$ so $17m = 51$ and $m = 3$.

(6) $\dfrac{x-1}{2} - \dfrac{x-2}{3} = 1$ Putting in brackets and multiplying both sides by 6 gives

$$6\left(\dfrac{(x-1)}{2} - \dfrac{(x-2)}{3}\right) = 6 \times 1 = 6 \quad \text{so} \quad 3(x-1) - 2(x-2) = 6$$

so $3x - 3 - 2x + 4 = 6$ and $x = 5$.

(7) $\dfrac{p+1}{p-1} = \dfrac{3}{4}$ Multiplying both sides by $4(p-1)$, and cancelling, gives

$$4(p+1) = 3(p-1) \quad \text{so} \quad 4p + 4 = 3p - 3 \quad \text{and} \quad p = -7.$$

(8) $\dfrac{2}{y} = \dfrac{3}{y+1}$ Multiplying both sides by $y(y+1)$, and cancelling, gives

$$2(y+1) = 3y \quad \text{so} \quad 2y + 2 = 3y \quad \text{and} \quad y = 2.$$

(9) $\dfrac{4}{2x+3} = \dfrac{3}{x-2}$ Multiplying both sides by $(2x+3)(x-2)$, and cancelling, gives

$$4(x-2) = 3(2x+3) \quad \text{so} \quad 4x - 8 = 6x + 9 \quad \text{so} \quad -17 = 2x \quad \text{and} \quad x = -\dfrac{17}{2}.$$

(10) $\dfrac{2x}{x+2} = \dfrac{3x}{x+5} - 1$ To get rid of fractions, we must multiply by $(x+2)(x+5)$.

Then, cancelling, we get

$$2x(x+5) = 3x(x+2) - (x+2)(x+5).$$

(Notice that the -1 has also been multiplied by $(x+2)(x+5)$.)
So $2x^2 + 10x = 3x^2 + 6x - (x^2 + 7x + 10) = 3x^2 + 6x - x^2 - 7x - 10$

so $11x = -10$ and $x = -\dfrac{10}{11}.$

(11) $\dfrac{2x+1}{3} + \dfrac{x+5}{2} = \dfrac{3x-1}{7}$

We get rid of the fractions first by multiplying both sides by $3 \times 2 \times 7 = 42$.
We have

$$42\left(\dfrac{2x+1}{3} + \dfrac{x+5}{2}\right) = 42\left(\dfrac{3x-1}{7}\right)$$

so $42\left(\dfrac{2x+1}{3}\right) + 42\left(\dfrac{x+5}{2}\right) = 42\left(\dfrac{3x-1}{7}\right).$

Notice that each separate chunk of the equation is getting multiplied by the 42.

We then cancel down each fraction in turn to obtain

$$14(2x + 1) + 21(x + 5) = 6(3x - 1)$$

$$28x + 14 + 21x + 105 = 18x - 6$$

$$28x + 21x - 18x = -14 - 105 - 6$$

so $\quad 31x = -125 \quad$ giving $\quad x = -\dfrac{125}{31}.$

(12) $\quad \dfrac{x + 3}{4} - \dfrac{x - 1}{5} = \dfrac{2x - 1}{10} \quad$ Putting in brackets and multiplying by 20 gives

$$20\left(\frac{x + 3}{4}\right) - 20\left(\frac{x - 1}{3}\right) = 20\left(\frac{2x - 1}{6}\right)$$

so $\quad 5(x + 3) - 4(x - 1) = 2(2x - 1) \quad$ so $\quad 21 = 3x \quad$ giving $\quad x = 7.$

Exercise 2.A.3

(1) (a) $S = 4\pi r^2 \quad$ so $\quad r^2 = \dfrac{S}{4\pi} \quad$ and $\quad r = \sqrt{\dfrac{S}{4\pi}}.$

(b) $V = \frac{4}{3}\pi r^3 \quad$ so $\quad \dfrac{3V}{4\pi} = r^3 \quad$ and $\quad r = \sqrt[3]{\dfrac{3V}{4\pi}}.$

(2) (a) $V = \pi r^2 h \quad$ so $\quad \dfrac{V}{\pi r^2} = h.$

(b) $V = \pi r^2 h \quad$ so $\quad \dfrac{V}{\pi h} = r^2 \quad$ and $\quad r = \sqrt{\dfrac{V}{\pi h}}.$

(c) $S = 2\pi r^2 + 2\pi rh \quad$ so $\quad S - 2\pi r^2 = 2\pi rh \quad$ and $\quad \dfrac{S - 2\pi r^2}{2\pi r} = h.$

(3) (a) $v^2 = u^2 + 2as \quad$ so $\quad v^2 - u^2 = 2as \quad$ and $\quad \dfrac{v^2 - u^2}{2s} = a.$

(b) $v^2 = u^2 + 2as \quad$ so $\quad v^2 - 2as = u^2 \quad$ and $\quad u = \sqrt{v^2 - 2as}$

(4) $\dfrac{1}{R} = \dfrac{1}{R_1} + \dfrac{1}{R_2} \quad$ so $\quad \dfrac{1}{R} - \dfrac{1}{R_1} = \dfrac{1}{R_2}.$

Multiplying by $R\,R_1\,R_2$ to get rid of the fractions, we get

$$R_1\,R_2 - RR_2 = RR_1 \quad \text{so} \quad R_2(R_1 - R) = RR_1 \quad \text{and} \quad R_2 = \frac{RR_1}{R_1 - R}.$$

If $R_1 = 3$ and $R = 2$ we have $R_2 = 6$. We should use a resistance R_2 of 6Ω

Exercise 2.B.1

(1) $\left(\dfrac{-3 + 1}{2}, \dfrac{2 - 6}{2}\right) = (-1, -2) \qquad$ (2) $\left(\dfrac{-2 + 3}{2}, \dfrac{-1 + 4}{2}\right) = (\frac{1}{2}, \frac{3}{2})$

(3) $\left(\dfrac{-1 + (-4)}{2}, \dfrac{-5 + (-6)}{2}\right) = (-\frac{5}{2}, -\frac{11}{2})$

Exercise 2.B.2

(1) $m = -5 \qquad$ (2) $y = \frac{3}{2}x + \frac{7}{2} \quad$ so $\quad m = \frac{3}{2} \qquad$ (3) $y = -\frac{1}{3}x + \frac{1}{3} \quad$ so $\quad m = -\frac{1}{3}$

(4) $y = \frac{5}{4}x + \frac{1}{2} \quad$ so $\quad m = \frac{5}{4}$

Exercise 2.B.3

(1) Sketch (c) (2) Sketch (b): $y + 4x = 4$ so $y = -4x + 4$.
(3) Sketch (a): $4y = x + 4$ so $y = \frac{1}{4}x + 1$. (4) Sketch (e) (5) Sketch (h)
(6) Sketch (g) (7) Sketch (d) (8) Sketch (f): $y + 2x = -2$ so $y = -2x - 2$.

Notice that appearances can be deceptive. For example, (c), (d) and (h) all look the same until you take account of the different scales marked on the axes.
Exercise 2.B.4, 2.B.5 and 2.B.6 all have answers given after Self-test 4.

Exercise 2.B.7

(1) The coordinates are $\left(\dfrac{1 \times (-1) + 2 \times 5}{2 + 1}, \dfrac{1 \times 2 + 2 \times 14}{2 + 1} \right) = (3,10)$.

(2) The coordinates are $\left(\dfrac{3 \times (-2) + 1 \times 6}{1 + 3}, \dfrac{3 \times (-3) + 1 \times 9}{1 + 3} \right) = (0,0)$.

Exercise 2.C.1

(1) Multiplying (2) by 2 gives $\begin{cases} 5a - 2b = 68 & \quad (1) \\ 6a + 2b = 20. & \quad (3) \end{cases}$

Adding (1) and (3) gives $11a = 88$ so $a = 8$.
Substituting in (1) gives $40 - 2b = 68$ so $b = -14$.
Check in (2): LHS $= 24 - 14 = 10 =$ RHS.

(2) Multiplying (1) by 5 and (2) by 2 gives $\begin{cases} 25p - 10q = 45 & \quad (3) \\ 4p + 10q = -16. & \quad (4) \end{cases}$

Adding (3) and (4) gives $29p = 29$ so $p = 1$.
Substituting in (1) gives $5 - 2q = 9$ so $q = -2$.
Check in (2): LHS $= 2 - 10 = -8 =$ RHS.

(3) First, get rid of fractions by writing $8 \times$ (1) and $3 \times$ (2).

This gives $\begin{cases} x - 8y = -20 & \quad (3) \\ 9x + y = 39. & \quad (4) \end{cases}$

Multiplying (4) by 8 gives $\begin{cases} x - 8y = -20 & \quad (3) \\ 72x + 8y = 312. & \quad (5) \end{cases}$

Adding (3) and (5) gives $73x = 292$ so $x = 4$.
Substituting in (2) gives $12 + y/3 = 13$ so $y/3 = 1$ and $y = 3$.
Check in (1): LHS $= \frac{4}{8} - 3 = -\frac{5}{2} =$ RHS.

(4) Use the same trick which we met earlier of letting $\dfrac{1}{x} = X$ and $\dfrac{1}{y} = Y$.

Then we have: $\begin{cases} 3X + 4Y = 0 & \quad (3) \\ 2X - 2Y = 7. & \quad (4) \end{cases}$

Multiplying (4) by 2 we have $\begin{cases} 3X + 4Y = 0 & \quad (3) \\ 4X - 4Y = 14. & \quad (5) \end{cases}$

Adding these two gives $7X = 14$ so $X = 2$.
Substituting in (3) gives $6 + 4Y = 0$ so $Y = -\frac{3}{2}$ so $x = \frac{1}{2}$ and $y = -\frac{2}{3}$.

Check in (1): LHS $= \dfrac{3}{\frac{1}{2}} + \dfrac{4}{-\frac{2}{3}} = 6 - 6 = 0 =$ RHS.

Exercise 2.D.1

(1) $x^2 + 9x + 14 = (x + 2)(x + 7) = 0$ so $x = -2$ or $x = -7$.

(2) $x^2 + 4x - 12 = (x - 2)(x + 6) = 0$ so $x = 2$ or $x = -6$.

(3) $x^2 - 11x + 18 = (x - 2)(x - 9) = 0$ so $x = 2$ or $x = 9$.

(4) $x^2 - x - 20 = (x + 4)(x - 5) = 0$ so $x = -4$ or $x = 5$.

(5) $2x^2 + 13x + 6 = (2x + 1)(x + 6) = 0$ so $x = -\frac{1}{2}$ or $x = -6$.

(6) $3x^2 - 7x - 6 = (3x + 2)(x - 3) = 0$ so $x = -\frac{2}{3}$ or $x = 3$.

Exercise 2.D.2

(1) $x = \pm 3$ (2) $x = \pm\frac{4}{5}$ (3) $x - 3 = \pm 2$ so $x = +1$ or $x = +5$.

(4) $2x - 3 = \pm 5$ so $x = 4$ or $x = -1$.

(5) $3x - 2 = \pm 6$ so $x = \frac{8}{3}$ or $x = -\frac{4}{3}$.

Exercise 2.D.3

(1) $x^2 + 4x = 21$ so $(x + 2)^2 - 4 = 21$ so $(x + 2)^2 = 25$ so $x + 2 = \pm 5$
and $x = 3$ or $x = -7$.

(2) $x^2 - 6x + 8 = 0$ so $x^2 - 6x = -8$ so $(x - 3)^2 - 9 = -8$ so $(x - 3)^2 = 1$
so $x - 3 = \pm 1$ and $x = 4$ or $x = 2$.

(3) $x^2 - 3x - 10 = 0$ so $x^2 - 3x = 10$ so $(x - \frac{3}{2})^2 - \frac{9}{4} = 10$ so $(x - \frac{3}{2})^2 = \frac{49}{4}$
so $x - \frac{3}{2} = \pm \frac{7}{2}$ and $x = 5$ or $x = -2$.

Exercise 2.D.4

(1) $y = x^2 - 4x + 3 = (x - 2)^2 - 4 + 3 = (x - 2)^2 - 1$ so the least value of y is -1 which is
when $x = 2$, that is, the lowest point on the curve is at $(2, -1)$.
     The y-intercept is $(0,3)$.
     When $y = 0$, $(x - 2)^2 - 1 = 0$ so $x - 2 = \pm 1$ and $x = 3$ or 1.
     Therefore the equation $x^2 - 4x + 3 = 0$ has the two roots $x = 1$ and $x = 3$, so the curve
$y = x^2 - 4x + 3$ cuts the x-axis at (0.1) and $(0,3)$.
     Curve (b) is just curve (a) turned upside down by being reflected in the x-axis, since the
sign for y is just the opposite way round.
     The sketch for this question is shown beside the one for question (2) below.

(2) $y = x^2 + 2x - 8 = (x + 1)^2 - 1 - 8$ so the least value of y is -9 when $x = -1$. The lowest
point on the curve is at $(-1, -9)$. The y-intercept is $(0, -8)$. When $y = 0$, $(x + 1)^2 - 9 = 0$ so
$(x + 1)^2 = 9$ and $x + 1 = \pm 3$. The roots of $x^2 + 2x - 8 = 0$ are $x = 2$ and $x = -4$, and the
curve $y = x^2 + 2x - 8$ cuts the x-axis at $(-4, 0)$ and $(2, 0)$. Again, curve (b) is just curve (a)
turned upside down.

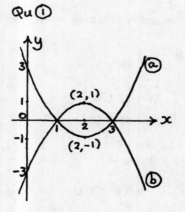

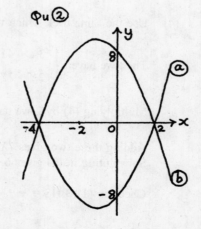

Exercise 2.D.5

(1) $x = \dfrac{-10 \pm \sqrt{100 - 64}}{2} = \dfrac{-10 \pm \sqrt{36}}{2} = \dfrac{-10 \pm 6}{2} = -8$ or -2

(2) $x = \dfrac{2 \pm \sqrt{4 + 32}}{2} = \dfrac{2 \pm 6}{2} = 4$ or -2

(3) $x = \dfrac{-5 \pm \sqrt{25 + 24}}{4} = \dfrac{-5 \pm 7}{4} = -3$ or $\frac{1}{2}$

(4) $x = \dfrac{-4 \pm \sqrt{16 - 8}}{2} = \dfrac{-4 \pm \sqrt{8}}{2} = \dfrac{-4 \pm 2\sqrt{2}}{2}$

$\quad = -2 + \sqrt{2}$ or $-2 - \sqrt{2} = -0.59$ or -3.41 to 2 d.p.

(5) $x = \dfrac{1 \pm \sqrt{1 + 24}}{6} = \dfrac{1 \pm 5}{6} = 1$ or $-\frac{2}{3}$

(6) $x = \dfrac{1 \pm \sqrt{1 + 56}}{4} = \dfrac{1 + \sqrt{57}}{4}$ or $\dfrac{1 - \sqrt{57}}{4} = 2.14$ or -1.64 to 2 d.p.

Exercise 2.D.6

(1) (a) $2x^2 + 7x + 3 = 0$ so $(2x + 1)(x + 3) = 0$ and $x = -\frac{1}{2}$ or -3.

(b) $3x^2 + 4x + 1 = 0$ so $(3x + 1)(x + 1) = 0$ and $x = -\frac{1}{3}$ or $x = -1$.

(c) $2x^2 + x - 4 = 0$ gives '$b^2 - 4ac$' $= 1 - 4 \times 2 \times -4 = 33$

so there is no whole number factorisation.

Using the formula, we have $x = \dfrac{-1 \pm \sqrt{33}}{4} = 1.186$ or -1.686 to 3 d.p.

(d) $6x^2 - 7x + 2 = 0$ so $(2x - 1)(3x - 2) = 0$ and $x = \frac{1}{2}$ or $x = \frac{2}{3}$.

(e) $x^2 - 5x + 3 = 0$ gives '$b^2 - 4ac$' $= 25 - 12 = 13$

so there is no whole number factorisation.

Using the formula, we have $x = \dfrac{5 \pm \sqrt{13}}{2} = 4.303$ or 0.697 to 3.d.p.

(f) $6x^2 + 5x - 6 = 0$ so $(3x - 2)(2x + 3) = 0$ and $x = \frac{2}{3}$ or $x = -\frac{3}{2}$.

(g) $x^2 - 81 = 0$ so $(x - 9)(x + 9) = 0$ and $x = 9$ or $x = -9$.

Or, you could say, $x^2 = 81$ so $x = \pm 9$.

(h) $6x^2 - x - 12 = 0$ so $(3x + 4)(2x - 3) = 0$ and $x = -\frac{4}{3}$ or $x = \frac{3}{2}$.

(i) $x^2 - 2 = 0$ so $x^2 = 2$ and $x = \pm\sqrt{2} = \pm 1.414$ to 3.d.p.

Or, you could say $(x + \sqrt{2})(x - \sqrt{2}) = 0$, factorising, which gives the same pair of answers as above.

(j) Factorising $x^2 - 5x = 0$ we have $x(x - 5) = 0$ so $x = 0$ or $x = 5$.

Don't be tempted to divide $x^2 - 5x = 0$ through by x.

If you do this, you lose the possible answer of $x = 0$. When $x = 0$, this division is actually impossible because we cannot divide by zero.

In all the questions above, where I have used factorisation, it is equally acceptable if you got your answers by using the formula.

(2) (a) $\dfrac{2x - 3}{2x + 3} = \dfrac{x - 1}{x + 1}$

Getting rid of fractions by multiplying by $(2x + 3)(x + 1)$, we have

$(2x - 3)(x + 1) = (2x + 3)(x - 1)$ so $2x^2 - x - 3 = 2x^2 + x - 3$

so $0 = 2x$ so $x = 0$.

(b) $\dfrac{2}{y+1} + \dfrac{1}{y-1} = \dfrac{3}{y}$

Getting rid of fractions by multiplying by $y(y+1)(y-1)$, we have

$$2y(y-1) + y(y+1) = 3(y+1)(y-1) \quad \text{so} \quad 2y^2 - 2y + y^2 + y = 3y^2 - 3$$

so $y = 3$.

(c) $\dfrac{2x+4}{x+1} = \dfrac{x-8}{2x-1}$

Getting rid of fractions by multiplying by $(x+1)(2x-1)$, we have:

$$4x^2 + 6x - 4 = x^2 - 7x - 8 \quad \text{so} \quad 3x^2 + 13x + 4 = 0$$

so $(3x+1)(x+4) = 0$ so $x = -\frac{1}{3}$ or $x = -4$.

Exercise 2.D.7

The sketches fit to the given equations as follows.

(1)   Sketch (e)   (4)   Sketch (d)   (6)   Sketch (c)   (7)   Sketch (b)
(8)   Sketch (f)   (10) Sketch (a)

The sketches for equations (2), (3), (5) and (9) are shown below.

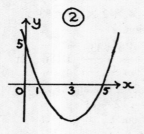 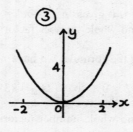 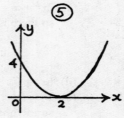 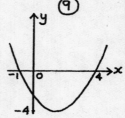

Exercise 2.E.1

(1) $y = f(x) = 3x^3 + 2x^2 - 3x - 2$

Guessing and substitution show that $f(1) = 0$ so $(x-1)$ is a factor, and $f(-1) = 0$ so $(x+1)$ is a factor.

Matching up the two sides, we have

$$f(x) = 3x^3 + 2x^2 - 3x - 2 = (x-1)(x+1)(3x+2).$$

The roots of $f(x) = 0$ are 1, -1 and $-\frac{2}{3}$. The y intercept is at $(0, -2)$.
The coefficient of x^3 is positive, so we have graph 1 below.

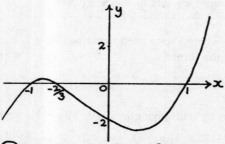 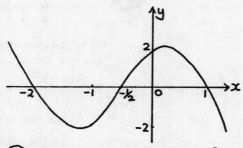

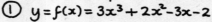

(2) $y = f(x) = 2 + 3x - 3x^2 - 2x^3$

Guessing and substitution show that $f(1)$ and $f(-2) = 0$ so $(x - 1)$ and $(x + 2)$ are both factors.

Matching up the two sides, we have

$$f(x) = 2 + 3x - 3x^2 - 2x^3 = (x - 1)\ (x + 2)(-2x - 1) = -(2x + 1)(x - 1)(x + 2)$$

taking out a factor of -1.

The roots of $f(x) = 0$ are $-\frac{1}{2}$, 1 and -2, and the y intercept is at $(0, 2)$.

The coefficient of x^3 is negative so we have graph 2 on the previous page.

(3) $y = f(x) = 4x^3 - 15x^2 + 12x + 4$

Guessing and substitution give $f(2) = 0$ so $(x - 2)$ is a factor. There is no obvious second root so matching up the two sides, we have

$$f(x) = 4x^3 - 15x^2 + 12x + 4 = (x - 2)(4x^2 + px - 2).$$

Matching the terms in x^2 gives $-15x^2 = -8x^2 + px^2$, so $p = -7$.
Checking, using the terms in x, gives $12x = -2px - 2x$, so $p = -7$ is correct.
Therefore

$$f(x) = (x - 2)\ (4x^2 - 7x - 2) = (x - 2)(x - 2)(4x + 1),$$

factorising the second bracket.

We see that $f(x) = 0$ has the root $x = -\frac{1}{4}$ and the double repeated root of $x = 2$. Just as we found with quadratic equations, this means that the curve of $y = f(x)$ *touches* the x-axis when $x = 2$.

The y intercept is at $(0,4)$ and the coefficient of x^3 is positive, so we get graph 3 below.

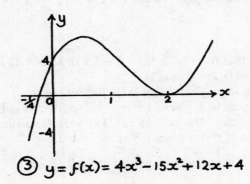

③ $y = f(x) = 4x^3 - 15x^2 + 12x + 4$

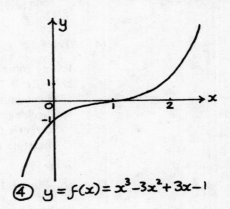

④ $y = f(x) = x^3 - 3x^2 + 3x - 1$

(4) $y = f(x) = x^3 - 3x^2 + 3x - 1$

Guessing and substituting shows that $f(1) = 0$, so $(x - 1)$ is a factor, and there is no obvious second root.

Matching up the two sides gives

$$x^3 - 3x^2 + 3x - 1 = (x - 1)(x^2 + px + 1).$$

Matching up the terms in x^2 gives

$$-3x^2 = -x^2 + px^2, \quad \text{so} \quad p = -2.$$

Checking, using the terms in x, gives $3x = x - px$, so $p = -2$ is correct.
Therefore,

$$y = f(x) = (x - 1)(x^2 - 2x + 1) = (x - 1)(x - 1)^2 = (x - 1)^3.$$

This time, we have a single triply repeated root at $x = 1$. The y intercept is at $(0, -1)$.

The coefficient of x^3 is positive, so we get graph 4 above.

If you look at this on a graph-sketching calculator, or plot values close to $x = 1$ for yourself, you will see that the curve flattens near $x = 1$ where the three roots are all bunched together.

Exercise 2.E.2

(1) $f(x) = x^3 + 2x^2 - 5x - 6$ so $f(2) = 8 + 8 - 10 - 6 = 0$.

Therefore $(x - 2)$ is a factor and we have $x^3 + 2x^2 - 5x - 6 = (x - 2)(x^2 + px + 3)$.
Matching the term in x^2 gives $2x^2 = -2x^2 + px^2$, so $p = 4$.
Checking, using the term in x gives $-5x = -2px + 3x$, so $p = 4$ is correct.
So $f(x) = (x - 2)(x^2 + 4x + 3) = (x - 2)(x + 1)(x + 3)$, factorising the second bracket.

(2) $f(x) = 2x^3 - 3x^2 - 8x - 3$ so $f(3) = 54 - 27 - 24 - 3 = 0$, so $(x - 3)$ is a factor.

We have $2x^3 - 3x^2 - 8x - 3 = (x - 3)(2x^2 + px + 1)$.
Matching up the terms in x^2 gives $-3x^2 = -6x^2 + px^2$, so $p = 3$.
Checking with the term in x gives $-8x = -3px + x$, so $p = 3$ is correct.
So $f(x) = (x - 3)(2x^2 + 3x + 1) = (x - 3)(2x + 1)(x + 1)$, factorising the second bracket.

(3) $f(x) = 3x^3 + x^2 - 12x - 4$

Testing some values for x, we find that $f(2) = 24 + 4 - 24 - 4 = 0$, so $(x - 2)$ is a factor. We have $3x^3 + x^2 - 12x - 4 = (x - 2)(3x^2 + px + 2)$.
Matching up the terms in x^2 gives $x^2 = -6x^2 + px^2$, so $p = 7$.
Checking, using the term in x, gives $-12x = -2px + 2x$, so $p = 7$ is correct.
So $f(x) = (x - 2)(3x^2 + 7x + 2) = (x - 2)(3x + 1)(x + 2)$, factorising the second bracket.
Therefore the solutions of $f(x) = 0$ are $x = 2$, $x = -\frac{1}{3}$ and $x = -2$.

(4) $f(x) = 2x^3 + 7x^2 + 2x - 3$

Testing some values, we find $f(-1) = -2 + 7 - 2 - 3 = 0$, so $(x + 1)$ is a factor. We have $2x^3 + 7x^2 + 2x - 3 = (x + 1)(2x^2 + px - 3)$.
Matching up the terms in x^2 gives $7x^2 = 2x^2 + px^2$, so $p = 5$
Checking, using the terms in x, gives $2x = px - 3x$, so $p = 5$ is correct.
So $f(x) = (x + 1)(2x^2 + 5x - 3) = (x + 1)(2x - 1)(x + 3)$, factorising the second bracket.
Therefore the solutions of $f(x) = 0$ are $x = -1$, $x = \frac{1}{2}$ and $x = -3$.

(5) $x^4 - 29x^2 + 100 = 0$.

We have a quadratic equation in a beard and dark glasses.
Putting $y = x^2$, we have $y^2 - 29y + 100 = 0$ so $(y - 25)(y - 4) = 0$.
So $y = 25$ which means that $x^2 = 25$, so $x = \pm 5$,   or $y = 4$ which means that $x^2 = 4$, so $x = \pm 2$.

(6) We have $f(x) = 5x^3 + ax^2 + bx - 6$. $(x - 3)$ is a factor, so $f(3) = 0$.
This gives $135 + 9a + 3b - 6 = 0$. Therefore

$$9a + 3b = -129 \quad \text{so} \quad 3a + b = -43. \tag{1}$$

Also,

$$f(-2) = -40 \quad \text{so} \quad -40 + 4a - 2b - 6 = -40 \quad \text{so} \quad 2a - b = 3. \tag{2}$$

(1) added to (2) gives $5a = -40$ so $a = -8$.
Substituting in (1) gives $-24 + b = -43$ so $b = -19$.
So $5x^3 - 8x^2 - 19x - 6 = (x - 3)(5x^2 + px + 2)$.
Matching terms in x^2 gives $-8x^2 = -15x^2 + px^2$, so $p = 7$.
Checking with the term in x gives $-19x = -3px + 2x$, so $p = 7$ is correct.

$$5x^2 + 7x + 2 = (5x + 2)(x + 1) \text{ therefore } f(x) = (x - 3)(5x + 2)(x + 1).$$

(7)   The working for the long division is shown below.

$$
\begin{array}{r}
4x^2 + 4x - 3 \\
3x-2\overline{\smash{\big)}\,12x^3+4x^2-17x+6} \\
\underline{12x^3 - 8x^2} \\
12x^2-17x \\
\underline{12x^2-\ 8x} \\
-9x+6 \\
\underline{-9x+6} \\
\end{array}
$$

Since the division process leaves no remainder, $(3x - 2)$ is a factor of $12x^3 + 4x^2 - 17x + 6$.

Alternatively, substituting $x = \frac{2}{3}$ gives $f(\frac{2}{3}) = 12(\frac{2}{3})^3 + 4(\frac{2}{3})^2 - 17(\frac{2}{3}) + 6 = 0$.

(8)   The working for the long division is shown below.

$$
\begin{array}{r}
3x^2 + 4x - 2 \\
2x-1\overline{\smash{\big)}\,6x^3+5x^2-8x+1} \\
\underline{6x^3 - 3x^2} \\
8x^2-8x \\
\underline{8x^2-4x} \\
-4x+1 \\
\underline{-4x+2} \\
-1 \\
\end{array}
$$

Alternatively, putting $x = \frac{1}{2}$ gives $f(\frac{1}{2}) = 6(\frac{1}{2})^3 + 5(\frac{1}{2})^2 - 8(\frac{1}{2}) + 1 = -1$.

Chapter 3
Exercise 3.A.1

(1)   In (a), the volumes are directly proportional to the heights, therefore

$$\frac{V_1}{h_1} = \frac{V_2}{h_2} \quad \text{so} \quad \frac{V_A}{4} = \frac{V_B}{1} \quad \text{so} \quad \frac{V_A}{V_B} = \frac{4}{1}.$$

We can see that the volume of A is 4 times the volume of B.
   In (b), the volumes are directly proportional to the (radius)^2, therefore

$$\frac{V_1}{r_1^2} = \frac{V_2}{r_2^2} \quad \text{so} \quad \frac{V_C}{1} = \frac{V_D}{16} \quad \text{so} \quad \frac{V_C}{V_D} = \frac{1}{16}.$$

Cylinder C has $\frac{1}{16}$ of the volume of cylinder D.

(2)   The kinetic energy is directly proportional to the (speed)^2, therefore

$$\frac{E_1}{v_1^2} = \frac{E_2}{v_2^2} \quad \text{so} \quad \frac{E_1}{25} = \frac{E_2}{900} \quad \text{so} \quad \frac{E_1}{E_2} = \frac{25}{900} = \frac{1}{36}.$$

(3)   The volume is directly proportional to the (radius)^3, therefore

$$\frac{V_1}{r_1^3} = \frac{V_2}{r_2^3} \quad \text{so} \quad \frac{V_1}{8} = \frac{V_2}{512} \quad \text{so} \quad \frac{V_1}{V_2} = \frac{8}{512} = \frac{1}{64}.$$

(4)   The time of the swing is directly proportional to the square root of the length, therefore

$$\frac{T_1}{\sqrt{l_1}} = \frac{T_2}{\sqrt{l_2}} \quad \text{so} \quad \frac{T_1}{\sqrt{9}} = \frac{T_2}{\sqrt{25}} \quad \text{so} \quad \frac{T_1}{T_2} = \frac{\sqrt{9}}{\sqrt{25}} = \frac{3}{5}.$$

The time of swing of the first pendulum is $\frac{3}{5}$ of the time of swing of the second pendulum.

Exercise 3.B.1

Here are the functions which you should have found.

(1)   (b) shows $f(x) + 2$.   (c) shows $f(x) - 2$.

(2)   (b) shows $g(x + 2)$.   (c) shows $g(x - 2)$.

Because we know $a = 2$, it must be a sideways shift which is happening in these two diagrams.

(3)   (b) shows $2h(x)$.   (c) shows $h(2x)$.

(4)   (b) shows $p(x) + 2$.   (c) shows $p(x + 2)$.

Exercise 3.B.2

(1)   (a) $f(g(x)) = f(2x) = 3(2x) - 5 = 6x - 5$ (b) $g(f(x)) = g(3x - 5) = 2(3x - 5) = 6x - 10$

(2)   (a) $f(g(x)) = f(4 - x) = (4 - x)^2$ (b) $g(f(x)) = g(x^2) = 4 - x^2$

(3)   (a) $f(g(x)) = f(x - 4) = \dfrac{1}{x - 4}, \quad x \neq 4$ (b) $g(f(x)) = g(1/x) = 1/x - 4, \quad x \neq 0$

If you are in doubt about any of these, replace x by 'lump' in the definition of the function to see what is happening. Notice that, in (3), we have to exclude the two values of x which would result in trying to divide by zero.

Exercise 3.B.3

(1) $f^{-1}(x) = \frac{1}{5}x$ (2) $f^{-1}(x) = x + 9$

(3) $y = 5x - 9$ so $y + 9 = 5x$ and $x = \frac{1}{5}(y + 9)$ so $f^{-1}(x) = \frac{1}{5}(x + 9)$.

(4) $f(x)$ is self-inverse so $f^{-1}(x) = 8 - x$.   (5) $f^{-1}(x) = 4x$

(6) $f(x)$ is self-inverse so $f^{-1}(x) = 4x$

(7) $y = 3 - 2x$ so $2x = 3 - y$ and $x = \frac{1}{2}(3 - y)$ so $f^{-1}(x) = \frac{1}{2}(3 - x)$.

(8)   Let $y = \dfrac{x - 3}{x + 2}$ so $xy + 2y = x - 3$ so $3 + 2y = x - xy = x(1 - y)$.

Notice the cunning choice of sides here to avoid lots of minuses. This gives

$$x = \frac{3 + 2y}{1 - y} \quad \text{so} \quad f^{-1}(x) = \frac{3 + 2x}{1 - x} \quad (x \neq 1).$$

(9)   Let $y = \dfrac{2x + 3}{x - 2}$ so $xy - 2y = 2x + 3$ so $xy - 2x = 2y + 3$

so $x(y - 2) = 2y + 3$ so $x = \dfrac{2y + 3}{y - 2}$ giving $f^{-1}(x) = \dfrac{2x + 3}{x - 2}$.

We see that this particular $f(x)$ is self-inverse.

Exercise 3.B.4

(1)   The facts we need for the graph sketch are as follows:
     (a) $g(x) = 0$ when $x = 2$.   (b) When $x = 0$, $g(x) = -\frac{1}{2}$.
     (c) The value of $g(x)$ is very large and positive if x is just less than -4.
         The value of $g(x)$ is very large and negative if x is just greater than -4.

     (d) $g(x) = \dfrac{x - 2}{x + 4} = \dfrac{1 - (2/x)}{1 + (4/x)}$,

         so, as x becomes large, the value of $g(x)$ approaches 1, since both $\dfrac{2}{x}$ and $\dfrac{4}{x}$
         become smaller and smaller.

The sketch is given in graph 1 below.

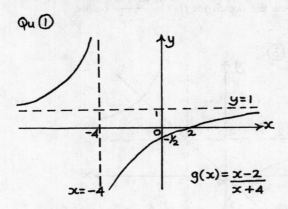

Qu ①

$g(x) = \dfrac{x-2}{x+4}$

$x = -4$

The working for the inverse function goes as follows.

Let $y = g(x) = \dfrac{x-2}{x+4}$ so $xy + 4y = x - 2$ so $4y + 2 = x - xy = x(1-y)$

so $x = \dfrac{4y+2}{1-y}$ and $g^{-1}(x) = \dfrac{4x+2}{1-x}$.

Check: $g(4) = \frac{1}{4}$ and $g^{-1}(\frac{1}{4}) = 4$.

(2)   The facts we need for the graph sketch are as follows:
(a) $h(x) = 0$ when $x = \frac{5}{2}$.   (b) When $x = 0$, $h(x) = -5$.
(c) The value of $h(x)$ is very large and negative if x is just less than -1.
    The value of $h(x)$ is very large and positive if x is just greater than -1.
(d) As x becomes large, the value of $h(x)$ approaches 2, because

$h(x) = \dfrac{2x-5}{x+1} = \dfrac{2-(5/2x)}{1+(1/x)}$ and both $5/2x$ and $1/x$ become very small.

The sketch is shown in graph 2 below.

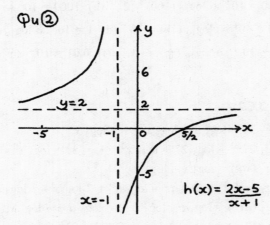

Qu ②

$y = 2$

$x = -1$

$h(x) = \dfrac{2x-5}{x+1}$

The working for the inverse function is as follows.

Let $y = h(x) = \dfrac{2x-5}{x+1}$ so $xy + y = 2x - 5$ so $y + 5 = 2x - xy = x(2-y)$

so $x = \dfrac{y+5}{2-y}$ and $h^{-1}(x) = \dfrac{x+5}{2-x}$.

Check: $h(3) = \frac{1}{4}$ and $h^{-1}(\frac{1}{4}) = 3$.

(3)   I show the sketch for $f(x) = \dfrac{2x + 3}{x - 2}$ below.

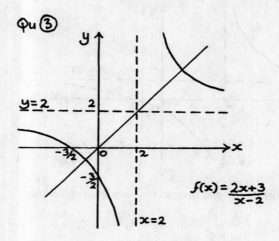

You can see that it is self-inverse because it is symmetrical about the line $y = x$.

Exercise 3.C.1

(1)   (a) $4 = 2^2$ so $\log_2 4 = 2$.   (b) $8 = 2^3$ so $\log_2 8 = 3$.   (c) $2 = 2^1$ so $\log_2 2 = 1$.

(d) $1 = 2^0$ so $\log_2 1 = 0$.   (e) $\frac{1}{2} = 2^{-1}$ so $\log_2 (\frac{1}{2}) = -1$.   (f) $\frac{1}{4} = 2^{-2}$ so $\log_2 (\frac{1}{4}) = -2$.

(2)   (a) $9 = 3^2$ so $\log_3 9 = 2$.   (b) $81 = 3^4$ so $\log_3 (81) = 4$.   (c) $\frac{1}{27} = 3^{-3}$ so $\log_3 (\frac{1}{27}) = -3$.

(d) $\frac{1}{3} = 3^{-1}$ so $\log_3 (\frac{1}{3}) = -1$.   (e) $1 = 3^0$ so $\log_3 1 = 0$.   (f) $3 = 3^1$ so $\log_3 3 = 1$.

(g) $\frac{1}{9} = 3^{-2}$ so $\log_3 (\frac{1}{9}) = -2$.   (h) $27 = 2^3$ so $\log_3 (27) = 3$.   (i) $\sqrt{3} = 3^{1/2}$ so $\log_3 (\sqrt{3}) = \frac{1}{2}$.

(3)   (a) $100 = 10^2$ so $\log_{10} (100) = 2$.   (b) $1000 = 10^3$ so $\log_{10} (1000) = 3$.

(c) $10 = 10^1$ so $\log_{10} (10) = 1$.   (d) $1 = 10^0$ so $\log_{10} (1) = 0$.

(e) $\frac{1}{10} = 10^{-1}$ so $\log_{10} (\frac{1}{10}) = -1$.   (f) $0.01 = 10^{-2}$ so $\log_{10} (0.01) = -2$.

Exercise 3.C.2

(1)   (a) $\log_3 3x = \log_3 3 + \log_3 x = 1 + \log_3 x$.

(b) $\log_3 27x^2 = \log_3 27 + \log_3 x^2 = \log_3 3^3 + \log_3 x^2 = 3 + 2 \log_3 x$.

(c) $\log_3 (x/y) = \log_3 x - \log_3 y$.

(d) $\log_3 (x^2/a^2) = \log_3 x^2 - \log_3 a^2 = 2 \log_3 x - 2 \log_3 a$.

(e) $\log_3 (ax^n) = \log_3 a + \log_3 (x^n) = \log_3 a + n \log_3 x$.

(f) $\log_3 (9a^x) = \log_3 9 + \log_3 (a^x) = \log_3 3^2 + x \log_3 a = 2 + x \log_3 a$.

(g) There is no possible change here.

(2)   (a) $\log_{10} x + \log_{10} (x - 1) = \log_{10} (x^2 - x)$.

(b) $2 \log_{10} x - \log_{10} y = \log_{10} (x^2) - \log_{10} y = \log_{10} (x^2/y)$.

(c) $\log_{10} (x + 1) - \log_{10} (x - 1) = \log_{10} \left(\dfrac{x + 1}{x - 1} \right)$.

(d) $3 \log_{10} x + 2 \log_{10} y = \log_{10} (x^3) + \log_{10} (y^2) = \log_{10} (x^3 y^2)$.

Chapter 4
Exercise 4.A.1

The following are the answers to part (A).

(1) $\sin 34° = \dfrac{x}{8}$ so $x = 8 \sin 34° = 4.47$ cm to 2 d.p.

(2) $\tan 38° = \dfrac{y}{5}$ so $y = 5 \tan 38° = 3.91$ cm to 2 d.p.

(3) $\cos 72° = \dfrac{x}{15}$ so $x = 15 \cos 72° = 4.64$ cm to 2 d.p.

(4) $\tan 54° = \dfrac{6}{x}$ so $x = \dfrac{6}{\tan 54°} = 4.36$ cm to 2 d.p.

(5) $\cos 48° = \dfrac{3}{x}$ so $x = \dfrac{3}{\cos 48°} = 4.48$ cm to 2 d.p.

(6) $\sin 28° = \dfrac{4}{y}$ so $y = \dfrac{4}{\sin 28°} = 8.52$ cm to 2 d.p.

These are the answers to part (B).

(1) $\tan a = \dfrac{7}{4}$ so $a = 60.3°$ to 1 d.p.   (2) $\sin b = \dfrac{5}{8}$ so $b = 38.7°$ to 1 d.p.

(3) $\cos c = \dfrac{6}{9}$ so $c = 48.2°$ to 1 d.p.   (4) $\sin d = \dfrac{8}{10}$ so $d = 53.1°$ to 1 d.p.

Exercise 4.A.2

Calling the length of the unknown side x in each case, the answers are as follows:

(1) $x^2 = 4^2 + 7^2 = 16 + 49 = 65$ so $x = 8.06$ to 2 d.p.

(2) $8^2 = 5^2 + x^2$ so $x^2 = 64 - 25 = 39$ and $x = 6.24$ to 2 d.p.

(3) $9^2 = x^2 + 6^2$ so $x^2 = 81 - 36 = 45$ and $x = 6.71$ to 2 d.p.

(4) $10^2 = 8^2 + d^2$ so $d^2 = 100 - 64 = 36$ and $d = 6$.

Exercise 4.B.1

(1) First, show the information on a sketch like the one below.

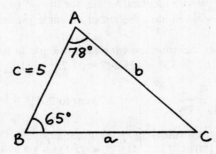

Then $\angle C = 180° - 78° - 65° = 37°$.

Also $\dfrac{a}{\sin 78°} = \dfrac{5}{\sin 37°}$ so $a = \dfrac{5 \sin 78°}{\sin 37°} = 8.13$ cm. to 2 d.p.

And $\dfrac{b}{\sin 65°} = \dfrac{5}{\sin 37°}$ so $b = \dfrac{5 \sin 65°}{\sin 37°} = 7.53$ cm. to 2 d.p.

(2)   First, we draw a sketch, which I've done below.

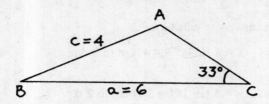

Now we can say

$$\frac{4}{\sin 33°} = \frac{6}{\sin A} \quad \text{so} \quad \sin A = \frac{6 \sin 33°}{4} \quad \text{so} \quad A = 54.7(8)° = 54.8° \text{ to 1 d.p.}$$

The only problem is that this looks wildly improbable from the sketch above, but the sketch does seem to fit the known facts quite well. What has gone wrong?

In fact, as you may already have realised, the known facts fit *two* possible triangles. Can you draw them both?

I've drawn sketches for both of them below in (a) and (b).

(I cheated by only giving one of them in my solution above; you may either have spotted the snag, or have sketched (b), or sketched (a) as I did.)

In (b), I have drawn a dotted line showing where the side *AB* of (a) would come, so you can see how it has been swung round from *B* to give the other possible position.

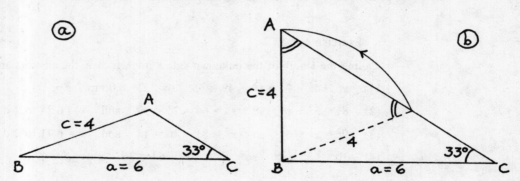

Now you have the right-hand sketch, you can see that the other possible answer for $\angle A$ is $180° - 54.78° = 125.2°$ to 1 d.p. and this is the value for triangle (a).

Your calculator will give you identical values for the sin of 54.8° and 125.2°.

How it is actually possible to have the sin of an angle greater than 90° will be explained later, in Section 5.A.(c).

Next, we find the other measurements for each triangle in turn.

In $\triangle ABC$ (a), $\angle B = 180° - 33° - 125.22° = 21.78° = 21.8°$ to 1 d.p.

and $$\frac{b}{\sin 21.78°} = \frac{4}{\sin 33°} \quad \text{so} \quad b = 2.73 \text{ cm to 2 d.p.}$$

(working with 2 d.p. to avoid rounding errors in the answer).

In $\triangle ABC$ (b), $\angle B = 180° - 33° - 54.78° = 92.22° = 92.2°$ to 1 d.p.

and $$\frac{b}{\sin 92.22°} = \frac{4}{\sin 33°} \quad \text{so} \quad b = 7.34 \text{ cm to 2 d.p.}$$

The two sets of answers now fit the two drawings in believable ways.

We met just this same situation of ambiguous information giving us two possible triangles in case (4) of Section 4.A.(e) on congruent triangles.

(3) First we draw a sketch like the one below.

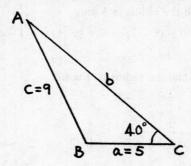

This time, there is only one possible diagram because if we swing AB around B it only cuts AC again the other side of C. Now

$$\frac{9}{\sin 40°} = \frac{5}{\sin A} \quad \text{so} \quad \sin A = \frac{5 \sin 40°}{9} \quad \text{so} \quad A = 20.9(2)° \text{ to 1 d.p.}$$

making a note of the second decimal place for use in further calculations. Therefore $B = 180° - 40° - 20.92° = 119.08° = 119.1°$ to 1 d.p.

Assuming from question (2) that it's all right to use $\sin 119.08°$ from your calculator to find b, you get

$$\frac{b}{\sin 119.08°} = \frac{9}{\sin 40°} \quad \text{so} \quad b = \frac{9 \sin 119.08°}{\sin 40°} = 12.24 \text{ cm to 2 d.p.}$$

Exercise 4.B.2

(1) (a) $\cos C = \dfrac{a^2 + b^2 - c^2}{2ab} = \dfrac{64 + 49 - 25}{112} \quad \text{so} \quad C = 38.2°$ to 1 d.p.

(b) $a^2 = b^2 + c^2 - 2bc \cos A = 25 + 64 - 80 \cos 72° \quad \text{so} \quad a = 8.02$ units to 2 d.p.

(c) (i) $\cos A = \dfrac{b^2 + c^2 - a^2}{2bc} = \dfrac{49 + 9 - 81}{42} \quad \text{so} \quad \cos A = -\dfrac{23}{42}$ and $A = 123.2(0)°$.

(ii) $\cos B = \dfrac{c^2 + a^2 - b^2}{2ca} = \dfrac{9 + 81 - 49}{54} = \dfrac{41}{54} \quad \text{so} \quad B = 40.6(0)°$).

(iii) $\angle C = 180° - 123.20° - 40.60° = 16.2°$ to 1 d.p.

(2) (a) $\angle Q = \angle R = 30°.$ (b) $\angle QPR = 120°.$

(c) $\cos \angle QPR = \dfrac{2^2 + 2^2 - (2\sqrt{3})^2}{2 \times 2 \times 2} = \dfrac{8 - 12}{8} = -\dfrac{1}{2}$

so $\quad \cos 120° = -\dfrac{1}{2} = -\cos 60°.$

(d) $\dfrac{2\sqrt{3}}{\sin 120°} = \dfrac{2}{\sin 30°} = \dfrac{2}{\frac{1}{2}} = 4$

so $\quad \sin 120° = \dfrac{2\sqrt{3}}{4} = \dfrac{\sqrt{3}}{2} = \sin 60°.$

Exercise 4.C.1

(1) The centre is at $(1, -2)$ and the radius is 4 units.

(2) $x^2 - 2x + y^2 - 4y = 0 = (x - 1)^2 - 1 + (y - 2)^2 - 4 \quad \text{so} \quad (x - 1)^2 + (y - 2)^2 = 5.$

The centre is at $(1,2)$ and the radius is $\sqrt{5}$ units.

(3) $x^2 - 8x + y^2 + 7 = 0 = (x - 4)^2 - 16 + y^2 + 7$ so $(x - 4)^2 + y^2 = 9$.

The centre is at $(4,0)$ and the radius is 3 units.

(4) $x^2 - 6x + y^2 + 2y - 6 = 0 = (x - 3)^2 - 9 + (y + 1)^2 - 1 - 6$

so $(x - 3)^2 + (y + 1)^2 = 16$.

The centre is at $(3, -1)$ and the radius is 4 units.

(5) $x^2 - x + y^2 + y = 0 = (x - \frac{1}{2})^2 - \frac{1}{4} + (y + \frac{1}{2})^2 - \frac{1}{4} = 0$

so $(x - \frac{1}{2})^2 + (y + \frac{1}{2})^2 = \frac{1}{2}$.

The centre is at $(\frac{1}{2}, -\frac{1}{2})$ and the radius is $1/\sqrt{2} = \sqrt{2}/2$ units.

(6) $x^2 + 3x + y^2 + 2y + 1 = 0$

so $(x + \frac{3}{2})^2 - \frac{9}{4} + (y + 1)^2 - 1 + 1 = 0$ so $(x + \frac{3}{2})^2 + (y + 1)^2 = \frac{9}{4}$.

The centre is at $(-\frac{3}{2}, -1)$ and the radius is $\frac{3}{2}$ units.

(7) The circle $x^2 + y^2 + 2x - 4y = 0$ can be rewritten as $(x + 1)^2 + (y - 2)^2 = 5^2$ so its centre is at the point $(-1, 2)$.

This is also the centre of the new circle, but the radius of the new circle is 5 units. Therefore its equation is $(x + 1)^2 + (y - 2)^2 = 5^2$ or $x^2 + 2x + y^2 - 4y = 20$.

(8) The equation of the circle is $x^2 - 2ax + y^2 - 2by + c = 0$.

It passes through the point $(0,0)$, therefore putting $x = 0$ and $y = 0$ must satisfy its equation. Doing this gives us $c = 0$.

The point $(3,0)$ also lies on the circle. Putting $x = 3$ and $y = 0$ gives $9 - 6a = 0$ so $a = \frac{3}{2}$.

The point $(0,4)$ also lies on the circle. Putting $x = 0$ and $y = 4$ gives $16 - 8b = 0$ so $b = 2$. Therefore the equation of the circle is $x^2 - 3x + y^2 - 4y = 0$.

Rewriting the equation as $(x - \frac{3}{2})^2 + (y - 2)^2 = \frac{25}{4}$ by completing the squares, gives its centre as the point $(\frac{3}{2}, 2)$ and its radius as $\frac{5}{2}$ units.

There is also a neat geometrical way to do this question.

The sketch below shows the three points A, O and B which the circle must pass through.

Now, $\angle AOB = 90°$ so it is an angle in a semicircle (Section 4.C.(c)) so AB is a diameter of the circle. Therefore, the centre C must be the point $(\frac{3}{2}, 2)$ and the radius must be $\frac{5}{2}$ units, since the length of AB is 5 units by Pythagoras' Theorem.

This then gives us the same equation as the method using algebra.

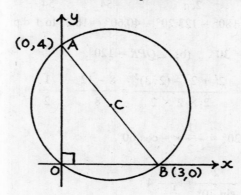

Exercise 4.C.2

(a) $3y = x - 5$ so $x = 3y + 5$.

Putting $x = 3y + 5$ in the equation of the circle gives

$$(3y + 5)^2 - 6(3y + 5) + y^2 - 2y + 5 = 0$$

so $9y^2 + 30y + 25 - 18y - 30 + y^2 - 2y + 5 = 0$.

Therefore $10y^2 + 10y = 0$ so $y^2 + y = 0$ so $y(y + 1) = 0$ so $y = 0$ or $y = -1$.

Remember not to divide through by y in the equation above of $y^2 + y = 0$. If you do this, you lose the answer of $y = 0$ (for which this division would have been impossible).

If $y = 0$ then $x = 5$ and if $y = -1$ then $x = 2$ so the line $3y = x - 5$ cuts the given circle at the two points (5, 0) and (2, –1).

(b)   Substituting $x = 2y - 4$ in the equation of the circle gives

$$(2y - 4)^2 - 6(2y - 4) + y^2 - 2y + 5 = 0$$

so $4y^2 - 16y + 16 - 12y + 24 + y^2 - 2y + 5 = 0$ so $5y^2 - 30y + 45 = 0$

so $y^2 - 6y + 9 = 0$ so $(y - 3)^2 = 0$.

The repeated root of $y = 3$ shows that the line $2y = x + 4$ is a tangent to the circle. When $y = 3$, $x = 2$ so its point of contact is (2, 3).

(c)   Substituting $y = 2x + 3$ in the equation of the circle gives

$$x^2 - 6x + (2x + 3)^2 - 2(2x + 3) + 5 = 0$$

so $x^2 - 6x + 4x^2 + 12x + 9 - 4x - 6 + 5 = 0$ so $5x^2 + 2x + 8 = 0$.

Putting $a = 5$ and $b = 2$ and $c = 8$ and using the quadratic equation formula gives $b^2 - 4ac = 2^2 - 160 = -156$.

Therefore this equation has no real roots and the line $y = 2x + 3$ does not cut this circle at all.

For the sketch, we write $x^2 - 6x + y^2 - 2y + 5 = 0$ as $(x - 3)^2 + (y - 1)^2 = 5$ so the centre of the circle is at the point (3, 1) and its radius is $\sqrt{5}$ units. I show the sketch of this circle and the three lines below.

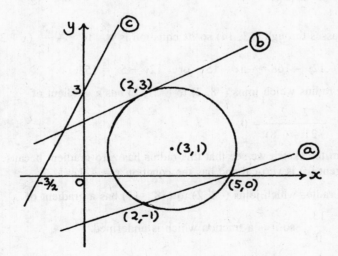

Exercise 4.C.3

$x^2 + 16x + y^2 - 4y - 101 = 0$ can be written as

$$(x + 8)^2 - 64 + (y - 2)^2 - 4 - 101 = 0   \quad \text{or} \quad   (x + 8)^2 + (y - 2)^2 = 169$$

so its centre is at the point (–8, 2) and its radius is 13 units. This makes it possible to draw the sketch on the next page.

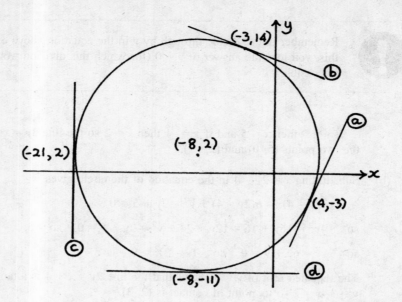

(a) The radius which joins (−8, 2) to (4, −3) has a gradient of

$$\frac{-3-2}{4-(-8)} = -\frac{5}{12} \text{ so the gradient of tangent (a) is } \frac{12}{5}.$$

It passes through (4, −3) so its equation is $y + 3 = \frac{12}{5}(x - 4)$

or $5y + 15 = 12x - 48$ or $5y = 12x - 63$.

(b) The radius which joins (−8, 2) to (−3, 14) has a gradient of

$$\frac{14-2}{-3-(-8)} = \frac{12}{5} \text{ so the gradient of tangent (b) is } -\frac{5}{12}.$$

It passes through (−3, 14) so its equation is $y - 14 = -\frac{5}{12}(x + 3)$

or $12y - 168 = -5x - 15$ or $12y + 5x = 153$.

(c) The radius which joins (−8, 2) to (−21, 2) has a gradient of

$$\frac{2-2}{-21-(-8)} = 0.$$

From the sketch we see that this radius has zero gradient because it is horizontal. Therefore tangent (c) is vertical and has the equation $x = -21$.

(d) The radius which joins (−8, 2) to (−8, −11) has a gradient of

$$\frac{13}{0} \text{ so it is a fraction which is undefined.}$$

Looking at the sketch shows us that this radius is vertical, so tangent (d) is horizontal and its equation is $y = -11$.

Exercise 4.D.1

The missing measurements are as follows:

$$30°, \quad 45°, \quad \frac{\pi}{3}, \quad 120°, \quad \frac{3\pi}{4}, \quad \frac{5\pi}{6}, \quad \pi, \quad 210°, \quad \frac{4\pi}{3}, \quad \frac{3\pi}{2}, \quad 315°, \quad 2\pi.$$

(1) (a) The arc length $= r\theta = 5 \times \pi/6 = 2.62$ cm to 2 d.p.

(b) The area of $\triangle AOB = \frac{1}{2}r^2 \sin \theta = \frac{1}{2} \times 25 \times \sin (\pi/6) = 6.25$ cm^2.

 Remember that you must set your calculator in radian mode before you find the sin of the angle.

(c) To find the area of the segment, we first find the area of the sector.
This is $\frac{1}{2} r^2 \theta = \frac{1}{2} \times 25 \times (\pi/6) = 6.545$ cm^2.
So the area of the segment is $6.545 - 6.25 = 0.30$ cm^2 to 2 d.p.

(2) I will follow my own recommendation here and work in radians. If you don't, you must use the formula for the area of a sector given in Section 4.D.(d).
I start by saying that $60°$ is the same as $\pi/3$ radians.
The area of the whole circle is $\pi(3^2) = 28.274$ m^2.
The area of the minor sector AOB is $\frac{1}{2}r^2 \ \theta = \frac{1}{2} \times 3^2 \times \pi/3 = 4.712$ m^2.
So the area of the coloured part of the circle is $28.274 - 4.712 = 23.56$ m^2 to 2 d.p.

The answer to the thinking point is $90°$ or $\pi/2$.
The reason for this is that the area of the triangle is given by $A = \frac{1}{2}r^2 \sin \theta$. The radius r is a fixed length, so the maximum area is obtained when $\sin \theta$ has its greatest value of one when $\theta = \pi/2$. The largest area the triangle can have is $\frac{1}{2} r^2$.

Chapter 5
Exercise 5.B.1

Here is the sketch of $y = \operatorname{cosec} x$ drawn using the graph of $y = \sin x$.
The two graphs touch each other whenever $\sin x = \pm 1$, and the graph shows that $\operatorname{cosec} x$ becomes very large whenever $\sin x$ approaches zero. The vertical lines are called asymptotes. The curve becomes very close to them near its jumps or discontinuities.

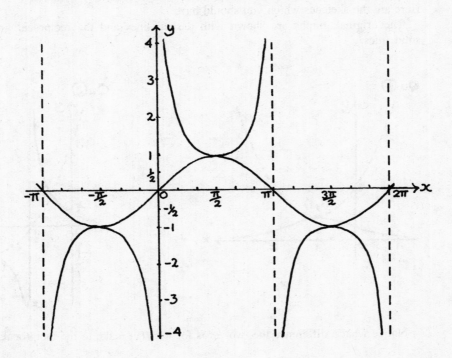

Exercise 5.B.2

Here is the sketch of $y = \cot x$ drawn using the graph of $y = \tan x$. I have shown $y = \tan x$ with a dashed line so that you can see $y = \cot x$ more easily.

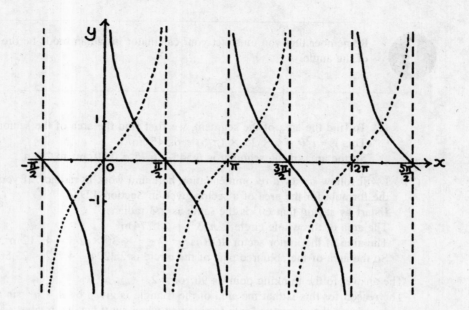

It is a kind of mirror image of $y = \tan x$ since it behaves in exactly the opposite way; going off to infinity when $\tan x = 0$, and itself equalling zero when $\tan x$ goes off towards infinity. (These last two properties are always true for reciprocal graphs, which helps when sketching them.) However, if you look at the graph of $y = \tan x$ in a mirror, you will see that you also have to slide the mirror image to the right (or left!) by $\pi/2$ in order to get the graph of $y = \cot x$.

Exercise 5.B.3

Here are the sketches which you should have.

The original graphs are shown with dashed lines and the reciprocal graphs are shown with solid lines.

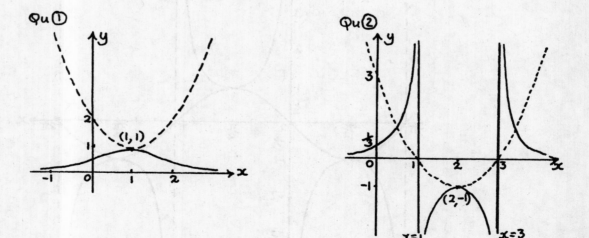

Notice what a difference the two zeros for $y = f(x)$ make to the reciprocal graph in (2).

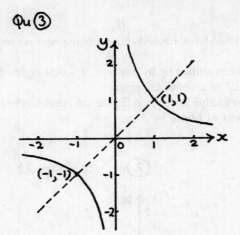

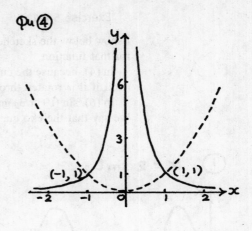

In (3), using the rules for sketching reciprocal graphs gives us the graph we already know the shape of from Section 3.B.(g).

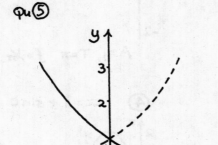

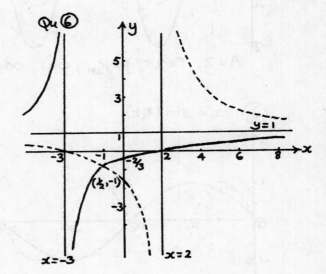

In (5), $y = 1/e^x = e^{-x}$.

In (6), the two graphs cross each other where $\dfrac{x+3}{x-2} = \dfrac{x-2}{x+3}$.

Therefore $(x+3)^2 = (x-2)^2$ so $x^2 + 6x + 9 = x^2 - 4x + 4$ so $10x = -5$ and $x = -\frac{1}{2}$.

Substituting $x = -\frac{1}{2}$ in either $f(x)$ or $1/f(x)$ gives $y = -1$, so the two graphs cut at the point $(-\frac{1}{2}, -1)$.

It's worth comparing the graph for this reciprocal function with the graph which we sketched in Section 3.B.(i) of the inverse function of $f(x)$.

Exercise 5.C.1

I show below the sketches you should have. Each sketch gives the answers to the individual questions for that function.

In (4), because the curve has been shifted up by one unit, it is no longer odd. It no longer fits onto itself if it is rotated through a half turn about the origin.

In (6), sin $(t + \pi/3)$ gets to every value faster by $\pi/3$. The sin curve has been shifted $\pi/3$ to the left. We say that the two curves are **out of phase** by $\pi/3$.

① $x = 2 \sin t$

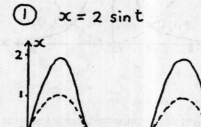

$A = 2$, $T = 2\pi$, $f = \frac{1}{2\pi}$, $\omega = 1$, odd.

② $x = \sin 2t$

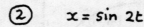
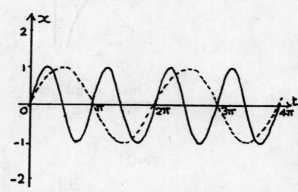

$A = 1$, $T = \pi$, $f = \frac{1}{\pi}$, $\omega = 2$, odd

③ $x = \sin(\frac{1}{2}t)$

$A = 1$, $T = 4\pi$, $f = \frac{1}{4\pi}$, $\omega = \frac{1}{2}$, odd

④ $x = 1 + \sin t$

$A = 1$, $T = 2\pi$, $f = \frac{1}{2\pi}$, $\omega = 1$, neither.

⑤ $x = \cos t$

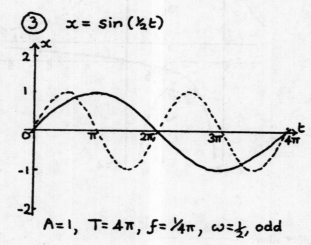

$A = 1$, $T = 2\pi$, $f = \frac{1}{2\pi}$, $\omega = 1$, odd.

⑥ $x = \cos(t + \frac{\pi}{2})$

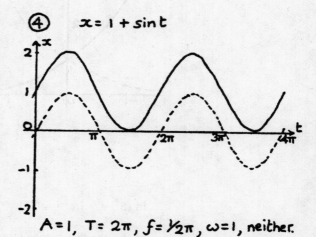

$A = 1$, $T = 2\pi$, $f = \frac{1}{2\pi}$, $\omega = 1$, odd.

Exercise 5.C.2

Here are the eight sketches which you should have drawn. Each sketch shows the position of P after time t and the corresponding length of x which I have drawn using a heavy black line. I have also shown the starting position of P when $t = 0$ on each sketch.

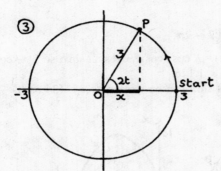

$x = \cos 3t \quad A = 1 \quad \omega = 3 \quad T = 2\pi/3$

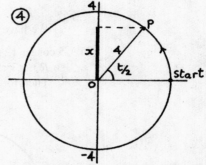

$x = 2 \sin t \quad A = 2 \quad \omega = 1 \quad T = 2\pi$

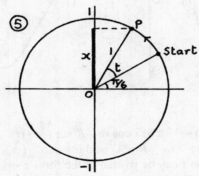

$x = 3 \cos 2t \quad A = 3 \quad \omega = 2 \quad T = \pi$

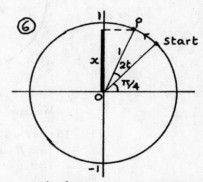

$x = 4 \sin(t/2) \quad A = 4 \quad \omega = \tfrac{1}{2} \quad T = 4\pi$

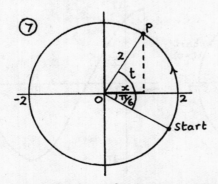

$x = \sin(t + \pi/6) \quad A = 1 \quad \omega = 1 \quad T = 2\pi$

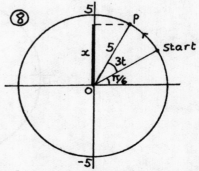

$x = \sin(2t + \pi/4) \quad A = 1 \quad \omega = 2 \quad T = \pi$

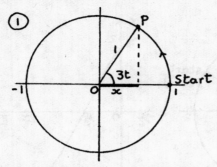

$x = 2 \cos(t - \pi/6) \quad A = 2 \quad \omega = 1 \quad T = 2\pi$

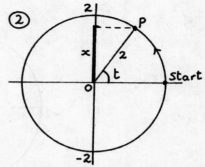

$x = 5 \sin(3t + \pi/6) \quad A = 5 \quad \omega = 3 \quad T = 2\pi/3$

Exercise 5.D.1

(1) $\sqrt{3} \cos t - \sin t = R \cos (t + \alpha) = R \cos t \cos \alpha - R \sin t \sin \alpha$
so $\sqrt{3} = R \cos \alpha$ and $1 = R \sin \alpha$ so $R = 2$ and $\tan \alpha = 1/\sqrt{3}$ giving $\alpha = \pi/6$.
Therefore $x = \sqrt{3} \cos t - \sin t$ can be written as $x = 2 \cos (t + \pi/6)$.

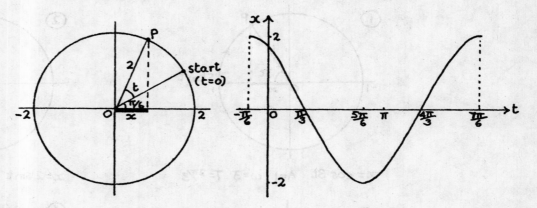

We have $A = 2$, $\omega = 1$ and $T = 2\pi$.

(2) $5 \cos t + 12 \sin t = R \cos (t - \alpha) = R \cos t \cos \alpha + R \sin t \sin \alpha$ so $5 = R\cos \alpha$ and $12 = R \sin \alpha$
so $R = 13$ and $\alpha = \tan^{-1} \left(\frac{12}{5}\right) = 1.176$.
Therefore $x = 5 \cos t + 12 \sin t$ can be written as $x = 13 \cos (t - 1.176)$.

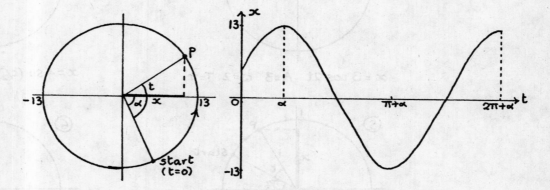

We have $A = 13$, $\omega = 1$ and $T = 2\pi$.

(3) $15 \cos t - 8 \sin t = R \cos (t + \alpha) = R \cos t \cos \alpha - R \sin t \sin \alpha$
so $15 = R \cos \alpha$ and $8 = R \sin \alpha$ so $R = 17$ and $\alpha = \tan^{-1} \left(\frac{8}{15}\right) = 0.490$.
Therefore $x = 15 \cos t - 8 \sin t$ can be written in the form $x = 17 \cos (t + 0.490)$.

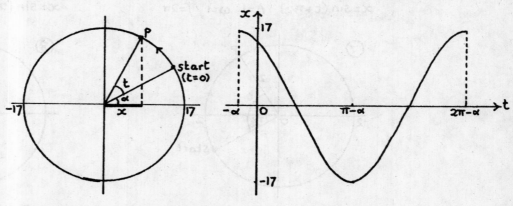

We have $A = 17$, $\omega = 1$ and $T = 2\pi$.

(4) $2 \cos t - 3 \sin t = R \cos (t + \alpha) = R \cos t \cos \alpha - R \sin t \sin \alpha$
so $2 = R \cos \alpha$ and $3 = R \sin \alpha$ so $R = \sqrt{13}$ and $\alpha = \tan^{-1}(\frac{3}{2}) = 0.983$.
Therefore $x = 2 \cos t - 3 \sin t$ can be written as $x = \sqrt{13} \cos (t + 0.983)$.

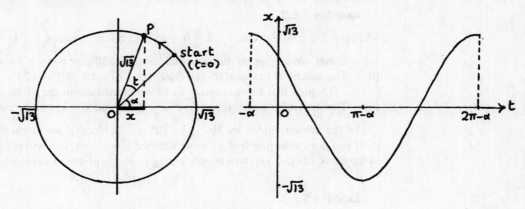

We have $A = \sqrt{13}$, $\omega = 1$ and $T = 2\pi$.

(5) $\cos 4t + \sin 4t = R \cos (4t - \alpha) = R \cos 4t \cos \alpha + R \sin 4t \sin \alpha$ so $1 = R \cos \alpha$ and $1 = R \sin \alpha$
so $R = \sqrt{2}$ and $\alpha = \tan^{-1} 1 = \pi/4$.
Therefore $x = \cos 4t + \sin 4t$ can be written as $x = \sqrt{2} \cos (4t + \pi/4)$.

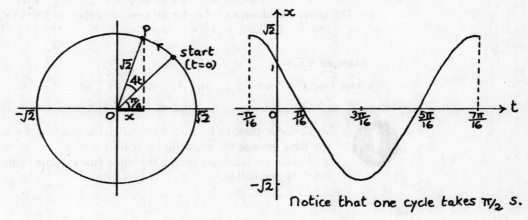

Notice that one cycle takes $\pi/2$ s.

We have $A = \sqrt{2}$, $\omega = 4$ and $T = 2\pi/4 = \pi/2$.

(6) $\sqrt{3} \sin 3t - \cos 3t = R \sin (3t - \alpha) = R \sin 3t \cos \alpha - R \cos 3t \sin \alpha$
so $\sqrt{3} = R \cos \alpha$ and $1 = R \sin \alpha$ so $R = 2$ and $\alpha = \tan^{-1} (1/\sqrt{3}) = \pi/6$.
Therefore $x = \sqrt{3} \sin 3t - \cos 3t$ can be written as $x = 2 \sin (3t - \pi/6)$.

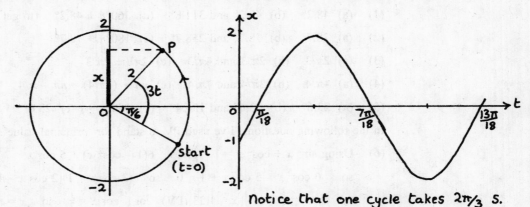

Notice that one cycle takes $2\pi/3$ s.

We have $A = 2$, $\omega = 3$ and $T = 2\pi/3$.

Exercise 5.E.1

This is checked by calculator.

Exercise 5.E.2

We have $2\cos^2 x + 3\cos x + 1 = 0$ so $(2\cos x + 1)(\cos x + 1) = 0$.

(a)   Either $\cos x = -\frac{1}{2}$ so the principal value is $120°$, or $\cos x = -1$ so the principal value is $180°$.

(b)   The solutions between $0°$ and $360°$ are $120°$ and $360° - 120° = 240°$, and $180°$.
     (Notice that there are only three solutions here in the given range!)

(c)   The general solution is $x = 360° \, n \pm 120°$ and $x = 360° \, n \pm 180°$.

All the answers given by $360°n \pm 180°$ are included if we just write $360° \, n + 180°$.
     If you sketch the graph of $y = \cos x$ with the line $y = -1$, you will see that the line $y = -1$ is a tangent to the curve of $\cos x$ and so it is only giving us single points of intersection on each cycle.

Exercise 5.E.3

Using $\tan^2 x + 1 = \sec^2 x$ gives us $\tan^2 x + 2\tan x - 3 = 0$, so $(\tan x - 1)(\tan x + 3) = 0$.

(a)   Either $\tan x = 1$ giving the principal value of $x = 45°$, or $\tan x = -3$ giving the principal value of $x = -71.57° = -71.6°$ to 1 d.p.

(b)   The solutions between $0°$ and $360°$ are $45°$, $180° + 45°$, $180° + (-71.57°)$ and $360° + (-71.57°)$ giving $45°$, $225°$, $108.4°$ and $288.4°$ to 1 d.p.

(c)   The general solution to 1 d.p. for all possible angles is given by $x = 180°n + 45°$ and $x = 180°n + (-71.6°) = 180°n - 71.6°$.

Exercise 5.E.4

We have $\cos^2 x + 2\sin x = 1$ so $(1 - \sin^2 x) + 2\sin x = 1$ so $0 = \sin^2 x - 2\sin x$.

> Don't divide through by $\sin x$ here, so giving yourself $0 = \sin x - 2$. If you do this, you have ignored the possibility that $\sin x = 0$.
>   Instead, we factorise, getting $0 = \sin x \, (\sin x - 2)$ so either $\sin x = 0$, or $\sin x = 2$ which is impossible.

(a) $\sin x = 0$ gives a principal value of 0 radians.

(b)   The solutions from 0 to 2π are 0, π and 2π.

(c)   The general solution is $x = n\pi$ radians where n is any whole number.

Exercise 5.E.5

(1)   (a) $48.2°$ (b) $48.2°$ and $311.8°$ (c) $360°n \pm 48.2°$ (to get (b), put $n = 0$ or 1)

(2)   (a) $78.7°$ (b) $78.7°$ and $258.7°$ (c) $180°n + 78.7°$

(3)   (a) $2\pi/3$ (b) $2\pi/3$ and $4\pi/3$ (c) $2\pi n \pm 2\pi/3$

(4)   (a) $-\pi/4$ (b) $3\pi/4$ and $7\pi/4$ (c) $n\pi + (-\pi/4) = n\pi - \pi/4$

(5)   (a) $23.6°$ (b) $23.6°$ and $156.4°$ (c) $180°n + (-1)^n \, 23.6°$

In the following questions, I've used PV to stand for 'principal value'.

(6)   Using $\sin^2 x + \cos^2 x = 1$ gives $6(1 - \cos^2 x) + 5\cos x = 7$

so $6\cos^2 x - 5\cos x + 1 = 0$ so $(3\cos x - 1)(2\cos x - 1) = 0$.

(a) $\cos x = \frac{1}{3}$ so $x = 1.23$ (PV)   or $\cos x = \frac{1}{2}$ so $x = \pi/3$ (PV).
(b) The solutions between 0 and 2π are 1.23 and 5.05 and $\pi/3$ and $5\pi/3$.
(c) The general solution is $x = 2n\pi \pm 1.23$ and $x = 2n\pi \pm \pi/3$.

(7) $\tan^2 x = \tan x$ so $\tan^2 x - \tan x = 0$ so $\tan x (\tan x - 1) = 0$.

Either $\tan x = 0$ giving a PV of 0 so that the general solution is $x = n\pi$.
   This gives the solutions 0, π or 2π if $0 \le x \le 2\pi$.
   Or $\tan x = 1$ giving a PV of $\pi/4$ so the general solution is $x = n\pi + \pi/4$.
   This gives the solutions $\pi/4$ and $3\pi/4$ if $0 \le x \le 2\pi$.

(8)   Using the identity $\tan^2 x + 1 = \sec^2 x$ we get $2 \tan^2 x = 1$ so $\tan x = \pm\sqrt{1/2}$.
   If $\tan x = 1/\sqrt{2}$, the PV is $35.3°$ and if $\tan x = -1/\sqrt{2}$, the PV is $-35.3°$.
   The general solution is $x = 180°n \pm 35.3°$. (This puts together both the principal values
which we have found.)
   The solutions between $0°$ and $360°$ are $35.3°$, $144.7°$, $215.3°$ and $324.7°$.

(9)   Using $\sin 2x = 2 \sin x \cos x$ gives $2 \sin x \cos x - 3 \cos x = 0$ so $\cos x (2 \sin x - 3) = 0$.
   Either $\sin x = \frac{3}{2}$ which has no solution, or $\cos x = 0$ giving a PV of $x = \pi/2$.
   This gives a general solution of $x = 2n\pi \pm \pi/2$ so there is just one solution of $\pi/2$ if
$0 \le x \le 2\pi$.

Exercise 5.E.6

(1)   Notice that we are working in radians here.
   In Section 5.D. (g) I showed that $3 \cos t - 2 \sin t = \sqrt{13} \cos(t + \alpha)$ where $\alpha = 0.588$
radians to 3 d.p.

   (a) This has no solutions since we can't have $\cos t > 1$.
   (b) This equation gives $\cos (t + \alpha) = 1$ so the principal value is $t + \alpha = 0$.
      This gives the general solution that $t + \alpha = 2n\pi$ so $t = 2n\pi - \alpha$, with $\alpha = 0.588$.
      If $0 \le t \le 2\pi$, we get $t = 5.70$ to 2 d.p.
   (c) The equation gives $\sqrt{13} \cos (t + \alpha) = 1$ so $\cos (t + \alpha) = 1/\sqrt{13}$ which gives the principal
      value for $(t + \alpha)$ of 1.290 radians to 3 d.p.
      The general solution for $(t + \alpha)$ is given by $t + \alpha = 2n\pi \pm 1.290$.
      Putting in $\alpha = 0.588$, the solutions between 0 and 2π are given by putting $n = 0$ and
   $n = 1$.
      These solutions are $t = 0.70$ and $t = 4.41$ to 2 d.p.
      I show all these answers in the sketch below.

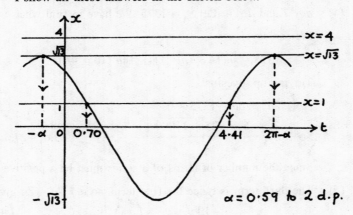

$\alpha = 0.59$ to 2 d.p.

(2)   In Section 5.D.(g) we showed that $3 \sin 2t + \cos 2t = \sqrt{10} \sin (2t + \alpha)$ where $\alpha = \tan^{-1} \frac{1}{3} =$
$18.43°$ to 2 d.p.
   We have $\sqrt{10} \sin (2t + \alpha) = 2$ so the principal value for $(2t + \alpha)$ is $39.23°$.
   The general solution for $2t + \alpha$ is $180°n + (-1)^n (39.23°)$, so

   $$2t = 180°n + (-1)^n (39.23°) - 18.43°$$

giving

   $$t = 90°n + (-1)^n (19.62°) - 9.22°.$$

   Putting $n = 0$, $n = 1$, $n = 2$ and $n = 3$ gives the solutions between $0°$ and $360°$ of
$t = 10.4°$, $61.2°$, $190.4°$ and $241.2°$ to 1 d.p.

Chapter 6
Exercise 6.B.1

(1) (a) $3 + 7 + 11 + \ldots + 79$
   (i) $a = 3$ and $d = 4$.
   (ii) $79 = a + (n - 1)d = 3 + (n - 1)4$, so $n - 1 = \frac{76}{4} = 19$, and $n = 20$.
   (iii) $S_{20} = \frac{20}{2}(3 + 79) = 820$.

   (b) $100 + 95 + 90 + \ldots + 15$
   (i) $a = 100$ and $d = -5$.
   (ii) $15 = a + (n - 1)d = 100 + (n - 1)(-5)$ so $5n = 105 - 15$, and $n = 18$.
   (iii) $S_{18} = \frac{18}{2}(100 + 15) = 1035$.

   (c) $6 + 6\frac{1}{4} + 6\frac{1}{2} + \ldots + 17\frac{3}{4}$
   (i) $a = 6$ and $d = \frac{1}{4}$.
   (ii) $17\frac{3}{4} = a + (n - 1)d = 6 + (n - 1)\frac{1}{4}$ so $11\frac{3}{4} = \frac{1}{4}(n - 1)$, and $47 = n - 1$,
         therefore $n = 48$.

(2) (a) $1 + 2 + 3 + \ldots + 100$

   $a = 1$ and $d = 1$ and $n = 100$.
   $S_{100} = \frac{100}{2}(1 + 100) = 5050$.

   (b) $2 + 4 + 6 + \ldots + 100$

   $a = 2$ and $d = 2$ and $n = 50$.
   $S_{50} = \frac{50}{2}(2 + 100) = 2550$.

   (c) The sum of the odd numbers up to 100 is $5050 - 2550 = 2500$.

   (d) $1 + 2 + 3 + 4 + \ldots + n$

   $a = 1$ and $l = n$ and the number of terms is n so $S_n = \frac{1}{2}n(1 + n)$.
   This is an often-used rule and it often appears in formula books.

(3) $a = 11$ and $S_{18} = 1269$ so we have $1269 = 9(22 + 17d)$.

   This gives $1269 = 198 + 153d$ so $d = 7$.

(4) $a = 7$ and $d = 4$. Let $S_n = 1375$. We have to find what n is.

   We can say

   $$1375 = \tfrac{1}{2}n(14 + (n - 1)4) = \tfrac{1}{2}n\,(10 + 4n).$$

   Tidying up gives us

   $$2n^2 + 5n - 1375 = 0.$$

   So $\quad n = \dfrac{-5 \pm \sqrt{25 + 11000}}{4} = \dfrac{-5 \pm 105}{4} = 25 \text{ or } -27.5.$

   Since the number of terms of a series must be a positive whole number, the answer is 25.

(5) The third term is twice the first term, so $a + 2d = 2a$ giving $d = \frac{1}{2}a$. Also

   $$S_{10} = 195 = \frac{10}{2}\left(2a + 9\left(\frac{a}{2}\right)\right) \quad \text{so} \quad 390 = 10\left(\frac{4a + 9a}{2}\right).$$

   This gives

   $$39 = \frac{13a}{2} \quad \text{so} \quad a = \frac{78}{13} = 6 \quad \text{and} \quad d = 3.$$

Exercise 6.C.2

(1) $0.7 = \dfrac{7}{10}$ (2) $0.25 = \dfrac{25}{100} = \dfrac{1}{4}$ (3) $0.401 = \dfrac{401}{1000}$ (4) $0.011 = \dfrac{11}{1000}$

(5)   Let $F = 0.7777\ldots$ Then $10F = 7.7777\ldots$

Subtracting, we have $9F = 7$ so $F = \frac{7}{9}$.

(6)   Let $F = 0.292929\ldots$ Then $100F = 29.292929\ldots$

Subtracting, we have $99F = 29$ so $F = \frac{29}{99}$.

(7)   Let $F = 2.5343434\ldots$ Then $100F = 253.4343434\ldots$

Subtracting, we have $99F = 250.9$ so $F = \dfrac{250.9}{99} = \dfrac{2509}{990}$.

(8)   If $F = 40.2\overline{106}$, then, multiplying by 1000, we have

$$1000F = 40\,210.6106106106\ldots$$
$$F = 40.2106106106\ldots$$

Subtracting, we get $999F = 40170.4$ so $F = \dfrac{40170.4}{999} = \dfrac{401704}{9990}$.

(9)   If $F = 0.\overline{142857}$, then multiplying by $1\,000\,000$, we have

$$1\,000\,000F = 142\,857.142857\ldots$$
$$F = 0.142857\ldots$$

Subtracting, we get $999\,999F = 142\,857$, so $F = \dfrac{142\,857}{999\,999} = \dfrac{1}{7}$ rather amazingly.

The digits of the decimal forms of $\frac{1}{7}, \frac{2}{7}, \frac{3}{7}$, etc. make interesting patterns. You might like to look at these for yourself.

Exercise 6.D.1

(1) $\displaystyle\sum_{r=1}^{9} r^2$ (2) $\displaystyle\sum_{r=1}^{11} \frac{r}{r+1}$ (3) $\displaystyle\sum_{r=1}^{29} \frac{1}{r(r+1)}$

(4)   We want the terms to alternate in sign, with the odd terms being negative and the even terms positive. We can make this happen by multiplying each term by something which flips sign in this way. $(-1)^r$ will fit our requirements exactly. (This is what we used when we wrote down the general solution for a sin in Section 5.E.(d).) So we write

$$\sum_{r=1}^{9} (-1)^r r^2.$$

Exercise 6.D.2

(1)   The first four terms are $5 + 7 + 9 + 11 = 32$. (An AP!)

The nth term is $2n + 3$, and the $(n + 1)$th term is $2(n + 1) + 3 = 2n + 5$.

(2)   The first four terms are $36 + 12 + 4 + \frac{4}{3} = 53\frac{1}{3}$, giving a GP this time. (Remember that $(\frac{1}{3})^0 = 1$ from Section 1.D.(b).)

The nth term is $36(\frac{1}{3})^{n-1}$. The $(n + 1)$th term is $36(\frac{1}{3})^n$.

(3)   The first four terms added are $1 + \frac{1}{2} + \frac{1}{6} + \frac{1}{24} = 1\frac{17}{24}$.

The nth term is $\dfrac{1}{n!}$ and the $(n + 1)$th term is $\dfrac{1}{(n + 1)!}$.

(I gave the meaning of $n!$ at the end of Section 6.A.(a).)

(4)   The first four terms added are $+\frac{1}{3} - \frac{2}{4} + \frac{3}{5} - \frac{4}{6} = -\frac{7}{30}$.

The nth term is $\left(\dfrac{n}{n + 2}\right)(-1)^{n+1}$, and the $(n + 1)$th term is $\left(\dfrac{n + 1}{n + 3}\right)(-1)^{n+2}$,

replacing n by $n + 1$ in the previous formula.

(5) The first four terms are $\dfrac{1}{1\times3} + \dfrac{1}{3\times5} + \dfrac{1}{5\times7} + \dfrac{1}{7\times9,}$ that is, $\dfrac{1}{3} + \dfrac{1}{15} + \dfrac{1}{35} + \dfrac{1}{63} = \dfrac{4}{9}.$

The nth term is $\dfrac{1}{(2n-1)(2n+1)}.$

The $(n+1)$th term is $\dfrac{1}{(2(n+1)-1)(2(n+1)+1)} = \dfrac{1}{(2n+1)(2n+3)}.$

Exercise 6.D.3

(1) $\displaystyle\sum_{r=1}^{n}(r-1)(r+3) = \sum_{r=1}^{n}(r^2 + 2r - 3)$

$$= \sum_{r=1}^{n}r^2 + 2\sum_{r=1}^{n}r - \sum_{r=1}^{n}3$$

$$= \tfrac{1}{6}n(n+1)(2n+1) + 2(\tfrac{1}{2}n(n+1)) - 3n \text{ using sums } (S\,2) \text{ and } (S1)$$

$$= \tfrac{1}{6}n((n+1)(2n+1) + 6(n+1) - 18)$$

$$= \tfrac{1}{6}n(2n^2 + 3n + 1 + 6n + 6 - 18)$$

$$= \tfrac{1}{6}n(2n^2 + 9n - 11).$$

Check: If $n = 3$

$$\text{LHS} = \sum_{r=1}^{3}(r-1)(r+3) = (0)(4) + (1)(5) + (2)(6) = 17.$$

Putting $n = 3$ in the RHS gives $\tfrac{1}{6}(3)(2\times3^2 + 9\times3 - 11) = \tfrac{1}{2}(18 + 27 - 11) = 17.$

(2) $\displaystyle\sum_{r=1}^{n}r(r-1)(r+1) = \sum_{r=1}^{n}r(r^2-1) = \sum_{r=1}^{n}(r^3 - r)$

$$= \sum_{r=1}^{n}r^3 - \sum_{r=1}^{n}r = \tfrac{1}{4}n^2(n+1)^2 - \tfrac{1}{2}n(n+1) \text{ using } (S3) \text{ and } (S\,1).$$

Factorising, we get

$$\tfrac{1}{4}n(n+1)(n(n+1) - 2) = \tfrac{1}{4}n(n+1)(n^2 + n - 2).$$

Check: If $n = 3$

$$\text{LHS} = \sum_{r=1}^{n}r(r-1)(r+1) = 1(0)(2) + 2(1)(3) + 3(2)(4) = 30.$$

Putting $n = 3$ in the RHS gives $\tfrac{1}{4}(3)(4)(9 + 3 - 2) = 30.$

These checks are useful not only to be confident that your working is correct, but also because they give you good practice in handling series, and seeing the terms building up into the sums.

Exercise 6.E.1

(1) $\dfrac{4}{(x+2)(x+3)} \equiv \dfrac{A}{x+2} + \dfrac{B}{x+3}$

so $\quad 4 \equiv A(x+3) + B(x+2).$

Putting $x = -3$, we get $4 = -B$, so $B = -4.$
Putting $x = -2$, we get $4 = A.$
Check with $x = 0$: LHS = 4 and RHS = $12 - 8 = 4.$

So $\quad \dfrac{4}{(x+2)(x+3)} \equiv \dfrac{4}{x+2} - \dfrac{4}{x+3}.$

(2) $\dfrac{6}{(2y-1)(2y+1)} \equiv \dfrac{A}{2y-1} + \dfrac{B}{2y+1}$

so $\quad 6 \equiv A(2y+1) + B(2y-1).$

Putting $y = \frac{1}{2}$, we get $6 = 2A$, so $A = 3$.
Putting $y = -\frac{1}{2}$, we get $6 = -2B$, so $B = -3$.
Check with $y = 0$: LHS = 6; RHS = 3 + 3 = 6.

So $\quad \dfrac{6}{(2y-1)(2y+1)} \equiv \dfrac{3}{2y-1} - \dfrac{3}{2y+1}.$

(3) $\dfrac{10}{x(x-1)(x+4)} \equiv \dfrac{A}{x} + \dfrac{B}{x-1} + \dfrac{C}{x+4}$

so $\quad 10 \equiv A(x-1)(x+4) + Bx(x+4) + Cx(x-1).$

Putting $x = 1$, we get $10 = 5B$, so $B = 2$.
Putting $x = -4$, we get $10 = 20C$, so $C = \frac{1}{2}$.
Putting $x = 0$, we get $10 = -4A$, so $A = \frac{-5}{2}$.
Check with $x = 2$, say. (We can't use $x = 0$ as we've used it already.)
We get LHS = 10 and RHS = $(-\frac{5}{2})6 + 2(12) + \frac{1}{2}(2) = 10.$

So $\quad \dfrac{10}{x(x-1)(x+4)} \equiv -\dfrac{\frac{5}{2}}{x} + \dfrac{2}{x-1} + \dfrac{\frac{1}{2}}{x+4}.$

 HELPFUL HINT When you use partial fractions for integrating, it is usually better to keep A, B and C as fractions on top of the original fractions, so I shall leave my answers in the form

$\dfrac{\frac{1}{2}}{x+4}$ rather than $\dfrac{1}{2(x+4)}$.

Exercise 6.E.2

(1) $\dfrac{5}{(x-2)(x+3)^2} \equiv \dfrac{A}{x-2} + \dfrac{B}{x+3} + \dfrac{C}{(x+3)^2}$

so $\quad 5 \equiv A(x+3)^2 + B(x-2)(x+3) + C(x-2).$

Putting $x = -3$, we get $5 = C(-5)$ so $C = -1$.
Putting $x = 2$, we get $5 = A(5^2)$ so $A = \frac{5}{25} = \frac{1}{5}$.
Comparing the terms in x^2, we have $0 = Ax^2 + Bx^2$ so $B = -A = -\frac{1}{5}$.
Checking with $x = 0$, we get the LHS = 5 and the RHS = $\frac{1}{5}(9) + (-\frac{1}{5})(-6) + (-1)(-2) = 5.$

So $\quad \dfrac{5}{(x-2)(x+3)^2} \equiv \dfrac{\frac{1}{5}}{x-2} - \dfrac{\frac{1}{5}}{x+3} - \dfrac{1}{(x+3)^2}.$

(2) $\dfrac{2}{y^2(y-1)} \equiv \dfrac{A}{y} + \dfrac{B}{y^2} + \dfrac{C}{y-1}$

so $\quad 2 = Ay(y-1) + B(y-1) + Cy^2.$

(y^2 works just like any other repeated factor.)
Putting $y = 0$, we get $2 = -B$ so $B = -2$.
Putting $y = 1$, we get $2 = C$.
Comparing the terms in y^2, we have $0 = Ay^2 + Cy^2$ so $A = -C$ so $A = -2$.
Checking with $y = 2$, we get the LHS = 2 and the RHS = $(-2)(2) + (-2)(1) + (2)(4) = 2.$

So $\quad \dfrac{2}{y^2(y-1)} \equiv -\dfrac{2}{y} - \dfrac{2}{y^2} + \dfrac{2}{y-1}.$

Exercise 6.E.3

(1) $\quad \dfrac{14}{(x^2 + 3)(x + 2)} \equiv \dfrac{Ax + B}{x^2 + 3} + \dfrac{C}{x + 2}$

so $\quad 14 \equiv (Ax + B)(x + 2) + C(x^2 + 3).$

Putting $x = -2$, we get $14 = 7C$ so $C = 2.$
Putting $x = 0$, we get $14 = 2B + 3C$ so $2B = 8$ and $B = 4.$
Comparing the terms in x^2, we have $0 = Ax^2 + Cx^2$ so $A = -C = -2.$
Checking with $x = 1$, the LHS $= 14$, and the RHS $= (-2 + 4)(3) + 2(4) = 14.$

So $\quad \dfrac{14}{(x^2 + 3)(x + 2)} \equiv \dfrac{(-2x + 4)}{x^2 + 3} + \dfrac{2}{x + 2} \equiv \dfrac{4 - 2x}{x^2 + 3} + \dfrac{2}{x + 2}.$

The second form looks a bit tidier.

(2) $\quad \dfrac{4}{y(y^2 + 1)} \equiv \dfrac{A}{y} + \dfrac{By + C}{y^2 + 1}$

so $\quad 4 \equiv A(y^2 + 1) + (By + C)y.$

Putting $y = 0$, we have $4 = A.$
Comparing the terms in y^2, we have $0 = Ay^2 + By^2$ so $B = -A = -4.$
Putting $y = 1$, we get $4 = 4(2) + (-4 + C)(1)$, so $4 = 8 - 4 + C$ so $C = 0.$
Checking with $y = 2$, we get the LHS $= 4$, and the RHS $- 4(5) + (-8)(2) = 20 - 16 = 4.$

So $\quad \dfrac{4}{y(y^2 + 1)} \equiv \dfrac{4}{y} - \dfrac{4y}{y^2 + 1}.$

Exercise 6.E.4

(I haven't filled in all the very straightforward parts of these questions.)

(1) $\quad \dfrac{4}{(x + 3)(x - 1)^2} \equiv \dfrac{A}{x + 3} + \dfrac{B}{x - 1} + \dfrac{C}{(x - 1)^2}$

so $\quad 4 \equiv A(x - 1)^2 + B(x - 1)(x + 3) + C(x + 3).$

Putting $x = 1$, we get $4 = 4C$ so $C = 1.$
Putting $x = -3$, we get $4 = 16A$ so $A = \frac{1}{4}.$
Matching the terms in x^2, we get $0 = Ax^2 + Bx^2$ so $B = -A = -\frac{1}{4}.$
Checking with $x = 0$ gives the LHS $= 4$, and the RHS $= \frac{1}{4} - \frac{1}{4}(-3) + 3 = 4.$

So $\quad \dfrac{4}{(x + 3)(x - 1)^2} \equiv \dfrac{\frac{1}{4}}{x + 3} + \dfrac{\frac{1}{4}}{x - 1} + \dfrac{1}{(x - 1)^2}.$

(2) $\quad \dfrac{3p + 1}{(2p - 1)(p + 2)^2} \equiv \dfrac{A}{2p - 1} + \dfrac{B}{p + 2} + \dfrac{C}{(p + 2)^2}$

so $\quad 3p + 1 \equiv A(p + 2)^2 + B(2p - 1)(p + 2) + C(2p - 1).$

Putting $p = -2$, we get $-5 = -5C$ so $C = 1.$
Putting $p = \frac{1}{2}$, we get $\frac{3}{2} + 1 = A(\frac{5}{2})^2 = \frac{25}{4}A$ so $A = \frac{2}{5}.$
Matching the terms in p^2, we get $0 = Ap^2 + 2Bp^2$ so $2B = -A$ so $B = -\frac{1}{5}.$
Checking with $p = 0$ gives the LHS $= 1$, and the RHS $= \frac{2}{5}(4) - \frac{1}{5}(-2) - 1 = 1.$

So $\quad \dfrac{3p + 1}{(2p - 1)(p + 2)^2} \equiv \dfrac{\frac{2}{5}}{2p - 1} - \dfrac{\frac{1}{5}}{p + 2} + \dfrac{1}{(p + 2)^2}.$

(3) $$\frac{4x-5}{(2x+1)(x^2-6x+9)} \equiv \frac{4x-5}{(2x+1)(x-3)^2} \equiv \frac{A}{2x+1} + \frac{B}{x-3} + \frac{C}{(x-3)^2}$$

so $\quad 4x - 5 \equiv A(x-3)^2 + B(2x+1)(x-3) + C(2x+1).$

Working in a similar way to (2), you should get

$$\frac{4x-5}{(2x+1)(x^2-6x+9)} \equiv \frac{-\frac{4}{7}}{2x+1} + \frac{\frac{2}{7}}{x-3} + \frac{1}{(x-3)^2}.$$

(4) $$\frac{10y}{(y-1)(y^2+9)} \equiv \frac{A}{y-1} + \frac{By+C}{y^2+9}$$

so $\quad 10y \equiv A(y^2+9) + (By+C)(y-1).$

Putting $y = 1$, we get $10 = 10A$ so $A = 1$.
Matching the terms in y^2, we get $0 = Ay^2 + By^2$, so $B = -A = -1$.
Matching the terms in y we get $10y = Cy - By$ so $C = 10 + B = 9$.
Checking with $y = 0$, the LHS = 0, and the RHS = $+1(9) + (9)(-1) = 0$.

So $\quad \dfrac{10y}{(y-1)(y^2+9)} \equiv \dfrac{1}{y-1} + \dfrac{(-y+9)}{y^2+9} \equiv \dfrac{1}{y-1} - \dfrac{y-9}{y^2+9}.$

Notice particularly here the rewriting of the second fraction with the minus sign outside, using the line of the fraction as a bracket.

(5) $$\frac{10x}{(x-1)(x^2-9)} \equiv \frac{10x}{(x-1)(x+3)(x-3)} \equiv \frac{A}{x-1} + \frac{B}{x+3} + \frac{C}{x-3}$$

which gives $\quad \dfrac{10x}{(x-1)(x^2-9)} \equiv \dfrac{-\frac{5}{4}}{x-1} + \dfrac{-\frac{5}{4}}{x+3} + \dfrac{\frac{5}{2}}{x-3}.$

(6) This one is top-heavy, so we rewrite it as

$$\frac{r^2+1}{r^2-1} = \frac{r^2-1+2}{r^2-1} = 1 + \frac{2}{r^2-1} = 1 + \frac{2}{(r-1)(r+1)}.$$

The partial fractions for $\quad \dfrac{2}{(r-1)(r+1)} \quad$ come out as $\quad \dfrac{1}{r-1} - \dfrac{1}{r+1}.$

The final complete answer is $\quad \dfrac{r^2+1}{r^2-1} \equiv 1 + \dfrac{1}{r-1} - \dfrac{1}{r+1}.$

(7) This is also top-heavy, so we write it as

$$\frac{x^4-1+2}{x^4-1} = 1 + \frac{2}{x^4-1}.$$

Did you spot how the $x^4 - 1$ could be factorised? It uses the difference of two squares twice. We can say $x^4 - 1 = (x^2 - 1)(x^2 + 1) = (x-1)(x+1)(x^2+1).$

So $\quad \dfrac{2}{x^4-1} \equiv \dfrac{2}{(x-1)(x+1)(x^2+1)} \equiv \dfrac{A}{x-1} + \dfrac{B}{x+1} + \dfrac{Cx+D}{x^2+1}$

so $\quad 2 \equiv A(x+1)(x^2+1) + B(x-1)(x^2+1) + (Cx+D)(x-1)(x+1).$

Putting $x = -1$ gives $2 = -4B$ so $B = -\frac{1}{2}.$
Putting $x = 1$ gives $2 = 4A$ so $A = \frac{1}{2}.$
Matching the terms in x^3 gives $0 = A + B + C$ so $C = 0.$
Putting $x = 0$ gives $2 = A - B - D$ so $D = -1.$
Putting $x = 2$ gives the LHS = 2 and the RHS = $\frac{1}{2}(3)(5) - \frac{1}{2}(1)(5) - (1)(3) = 2.$

The final answer is $\quad \dfrac{x^4+1}{x^4-1} \equiv \dfrac{\frac{1}{2}}{x-1} - \dfrac{\frac{1}{2}}{x+1} - \dfrac{1}{x^2+1}.$

(8) $\dfrac{u^2 - 1}{u^2(2u + 1)} \equiv \dfrac{A}{u} + \dfrac{B}{u^2} + \dfrac{C}{2u + 1}.$

 Students sometimes leave out the first of these fractions, forgetting that u^2 is a repeated factor of u times u.

Now we have, getting rid of fractions,

$$u^2 - 1 \equiv Au(2u + 1) + B(2u + 1) + Cu^2.$$

Putting $u = 0$ we get $-1 = B$.
Putting $u = -\frac{1}{2}$ we get $\frac{1}{4} - 1 = \frac{1}{4} C$ so $1 - 4 = C$ and $C = -3$.
Matching the terms in u^2 gives us $u^2 = 2Au^2 + Cu^2$ so $1 = 2A + C$ so $A = 2$.
Putting $u = 1$ gives the LHS = 0 and the RHS = $2(1)(3) - 1(3) - 3(1) = 0$.

so $\dfrac{u^2 - 1}{u^2(2u + 1)} \equiv \dfrac{2}{u} - \dfrac{1}{u^2} - \dfrac{3}{2u + 1}.$

(9) $\dfrac{x^2 + 1}{(x + 2)(x + 4)}$

This one is also top-heavy and the rearranging is a bit tricky. You may prefer to use long division. If not, it can be rearranged this way:

$$\dfrac{x^2 + 1}{x^2 + 6x + 8} = \dfrac{x^2 + 6x + 8 - 6x - 7}{x^2 + 6x + 8} = 1 - \dfrac{6x + 7}{x^2 + 6x + 8}.$$

Notice that the line of the fraction is acting as a bracket, again.

The partial fractions for $\dfrac{6x + 7}{x^2 + 6x + 8} = \dfrac{6x + 7}{(x + 2)(x + 4)}$ are $\dfrac{-\frac{5}{2}}{x + 2} + \dfrac{\frac{17}{2}}{x + 4}.$

So $\dfrac{x^2 + 1}{(x + 2)(x + 4)} \equiv 1 - \left(\dfrac{-\frac{5}{2}}{x + 2} + \dfrac{\frac{17}{2}}{x + 4} \right) \equiv 1 + \dfrac{\frac{5}{2}}{x + 2} - \dfrac{\frac{17}{2}}{x + 4}.$

Notice the signs!

(10) (a) The first four terms of $\displaystyle\sum_{r=1}^{n} \dfrac{2}{4r^2 - 1}$ are $\dfrac{2}{3}, \dfrac{2}{15}, \dfrac{2}{35},$ and $\dfrac{2}{63}.$

(b) $4r^2 - 1 = (2r - 1)(2r + 1)$ (the difference of two squares again!)

$$\dfrac{2}{4r^2 - 1} = \dfrac{2}{(2r - 1)(2r + 1)} = \dfrac{1}{2r - 1} - \dfrac{1}{2r + 1}.$$

(c) $\displaystyle\sum_{r=1}^{n} \dfrac{2}{4r^2 - 1} = \sum_{r=1}^{n} \left(\dfrac{1}{2r - 1} - \dfrac{1}{2r + 1} \right) = \sum_{r=1}^{n} \dfrac{1}{2r - 1} - \sum_{r=1}^{n} \dfrac{1}{2r + 1}$

$$= \left(1 + \dfrac{1}{3} + \dfrac{1}{5} + \ldots + \dfrac{1}{2n - 1} \right) - \left(\dfrac{1}{3} + \dfrac{1}{5} + \ldots + \dfrac{1}{2n - 1} + \dfrac{1}{2n + 1} \right)$$

The second bracket has just been slid along one space, so we get $1 - \dfrac{1}{2n + 1}.$

(d) As $n \to \infty$, $\dfrac{1}{2n + 1} \to 0$ so $\displaystyle\sum_{r=1}^{n} \dfrac{2}{4r^2 - 1} \to 1.$

The sum to infinity of this series is 1.

Chapter 7
Exercise 7.A.2

(1)   The sixth row of Pascal's Triangle is:

$$1 \quad\quad 6 \quad\quad 15 \quad\quad 20 \quad\quad 15 \quad\quad 6 \quad\quad 1 \qquad\qquad\qquad (P6)$$

so the expansion of $(x - 2y)^6$ is given by

$$x^6 + 6(x^5)(-2y)^1 + 15(x^4)(-2y)^2 + 20(x^3)(-2y)^3 + 15(x^2)(-2y)^4 + 6(x)(-2y)^5 + (-2y)^6$$
$$= x^6 - 12x^5y + 60x^4 y^2 - 160x^3y^3 + 240x^2 y^4 - 192xy^5 + 64y^6.$$

(2)   The fifth row of Pascal's Triangle is given by:

$$1 \quad\quad 5 \quad\quad 10 \quad\quad 10 \quad\quad 5 \quad\quad 1 \qquad\qquad\qquad (P5)$$

so the expansion of $(2x^2 - y^2)^5$ is

$$(2x^2)^5 + 5(2x^2)^4 (-y^2) + 10(2x^2)^3 (-y^2)^2 + 10(2x^2)^2 (-y^2)^3 + 5(2x^2) (-y^2)^4 + (-y^2)^5$$
$$= 32x^{10} - 80 x^8y^2 + 80x^6 y^4 - 40x^4 y^6 + 10x^2y^8 - y^{10}.$$

(3)   The fourth row of Pascal's Triangle is:

$$1 \quad\quad 4 \quad\quad 6 \quad\quad 4 \quad\quad 1 \qquad\qquad\qquad (P4)$$

so the expansion of $\left(2x - \dfrac{1}{x}\right)^4$ is

$$(2x)^4 + 4(2x)^3 \left(-\dfrac{1}{x}\right) + 6(2x)^2 \left(-\dfrac{1}{x}\right)^2 + 4(2x) \left(-\dfrac{1}{x}\right)^3 + \left(-\dfrac{1}{x}\right)^4$$

$$= 16x^4 - 32x^2 + 24 - \dfrac{8}{x^2} + \dfrac{1}{x^4}.$$

(4)   The third row of Pascal's Triangle is

$$1 \quad\quad 3 \quad\quad 3 \quad\quad 1 \qquad\qquad\qquad (P3)$$

so the expansion of $\left(\dfrac{3}{x} + 4x^2\right)^3$ is

$$\left(\dfrac{3}{x}\right)^3 + 3 \left(\dfrac{3}{x}\right)^2 (4x^2) + 3 \left(\dfrac{3}{x}\right) (4x^2)^2 + (4x^2)^3 = \dfrac{27}{x^3} + 108 + 144x^3 + 64x^6.$$

 It is very easy to make mistakes with complicated terms like we have in these questions. It is safest always to put in the working step as I have done, rather than trying to do it in your head.

Exercise 7.A.3

(1) $\dfrac{16!}{16! \, 0!} = 1$

(which you can see must be the case since you are choosing all as)
    We define 0! to be equal to 1 to make the formula work in this case.

(2) $\dfrac{16!}{15! \, 1!} = 16.$ The term is $16a^{15}b$.

(3) $\dfrac{16!}{14! \, 2!} = \dfrac{16 \times 15}{2!} = 120.$ The term is $120a^{14}b^2$.

(4) $\dfrac{16!}{12!\;4!} = \dfrac{16 \times 15 \times 14 \times 13}{4!} = 1820.$ The term is $1820a^{12}b^4$.

(5) $\dfrac{16!}{8!\;8!} = 12870.$ The term is $12870a^8b^8$.

(6) $\dfrac{16!}{4!\;12!} = 1820.$ The term is $1820a^4b^{12}$.

(7) $\dfrac{16!}{2!\;14!} = 120.$ The term is $120a^2b^{14}$.

(8) $\dfrac{16!}{0!\;16!} = 1.$ The term is b^{16}.

(9)   This works in exactly the same way as the others.

r is just standing for whichever power of a we might be interested in.

We get $\dfrac{16!}{r!\;(16-r)!}$ and the term is $\dfrac{16!}{r!\;(16-r)!}\,a^r b^{16-r}$

Notice the symmetry of the pairs (a) and (h), (c) and (g), and (d) and (f). This is the same symmetry which we say in Pascal's Triangle.

Exercise 7.A.4

(1)   The first four terms of $(2x - y)^{12}$.

Using $(B1)$, with 'a' $= 2x$ and 'b' $= -y$ and $n = 12$, we get

$$(2x)^{12} + 12(2x)^{11}\,(-y) + \frac{12 \times 11}{2 \times 1}\,(2x)^{10}\,(-y)^2 + \frac{12 \times 11 \times 10}{3 \times 2 \times 1}\,(2x)^9\,(-y)^3$$

$$= 4096x^{12} - 24576x^{11}y + 67584x^{10}y^2 - 112640x^9y^3.$$

(2)   The first four terms of $(1 - 2x)^{18}$.

Using $(B\,2)$ with 'x' $= -2x$ and $n = 18$, we get

$$1 + 18(-2x) + \frac{18 \times 17}{2 \times 1}\,(-2x)^2 + \frac{18 \times 17 \times 16}{3 \times 2 \times 1}\,(-2x)^3 = 1 - 36x + 612x^2 - 6528x^3.$$

(3)   The first four terms in the expansion of $(1 + x^2)^{10}$.

Using $(B\,2)$ with 'x' $= x^2$ and $n = 10$, we get

$$1 + 10(x^2) + \frac{10 \times 9}{2 \times 1}\,(x^2)^2 + \frac{10 \times 9 \times 8}{3 \times 2 \times 1}\,(x^2)^3 = 1 + 10x^2 + 45x^4 + 120x^6.$$

(4)   The first four terms in the expansion of $(\tfrac{1}{2}x + 3y)^{16}$.

Using $(B\ 1)$ with 'a' $= x/2$ and 'b' $= 3y$ and $n = 16$ we get

$$\left(\frac{x}{2}\right)^{16} + 16\left(\frac{x}{2}\right)^{15}(3y) + \frac{16 \times 15}{2 \times 1}\left(\frac{x}{2}\right)^{14}(3y)^2 + \frac{16 \times 15 \times 14}{3 \times 2 \times 1}\left(\frac{x}{2}\right)^{13}(3y)^3$$

$$= \frac{x^{16}}{65\,536} + \frac{3}{2048}\,x^{15}y + \frac{135}{2048}x^{14}y^2 + \frac{945}{512}\,x^{13}y^3.$$

Exercise 7.A.5

(1) (b) is $\dfrac{9!}{4!\,5!}$.   (c) is $\dfrac{10!}{4!\,6!}$.

Then (a) + (b) $= \dfrac{9!}{3!\,6!} + \dfrac{9!}{4!\,5!} = \dfrac{9!}{3!\,5!}\left(\dfrac{1}{6} + \dfrac{1}{4}\right)$

$$= \dfrac{9!}{3!\,5!}\left(\dfrac{4+6}{6\times4}\right) = \dfrac{9! \times 10}{(3!\times4)(5!\times6)} = \dfrac{10!}{4!\,6!} = \text{(c)}.$$

This rearrangement is possible because it does not matter which order we multiply the numbers in.

   The answer that (a) + (b) = (c) is another way of saying that 84 + 126 in (P 9) gives 210 in (P 10).

(2) (a) is $\dfrac{12!}{3!\,9!}$.   (b) is $\dfrac{12!}{4!\,8!}$.   (c) is $\dfrac{13!}{4!\,9!}$.

Then (a) + (b) $= \dfrac{12!}{3!\,9!} + \dfrac{12!}{4!\,8!} = \dfrac{12!}{3!\,8!}\left(\dfrac{1}{9} + \dfrac{1}{4}\right)$

$$= \dfrac{12!}{3!\,8!}\left(\dfrac{4+9}{9\times4}\right) = \dfrac{12! \times 13}{(3!\times4)(8!\times9)} = \dfrac{13!}{4!\,9!} = \text{(c)}.$$

This answer corresponds to 220 + 495 in (P 12) giving 715 in (P 13).

(3) This last question looks horrible because it has letters instead of numbers, but it works in just the same way as (1) and (2).

(b) is $\dfrac{k!}{r!\,(k!-r)!}$.   (c) is $\dfrac{(k+1)!}{r!\,(k+1-r)!}$.

So   (a) + (b) $= \dfrac{k!}{(r-1)!\,(k-r+1)!} + \dfrac{k!}{r!\,(k-r)!}$

$$= \dfrac{k!}{(r-1)!\,(k-r)!}\left(\dfrac{1}{k-r+1} + \dfrac{1}{r}\right) = \dfrac{k!}{(r-1)!\,(k-r)!}\left(\dfrac{r+k-r+1}{r(k-r+1)}\right)$$

$$= \dfrac{k! \times (k+1)}{((r-1)!\times r)((k-r)!\times(k-r+1))} = \dfrac{(k+1)!}{r!\,(k-r+1)!} = \text{(c)}.$$

$k + 1 - r$ is the same as $k - r + 1$. The numbers are just being put together in a different order. Having this result will make it very easy for us to prove the binomial theorem by induction in Section 7.D.(b).

Exercise 7.A.6

(1) (a) The term in x^6 in the expansion of $(2 - 3x)^{11}$ is

$$\dfrac{11!}{5!\,6!}\,(2)^5\,(-3x)^6 = 10\,777\,536x^6.$$

(b) The term in x^6 in the expansion of $(2x - y)^8$ is

$$\dfrac{8!}{6!\,2!}\,(2x)^6\,(-y)^2 = 1792x^6\,y^2.$$

(c) The term in x^6 in the expansion of $(y^2 - 2x^2)^{10}$ is

$$\dfrac{10!}{7!\,3!}\,(y^2)^7\,(-2x^2)^3 = -960y^{14}x^6.$$

(2) (a) The constant term in the expansion of

$$\left(2x - \frac{3}{x}\right)^{10} \quad \text{is} \quad \frac{10!}{5!\,5!}\,(2x)^5\left(-\frac{3}{x}\right)^5 = -1\,959\,552.$$

(b) The constant term in the expansion of

$$\left(x + \frac{1}{x^2}\right)^9 \quad \text{is} \quad \frac{9!}{6!\,3!}\,x^6\left(\frac{1}{x^2}\right)^3 = 84.$$

(c) The constant term in the expansion of

$$\left(2x^3 + \frac{1}{x}\right)^{16} \quad \text{is} \quad \frac{16!}{4!\,12!}\,(2x^3)^4\left(\frac{1}{x}\right)^{12} = 29\,120.$$

(3) The term in x^{10} in the expansion of $(1 + x)^7\,(2 - 3x)^5$ comes from $x^7 \times x^3$ and $x^6 \times x^4$ and $x^5 \times x^5$. So it is:

$$[x^7]\left[\frac{5!}{2!\,3!}\,(2)^2\,(-3x)^3\right] + [7x^6]\left[\frac{5!}{1!\,4!}\,(2)\,(-3x)^4\right] + \left[\frac{7!}{2!\,5!}\,x^5\right][(-3x)^5]$$

$$= -1080x^{10} + 5670x^{10} - 5103x^{10} = -513x^{10}.$$

Exercise 7.B.1

(1) Most people try to spread the numbers much too evenly when they do this. To show you by example that such an even spread isn't to be expected, I have written out for you two lists of numbers below. The first list gives ten sets of six numbers chosen by using the random number generator on my calculator and taking the first two digits, discarding anything over 49. The second list gives the results of the draws from ten consecutive weeks in a lottery similar to the one I described before this exercise.

List (1)

6,	9,	18,	29,	35,	46		8,	28,	36,	40,	41,	46
7,	12,	28,	31,	44,	46		5,	7,	32,	34,	41,	47
7,	18,	24,	26,	27,	36		8,	21,	25,	29,	36,	43
13,	20,	31,	32,	38,	42		1,	11,	18,	34,	46,	48
2,	17,	22,	25,	35,	46		7,	9,	18,	19,	27,	31

List (2)

11,	25,	28,	33,	34,	47		5,	8,	23,	24,	28,	48
16,	18,	21,	27,	38,	41		1,	15,	22,	28,	40,	49
2,	12,	20,	22,	41,	45		2,	10,	14,	25,	37,	41
5,	10,	19,	24,	34,	46		10,	11,	29,	32,	33,	40
10,	22,	28,	30,	36,	37		4,	5,	9,	25,	30,	47

You can see from these lists that two consecutive numbers have come up surprisingly often, in fact ten times altogether. Also, runs of the same number occur, with 10 coming up in four consecutive lottery draws, for example. There are also corridors with no numbers in. For example, there are no 42, 43 or 44 in the ten consecutive lottery draws. You can see the uneven spread of the numbers even more clearly if you show these lists on squared paper also.

(2) Exactly the same rules of probability apply to each set of numbers, so each of them have an equal chance of

$$\frac{1}{13\,983\,816}$$

of being the correct choice.

(3) You would be wiser to pick the third choice. The reason for this is that many people think that 'random' means more or less evenly spread. This is not so. 'Random' means just that; the choices can happen anywhere, and therefore there will sometimes be bunching, and

sometimes large gaps, as you can see in the two lists above. The behaviour of such random distributions of numbers is explained mathematically in courses on statistics. These mathematical descriptions can then be used to answer questions such as how much concern should be felt about clusters of cases of particular illnesses such as childhood leukaemia; is there a local cause or are such clusters to be expected anyway?

Since most people would probably think that 44, 45, 46, 47, 48, 49, would be very unlikely, if these numbers did come up, you would be less likely to have to share your winnings! The choice of 1, 2, 3, 4, 5, 6 would be less satisfactory since many people have favourite 'lucky' numbers which they choose, and such numbers are usually small. Other people are likely to choose birthdays, so they will be restricted to 31 or less. It is also likely that many people would be very doubtful about choosing even two numbers together, never mind the whole six. In fact, in a situation like this, the probability of getting at least one pair of consecutive numbers is about one half; much higher than you would probably guess.

(4) On any particular draw of this lottery, there will be 43 numbers which are not chosen. The number of different possibilities if six numbers are chosen from these 43 numbers and the order of choice doesn't matter, is given by

$$\frac{43!}{37!\ 6!} = 6\ 096\ 454.$$

We know already that the total number of ways of picking 6 numbers from 49 numbers is 13 983 816.

This means that there are 13 983 816 − 6 096 454 = 7 887 362 ways of choosing at least one number correctly. Therefore the probability of doing this is

$$\frac{7\ 887\ 362}{13\ 983\ 816} = 0.564 \text{ to 3 d.p.}$$

We see that you should expect to have at least one number correct more often than not, which makes it not quite as encouraging for future prize-winning as it might seem.

Exercise 7.C.1

(1) $(1 + 2x)^{-3} = 1 + (-3)(2x) + \dfrac{(-3)(-4)}{2 \times 1}(2x)^2 + \dfrac{(-3)(-4)(-5)}{3 \times 2 \times 1}(2x)^3 \ldots$

$= 1 - 6x + 24x^2 - 80x^3 \ldots$

The expansion is valid if $|2x| < 1$ or $-\frac{1}{2} < x < +\frac{1}{2}$.

(2) $(1 - 3x)^{-1} = 1 + (-1)(-3x) + \dfrac{(-1)(-2)}{2 \times 1}(-3x)^2 + \dfrac{(-1)(-2)(-3)}{3 \times 2 \times 1}(-3x)^3 \ldots$

$= 1 + 3x + 9x^2 + 27x^3 + \ldots$

This is just the same as the first expansion in this section except that x has been replaced by $-3x$, so it tidies up particularly nicely.

HELPFUL HINT

It's worth remembering that

$$(1 - \spadesuit)^{-1} = 1 + \spadesuit + \spadesuit^2 + \spadesuit^3 + \ldots \text{ and } (1 + \spadesuit)^{-1} = 1 - \spadesuit + \spadesuit^2 - \spadesuit^3 + \spadesuit^4 - \ldots$$

where $\spadesuit$ stands for whatever we have in this position in the bracket, and $|\spadesuit| < 1$.

Here, $\spadesuit = 3x$ so we must have $|3x| < 1$ or $-\frac{1}{3} < x < +\frac{1}{3}$ for the expansion to be valid.

(3) $(1 + \frac{1}{3}x)^{-2} = 1 + (-2)(\frac{1}{3}x) + \dfrac{(-2)(-3)}{2 \times 1}(\frac{1}{3}x)^2 + \dfrac{(-2)(-3)(-4)}{3 \times 2 \times 1}(\frac{1}{3}x)^3 + \ldots$

$= 1 - \frac{2}{3}x + \frac{1}{3}x^2 - \frac{4}{27}x^3 + \ldots$

For the expansion to be valid, we must have $|\frac{1}{3}x| < 1$ or $-3 < x < +3$.

(1) (a) Here, $\spadesuit = -3x$. For the expansion to be valid, we must have $|\spadesuit| < 1$, so $|3x| < 1$, so $|x| < \frac{1}{3}$ or $-\frac{1}{3} < x < \frac{1}{3}$. (Remember that $|-3x| = |3x|$.) Then:

$$(1 - 3x)^{1/3} = 1 + \left(\tfrac{1}{3}\right)(-3x) + \frac{\left(\tfrac{1}{3}\right)\left(-\tfrac{2}{3}\right)(-3x)^2}{2 \times 1} + \frac{\left(\tfrac{1}{3}\right)\left(-\tfrac{2}{3}\right)\left(-\tfrac{5}{3}\right)}{3 \times 2 \times 1}(-3x)^3 + \dots$$

$$= 1 - x - x^2 - \tfrac{5}{3}x^3, \text{ giving the first four terms.}$$

(b) Here, $\spadesuit = \dfrac{x}{2}$, so we must have $\left|\dfrac{x}{2}\right| < 1$ that is $|x| < 2$ or $-2 < x < 2$. Then:

$$\left(1 + \frac{x}{2}\right)^{2/3} = 1 + \left(\frac{2}{3}\right)\left(\left(\frac{x}{2}\right)\right) + \frac{\left(\tfrac{2}{3}\right)\left(-\tfrac{1}{3}\right)}{2 \times 1}\left(\frac{x}{2}\right)^2 + \frac{\left(\tfrac{2}{3}\right)\left(-\tfrac{1}{3}\right)\left(-\tfrac{4}{3}\right)}{3 \times 2 \times 1}\left(\frac{x}{2}\right)^3 + \dots\right)$$

$$= 1 + \frac{x}{3} - \frac{x^2}{36} + \frac{x^3}{648}, \text{ giving the first four terms.}$$

(c) This time, we must rearrange to get the $(1 + \spadesuit)$ form.

$$(16 - 3x)^{1/4} = \left[16^{1/4}\left(1 - \frac{3x}{16}\right)^{1/4}\right] = 2\left(1 - \frac{3x}{16}\right)^{1/4} \text{ because } 16^{1/4} = 2.$$

$$= 2\left[1 + \left(\tfrac{1}{4}\right)\left(-\frac{3x}{16}\right) + \frac{\left(\tfrac{1}{4}\right)\left(-\tfrac{3}{4}\right)}{2 \times 1}\left(-\frac{3x}{16}\right)^2 + \frac{\left(\tfrac{1}{4}\right)\left(-\tfrac{3}{4}\right)\left(-\tfrac{7}{4}\right)}{3 \times 2 \times 1}\left(-\frac{3x}{16}\right)^3 + \dots\right]$$

$$= 2\left[1 - \frac{3x}{64} - \frac{27x^2}{8192} - \frac{189x^3}{524288} - \dots\right]$$

$$= 2 - \frac{3x}{32} - \frac{27x^2}{4096} - \frac{189x^3}{262144}, \text{ giving the first four terms.}$$

Here, $\spadesuit = -\dfrac{3x}{16}$, so we must have $\left|\dfrac{3x}{16}\right| < 1$ so $-16 < 3x < 16$ so $-\dfrac{16}{3} < x < \dfrac{16}{3}$.

(d) Rearranging again, we have

$$(4 + x)^{-1/2} = \left[4\left(1 + \frac{x}{4}\right)\right]^{-1/2} = 4^{-1/2}\left(1 + \frac{x}{4}\right)^{-1/2}$$

$$= \frac{1}{4^{\frac{1}{2}}}\left(1 + \frac{x}{4}\right)^{-1/2} = \frac{1}{2}\left(1 + \frac{x}{4}\right)^{-1/2}$$

$$= \frac{1}{2}\left[1 + \left(-\tfrac{1}{2}\right)\left(\frac{x}{4}\right) + \frac{\left(-\tfrac{1}{2}\right)\left(-\tfrac{3}{2}\right)}{2 \times 1}\left(\frac{x}{4}\right)^2 + \frac{\left(-\tfrac{1}{2}\right)\left(-\tfrac{3}{2}\right)\left(-\tfrac{5}{2}\right)}{3 \times 2 \times 1}\left(\frac{x}{4}\right)^3 + \dots\right]$$

$$= \frac{1}{2}\left[1 - \frac{x}{8} + \frac{3x^2}{128} - \frac{5x^3}{1024}\dots\right]$$

$$= \frac{1}{2} - \frac{x}{16} + \frac{3x^2}{256} - \frac{5x^3}{2048}, \text{ giving the first four terms.}$$

Here, $\spadesuit = \dfrac{x}{4}$, so we require $\left|\dfrac{x}{4}\right| < 1$, so $-4 < x < 4$.

(e) Again, rearranging is necessary.

$$(-2 + x)^{-2} = \left[-2\left(1 - \frac{x}{2}\right)\right]^{-2} = -2^{-2}\left(1 - \frac{x}{2}\right)^{-2}$$

We must take out −2 to make the bracket start with +1. Taking out +2 will not do what we want.

$$= -\frac{1}{2^2}\left[1 + (-2)\left(-\frac{x}{2}\right) + \frac{(-2)(-3)}{2\times 1}\left(-\frac{x}{2}\right)^2 + \frac{(-2)(-3)(-4)}{3\times 2\times 1}\left(-\frac{x}{2}\right)^3 \dots\right]$$

$$= -\frac{1}{4}\left[1 + x + \frac{3x^2}{4} + \frac{x^3}{2} + \dots\right]$$

$$= -\frac{1}{4} - \frac{x}{4} - \frac{3x^2}{16} - \frac{x^3}{8}, \text{ giving the first four terms.}$$

Here, $\spadesuit = \dfrac{x}{2}$, so we require $\left|\dfrac{x}{2}\right| < 1$ so $-1 < \dfrac{x}{2} < 1$ so $-2 < x < 2.$

(f) Rearranging yet again, we have

$$(27 - 4x)^{-2/3} = 27^{-2/3}\left(1 - \frac{4x}{27}\right)^{-2/3} = \frac{1}{27^{2/3}}\left(1 - \frac{4x}{27}\right)^{-2/3}$$

The first four terms are given by

$$\frac{1}{9}\left[1 + \left(-\frac{2}{3}\right)\left(-\frac{4x}{27}\right) + \frac{(-\frac{2}{3})(-\frac{5}{3})}{2\times 1}\left(-\frac{4x}{27}\right)^2 + \frac{(-\frac{2}{3})(-\frac{5}{3})(-\frac{8}{3})}{3\times 2\times 1}\left(-\frac{4x}{27}\right)^3\right]$$

(NB $27^{2/3} = (27^{1/3})^2 = 3^2 = 9$. See Section 1.D.(b) if you need help with this.)

$$= \frac{1}{9}\left[1 + \frac{8x}{81} + \frac{80x^2}{6561} + \frac{2560x^3}{1\,594\,323} + \dots\right]$$

$$= \frac{1}{9} + \frac{8x}{729} + \frac{80x^2}{59\,049} + \frac{2560x^3}{14\,348\,907}.$$

This time, $\spadesuit = \dfrac{4x}{27}$, so we require $\left|\dfrac{4x}{27}\right| < 1$ so $-\frac{27}{4} < x < \frac{27}{4}.$

(2) METHOD (1) $(3 - 2x)^{-1}(1 + 3x)^{-1} = \frac{1}{3}(1 - \frac{2}{3}x)^{-1}(1 + 3x)^{-1}$

$$= \frac{1}{3}(1 + \frac{2}{3}x + \frac{4}{9}x^2 + \dots)(1 - 3x + 9x^2 - \dots)$$

Now we take just the multiplications giving the terms up to x^2, giving

$$\frac{1}{3}\left[1 - 3x + 9x^2\right.$$
$$+ \frac{2}{3}x - 2x^2$$
$$\left.+ \frac{4}{9}x^2\right]$$

$$= \frac{1}{3}(1 - \frac{7}{3}x + \frac{67}{9}x^2) = \frac{1}{3} - \frac{7}{9}x + \frac{67}{27}x^2.$$

METHOD (2)   Writing $\dfrac{1}{(3 - 2x)(1 + 3x)}$ in partial fractions, we get

$$\frac{\frac{2}{11}}{3 - 2x} + \frac{\frac{3}{11}}{1 + 3x} = \frac{2}{11}(3 - 2x)^{-1} + \frac{3}{11}(1 + 3x)^{-1}$$

$$= \frac{2}{11}\left[3^{-1}(1 - \frac{2}{3}x)^{-1}\right] + \frac{3}{11}(1 + 3x)^{-1}$$

$$= \frac{2}{33}\left[1 + 2x/3 + (2x/3)^2 + \dots\right] + \frac{3}{11}\left[1 - 3x + (3x)^2 - \dots\right]$$

$$= \text{(tidied up) } \frac{1}{3} - \frac{7}{9}x + \frac{67}{27}x^2 \text{ giving just the first three terms.}$$

In order for the first expansion to be valid, we require that $\left|\dfrac{2x}{3}\right| < 1$ so $-\frac{3}{2} < x < \frac{3}{2}$.

For the second expansion to be valid, we must have $|3x| < 1$ so $-\frac{1}{3} < x < \frac{1}{3}$.

Therefore, in order to fit both requirements simultaneously, we must take the tighter of the two restrictions which is that $-\frac{1}{3} < x < \frac{1}{3}$.

Exercise 7.D.1

(1)   We want to show that $1 + 2 + 3 + 4 + \ldots + n = \frac{1}{2}n(n + 1)$.

First, check that the statement is true when $n = 1$.
LHS = 1. RHS $= \frac{1}{2} \times 1 \times 2 = 1$, so, yes, St[1] is true.
Then, suppose that the statement *is* true for $n = $ a particular value, k, so

$$1 + 2 + 3 + \ldots + k = \tfrac{1}{2}k(k + 1). \qquad\qquad \text{St}[k]$$

Now we must show that this means that it is also true for $n = k + 1$. That is, we must show that

$$1 + 2 + 3 + \ldots + k + (k + 1) = \tfrac{1}{2}(k + 1)(k + 2). \qquad\qquad \text{St}[k + 1]$$

(The extra term has been added on to the LHS and k has been replaced by $k + 1$ in the formula on the RHS.)

Now, the whole bunch of $1 + 2 + 3 + \ldots + k$ on the LHS of St[$k + 1$] can be replaced by $\frac{1}{2}k(k + 1)$, using St[k]. So we have that

$$\text{the LHS of St}[k + 1] = \tfrac{1}{2}k(k + 1) + (k + 1).$$

But, factorising this, we have

$$\tfrac{1}{2}k(k + 1) + (k + 1) = \tfrac{1}{2}(k + 1)(k + 2) = \text{the RHS of St}[k + 1].$$

We have taken out the factor of $(k + 1)$, and also the $\frac{1}{2}$, because we know we want it at the front. This means that we must have the 2 in the second bracket to make the whole thing multiply out correctly.

So, *if* the statement is true when $n = k$, *then* it is also true when $n = k + 1$.

But it is true when $n = 1$, so therefore it is true when $n = 2$, and so on through all the counting numbers.

(2)   (a) $S_1 = 1 = 1^2$ (b) $S_2 = 4 = 2^2$ (c) $S_3 = 9 = 3^2$ (d) $S_4 = 16 = 4^2$

So it looks as if the sum of n odd numbers is n^2.
Suppose this statement is true when $n = $ a particular value, k. Then

$$1 + 3 + 5 + \ldots + (2k - 1) = k^2. \qquad\qquad \text{St}[k]$$

Nasty pitfall. The nth odd number isn't n, but $2n - 1$. (For instance, the third odd number isn't 3 but 5.)

We must now show that if St[k] is true, then St[$k + 1$] is also true, so

$$1 + 3 + 5 + \ldots + (2k - 1) + (2k + 1) = (k + 1)^2. \qquad\qquad \text{St}[k + 1]$$

Using St[k], the LHS of St[$k + 1$] $= k^2 + (2k + 1) = (k + 1)^2 = $ RHS of St[$k + 1$].

But the statement is true when $n = 1$, so therefore it is true through all the counting numbers.

(3)   We have to show that

$$1 \times 2 + 2 \times 3 + 3 \times 4 + \ldots + n(n + 1) = \tfrac{1}{3}n(n + 1)(n + 2).$$

First, we check that this formula is true when $n = 1$.
The LHS is $1 \times 2 = 2$ and the RHS is $1(2) = 2$, so St[1] is true.
Then we assume the statement is true for $n = $ a particular value, k, so that

$$1 \times 2 + 2 \times 3 + 3 \times 4 + \ldots + k(k + 1) = \tfrac{1}{3}k(k + 1)(k + 2). \qquad\qquad \text{St}[k]$$

We must then show that St$[k + 1]$ is true, that is, that

$$1 \times 2 + 2 \times 3 + 3 \times 4 + \ldots + k(k + 1) + (k + 1)(k + 2) = \tfrac{1}{3}(k + 1)(k + 2)(k + 3),$$

adding the extra term to the LHS and putting $n = k + 1$ on the RHS.

Using St$[k]$, the LHS of St$[k + 1]$ is

$$\tfrac{1}{3}k(k + 1)(k + 2) + (k + 1)(k + 2).$$

Now, we factorise by taking out $\tfrac{1}{3}(k + 1)(k + 2)$. We get

$$\tfrac{1}{3}k(k + 1)(k + 2)(k + 3)$$

which is the RHS of St$[k + 1]$. (Check by multiplying out this last bracket if you are unsure about this step.)

So, if St$[k]$ is true, then St$[k + 1]$ is true.

But St$[1]$ is true, and so St$[2]$ is true, and so on through all the counting numbers.

Chapter 8
Exercise 8.A.1

(1) $dy/dx = 14x + 12x^3$ (2) $dx/dt = 5 - \tfrac{3}{2}t^2$

(3) $y = 3 - 2x^{-3}$ so $dy/dx = 0 + 6x^{-4} = 6/x^4$.

(4) $dx/dt = 2 \times \tfrac{1}{2} \times t^{-1/2} - 3 \times \tfrac{1}{2} \times t^{-3/2} = t^{-1/2} - \tfrac{3}{2}t^{-3/2}$

(5) (a) If x is increased by a small amount δx, so that y gets increased by a correspondingly small amount δy, we have

$$y + \delta y = (x + \delta x)^3 = x^3 + 3x^2(\delta x) + 3x(\delta x)^2 + (\delta x)^3$$

But $y = x^3$

so $\delta y = 3x^2(\delta x) + 3x(\delta x)^2 + (\delta x)^3$ so $\delta y/\delta x = 3x^2 + 3x(\delta x) + (\delta x)^2$.

Now, when we let $\delta x \to 0$, we can ignore the terms $3x(\delta x)$ and $(\delta x)^2$, so the limit of $\delta y/\delta x$ as $\delta x \to 0$ is $dy/dx = 3x^2$.

(b) Here, we are taking the particular case of $x = 2$ and $\delta x = 0.001$ so we have

$$y + \delta y = (x + \delta x)^3 = (2 + 0.001)^3 = 2^3 + 3(2^2)(0.001) + 3(2)(0.001)^2 + (0.001)^3$$

But $y = 2^3$ so

$$\delta y = 3(2)^2(0.001) + 3(2)(0.001)^2 + (0.001)^3$$

and since $\delta x = 0.001$, we have

$$\delta y/\delta x = 3(2^2) + 3(2)(0.001) + (0.001)^2 = 12.006001.$$

If we make δx even smaller, $\delta y/\delta x$ will approach even more closely to the value of $3(2^2) = 12$ given by $dy/dx = 3x^2$ when $x = 2$.

Exercise 8.A.2

(1) We have $y = \sin t$ so $dy/dt = \cos t$ and $d^2y/dx^2 = -\sin t = -y$

so we *do* get the same kind of link between the acceleration and the distance of Y from O. The point Y is also moving in SHM, but it is starting from the central position on its path whereas X started from its most extreme positive position.

(When $t = 0$ we have $y = 0$ too, but if $x = \cos t$, then $x = 1$ when $t = 0$.)

(2) For this new object, $x = 3 \cos t + 4 \sin t$.

$$dx/dt = -3 \sin t + 4 \cos t \quad \text{and} \quad d^2y/dt^2 = -3 \cos t - 4 \sin t = -x.$$

So this object is also moving in SHM. This equation describes a motion which starts from an intermediate point when the time $t = 0$.

We know from Section 5.D.(f) that if $x = 3 \cos t + 4 \sin t$ then the two most extreme values of x are $x = +5$ and $x = -5$. When $t = 0$, $x = 3$.

Exercise 8.C.1

Here are the answers you should have. To help you, I have shown the two separate bits in square brackets, before tidying up, and also put what I chose for X at the side of each answer.

(1) $dy/dx = [4(2x^2 + 3)^3] [4x] = 16x(2x^2 + 3)^3$ with $X = 2x^2 + 3$.

(2) $dx/dt = [5(t^3 + 2)^4] [3t^2] = 15t^2(t^3 + 2)^4$ with $X = t^3 + 2$.

(3) $dy/dx = [4(3x^2 - 2x)^3] [6x - 2] = 8(3x - 1)(3x^2 - 2x)$ with $X = 3x^2 - 2x$.

(4) $dx/dt = [\frac{1}{2}(3t + 4)^{-1/2}] [3] = \frac{3}{2}(3t + 4)^{-1/2}$ with $X = 3t + 4$.

(5) $dy/dx = [3e^{4x}] [4] = 12e^{4x}$ with $X = 4x$.

(6) $dx/dt = [e^{t^2+1}] [2t] = 2te^{t^2+1}$ with $X = t^2 + 1$.

(7) $dy/dx = [2e^{x^2+x}] [2x + 1] = 2(2x + 1)e^{x^2 + x}$ with $X = x^2 + x$.

(8) $dx/dt = [-\sin(4t + \pi/3)] [4] = -4\sin(4t + \pi/3)$ with $X = 4t + \pi/3$.

(9) $dx/dt = \cos t + [\cos 2t] [2] = \cos t + 2 \cos 2t$ with $X = 2t$.

(10) $dy/dx = [\cos(x^2)] [2x] = 2x(\cos(x^2))$ with $X = x^2$.

(11) $dy/dx = [2 \sin x] [\cos x] = 2 \sin x \cos x$ with $X = \sin x$.

(12) $dy/dx = [3 \cos^2 x] [- \sin x] = -3 \sin x \cos^2 x$ with $X = \cos x$.

(13) $\dfrac{dy}{dx} = \left[\dfrac{1}{4x}\right] [4] = \dfrac{4}{4x} = \dfrac{1}{x}$ with $X = 4x$.

 Students often give the answer of $\dfrac{1}{4x}$ for this one!

(14) $\dfrac{dy}{dx} = \left[\dfrac{1}{3x + 1}\right] [3] = \dfrac{3}{3x + 1}$ with $X = 3x + 1$.

(15) $\dfrac{dx}{dt} = \left[\dfrac{1}{2t^2 + 1}\right] [4t] = \dfrac{4t}{2t^2 + 1}$ with $X = 2t^2 + 1$.

(16) $dy/dx = [-\sin(5x^2 + \pi)] [10x] = -10x \sin(5x^2 + \pi)$ with $X = 5x^2 + \pi$.

(17) $dx/dt = [\cos(2t^2 + 3)] [4t] = 4t \cos(2t^2 + 3)$ with $X = 2t^2 + 3$.

(18) $\dfrac{dy}{dx} = \left[\dfrac{1}{\sin x}\right] [\cos x] = \dfrac{\cos x}{\sin x} = \cot x$ with $X = \sin x$.

Exercise 8.C.2

These are the answers you should have for dy/dx in each case.

(1) $5e^{5x}$ (2) $-2e^{-2x}$ (3) $2xe^{x^2}$ (4) $\dfrac{2}{2x + 3}$ (5) $\dfrac{1}{1 + x}$

(6) $\dfrac{-1}{1 - x} = \dfrac{1}{x - 1}$ (7) $7 \cos 7x$ (8) $-4 \sin x \cos^3 x$ (9) $2 \cos(2x + \pi)$

(10) $-3 \sin(3x + 4)$.

Exercise 8.C.3

I have used the square brackets again to make the working clear for you.

(1) $dy/dx = [5 \cos^4 2x] [-2 \sin 2x] = -10 \sin 2x \cos^4 2x.$

(2) $dy/dx = [3 \sin^2 (4x + 1)] [4 \cos (4x + 1)] = 12 \cos (4x + 1) \sin^2 (4x + 1).$

(3) $\dfrac{dx}{dt} = \left[\dfrac{1}{\sin (2t + 3)} \right] [2 \cos (2t + 3)] = 2 \cot (2t + 3).$

(4) $dx/d\theta = [3 (2 \cos 2\theta + 5)^2] [-4 \sin 2\theta] = -12 \sin 2\theta (2 \cos 2\theta + 5)^2.$

(5) $\dfrac{dy}{dx} = \left[\dfrac{1}{1 + \cos x} \right] [- \sin x] = \dfrac{- \sin x}{1 + \cos x}.$

(6) $\dfrac{dx}{dt} = \left[\dfrac{1}{3t + \sin^2 3t} \right] [3 + 2 \sin 3t \, (3 \cos 3t)] = \dfrac{3 + 6 \sin 3t \cos 3t}{3t + \sin^2 3t}.$

(7) $\dfrac{dx}{dt} = \left[\dfrac{1}{2 + e^{t^2 + 1}} \right] [2te^{t^2 + 1}] = \dfrac{2te^{t^2 + 1}}{2 + e^{t^2 + 1}}.$

(8) $dy/dx = [\cos (\cos 4x)] [-4 \sin 4x] = -4 \sin 4x \cos (\cos 4x).$

(9) $dy/dt = [\tfrac{1}{2} (1 + \sin^2 t)^{-1/2}] [2 \sin t \cos t] = \sin t \cos t (1 + \sin^2 t)^{-1/2}.$

(10) Using the third law of logs to make this question easier, we have

$$y = \tfrac{1}{2} \ln (1 + \sin^2 t) \quad \text{so} \quad \dfrac{dy}{dt} = \tfrac{1}{2} \left[\dfrac{1}{1 + \sin^2 t} \right] [2 \sin t \cos t] = \dfrac{\sin t \cos t}{1 + \sin^2 t}.$$

Exercise 8.C.4

Again, I have sometimes used square brackets to show the way in which the answers build up more clearly.

(1) $u = 7x^2$ and $v = \cos 3x$ so

$$\dfrac{dy}{dx} = (\cos 3x) (14x) + (7x^2) (-3 \sin 3x) = 14x \cos 3x - 21x^2 \sin 3x.$$

(2) $u = e^{3x}$ and $v = \sin 2x$ so

$$\dfrac{dy}{dx} = (\sin 2x) (3e^{3x}) + e^{3x} (2 \cos 2x) = e^{3x} (3 \sin 2x + 2 \cos 2x).$$

(3) $u = 4x^5$ and $v = (x^2 + 3)^3$ so

$$\dfrac{dy}{dx} = [(x^2 + 3)^3] [20x^4] + [4x^5] [3(x^2 + 3)^2 (2x)]$$

$$= 4x^4 (x^2 + 3)^2 [5(x^2 + 3) + 6x^2] = 4x^4 (x^2 + 3)^2 (11x^2 + 15).$$

(4) $u = e^{2t + 1}$ and $v = \cos (2t + 1)$ so

$$\dfrac{dx}{dt} = [\cos (2t + 1)] [2e^{2t + 1}] + [e^{2t + 1}] [-2 \sin (2t + 1)]$$

$$= 2e^{2t + 1} (\cos (2t + 1) - \sin (2t + 1)).$$

(5) $u = 7t^2$ and $v = \ln (2t - 1)$ so

$$\dfrac{dx}{dt} = [\ln (2t - 1)] [14t] + [7t^2] \left[\dfrac{2}{2t - 1} \right] = 14t \left(\ln (2t - 1) + \dfrac{t}{2t - 1} \right).$$

(6) $u = (t^2 + 1)^{1/2}$ and $v = \sin(2t + \pi)$ so

$$\frac{dx}{dt} = [\sin(2t + \pi)][\tfrac{1}{2}(t^2 + 1)^{-1/2}(2t)] + [(t^2 + 1)^{1/2}][2\cos(2t + \pi)]$$

$$= \frac{t\sin(2t + \pi)}{(t^2 + 1)^{1/2}} + 2(t^2 + 1)^{1/2}\cos(2t + \pi).$$

(7) $y = [(x^2 + 1)^5][e^{3x}\cos 2x]$ so $u = (x^2 + 1)^5$ and $v = e^{3x}\cos 2x$.

First, using the Product Rule to find dv/dx, we have

$$\frac{dv}{dx} = [\cos 2x][3e^{3x}] + [e^{3x}][-2\sin 2x] = e^{3x}(3\cos 2x - 2\sin 2x).$$

Now, using the Product Rule again to find dy/dx, we have

$$\frac{dy}{dx} = [e^{3x}\cos 2x][5(x^2 + 1)^4(2x)] + [(x^2 + 1)^5][e^{3x}(3\cos 2x - 2\sin 2x)]$$

$$= e^{3x}(x^2 + 1)^4[10x\cos 2x + (x^2 + 1)(3\cos 2x - 2\sin 2x)].$$

(8) $x = (2 + t)e^{3t}$ so $dx/dt = (e^{3t})(1) + (2 + t)(3e^{3t}) = (7 + 3t)e^{3t}$.

Using the Product Rule again, we get

$$\frac{d^2x}{dt^2} = (7 + 3t)(3e^{3t}) + (e^{3t})(3) = (24 + 9t)e^{3t}$$

so $\dfrac{d^2x}{dt^2} - 6\,dx/dt + 9x = (24 + 9t)e^{3t} - 6(7 + 3t)e^{3t} + 9(2 + t)e^{3t} = (0 + 0t)e^{3t} = 0.$

(9) $x = e^{kt}$ so $dx/dt = k\,e^{kt}$ and $d^2x/dt^2 = k^2e^{kt}$.

$$\frac{d^2x}{dt^2} - 2\frac{dx}{dt} - 3x = 0 \quad \text{so} \quad k^2e^{kt} - 2ke^{kt} - 3e^{kt} = 0.$$

Therefore $e^{kt}(k^2 - 2k - 3) = 0$ but e^{kt} is never equal to zero.
So we have

$$k^2 - 2k - 3 = 0 \quad \text{so} \quad (k - 3)(k + 1) = 0 \quad \text{so} \quad k = 3 \quad \text{or} \quad k = -1.$$

(10)   We know from the last question that both e^{3t} and e^{-t} fit this equation. Multiplying them by constant numbers will make no difference to this fit.
     Therefore $x = Ae^{3t} + Be^{-t}$ is a solution to this equation.
     (In fact, it is the solution to this equation which covers all possible situations.)

(11) $x = e^{-t}\ln(1 + e^t)$.

Using the Product Rule gives

$$\frac{dx}{dt} = \ln(1 + e^t)(-e^{-t}) + e^{-t}\left(\frac{e^t}{1 + e^t}\right) = -e^{-t}\ln(1 + e^t) + \frac{1}{1 + e^t}$$

because $(e^t)(e^{-t}) = e^0 = 1$.

Therefore $\dfrac{dx}{dt} + x = -e^{-t}\ln(1 + e^t) + \dfrac{1}{1 + e^t} + e^{-t}\ln(1 + e^t) = \dfrac{1}{1 + e^t}.$

Exercise 8.C.5

(1) $\dfrac{dy}{dx} = \dfrac{(\sin x)(-\sin x) - (\cos x)(\cos x)}{\sin^2 x} = -\dfrac{\sin^2 x + \cos^2 x}{\sin^2 x} = -\operatorname{cosec}^2 x.$

(2)   If $y = (\cos x)^{-1}$ then $\dfrac{dy}{dx} = -(\cos x)^{-2}(-\sin x) = \dfrac{\sin x}{\cos^2 x} = \sec x \tan x.$

(3)   If $y = \dfrac{1}{\sin x}$ then $dy/dx = \dfrac{(\sin x)(0) - (1)(\cos x)}{\sin^2 x} = -\dfrac{\cos x}{\sin^2 x} = -\operatorname{cosec} x \cot x.$

These last two questions show how you can use either the Product Rule or the Quotient Rule to answer them.

(4)   If $x = \dfrac{\sin t}{1 + \cos t}$ then $\dfrac{dx}{dt} = \dfrac{(1 + \cos t)\cos t - \sin t\,(-\sin t)}{(1 + \cos t)^2} = \dfrac{2}{(1 + \cos t)^2}.$

(5)   If $y = \ln(\sec x + \tan x)$ then we know from Example (1) in Section 8.C.(e) that $d/dx\,(\tan x) = \sec^2 x$ and from question (2) above that $d/dx\,(\sec x) = \sec x \tan x$. Therefore

$$\frac{d}{dx}\,(\ln(\sec x + \tan x)) = \frac{\sec x \tan x + \sec^2 x}{\sec x + \tan x} = \sec x$$

taking out $\sec x$ as a factor on the top of this fraction and then cancelling.

(6) $y = \dfrac{e^x - e^{-x}}{e^x + e^{-x}}$, so

$$\frac{dy}{dx} = \frac{(e^x + e^{-x})(e^x + e^{-x}) - (e^x - e^{-x})(e^x - e^{-x})}{(e^x + e^{-x})^2}$$

$$= \frac{(e^{2x} + 2 + e^{-2x}) - (e^{2x} - 2 + e^{-2x})}{(e^x + e^{-x})^2} = \frac{4}{(e^x + e^{-x})^2}$$

(remembering that $e^x \times e^{-x} = e^{x-x} = e^0 = 1$ and $d/dx(e^{-x}) = -e^{-x}$).

(7) $y = \ln\left(\dfrac{1 + x}{1 - x}\right) = \ln(1 + x) - \ln(1 - x)$ using the second law of logs, so

$$\frac{dy}{dx} = \frac{1}{1 + x} - \left(\frac{-1}{1 - x}\right) = \frac{(1 - x) + (1 + x)}{(1 + x)(1 - x)} = \frac{2}{1 - x^2}.$$

(8)   Here, we need to use the Product Rule to differentiate the 'u' of $x^2 \sin x$ on the top. Doing this first gives

$$\frac{dy}{dx} = (\sin x)(2x) + x^2(\cos x).$$

So $$\frac{dy}{dx} = \frac{(3 - x)(2x \sin x + x^2 \cos x) - (x^2 \sin x)(-1)}{(3 - x)^2}$$

$$= \frac{x(3x \cos x - x^2 \cos x - x \sin x + 6 \sin x)}{(3 - x)^2}$$

(9) $y = \dfrac{3x - 2}{2x + 3}$ so $\dfrac{dy}{dx} = \dfrac{(2x + 3)(3) - (3x - 2)(2)}{(2x + 3)^2} = \dfrac{13}{(2x + 3)^2}.$

(Notice that this time we have found that the gradient of the curve is always positive, except when $x = -\frac{3}{2}$, when the curve itself is undefined.)

(10) $y = \dfrac{ax + b}{cx + d}$ so $\dfrac{dy}{dx} = \dfrac{(cx + d)(a) - (ax + b)(c)}{(cx + d)^2} = \dfrac{ad - bc}{(cx + d)^2}.$

The bottom of this fraction is always positive, and $ad - bc$ is a constant number itself since it is made up of constant numbers.

Therefore, dy/dx will remain always either positive or negative depending on the particular values of a, b, c, and d, unless $x = -d/c$, when the function y itself is undefined.

The only exception to this is the special case when $ad - bc = 0$ when

$$\frac{dy}{dx} = 0$$ everywhere, so $y = \dfrac{ax + b}{cx + d}$ must be totally flat.

How could this be?

If we put $a = 6$, $d = 1$, $b = 3$ and $c = 2$, we would have an example of this. We then get

$$y = \frac{6x + 3}{2x + 1} = \frac{3(2x + 1)}{2x + 1} = 3.$$

It *is* totally flat!

THINKING POINT

See if you can show for yourself that, if $ad - bc = 0$, then

$$y = \frac{ax + b}{cx + d}$$

must simplify down to the form $y = $ a constant, so giving a horizontal line with a gradient of zero.

Hint: if $ad - bc = 0$ then $ad = bc$. Put $ad = bc = k$ and substitute for d and c in the equation for y. Then tidy this up.

Exercise 8.D.1

(1) If $e^x = 2$ then $e^{-x} = 1/e^x = \frac{1}{2}$.

Using the definitions for $\sinh x$, $\cosh x$ and $\tanh x$ gives us

$$\sinh x = \frac{1}{2}\left(2 - \frac{1}{2}\right) = \frac{3}{4},$$

$$\cosh x = \frac{1}{2}\left(2 + \frac{1}{2}\right) = \frac{5}{4},$$

$$\tanh x = \left(\frac{3}{4}\right) \div \left(\frac{5}{4}\right) = \frac{3}{5}.$$

(2) $e^0 = 1$ so $\sinh x = 0$ and $\cosh x = 1$. This gives $\tanh x = 0/1 = 0$.

(3) (a) $2 \sinh 2x$ (b) $3 \cosh (3x + 5)$

(c) Using the Product Rule, we get

$$2e^{2x} \sinh 3x + 3e^{2x} \cosh 3x = e^{2x} (2 \sinh 3x + 3 \cosh 3x).$$

(d) $5 \operatorname{sech}^2 5x$ (e) $\dfrac{\sinh x}{\cosh x} = \tanh x$ (f) $2 \cosh 3x (3 \sinh 3x) = 6 \sinh 3x \cosh 3x$

Exercise 8.D.2

(1) $dy/dx = \cosh x$ and $\cosh(0) = 1$, so the gradient of the curve $y = \sinh x$ at the origin is 1.

(2) The line $y = x$ is the tangent to the curve $y = \sinh x$ when it passes through the origin, since this line also passes through the origin and has a gradient of 1.

(3) Since $y = \sinh x$ and $y = \sinh^{-1} x$ are symmetrically placed either side of $y = x$, this line is also the tangent to $y = \sinh^{-1} x$ at the origin, and the gradient of $\sinh^{-1} x$ here must also be 1.

Exercise 8.D.3

(1) (a) If $y = \tan^{-1}\left(\dfrac{x}{3}\right)$ then $a = 3$ so $\dfrac{dy}{dx} = \dfrac{3}{x^2 + 9}$

(b) If $y = \sinh^{-1}\left(\dfrac{x}{5}\right)$ then $a = 5$ so $\dfrac{dy}{dx} = \dfrac{1}{\sqrt{x^2 + 25}}$.

(c) If $y = \tan^{-1}\left(\dfrac{3x}{2}\right)$ then $a = 2/3$ and $\dfrac{dy}{dx} = \dfrac{\frac{2}{3}}{x^2 + 4/9} = \dfrac{6}{9x^2 + 4}$.

(d) If $y = \sinh^{-1}\left(\dfrac{2x}{3}\right)$ then $a = 3/2$ and $\dfrac{dy}{dx} = \dfrac{1}{\sqrt{x^2 + 9/4}} = \dfrac{2}{\sqrt{4x^2 + 9}}$.

(2) (a) If $y = \tan^{-1}(5x)$ then $\dfrac{dy}{dx} = 5 \times \dfrac{1}{(5x)^2 + 1} = \dfrac{5}{25x^2 + 1}$

(b) If $y = \sinh^{-1}(3x)$ then $\dfrac{dy}{dx} = 3 \times \dfrac{1}{\sqrt{(3x)^2 + 1}} = \dfrac{3}{\sqrt{9x^2 + 1}}$.

(c) If $y = \tan^{-1}(x + 3)$ then $\dfrac{dy}{dx} = \dfrac{1}{(x + 3)^2 + 1} = \dfrac{1}{x^2 + 6x + 10}$.

(d) If $y = \tan^{-1}(3x + 4)$ then $\dfrac{dy}{dx} = \dfrac{3}{(3x + 4)^2 + 1} = \dfrac{3}{9x^2 + 24x + 17}$.

(e) If $y = \tan^{-1}(1 - x)$ then $\dfrac{dy}{dx} = \dfrac{-1}{(1 - x)^2 + 1} = \dfrac{-1}{x^2 - 2x + 2}$.

(f) If $y = \sinh^{-1}(2x + 1)$ then $\dfrac{dy}{dx} = \dfrac{2}{\sqrt{(2x + 1)^2 + 1}} = \dfrac{2}{\sqrt{4x^2 + 4x + 2}}$.

(g) If $y = \sinh^{-1}(3 - 2x)$ then $\dfrac{dy}{dx} = \dfrac{-2}{\sqrt{(3 - x)^2 + 1}} = \dfrac{-2}{\sqrt{x^2 - 6x + 10}}$.

I'll include more details for the last three questions, by putting $X = $ 'lump', as these are slightly more complicated.

(h) $y = \tan^{-1}\left(\dfrac{x + 3}{4}\right)$ so $X = \dfrac{x + 3}{4}$ and $dX/dx = \dfrac{1}{4}$, so

$$\frac{dy}{dx} = \frac{1}{X^2 + 1} = \frac{1}{\left(\frac{x+3}{4}\right)^2 + 1} = \frac{16}{(x + 3)^2 + 16} = \frac{16}{x^2 + 6x + 25}$$

so $\dfrac{dy}{dx} = \dfrac{dy}{dX}\dfrac{dX}{dx} = \dfrac{4}{x^2 + 6x + 25}$.

(i) $y = \tan^{-1}\left(\dfrac{2x + 1}{3}\right)$ so $X = \dfrac{2x + 1}{3}$ and $\dfrac{dX}{dx} = \dfrac{2}{3}$.

$$\frac{dy}{dX} = \frac{1}{\left(\frac{2x+1}{3}\right)^2 + 1} = \frac{9}{(2x + 1)^2 + 9} = \frac{9}{4x^2 + 4x + 10}$$

so $\dfrac{dy}{dx} = \dfrac{2}{3} \times \dfrac{9}{4x^2 + 4x + 10} = \dfrac{3}{2x^2 + 2x + 5}$.

(j) $y = \tan^{-1}\left(\dfrac{3x + 2}{4}\right)$ so $X = \dfrac{3x + 2}{4}$ and $\dfrac{dX}{dx} = \dfrac{3}{4}$.

$$\frac{dy}{dX} = \frac{1}{\left(\frac{3x+2}{4}\right)^2 + 1} = \frac{16}{(3x + 2)^2 + 16} = \frac{16}{9x^2 + 12x + 20}$$

so $\dfrac{dy}{dx} = \dfrac{dy}{dX}\dfrac{dX}{dx} = \dfrac{12}{9x^2 + 12x + 20}$.

(3) We have $y = \tanh^{-1} x = \frac{1}{2}\ln\left(\dfrac{1 + x}{1 - x}\right)$.

The first thing to do here is to use the second rule of logs to write y more simply. (This saves a huge amount of work.) This gives us

$$y = \tfrac{1}{2} \ln (1 + x) - \tfrac{1}{2} \ln (1 - x)$$

so

$$\frac{dy}{dx} = \tfrac{1}{2} \left(\frac{1}{1 + x} - \frac{(-1)}{1 - x} \right) = \tfrac{1}{2} \left(\frac{1 - x + 1 + x}{1 - x^2} \right) = \frac{1}{1 - x^2}.$$

(4)   We can solve $8 \sinh x = 3 \operatorname{sech} x$ in two ways.

METHOD (1)   Do it in the same way you would solve a trig equation.

$$8 \sinh x = 3/\cosh x \quad \text{so} \quad 8 \sinh x \cosh x = 3 \quad \text{so} \quad 4 \sinh 2x = 3$$

using the rule of Section 8.D.(d) that $\sinh 2x = 2 \sinh x \cosh x$.

We now have $\sinh 2x = \tfrac{3}{4}$ and $\sinh^{-1} (\tfrac{3}{4}) = 0.6931$ so $x = 0.347$ to 3 d.p.

Notice that you don't have the infinite numbers of solutions that you get with trig equations to worry about here! This is one way in which sinh and cosh *don't* behave like sin and cos.

I show a sketch of what's happening graphically here below.

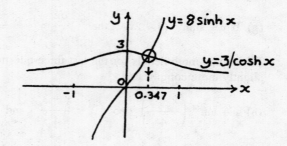

METHOD (2)   Write everything in terms of e^x. This gives

$$8 \left(\frac{e^x - e^{-x}}{2} \right) = 3 \left(\frac{2}{e^x + e^{-x}} \right) \quad \text{so} \quad 4(e^x - e^{-x})(e^x + e^{-x}) = 6$$

so $2(e^{2x} - e^{-2x}) = 3 \quad \text{so} \quad 2e^{4x} - 3e^{2x} - 2 = 0$

so $e^{2x} = \dfrac{3 \pm \sqrt{9 + 16}}{4} = \dfrac{3 \pm 5}{4} = 2$

since e^{2x} is always positive. This gives $2x = \ln 2 \quad \text{so} \quad x = 0.347$ as before.

(5)   (a)   We solve $2 \sinh^2 x - 5 \cosh x - 1 = 0$ in a very similar way to the trig equations in Section 5.E.(b).

$$\sinh^2 x = \cosh^2 x - 1 \quad \text{so} \quad 2(\cosh^2 x - 1) - 5 \cosh x - 1 = 0$$

so $2 \cosh^2 x - 5 \cosh x - 3 = 0 \quad \text{so} \quad (2 \cosh x + 1)(\cosh x - 3) = 0$

so $\cosh x = -\tfrac{1}{2}$ (which is impossible as you can see from the graph in Figure 8.D.5) or $\cosh x = 3$. This gives $x = \cosh^{-1} (3) = 1.763$ to 3 d.p.

(x could also be -1.763 from the symmetry of the cosh graph. Excellent, if you also got this answer!)

(b) $3 \operatorname{sech}^2 x + 8 \tanh x - 7 = 0$ so, using $1 - \tanh^2 x = \operatorname{sech}^2 x$, we have

$$3 (1 - \tanh^2 x) + 8 \tanh x - 7 = 0 \quad \text{so} \quad 0 = (3 \tanh x - 2)(\tanh x - 2)$$

so $\tanh x = \tfrac{2}{3}$ giving $x = 0.805$ to 3 d.p. or $\tanh x = 2$ which is impossible since the values of $\tanh x$ lie between -1 and $+1$ as you can see in Figure 8.D.7.

Exercise 8.E.1

(1) $y = e^x$ so $dy/dx = e^x$.

 (a) $x = 0$ so $dy/dx = e^0 = 1$ so the gradient of the tangent at $(0, 1)$ is 1.

  The equation of this tangent is $y - 1 = 1(x - 0)$ so $y = x + 1$.

 (b) $x = 1$ so $dy/dx = e^1 = e$.

  The equation of the tangent at $(1, e)$ is $y - e = e(x - 1)$ so $y = ex$.
  Notice that this particular tangent passes through the origin, and has a gradient of e. The lines $y = x$ and $y = 3x$ also pass through the origin. Since $1 < e$, $y = x$ doesn't cut the curve of $y = e^x$ at all – there are no solutions to the equation $x = e^x$. However, the line $y = 3x$ cuts the curve $y = e^x$ twice, because $3 > e$. This means that the equation $3x = e^x$ has two solutions.

 (c) $x = 2$ so $dy/dx = e^2$.

  The equation of the tangent at $(2, e^2)$ is $y - e^2 = e^2 (x - 2)$ so $y = e^2 x - e^2$, so it cuts the y-axis at about -7.4.

(2) $y = \tan x$ so $dy/dx = \sec^2 x$. (I showed this result in Section 8.C.(e).)

 (a) $x = 0$ so $dy/dx = 1$.

  When $x = 0$, $y = 0$ also, so the equation of the tangent at $(0,0)$ is $y = x$.

 (b) If $x = \pi/4$, then $y = 1$ and

  $$\frac{dy}{dx} = \sec^2 (\pi/4) = \frac{1}{\cos^2 (\pi/4)} = \frac{1}{(\frac{1}{\sqrt{2}})^2} = 2.$$

  The equation of the tangent at $(\pi/4, 1)$ is $y - 1 = 2(x - \pi/4)$ so $y = 2x + (1 - \pi/2)$. This tangent cuts the y-axis at about -0.57.

(3) Tangent (a) in (2) actually *crosses* the curve at the origin. It is behaving just like the tangents at the origin to $y = \sinh x$ and $y = \tanh x$ which we found in the last section. Each of these tangents happens to be the line $y = x$, but other tangents can also cross their curves. An example of this is tangent (c) to $y = \cos x$ at $x = \pi/2$, just before this exercise.

Table 8.E.2

Function	$\dfrac{dy}{dx}$	Value of $\dfrac{dy}{dx}$	$\dfrac{dy^2}{dx^2}$	Is $\dfrac{dy^2}{dx^2}$ +, – or 0?
(a) $y = x^2$ at O	$2x$	0	2	+
(b) $y = x^3$ at O	$3x^2$	0	$6x$	0
(c) $y = -x^2$ at O	$-2x$	0	-2	–
(d) $y = \tan x$ at O	$\sec^2 x$	1	$2 \sec x (\sec x \tan x)$ $= 2 \sec^2 x \tan x$ (Chain Rule)	0
(e) $y = \cos x$ (i) at A (ii) at B (iii) at C (iv) at D	$-\sin x$	(i) 0 (ii) -1 (iii) 0 (iv) 1	$-\cos x$	(i) – (ii) 0 (iii) + (iv) 0
(f) $y = x^4$ at O	$4x^3$	0	$12x^2$	0

Exercise 8.E.2

(a) $y = f(x) = \dfrac{x-1}{x^2-4} = \dfrac{x-1}{(x-2)(x+2)}$

(1) $f(0) = \frac{1}{4}$ so $y = f(x)$ cuts the y-axis at $(0, \frac{1}{4})$.

(2) $f(x) = 0$ when $x = 1$ so $y = f(x)$ cuts the x-axis at $(1, 0)$.

(3) We cannot divide by zero so we must exclude the values $x = -2$ and $x = +2$.

    The straight lines $x = -2$ and $x = +2$ are vertical asymptotes for $y = f(x)$.
    The value of $f(x)$ is large and negative if x is just less than -2.
    The value of $f(x)$ is large and positive if x is just greater than -2.
    The value of $f(x)$ is large and negative if x is just less than $+2$.
    The value of $f(x)$ is large and positive if x is just greater than $+2$.

(4) $f(x) = \dfrac{x-1}{x^2-4} = \dfrac{(1/x) - (1/x^2)}{1 - (4/x^2)}$ dividing top and bottom by x^2.

$1/x$, $1/x^2$ and $4/x^2$ all tend to zero when x becomes large, so $f(x) \to 0/1 = 0$ for both large positive and negative values of x. This means that $y = 0$ is a horizontal asymptote.

(5) $\dfrac{dy}{dx} = \dfrac{(x^2-4)(1) - (x-1)(2x)}{(x^2-4)^2} = \dfrac{2x - 4 - x^2}{(x^2-4)^2}$

so $dy/dx = 0$ if $x^2 - 2x + 4 = 0$.

There is no value of x on the x-axis which fits this equation since '$b^2 - 4ac$' $= -12$.
    In other words, this equation has no real roots.
    Therefore, there are no turning points. This curve never has a horizontal tangent.
    I show a sketch of it at the end of the answers to this exercise.

(b) $y = g(x) = \dfrac{x}{1+x^2}$

(1) $g(0) = 0$ so $y = g(x)$ cuts the x-axis at $(0, 0)$. This answers (2) as well.

(3) $1 + x^2$ is always positive so no value of x has to be excluded.

(4) $g(x) = \dfrac{x}{1+x^2} = \dfrac{1/x}{(1/x^2) + 1}$ dividing top and bottom by x^2.

$1/x$ and $1/x^2$ both tend to zero when x becomes large, so again $y = 0$ is a horizontal asymptote.

(5) $\dfrac{dy}{dx} = \dfrac{(1 + x^2)(1) - (x)(2x)}{(1 + x^2)^2} = \dfrac{1 - x^2}{(1 + x^2)^2} = 0$ if $1 - x^2 = 0$

so $dy/dx = 0$ if $x = -1$ or $x = +1$ since $1 - x^2 = (1 - x)(1 + x)$.

    Passing through $x = -1$, the sequence of values of dy/dx goes $- 0 +$ so there is a local maximum of $g(-1) = -\frac{1}{2}$ at the point $(-1, -\frac{1}{2})$.
    Passing through $x = +1$, the sequence of values of dy/dx goes $+ 0 -$ so there is a local maximum of $g(+1) = \frac{1}{2}$ at the point $(+1, \frac{1}{2})$.
    I show a sketch of this curve at the end of the answers to this exercise.
    It is another example of an odd function. If you turn the page upside down, the curve looks unchanged. It has its one moment of glory near the origin, and then sinks back to the x-axis.

(c) $y = h(x) = x + \dfrac{4}{x}$

(1) $h(0)$ is undefined. There is a vertical asymptote of $x = 0$.

(2) $h(x) = 0$ if $x + 4/x = 0$ or $x^2 = -4$.

There are no values of x on the x-axis which fit this equation, so $h(x)$ never cuts the x-axis.

(3) We have already decided in (1) that we must exclude $x = 0$.

(4) As $x \to \infty$ we have $4/x \to 0$ so $h(x)$ gets closer and closer to the line $y = x$.

The same thing is true if $x \to -\infty$, so $y = x$ is an asymptote of $y = h(x)$.

(5) $dy/dx = h'(x) = 1 - 4/x^2$ so $h'(x) = 0$ if $x = -2$ or $x = +2$.

$d^2y/dx^2 = 8/x^3$ so if $x = -2$, d^2y/dx^2 is negative.

Therefore, there is a local maximum of $h(-2) = -4$ at the point $(-2, -4)$.
If $x = +2$, d^2y/dx^2 is positive so there is a local minimum of $h(2) = 4$ at the point $(2,4)$.
Notice that the local minimum value is larger than the local maximum value! This is possible because the function $h(x)$ has a discontinuity at $x = 0$. The curve does a jump there.
I show a sketch of this curve at the end of the answers to this exercise.

(d) $y = p(x) = x - \dfrac{9}{x}$

This looks quite like (c) but the change in sign makes an enormous difference to the behaviour of the function and the shape of its curve.

(1) $p(0)$ is undefined. There is a vertical asymptote of $x = 0$.

(2) $p(x) = 0$ if $x - (9/x) = 0$ so $x = 9/x$ so $x^2 - 9 = 0$ so $x = -3$

or $x = +3$.

The curve cuts the x-axis at $(-3, 0)$ and $(+3, 0)$.

(3) We have already decided that we must exclude $x = 0$.

(4) As $x \to \infty$, $p(x) \to x$ just like example (c) above.

(5) $dy/dx = p'(x) = 1 + 9/x^2 = 0$ if $9/x^2 = -1$ or $x^2 = -9$.

There are no values of x on the x-axis which fit this equation therefore this curve has no turning points.
I show a sketch of it at the end of the answers to this exercise.
Both (c) and (d) are also odd functions. Try turning the page upside down.

(e) $y = f(x) = x^2\, e^x$

(1) $f(0) = 0$ so the curve passes through the origin. In fact, it has the repeated root $x = 0$ there, so it just touches the x-axis.

(2) has been answered by (1). There are no other solutions of $f(x) = 0$ since e^x is never equal to zero.

(3) There are no values of x which must be excluded.

(4) $x^2 e^x$ becomes very large, as x becomes very large and positive.
When x becomes large and negative, e^x gets smaller faster than x^2 gets larger, so $x^2 e^x$ becomes very small.

(5) $dy/dx = e^x(2x + x^2) = 0$ if $x = 0$ or $x = -2$.

$$d^2y/dx^2 = (2x + x^2)e^x + e^x(2 + 2x) = e^x(x^2 + 4x + 2).$$

If $x = 0$, d^2y/dx^2 is positive, so $y = 0$ is a local minimum.

If $x = -2$, d^2y/dx^2 is negative, so $y = 4/e^2$ is a local maximum.

This gives the sketch which I show below.

(a)
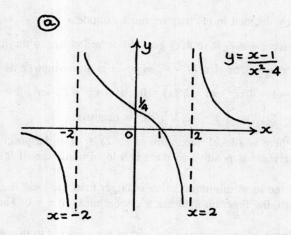

(b)
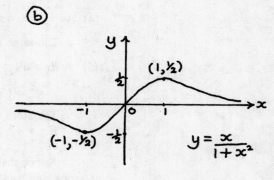

(c)
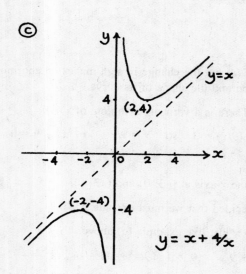

(d)
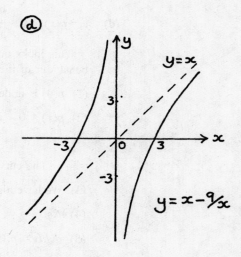

(e)
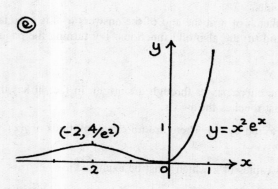

(1)   I show a drawing for this question in diagram (1) below.

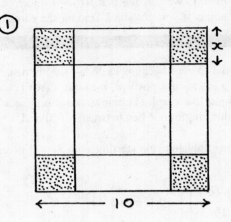

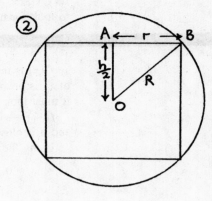

The volume of the box is given by $V = (10 - 2x)(10 - 2x)x = 100x - 40x^2 + 4x^3$.

$dV/dx = 100 - 80x + 12x^2$ so $dV/dx = 0$ if $3x^2 - 20x + 25 = 0$

or $(3x - 5)(x - 5) = 0$. The solutions to this equation are $x = 5$ or $x = \frac{5}{3}$.

Clearly, $x = 5$ gives the minimum value. After removing these squares there would be no cardboard left!

Checking for absolute certain that $x = \frac{5}{3}$ does give a maximum value for V, we have $d^2V/dx^2 = 6x - 20$ which is negative if $x = \frac{5}{3}$.

So the maximum volume of the box is $(\frac{20}{3})(\frac{20}{3})(\frac{5}{3}) = 2000/27 = 74.1 \text{ cm}^3$ to 1 d.p.

(2)   I show a cross-section along the axis of the cylinder in diagram (2) above.

If V is the volume of the cylinder, then $V = \pi r^2 h$.

We now need to use the physical relationship of the cylinder to the sphere to find r in terms of h and R.

We can do this by using Pythagoras' Theorem in triangle AOB. This gives us

$$R^2 = r^2 + \left(\frac{h}{2}\right)^2 = r^2 + \frac{h^2}{4} \quad \text{so} \quad r^2 = R^2 - \frac{h^2}{4}.$$

It is easier this time to substitute for r than h, so we have

$$V = \pi r^2 h = \pi \left(R^2 - \frac{h^2}{4}\right) h = \pi R^2 h - \pi \frac{h^3}{4}.$$

Differentiating V with respect to h we get

$$\frac{dV}{dh} = \pi R^2 - 3\pi \frac{h^2}{4} \quad \text{so} \quad \frac{dV}{dh} = 0 \quad \text{if} \quad 3h^2 = 4R^2 \quad \text{or} \quad h = \frac{2R}{\sqrt{3}}.$$

Differentiating again, we get $d^2V/dh^2 = -6\pi h/4$ which is negative since h is positive, so this value of h does give us the maximum volume of the cylinder.

Also, $R^2 = r^2 + (h/2)^2$ so if $h = 2R/\sqrt{3}$ we have $R^2 = r^2 + R^2/3$. This gives

$$r = \sqrt{\tfrac{2}{3}}\, R.$$

The volume of the cylinder $= \pi r^2 h = \pi \left(\dfrac{2R^2}{3}\right)\left(\dfrac{2R}{\sqrt{3}}\right) = \dfrac{4\pi R^3}{3\sqrt{3}}.$

The volume of the sphere is $\frac{4}{3} \pi R^3$ so the largest cylinder fills $1/\sqrt{3}$ of the hollow inside the sphere. $1/\sqrt{3}$ is 0.577 to 3 d.p. so it *does* fill more than half of it.

(3)   In Section 5.D.(f) we showed that $4 \sin t + 3 \cos t = 5 \sin(t + 0.6435)$.

The maximum value for a sin is one, so the maximum distance of the particle from the origin is 5 units.

Exercise 8.E.4

(1) We want to solve $2x^3 - 3x^2 + 6x + 1 = 0$, so we need to have some idea of the shape of $f(x) = 2x^3 - 3x^2 + 6x + 1$. We can see that $f(0) = 1$ and $f'(x) = 6x^2 - 6x + 6$. Try using this to look for turning points if you got stuck finding the roots of this equation.

$f'(x)$ is never equal to zero because its '$b^2 - 4ac$' is negative. Also, since the coefficient of x^2 is positive, $f'(x)$ is always positive, meaning that $f(x)$ always has a positive gradient. It must look something like graph (1) below, and it only *has* one root.

$f(-1) = -10$, so this single root lies between $x = 0$ and $x = -1$ since $f(x)$ is continuous, and it is closer to $x = 0$.

Using the Newton–Raphson rule starting with $x_1 = 0$ gives us

$$x_2 = 0 - \tfrac{1}{6} = -\tfrac{1}{6}.$$

$$x_3 = -\tfrac{1}{6} - \left(\frac{-0.0741}{7.1667}\right) = -0.156 \text{ to 3 d.p.}$$

Repeating the process gives $x = -0.156$ to 3 d.p. again, so this is the required solution.

(2) The sketch of $e^x = 3 - x$, in graph (2) below, shows the single solution.

We have to solve $e^x - 3 + x = 0$, so we put $f(x) = e^x - 3 + x$, giving $f'(x) = e^x + 1$.

$f(1) = e - 2 = 0.718$ and $f(0.5) = -0.851$ so the solution lies between these two values since $f(x)$ is continuous.

Starting with $x_1 = 1$ gives $\quad x_2 = 1 - \left(\dfrac{0.718}{3.718}\right) = 0.807$ to 3 d.p.

$$x_3 = 0.807 - \left(\frac{0.0482}{3.2412}\right) = 0.792 \text{ to 3 d.p.}$$

Repeating the process confirms this answer to 3 d.p.

(3) We can see from sketch (3) below that (i) has just the single solution of $x = 0$ while (ii) has three solutions symmetrically placed. ($y = x$ is a tangent here, just like it was to $y = \tanh x$.)

Let $f(x) = \sinh x - 2x$ so $f'(x) = \cosh x - 2$.

Using a calculator, $f(1) = -0.825$, $f(2) = -0.373$ and $f(3) = 4.018$ so there is a solution between $x = 2$ and $x = 3$ since $f(x)$ is continuous.

Starting with $x_1 = 2$ gives $x_2 = 2 - \left(\dfrac{-0.373}{1.762}\right) = 2.212$ to 3 d.p.

Repeating the process three more times confirms the answer as $x = 2.177$ to 3 d.p.

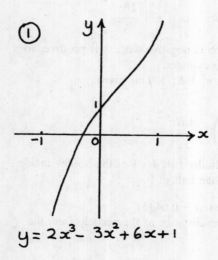

$y = 2x^3 - 3x^2 + 6x + 1$

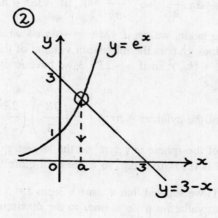

$y = e^x$

$y = 3 - x$

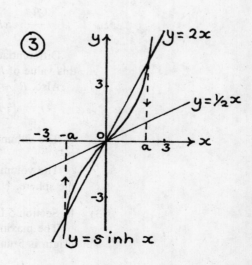

$y = 2x$

$y = \tfrac{1}{2}x$

$y = \sinh x$

Exercise 8.F.2

In these answers, I shall use the shorthand 'w.r.t.' to mean 'with respect to'. (This is quite commonly done in maths.)

(1) Differentiating $x^2 y^2 = 25$ implicitly w.r.t. x gives

$$2xy^2 + x^2 \left(2y \frac{dy}{dx}\right) = 0 \quad \text{so} \quad \frac{dy}{dx} = -\frac{2xy^2}{2x^2y} = -\frac{y}{x}.$$

The gradient of $x^2 y^2 = 25$ at $(1, 5)$ is -5.

The equation of the tangent here is $y - 5 = -5(x - 1)$ or $y + 5x = 10$.

(2) Differentiating $x^2 + 3xy - y^2 = 2$ implicitly w.r.t x gives

$$2x + 3y + 3x \frac{dy}{dx} - 2y \frac{dy}{dx} = 0$$

so $\dfrac{dy}{dx} = \dfrac{-2x - 3y}{3x - 2y} = \dfrac{2x + 3y}{2y - 3x}.$

The gradient of the curve at $(1, 2)$ is $\frac{8}{1} = 8$.

The equation of the tangent here is $y - 2 = 8(x - 1)$ or $y = 8x - 6$.

(3) Differentiating $1/y + 2/x = 1$ implicitly w.r.t. x gives

$$-\frac{1}{y^2} \frac{dy}{dx} - \frac{2}{x^2} = 0 \quad \text{so} \quad \frac{dy}{dx} = \frac{-2/x^2}{1/y^2} = -2y^2/x^2.$$

The gradient of the curve at $(4, 2)$ is $-\frac{8}{16} = -\frac{1}{2}$.

The equation of the tangent here is $y - 2 = -\frac{1}{2}(x - 4)$ or $2y + x = 8$.

Alternatively, multiplying the original equation by xy gives $x + 2y = xy$.

Differentiating this implicitly w.r.t. x gives

$$1 + 2\frac{dy}{dx} = y + x\frac{dy}{dx} \quad \text{so} \quad \frac{dy}{dx} = \frac{y - 1}{2 - x}.$$

This looks alarmingly different from the first answer, but substituting $x = 4$ and $y = 2$ gives $dy/dx = -\frac{1}{2}$ as before, and therefore, of course, the equation of the tangent is the same. Because x and y are related by fitting the equation of the given curve, dy/dx will give the same answer for any particular point, using either form.

(4) (a) Putting $X = 2x - 5$ and $y = \sin^{-1} X$, we have

$$\frac{dy}{dx} = \frac{1}{\sqrt{1 - (2x - 5)^2}} \times 2 = \frac{2}{\sqrt{20x - 4x^2 - 24}} = \frac{1}{\sqrt{5x - x^2 - 6}}.$$

We must have $-1 < 2x - 5 < 1$ so $4 < 2x < 6$ or $2 < x < 3$.

(b) Putting $X = 3x + 1$ and $y = \cosh^{-1} X$, we have

$$\frac{dy}{dx} = \frac{1}{\sqrt{(3x + 1)^2 - 1}} \times 3 = \frac{3}{\sqrt{9x^2 + 6x}}.$$

We must have $3x + 1 > 1$ so $x > 0$.

(5) $y = \tanh^{-1} x$ so $\tanh y = x$ and differentiating implicitly w.r.t. x gives

$$1 = \text{sech}^2 y \frac{dy}{dx} \quad \text{so} \quad \frac{dy}{dx} = \frac{1}{\text{sech}^2 y} = \frac{1}{1 - \tanh^2 y} = \frac{1}{1 - x^2}.$$

(This working uses results from Section 8.D.(c).)

The answer we have found agrees with the answer to question (3) of Exercise 8.D.3 where we used

$$\tanh^{-1} x = \frac{1}{2} \ln \left(\frac{1 + x}{1 - x}\right).$$

From Figure 8.D.8 we see that $-1 < x < 1$, and $y = \tanh^{-1} x$ has a positive gradient.

(6) (a) If $y = x^x$ then $\ln y = \ln (x^x) = x \ln x$ so $\frac{1}{y} \, dy/dx = \ln x + x \, (1/x) = \ln x + 1$
Therefore $dy/dx = y(\ln x + 1) = (\ln x + 1) \, x^x$.

(b) If $y = 3x^{(x+2)}$ then $\ln y = \ln (3x^{(x+2)}) = \ln 3 + \ln (x^{(x+2)}) = \ln 3 + (x+2) \ln x$.

(The splitting up is a bit tricky here! It uses both the first and third rules of logs.)
Differentiating implicitly w.r.t. x gives

$$\frac{1}{y} \frac{dy}{dx} = 0 + \ln x + (x+2)\left(\frac{1}{x}\right) = \ln x + 1 + \frac{2}{x}$$

so $\quad \dfrac{dy}{dx} = \left(\ln x + 1 + \dfrac{2}{x}\right) 3x^{(x+2)}.$

(c) $y = (3x)^{(x+2)}$ so $\ln y = \ln ((3x)^{(x+2)}) = (x+2) \ln 3x.$

(This is what you also get for (b) if you aren't careful!)
Differentiating implicitly w.r.t. x gives

$$\frac{1}{y} \frac{dy}{dx} = \ln 3x + (x+2)\left(\frac{3}{3x}\right) = \ln 3x + 1 + \frac{2}{x}$$

so $\quad \dfrac{dy}{dx} = \left(\ln 3x + 1 + \dfrac{2}{x}\right) (3x)^{(x+2)}.$

Exercise 8.G.1

(1) Putting 't' $= 2t$, we get $\quad e^{2t} = 1 + \dfrac{2t}{1!} + \dfrac{4t^2}{2!} + \dfrac{8t^3}{3!} + \dots + \dfrac{(2t)^r}{r!} + \dots$

Putting 't' $= -t$, we get $\quad e^{-t} = 1 - \dfrac{t}{1!} + \dfrac{t^2}{2!} - \dfrac{t^3}{3!} + \dots + (-1)^r \dfrac{t^r}{r!} + \dots$

(2) (a) Putting 't' $= 2t$, we get

$$\sin 2t = \frac{2t}{1!} - \frac{8t^3}{3!} + \frac{32t^5}{5!} - \dots + (-1)^r \frac{(2t)^{2r+1}}{(2r+1)!} + \dots = \sum_{r=0}^{\infty} (-1)^r \frac{(2t)^{2r+1}}{(2r+1)!}.$$

(b) Putting 't' $= -t$, we have $\sin (-t) = -\dfrac{t}{1!} + \dfrac{t^3}{3!} - \dfrac{t^5}{5!} + \dfrac{t^7}{7!} - \dots = -\sin t$

which is what we expect from the shape of its graph in Section 5.A.(a).

(3) $f(t) = \cos t, f'(t) = -\sin t, f''(t) = -\cos t$ and $f'''(t) = \sin t.$

Differentiating again brings us back to where we started.
We have $f(0) = 1, f'(0) = 0, f''(0) = -1$ and $f'''(0) = 0$, giving us the series

$$\cos t = 1 - \frac{t^2}{2!} + \frac{t^4}{4!} - \frac{t^6}{6!} + \dots + (-1)^r \frac{t^{2r}}{(2r)!} + \dots = \sum_{r=0}^{\infty} (-1)^r \frac{t^{2r}}{(2r)!}$$

starting the count with $r = 0$. It's a good idea to check that you've written down the
rth term correctly by putting $r = 0$, 1 and 2 and checking that you do actually get the first
three terms of the series. Then you can do any necessary tweakings to make it right.

(a) Putting 't' $= 2t$ in the series above, we get

$$\cos 2t = 1 - \frac{4t^2}{2!} + \frac{16t^4}{4!} - \frac{64t^6}{6!} + \dots + (-1)^r \frac{(2t)^{2r}}{(2r)!} + \dots$$

(b) Putting 't' = $-t$ gives us $\cos(-t) = 1 - \dfrac{t^2}{2!} + \dfrac{t^4}{4!} - \dfrac{t^6}{6!} + \ldots = \cos t$.

as we expect it to be, from the shape of the graph in Section 5.A.(a).

(c) Similarly, putting 't' = nt gives us

$$\cos nt = 1 - \frac{n^2 t^2}{2!} + \frac{n^4 t^4}{4!} - \frac{n^6 t^6}{6!} + \ldots + (-1)^r \frac{(nt)^{2r}}{(2r)!} + \ldots$$

(4) $f(t) = \sinh t$ so $f'(t) = \cosh t$ and $f''(t) = \sinh t$ and so on.

From this, we have $f(0) = 0$, $f'(0) = 1$ and $f''(0) = 0$, and so on. This gives

$$\sinh t = \frac{t}{1!} + \frac{t^3}{3!} + \frac{t^5}{5!} + \ldots + \frac{t^{2r+1}}{(2r+1)!} + \ldots = \sum_{r=0}^{\infty} \frac{t^{2r+1}}{(2r+1)!}$$

Adding the series gives $\sinh t + \cosh t = 1 + \dfrac{t}{1!} + \dfrac{t^2}{2!} + \dfrac{t^3}{3!} + \dfrac{t^4}{4!} + \ldots = e^t$.

Chapter 9
Exercise 9.A.1

(1) If $f'(x) = e^x$ then $y = f(x) = e^x + C$. Putting $y = 2$ and $x = 0$ gives $2 = e^0 + C = 1 + C$ so $C = 1$, giving us $y = e^x + 1$.

(2) If $g'(x) = \sinh x$ then $y = g(x) = \cosh x + C$. (See Sections 8.D.(a) and (b) if necessary.) Putting $y = x = 0$ gives $0 = \cosh(0) + C = 1 + C$ so $C = -1$ giving $y = \cosh x - 1$.

(3) If $h'(x) = 3x^2$ then $y = h(x) = x^3 + C$. Putting $y = 2$ and $x = 1$ gives $2 = 1 + C$ so $C = 1$ giving $y = x^3 + 1$.

The three curves look like my drawings below.

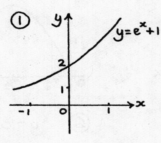
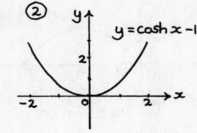
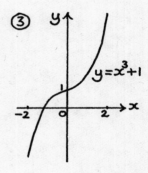

Exercise 9.A.2

(1) If $x = 5t^3 + 3t^2$ then   (a) $v = dx/dt = 15t^2 + 6t$ and   (b) $a = d^2 x/dt^2 = 30t + 6$.

(2) If $v = 3t^2 + 2t$ then

(a) $x = t^3 + t^2 + 5$ since $x = 5$ when $t = 0$.
(b) $a = dv/dt = 6t + 2$.

(3) If $a = d^2 x/dt^2 = 2t + 1$ then

(a) $v = dx/dt = t^2 + t + 5$ since $v = 5$ when $t = 0$.
(b) $x = \frac{1}{3}t^3 + \frac{1}{2}t^2 + 5t$ since $x = 0$ when $t = 0$.

(4) If $v = 4e^{2t}$ then

(a) $x = 2e^{2t} - 1$ since $x = 1$ when $t = 0$ and $e^0 = 1$.
(b) $a = dv/dt = 8e^{2t}$.

(5) If $v = dx/dt = - \sin t$ then

    (a) $x = \cos t$ since $x = 1$ when $t = 0$.
    (b) $a = dv/dt = d^2x/dt^2 = - \cos t$ so we have $d^2x/dt^2 = - x$.

The acceleration is towards the origin and equal to the distance from the origin so this is an example of SHM. (For more information on SHM see Sections 5.C.(b) and 8.A.(e).)

(6) $v = dx/dt = \cos (t + \pi/6)$ so

    (a) $x = \sin (t + \pi/6)$ since $x = \frac{1}{2}$ when $t = 0$.
    (b) $a = dv/dt = d^2x/dt^2 = - \sin (t + \pi/6)$ so, again, $d^2x/dt^2 = - x$ giving another example of SHM.

Exercise 9.A.3

(1) $\int_1^3 (2t^2 + 3t + 1)dt = \left[\frac{2}{3} t^3 + \frac{3}{2}t^2 + t\right]_1^3 = (\frac{2}{3}(27) + \frac{3}{2}(9) + 3) - (\frac{2}{3} + \frac{3}{2} + 1) = 31\frac{1}{3}.$

(2) $\int_0^4 (2t^3 - t)\, dt = \left[\frac{1}{2} t^4 - \frac{1}{2} t^2\right]_0^4 = (128 - 8) - (0) = 120$

(3) $\int_0^{\pi/6} \sin x\, dx = \left[- \cos x\right]_0^{\pi/6} = - \cos (\pi/6) + \cos (0) = - \sqrt{3}/2 + 1.$

The values of the sin, cos and tan of certain particular angles are often useful in integration. You will find these special values in Section 4.A.(g). Also, if you have trouble remembering which way the signs go when you differentiate and integrate sin and cos, remember the rule 'Solve **D**amn **P**roblem or **S**in **D**ifferentiates **P**lus'. The other three are then easy to get.

The sketches for questions (3), (4), (5) and (6) are shown below.

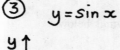

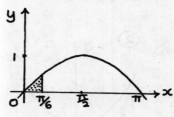

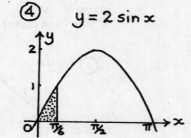

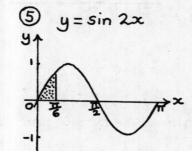

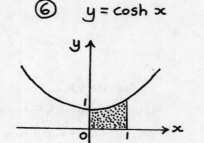

(4) $\int_0^{\pi/6} 2 \sin x\, dx = \left[-2 \cos x\right]_0^{\pi/6} = 2(- \sqrt{3}/2 + 1) = 2 - \sqrt{3}.$

The effect of the 2 is just to double the integral so doubling the area, as you can see above. It can be taken outside as a factor, giving $2\int_0^{\pi/6} \sin x\, dx$ if we want.

(5) $\int_0^{\pi/6} \sin 2x\, dx = \left[-\frac{1}{2} \cos 2x\right]_0^{\pi/6} = -\frac{1}{2} \cos(\pi/3) + \frac{1}{2} \cos(0) = -\frac{1}{4} + \frac{1}{2} = \frac{1}{4}$

See the sketch above. Notice that this *isn't* twice (3)! If you had any trouble with these sketches, you should go back to Section 5.C.(a).

(6) $\int_0^1 \cosh x \, dx = \left[\sinh x \right]_0^1 = \sinh(1) - \sinh(0) = 1.175 - 0 = 1.175$ to 3 d.p.

using a calculator to evaluate sinh(1). The sketch is on the previous page. If you need help with this, go back to Sections 8 D.(a) and (b).

(7) $\int_0^{\pi/4} 3 \cos 2u \, du = \left[\frac{3}{2} \sin 2u \right]_0^{\pi/4} = \frac{3}{2} \sin(\pi/2) - \frac{3}{2} \sin(0) = \frac{3}{2}$

(8) $\int_1^2 e^{2x} \, dx = \left[\frac{1}{2} e^{2x} \right]_1^2 = \frac{1}{2} e^4 - \frac{1}{2} e^2 = 23.6$ to 1 d.p.

(9) $\int_0^1 e^{3x+1} \, dx = \left[\frac{1}{3} e^{3x+1} \right]_0^1 = \frac{1}{3} e^4 - \frac{1}{3} e^1 = 17.3$ to 1 d.p.

(10) $\int (t^2 + 2t) dt = \frac{1}{3}t^3 + t^2 + C$ (11) $\int \sin 3x \, dx = -\frac{1}{3} \cos 3x + C$

(12) $\int \cos (2x + \pi/2) \, dx = \frac{1}{2} \sin (2x + \pi/2) + C$ (13) $\int \sinh 2u \, du = \frac{1}{2} \cosh 2u + C$

(14) $\int e^{5x} \, dx = \frac{1}{5} e^{5x} + C$ (15) $\int e^{2x+3} \, dx = \frac{1}{2} e^{2x+3} + C$

(16) $x = \int (8t^3 + t) \, dt = 2t^4 + \frac{1}{2} t^2 + C$, but $x = 2$ when $t = 0$ so $C = 2$.

(17) $x = \int \sin 2t \, dt = -\frac{1}{2} \cos 2t + C$, but $x = 1$ when $t = 0$ so $-\frac{1}{2} + C = 1$ and $C = \frac{3}{2}$.

(18) $x = \int (t + e^{2t}) \, dt = \frac{1}{2} t^2 + \frac{1}{2} e^{2t} + C$, but $x = 0$ when $t = 0$ so $\frac{1}{2} + C = 0$ and $C = -\frac{1}{2}$.

Thinking point after Exercise 9.A.3

In graph (c), $a = 0$ when $t = \frac{1}{3}$. At this instant, the particle's acceleration is zero. It changes from negative (the particle was under some kind of decreasing drag) to positive. (The particle is now receiving some kind of increasing forwards push.)

This is reflected in graph (b). We see the particle's velocity decreasing up to $t = \frac{1}{3}$ and then increasing again. The curve has a negative slope up to $t = \frac{1}{3}$ (dv/dt is negative), zero slope at $t = \frac{1}{3}$ ($dv/dt = 0$), and a positive slope when $t > \frac{1}{3}$ (dv/dt is positive.)

Graph (c) shows how this affects the distance travelled.

We have $d^2x/dt^2 = dv/dt = a = 0$ at $t = \frac{1}{3}$. The graph shows how the distance is being covered increasingly slowly as the time gets closer to $t = \frac{1}{3}$, and then faster again as the velocity picks up afterwards. The flattening in the graph (which gives the point of inflection where the tangent crosses the curve), shows that the change in the change of distance (that is, $d/dt \, (dx/dt)$), is happening increasingly slowly up to $t = \frac{1}{3}$. It is zero at $t = \frac{1}{3}$ and then becomes positive afterwards. (Remember that d^2x/dt^2 is the same thing as dv/dt.)

You could imagine that the particle puts its brakes on at $t = 0$, gradually raising the pedal until $t = \frac{1}{3}$, when both its feet are instantaneously off the pedals. After $t = \frac{1}{3}$, it gradually increases its pressure on the accelerator, and vrooms off along the positive x-axis.

If we know what both x and v are when $t = 0$, we can move both forwards (differentiating), and backwards (integrating), between these particular curves. If we don't have this specific knowledge then moving backwards from (c) would give $v = 3t^2 - 2t + C$ and $x = t^3 - t^2 + Ct + K$ where C and K are constants representing the unknown amount of shift from the horizontal axis in both (b) and (a).

Exercise 9.A.4

(1) $\ln |x + 3| + C$ (2) $2\ln |t - 4| + C$ (3) $\frac{1}{3} \ln |3x + 2| + C$

(4) $-\frac{1}{2} \ln |5 - 2x| + C$ (5) $-4 \ln |3 - t| + C$ (6) $-\frac{2}{5} \ln |5t + 4| + C$

(7) $\left[-\frac{1}{2} \ln |5 - 2x| \right]_1^2 = -\frac{1}{2} (\ln 1 - \ln 7) = 0.973$

(8) This one can't be done because $x = \frac{4}{3}$ (which gives a zero on the bottom of the fraction) lies between the limits of 1 and 3.

(9) $\left[\frac{1}{2} \ln |3 + 2x| \right]_{-4}^{-2} = \frac{1}{2} (\ln |-1| - \ln |-5|) = \frac{1}{2} (\ln 1 - \ln 5) = -0.805$.

Exercise 9.B.1

(A) (1) $8t$ (2) $\frac{9}{2}x^{1/2}$ (3) $2e^{2t}$ (4) $5e^{5x+2}$ (5) $-3e^{-3t}$

(6) $\frac{3}{3t} = \frac{1}{t}$ (7) $\frac{1}{3+x}$ (8) $\frac{-1}{2-x} = \frac{1}{x-2}$ (9) $\frac{-5}{3-5t} = \frac{5}{5t-3}$

(10) $3\cos 3t$ (11) $-\sin(x+\pi/3)$ (12) $2\cos(2x+\pi/6)$

(13) $2\sinh(2t+5)$ (14) $-5\cosh(8-5x)$ (15) $3\sin^2 x \cos x$

(16) $\sec^2 x$

(B) (1) $\frac{1}{2}x^4 + C$ (2) $\frac{2}{3}t^{3/2} + C$ (3) $\frac{1}{3}e^{3t} + C$ (4) $\frac{1}{2}e^{2x+3} + C$

(5) $-\frac{1}{2}e^{-2t} + C$ (6) $-\frac{1}{4}e^{3-4t} + C$ (7) $\ln|x+2| + C$ (8) $\frac{1}{3}\ln|3x+2| + C$

(9) $-\ln|1-x| + C$ (10) $-\frac{1}{5}\cos 5t + C$ (11) $\sin(x+\pi/6) + C$

(12) $-\frac{1}{2}\cos(2t+\pi/4) + C$ (13) $\frac{1}{3}\sinh 3x + C$ (14) $\frac{1}{2}\cosh(2x+3) + C$

(15) $\sin^4 x + C$ (16) $\frac{1}{2}\tan 2x + C$

(C) (1) $[\frac{1}{3}u^3 + \frac{3}{2}u^2]_0^2 = \frac{26}{3}$ (2) $[\frac{5}{3}t^3 + \ln t]_1^3 = \frac{130}{3} + \ln 3$

(3) $[\frac{1}{4}\sin 4t]_0^{\pi/8} = \frac{1}{4}(\sin \pi/2 - \sin 0) = \frac{1}{4}$

(4) $[-\frac{1}{2}\cos 2x]_0^{\pi/6} = -\frac{1}{2}\cos \pi/3 + \frac{1}{2}\cos 0 = -\frac{1}{4} + \frac{1}{2} = \frac{1}{4}$

(Remember $\cos 0 \neq 0$.)

(5) $[-\frac{1}{2}\cos(2x+\pi/3)]_0^{\pi/2} = -\frac{1}{2}\cos(4\pi/3) + \frac{1}{2}\cos(\pi/3) = \frac{1}{4} + \frac{1}{4} = \frac{1}{2}$

(6) $[\frac{1}{5}e^{5t}]_{-1}^1 = \frac{1}{5}(e^5 - e^{-5}) = 29.68$ to 2 d.p.

(7) $[\frac{1}{2}e^{2t+3}]_{-1}^0 = \frac{1}{2}(e^3 - e^1) = 8.68$ to 2 d.p.

(8) $[\frac{1}{2}\sinh 2x]_{-1}^1 = 0$

(You can see it will be this from Section 8.D.(a) and Section 9.A.(d).)

(9) $[\tan x]_0^{\pi/4} = 1$ (10) $[\frac{1}{2}\ln|2u+3|]_0^3 = \frac{1}{2}(\ln 9 - \ln 3) = 0.55$ to 2 d.p.

(11) $[\ln|x+2| + 2\ln|2x+5|]_0^1 = \ln 3 + 2\ln 7 - \ln 2 - 2\ln 5 = 1.08$ to 2 d.p.

(12) $[-\ln|2-x| - \frac{1}{2}\ln|2x+3|]_{-1}^1 = -\ln 1 - \frac{1}{2}\ln 5 + \ln 3 + \frac{1}{2}\ln 1 = 0.29$ to 2 d.p.

I'll give you the answers to each of the thinking point integrals when we look at the technique for that particular one.

Exercise 9.B.2

(1) Let $t = 2x^2 - 3$ so $dt/dx = 6x$ so we can replace $3x\, dx$ by $\frac{1}{2}dt$. Then

$$\int \frac{3x}{(2x^2-3)^4}\, dx = \int \frac{1}{2}t^{-4}\, dt = -\frac{1}{6}t^{-3} + C = \frac{-1}{6(2x^2-3)^3} + C.$$

(2) Let $t = 2x^3 + 9$ so $dt/dx = 6x^2$.

Also, if $x = 0$ then $t = 9$ and if $x = 2$ then $t = 25$. So

$$\int_0^2 \frac{6x^2}{\sqrt{2x^3+9}}\, dx = \int_9^{25} \frac{dt}{t^{1/2}} = \left[2t^{1/2}\right]_9^{25} = 10 - 6 = 4.$$

(3) Let $t = x - 4$ so $dt/dx = 1$ and $x = t + 4$.

Also, if $x = 4$ then $t = 0$ and if $x = 5$ then $t = 1$. So

$$\int_4^5 x(x-4)^9\, dx = \int_0^1 t^9(t+4)\, dt = \left[t^{11}/11 + 4t^{10}/10\right]_0^1 = 27/55.$$

Exercise 9.B.3

(1) $I = \ln|x^2 + 3x + 9| + C$ (2) $I = \ln|3x^2 + 2x + 7| + C$

(3) $I = 2\ln|x^2 + 4x + 7| + C$

(4) Let $u = 3x^2 + 4$ so $du/dx = 6x$. Then

$$I = \tfrac{1}{6}\int \frac{du}{u^3} = \tfrac{1}{6}\left(\frac{u^{-2}}{-2}\right) + C = -\tfrac{1}{12}(3x^2 + 4)^{-2} + C.$$

(5) Let $u = x^3 - 1$ so $du/dx = 3x^2$. Then

$$I = \tfrac{1}{3}\int \frac{du}{\sqrt{u}} = \tfrac{1}{3}\left(\frac{u^{1/2}}{\tfrac{1}{2}}\right) = \tfrac{2}{3}\sqrt{x^3 - 1} + C.$$

(6) Let $u = 1 - 4x^2$ so $du/dx = -8x$. Then

$$I = -\tfrac{1}{8}\int \frac{du}{u^{1/2}} = -\tfrac{1}{4}u^{1/2} + C = -\tfrac{1}{4}\sqrt{1 - 4x^2} + C.$$

(7) Let $u = 1 - x^2$ so $du/dx = -2x$. Then

$$I = -\tfrac{1}{2}\int u^{1/2}\, du = -\tfrac{1}{2}\left(\frac{u^{3/2}}{\tfrac{3}{2}}\right) = -\tfrac{1}{3}(1 - x^2)^{3/2} + C.$$

(8) Let $u = x + x^2$ so $du/dx = 1 + 2x$. Then

$$I = \int \frac{du}{\sqrt{u}} = 2u^{1/2} + C = 2\sqrt{(x + x^2)} + C.$$

(9) Let $u = 2x - 7$ so $du/dx = 2$. Then

$$I = \tfrac{1}{2}\int \frac{du}{u^2} = -\tfrac{1}{2}u^{-1} + C = \frac{-1}{2(2x - 7)} + C.$$

Exercise 9.B.4

(1) $I = -\tfrac{1}{6}\cos^6 x + C$ (2) $I = -\tfrac{1}{8}\cos^4 2x + C$ (3) $I = \tfrac{1}{3}\sin^3(x + \pi/4) + C$

(4) Let $u = \cos 3x$ so $du/dx = -3\sin 3x$ and $-\tfrac{1}{3}du = \sin 3x\, dx$. Then

$$I = -\tfrac{1}{3}\int \frac{du}{u^4} = \tfrac{1}{9}u^{-3} + C = \frac{1}{9\cos^3 3x} + C.$$

(5) Let $u = 1 + \sin x$ so $du/dx = \cos x$. Then

$$I = \int \frac{du}{u^3} = -\tfrac{1}{2}u^{-2} + C = -\frac{1}{2(1 + \sin x)^2} + C.$$

(6) Let $u = \sin x$ so $du/dx = \cos x$. Then

$$I = \int \frac{du}{u} = \ln|u| + C = \ln|\sin x| + C.$$

(7) $I = \tfrac{1}{2}\int (\cos 2x + 1)\, dx = \tfrac{1}{4}\sin 2x + \tfrac{1}{2}x + C$ using $\cos 2x = 2\cos^2 x - 1$.

(8) $\cos 6x = 1 - 2\sin^2 3x$ (using $\cos 2A = 1 - 2\sin^2 A$) so $\sin^2 3x = \tfrac{1}{2}(1 - \cos 6x)$, therefore

$$I = \tfrac{1}{2}\int 1 - \cos 6x\, dx = \tfrac{1}{2}(x - \tfrac{1}{6}\sin 6x) + C.$$

(9) $\int \cos^3 x\, dx = \int \cos x\,(1 - \sin^2 x)\, dx = \sin x - \tfrac{1}{3}\sin^3 x + C.$

(10) $I = \int \cos 5x\,(1 - \sin^2 5x)\, dx = \tfrac{1}{5}\sin 5x - \tfrac{1}{15}\sin^3 5x + C$

(11) $I = \tfrac{1}{2}\int (\cosh 2x - 1)\, dx = \tfrac{1}{4}\sinh 2x - \tfrac{1}{2}x + C$

using $\cosh 2x = \cosh^2 x + \sinh^2 x = 2\cosh^2 x - 1$ from Section 8.D.(d).

(12) $I = \frac{1}{5} \cosh^5 (x + 1) + C$

(13) If $t = \sin x$ then $dt/dx = \cos x$.

If $x = \pi/6$ then $t = \frac{1}{2}$ and if $x = 0$ then $t = 0$, so

$$I = \int_0^{1/2} \frac{3 \, dt}{t^2 + 5t + 6} = \int_0^{1/2} \frac{3 \, dt}{(t + 3)(t + 2)} = \int_0^{1/2} \left(\frac{-3}{t + 3} + \frac{3}{t + 2} \right) dt$$

so we have $I = [-3 \ln |t + 3| + 3 \ln |t + 2|]_0^{1/2} = 0.207$ to 3 d.p.

(14) If $t = \cosh x$ then $dt/dx = \sinh x$, so

$$I = \int \frac{7 \, dt}{2t^2 + 5t - 3} = \int \left(\frac{2}{2t - 1} - \frac{1}{t + 3} \right) dt$$

$$= \ln |2t - 1| - \ln |t + 3| + \ln A$$

$$= \ln \left| \frac{A (2t - 1)}{t + 3} \right| = \ln \left| \frac{A (2 \cosh x - 1)}{\cosh x + 3} \right|.$$

(15) If $x = 2 \sin t$ then $dx/dt = 2 \cos t$.

If $x = 1$ then $\sin t = \frac{1}{2}$ so $t = \pi/6$. If $x = 0$ then $\sin t = 0$ so $t = 0$.

$$I = \int_0^{\pi/6} \frac{2 \cos t}{\sqrt{4 - 4 \sin^2 t}} \, dt = \int_0^{\pi/6} \frac{2 \cos t}{2 \cos t} \, dt = \left[t \right]_0^{\pi/6} = \pi/6$$

(16) If $x = \sin t$ then $dx/dt = \cos t$.

If $x = \frac{1}{2}$ then $t = \pi/6$ and if $x = 0$ then $t = 0$.

$$I = \int_0^{\pi/6} \cos t \cos t \, dt = \int_0^{\pi/6} \frac{1}{2} (\cos 2t + 1) \, dt$$

$$= \frac{1}{2} \left[\frac{1}{2} \sin 2t + t \right]_0^{\pi/6} = \frac{1}{2} [\sqrt{3}/4 + \pi/6] = 0.478 \text{ to 3 d.p.}$$

(17) Let $t = 4 + \sin x$ so $dt/dx = \cos x$.

If $x = \pi/2$ then $t = 5$ and if $x = 0$ then $t = 4$.

$$I = \int_4^5 1/t^2 \, dt = \left[-1/t \right]_4^5 = -\frac{1}{5} + \frac{1}{4} = \frac{1}{20}$$

(18) If $x = 2 \cos t$ then $dx/dt = -2 \sin t$.

If $x = \sqrt{2}$ then $t = \pi/4$ and if $x = 1$ then $t = \pi/3$.

$$I = \int_{\pi/3}^{\pi/4} \frac{-2 \sin t}{(4 \cos^2 t)(2 \sin t)} \, dt = \frac{1}{4} \int_{\pi/4}^{\pi/3} \sec^2 t \, dt = \frac{1}{4} \left[\tan t \right]_{\pi/4}^{\pi/3} = \frac{1}{4} (\sqrt{3} - 1)$$

Exercise 9.B.5

(1) 'a' = 5   so $I = 3 \sinh^{-1} (x/5) + C$.   (2) '$a$' = 3   so $I = 3 \sin^{-1} (x/3) + C$.

(3) 'a' = 4   so $I = [\frac{1}{4} \tan^{-1} (x/4)]_0^\pi = \frac{1}{4} - 0 = \frac{1}{4}$.

Exercise 9.B.6

(1) $I = \frac{1}{3} \tan^{-1} (t/3) + C$

(2) $I = \int \frac{dx}{(x + 3)^2 + 1}$ so 'x' = $x + 3$ and 'a' = 1, giving $\tan^{-1} (x + 3) + C$.

(3) $\quad I = \displaystyle\int \frac{2x+4}{x^2+4x+13}\, dx + \int \frac{dx}{x^2+4x+13}$

$\qquad = \ln|x^2+4x+13| + \displaystyle\int \frac{dx}{(x+2)^2+9}$

$\qquad = \ln|x^2+4x+13| + \frac{1}{3}\tan^{-1}\left(\dfrac{x+2}{3}\right) + C$

(4) $\quad I = \displaystyle\int \frac{4t+16-13}{t^2+8t+17}\, dt = 2\int \frac{2t+8}{t^2+8t+17}\, dt - 13\int \frac{dt}{(t+4)^2+1}$

$\qquad = 2\ln|t^2+8t+17| - 13\tan^{-1}(t+4) + C$

(5) $\quad I = \displaystyle\int \frac{dx}{4(x^2+\frac{9}{4})} = \frac{1}{4}\int \frac{dx}{x^2+\frac{9}{4}}$

$\qquad = \frac{1}{4} \times \frac{2}{3}\tan^{-1}(2x/3) + C = \frac{1}{6}\tan^{-1}(2x/3) + C$ with 'a' $= \frac{3}{2}$

(6) $\quad I = \displaystyle\int \frac{3dx}{9(x^2+\frac{16}{9})} = \frac{1}{3}\int \frac{dx}{x^2+\frac{16}{9}}$

$\qquad = \frac{1}{3} \times \frac{3}{4}\tan^{-1}(3x/4) + C = \frac{1}{4}\tan^{-1}(3x/4) + C$ with 'a' $= \frac{4}{3}$

(7) $\quad I = \frac{1}{2}\displaystyle\int \frac{6u+4}{u^2+6u+13}\, du = \frac{1}{2}\int \frac{6u+18-14}{u^2+6u+13}\, du$

$\qquad = \frac{3}{2}\displaystyle\int \frac{2u+6}{u^2+6u+13}\, du - 7\int \frac{du}{(u+3)^2+4}$

$\qquad = \frac{3}{2}\ln|u^2+6u+13| - \frac{7}{2}\tan^{-1}\left(\dfrac{u+3}{2}\right) + C$

(8) $\quad I = \displaystyle\int \frac{3du}{(3u+4)^2+1}$ so 'x' $= 3u+4$ and 'a' $= 1$ so $dx/du = 3$.

$\qquad$ This gives $I = \displaystyle\int \frac{dx}{x^2+1} = \tan^{-1}x + C = \tan^{-1}(3u+4) + C$

(9) $\quad$ I will do this one using the second method I gave you where we put up with the fractions in order to use the $\tan^{-1}$ rule directly. We have

$$I = \frac{1}{2}\int \frac{3du}{u^2+u+\frac{5}{2}} = \frac{3}{2}\int \frac{du}{(u+\frac{1}{2})^2+\frac{9}{4}}.$$

Now we put $u+\frac{1}{2} = x$ and $\frac{3}{2} = a$ which gives us

$$I = \frac{3}{2} \times \frac{2}{3}\tan^{-1}\left(\frac{u+\frac{1}{2}}{\frac{3}{2}}\right) + C = \tan^{-1}\left(\frac{2u+1}{3}\right) + C$$

You might like to look at questions (8) and (9) being worked out the opposite way round in (2)(d) and (2)(i) of exercise 8.D.3.

Exercise 9.B.7

(1) $\quad I = \displaystyle\int \left(\frac{4}{x+2} - \frac{4}{x+3}\right) dx = 4\ln|x+2| - 4\ln|x+3| + C$

(2) $\quad I = \int \left(\dfrac{\frac{1}{5}}{x-2} - \dfrac{\frac{1}{5}}{x+3} - \dfrac{1}{(x+3)^2} \right) dx$

$\qquad = \frac{1}{5} \ln |x-2| - \frac{1}{5} \ln |x+3| + \dfrac{1}{x+3} + C$

(3) This one often causes problems although it looks so simple. The important thing to notice about it is that it counts as top heavy, so we say

$$\int \frac{x-1}{x+1} \, dx = \int \frac{x+1-2}{x+1} \, dx = \int 1 - \frac{2}{x+1} \, dx = x - 2 \ln |x+1| + C$$

(4) $\quad I = \int \left(\dfrac{4}{y} - \dfrac{4y}{y^2+1} \right) dy = 4 \ln |y| - 2 \ln |y^2 + 1| + C$

(5) $\quad I = \int \left(\dfrac{\frac{2}{5}}{2t-1} - \dfrac{\frac{1}{5}}{t+2} + \dfrac{1}{(t+2)^2} \right) dt$

$\qquad = \frac{1}{5} \ln |2t-1| - \frac{1}{5} \ln |t+2| - \dfrac{1}{t+2} + C$

(6) $\quad I = \int \left(\dfrac{1}{y-1} - \dfrac{y-9}{y^2+9} \right) dy = \ln |y-1| - \int \dfrac{y}{y^2+9} \, dy + \int \dfrac{9}{y^2+9} \, dy$

$\qquad = \ln |y-1| - \frac{1}{2} \ln |y^2+9| + 3 \tan^{-1} (y/3) + C$

(7) $\quad I = \int \left(\dfrac{-\frac{5}{4}}{x-1} - \dfrac{\frac{5}{4}}{x+3} + \dfrac{\frac{5}{2}}{x-3} \right) dx$

$\qquad = -\frac{5}{4} \ln |x-1| - \frac{5}{4} \ln |x+3| + \frac{5}{2} \ln |x-3| + C$

(8) $\quad I = \int \dfrac{t^2+1}{t^2-1} \, dt = \int \dfrac{t^2-1+2}{t^2-1} \, dt = \int \left(1 + \dfrac{2}{t^2-1} \right) dt$

$\qquad = \int \left(1 + \dfrac{1}{t-1} - \dfrac{1}{t+1} \right) dt = t + \ln |t-1| - \ln |t+1| + C$

(9) $\quad I = \int \dfrac{x^2+1}{x^2+6x+8} \, dx = \int \dfrac{x^2+6x+8-6x-7}{x^2+6x+8} \, dx = \int 1 - \dfrac{(6x+7)}{x^2+6x+8} \, dx$

$\qquad = \int \left(1 + \dfrac{\frac{5}{2}}{x+2} - \dfrac{\frac{17}{2}}{x+4} \right) dx = x + \frac{5}{2} \ln |x+2| - \frac{17}{2} \ln |x+4| + C$

 Don't forget the 1 when you rewrite the integral after finding the partial fractions.

(10) First, factorise $y^3 - y^2$ as $y^2(y-1)$. Then you will find

$$I = \int \left(-\dfrac{1}{y} - \dfrac{1}{y^2} + \dfrac{1}{y-1} \right) dy = -\ln |y| + 1/y + \ln |y-1| + C.$$

(11) $\quad I = \int \left(\dfrac{2}{u} - \dfrac{1}{u^2} - \dfrac{3}{2u+1} \right) du = 2 \ln |u| + 1/u - \frac{3}{2} \ln |2u+1| + C$

(12) Let $t = e^x$ so that $dt/dx = e^x$. Remembering that $e^{2x} = (e^x)^2$, we can then rewrite I in the form

$$I = \int \frac{dt}{t^2 + 5t + 6} = \int \left(\frac{1}{t + 2} - \frac{1}{t + 3} \right) dt = \ln |t + 2| - \ln |t + 3| + \ln A.$$

Now we can combine this answer into a single log, and we must also replace t by e^x. Doing this gives

$$I = \ln \left[A \left(\frac{e^x + 2}{e^x + 3} \right) \right].$$

If you had any difficulties finding any of these partial fractions, you will find that all of them come in the answers to the exercises of Section 6.E.

Exercise 9.B.8

(1) (a) $du/dx = 5e^{5x}$ (b) $du/dt = 4\cos 4t$ (c) $v = \frac{1}{4} e^{4x}$ (d) $v = -e^{-t}$

   (e) $du/dx = 2/2x = 1/x$

   (Watch out for this one! The usual mistake is to say it is $1/2x$.)

   (f) $v = \frac{1}{2} \cosh 2t$ (Section 8.D)   (g) $v = -\frac{1}{3} \cos 3x$ (h) $du/dx = -2 \sin 2x$

(2) $I = \int 2t \sin 3t \, dt$

   Let $u = 2t$ so $du/dt = 2$ and $dv/dt = \sin 3t$ so $v = -\frac{1}{3} \cos 3t$.
   This gives $I = -\frac{2}{3} t \cos 3t + \frac{2}{3} \int \cos 3t \, dt = -\frac{2}{3}t \cos 3t + \frac{2}{9} \sin 3t + C$.

(3) $I = \int xe^x \, dx$

   Let $u = x$ so $du/dx = 1$ and $dv/dx = e^x$ so $v = e^x$.
   This gives $I = xe^x - \int e^x \, dx = xe^x - e^x + C$.

(4) $I = \int x^2 \sin x \, dx$

   Let $u = x^2$ so $du/dx = 2x$ and $dv/dx = \sin x$ so $v = -\cos x$.
   This gives $I = -x^2 \cos x - \int -2x \cos x dx = -x^2 \cos x + 2 \int x \cos x \, dx$.
   For $I_1 = \int x \cos x \, dx$, we let $u = x$ so $du/dx = 1$ and $dv/dx = \cos x$ so $v = \sin x$.
   This gives $I = -x^2 \cos x + 2(x \sin x - \int \sin x \, dx) = -x^2 \cos x + 2x \sin x + 2 \cos x + C$.

(5) (a) $I = \int_1^2 x^2 \ln x \, dx$

   Let $u = \ln x$ so $du/dx = 1/x$ and $dv/dx = x^2$ so $v = \frac{1}{3}x^3$.
   This gives $I = [\frac{1}{3} x^3 \ln x]_1^2 - \int_1^2 \frac{1}{3}x^3 . \ 1/xdx = \frac{8}{3} \ln 2 - \frac{1}{3} [x^3/3]_1^2 = \frac{8}{3} \ln 2 - \frac{7}{9}$

   (b) $I = \int_1^2 1/x \ln x \, dx$

   Let $u = \ln x$ so $du/dx = 1/x$ and $dv/dx = 1/x$ so $v = \ln x$. Doing this gives

   $$I = [(\ln x)^2]_1^2 - \int_1^2 1/x \ln x \, dx = [(\ln x)^2]_1^2 - I \text{ so } I = \frac{1}{2} (\ln 2)^2.$$

   This one can also be done by substitution.
   If we let $\ln x = t$ then $dt/dx = 1/x$ and $I = \int_0^{\ln 2} t \, dt = [\frac{1}{2} t^2]_0^{\ln 2} = \frac{1}{2} (\ln 2)^2$ as before.

   (c) $I = \int_1^2 x^2 (\ln x)^2 \, dx$

   Putting $u = (\ln x)^2$ and $dv/dx = x^2$ gives $du/dx = 2/x \ln x$ and $v = \frac{1}{3} x^3$.
   This then gives

   $$I = [\frac{1}{3} x^3 (\ln x)^2]_1^2 - \int_1^2 \frac{1}{3} x^3 . \ 2/x \ln x \, dx = \frac{8}{3} (\ln 2)^2 - \frac{2}{3} \int_1^2 x^2 \ln x \, dx$$

   $$= \frac{8}{3} (\ln 2)^2 - \frac{2}{3} (\frac{8}{3} \ln 2 - \frac{7}{9}) \text{ from (a), giving } I = \frac{8}{3} (\ln 2)^2 - \frac{16}{9} \ln 2 + \frac{14}{27}.$$

(6) $I = \int_0^1 x^2 \tan^{-1} x \, dx$

Let $u = \tan^{-1} x$ so $du/dx = \dfrac{1}{1 + x^2}$ and $dv/dx = x^2$ so $v = \frac{1}{3} x^3$.

This gives $I = \left[\frac{1}{3} x^3 \tan^{-1} x \right]_0^1 - \int_0^1 \frac{1}{3} x^3 \cdot \dfrac{1}{1 + x^2} \, dx = \frac{1}{3} (\pi/4) - \frac{1}{3} \int_0^1 \dfrac{x^3}{1 + x^2} \, dx$.

This fraction is top-heavy so we cunningly rewrite the top as $x(1 + x^2) - x$. This then gives us

$$I = \pi/12 - \frac{1}{3} \int_0^1 x - \dfrac{x}{1 + x^2} \, dx = \pi/12 - \frac{1}{3} \left[x^2/2 - \frac{1}{2} \ln (1 + x^2) \right]_0^1 = \pi/12 - \frac{1}{6} (1 - \ln 2).$$

(7) $I = \int_0^\pi e^{-x} \sin x \, dx$

If we let $u = e^{-x}$ and $dv/dx = \sin x$, we have $du/dx = -e^{-x}$ and $v = - \cos x$. This then gives us

$$I = \left[e^{-x} (- \cos x) \right]_0^\pi - \int_0^\pi (- \cos x)(-e^{-x}) \, dx = e^{-\pi} + 1 - \int_0^\pi e^{-x} \cos x \, dx.$$

Using integration by parts again on this second integral gives

$$I = e^{-\pi} + 1 - \left(\left[e^{-x} \sin x \right]_0^\pi - \int_0^\pi - e^{-x} \sin x \, dx \right) = e^{-\pi} + 1 - I$$

so $I = \frac{1}{2} (e^{-\pi} + 1)$.

(8) $I = \int e^{-t} \cosh 2t \, dt$

I'll let $u = e^{-t}$ so $du/dt = -e^{-t}$ and $dv/dt = \cosh 2t$ so $v = \frac{1}{2} \sinh 2t$.
This gives $I = \frac{1}{2} e^{-t} \sinh 2t + \frac{1}{2} \int e^{-t} \sinh 2t \, dt$.
Now I integrate $\int e^{-t} \sinh 2t \, dt$ by parts, and I must have $u = e^{-t}$ and $dv/dt = \sinh 2t$ so $du/dt = -e^{-t}$ and $v = \frac{1}{2} \cosh 2t$. This then gives me $I = \frac{1}{2} e^{-t} \sinh 2t + \frac{1}{2} (\frac{1}{2} e^{-t} \cosh 2t + \frac{1}{2} I)$
so $I = \frac{1}{3} e^{-t} (2 \sinh 2t + \cosh 2t) + C$.

(9) $I = \int \sin 4x \sin 3x dx = \frac{1}{2} \int (\cos x - \cos 7x) \, dx = \frac{1}{2} (\sin x - \frac{1}{7} \sin 7x) + C$ using
$\sin A \sin B = \frac{1}{2} [\cos (A - B) - \cos (A + B)]$.

(10) $I = \int \cos 3t \cos 6t \, dt = \int \cos 6t \cos 3t \, dt$

I've swapped the order here to avoid minus signs. I can do this because the multiplication is the same either way round. This then gives me

$$I = \frac{1}{2} \int (\cos 9t + \cos 3t) \, dt = \frac{1}{18} (\sin 9t + 3 \sin 3t) + C$$

using $\cos A \cos B = \frac{1}{2} [\cos (A + B) + \cos (A - B)]$.

Exercise 9.B.9

(1) (a) $\int_0^{\pi/2} \cos^n x \, dx = \int_0^{\pi/2} \cos x \cos^{n-1} x \, dx$

Let $u = \cos^{n-1} x$ so $du/dx = -(n - 1) \cos^{n-2} x \sin x$ and $dv/dx = \cos x$ so $v = \sin x$.
Using integration by parts gives

$$I_n = \left[\cos^{n-1} x \sin x \right]_0^{\pi/2} + \int_0^{\pi/2} (n - 1) \sin^2 x \cos^{n-2} x \, dx$$

so $I_n = 0 + (n - 1) \int_0^{\pi/2} (1 - \cos^2 x) \cos^{n-2} x \, dx = (n - 1)(I_{n-2} - I_n)$

so $n I_n = (n - 1) I_{n-2}$ and $I_n = \dfrac{n - 1}{n} I_{n-2}$.

(b) $I_0 = \int_0^{\pi/2} dx = \left[x \right]_0^{\pi/2} = \pi/2$ and $I_1 = \int_0^{\pi/2} \cos x \, dx = \left[\sin x \right]_0^{\pi/2} = 1$.

$I_2 = \frac{1}{2} I_0 = \pi/4$ and $I_4 = \frac{3}{4} I_2 = 3\pi/16$. $I_3 = \frac{2}{3} I_1 = \frac{2}{3}$.

(c) The answers are the same as Example (1) because the curves for $y = \sin^n x$ and $y = \cos^n x$ from 0 to $\pi/2$ are mirror images of each other in the line $x = \pi/4$. I show this for the $\sin x/\cos x$ and $\sin^2 x/\cos^2 x$ pairs in the graphs below.

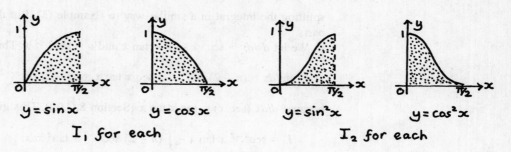

$y = \sin x$ $y = \cos x$ $y = \sin^2 x$ $y = \cos^2 x$

I_1 for each I_2 for each

(2)   (a) $\int_{-\pi}^{\pi} \sin nx \, dx = dx \left[-1/n \cos nx \right]_{-\pi}^{\pi} = -1/n \left[\cos n\pi - \cos(-n\pi) \right] = 0$

because $\cos(-n\pi) = \cos n\pi$ since the cos graph is even.

(b) $\int_{-\pi}^{\pi} \cos nx \, dx = \left[1/n \sin nx \right]_{-\pi}^{\pi} = 1/n \left[\sin nx - \sin (-n\pi) \right] = 1/n \left[0 - 0 \right] = 0$

(c) $\int_{-\pi}^{\pi} \sin^2 nx \, dx = \frac{1}{2} \int_{-\pi}^{\pi} (1 - \cos (2nx)) \, dx = \frac{1}{2} \left[x - \frac{1}{2}n \sin (2nx) \right]_{-\pi}^{\pi}$

$$= \pi/2 + \pi/2 = \pi$$

(See Example (4) in Section 9.B.(c) for the method here.)

(d)   (i) $\int_{-\pi}^{\pi} \sin nx \cos mx \, dx = \frac{1}{2} \int_{-\pi}^{\pi} \sin [(n + m)x] + \sin [(n - m)x] \, dx$

(see Example (8) in Section 9.B.(f) for the method here.)

$$= \frac{1}{2} \left[-\frac{1}{(n + m)} \cos (n + m)x - \frac{1}{(n - m)} \cos (n - m)x \right]_{-\pi}^{\pi} = 0$$

because $\cos (n + m)x = \cos [-(n + m)x]$, and the same for $(n - m)$.

(d)   (ii)   If $n = m$, we have

$$I = \int_{-\pi}^{\pi} \sin nx \cos nx \, dx = \frac{1}{2} \int_{-\pi}^{\pi} \sin 2nx \, dx$$

$$= \frac{1}{2} \left[-\frac{1}{2n} \cos 2nx \right]_{-\pi}^{\pi} = 0.$$

This uses $\sin 2A = 2 \sin A \cos A$ from Section 5.D.(d).

(3)   Let $I_n = \int \cosh x \cosh^{n-1} x \, dx$ because we can easily integrate $\cosh x$.

Now put $u = \cosh^{n-1} x$ so $du/dx = (n - 1) \cosh^{n-2} x \sinh x$, and $dv/dx = \cosh x$ so $v = \sinh x$. This gives

$$I_n = (\cosh^{n-1} x) (\sinh x) - \int (n - 1) \sinh^2 x \cosh^{n-2} x \, dx$$

$$= \cosh^{n-1} x \sinh x - (n - 1) \int \cosh^{n-2} x (\cosh^2 x - 1) \, dx$$

using $\cosh^2 x - \sinh^2 x = 1$ from Section 8.D.(a), so

$$I_n = \cosh^{n-1} x \sinh x - (n - 1) I_n + (n - 1) I_{n-2}$$

and, rearranging, we have $nI_n = \cosh^{n-1} x \sinh x + (n - 1) I_{n-2}$.

(4)   We say

$$I_n = \int \sec^2 x \; \sec^{\,n-2} x \; dx,$$

splitting the integral in a similar way to Example (3). But this time we *do* use integration by parts.

Now we let $dv/dx = \sec^2 x$ so $v = \tan x$ and $u = \sec^{n-2} x$. Then

$$du/dx = (n-2)\sec^{n-3} x \; \sec x \tan x = (n-2)\sec^{n-2} x \tan x$$

because $d/dx \, (\sec x) = \sec x \tan x$ (Section 8.C.(e)). This gives us

$$I_n = \sec^{n-2} x \tan x - \int (n-2)\sec^{n-2} x \tan^2 x dx.$$

Now we use the identity $\tan^2 x + 1 = \sec^2 x$ to get the integral all back in terms of $\sec x$. This gives us

$$I_n = \sec^{n-2} x \tan x - (n-2)\int (\sec^n x - \sec^{n-2} x) \, dx$$

$$= \sec^{n-2} x \tan x - (n-2) I_n + (n-2) I_{n-2}$$

so $(n-1) I_n = \sec^{n-2} x \tan x + (n-2) I_{n-2}.$

Now,

$$I_6 = \int \sec^6 dx \quad \text{and} \quad 5I_6 = \sec^4 x \tan x + 4I_4,$$

$$4I_4 = 4 \times \tfrac{1}{3}(\sec^2 x \tan x + 2I_2) \quad \text{and} \quad I_2 = \int \sec^2 x \; dx = \tan x$$

so, tidied up, we have

$$I_6 = \tfrac{1}{15} \tan x \, (3 \sec^4 x + 4 \sec^2 x + 8).$$

Finally, $\tan(\pi/4) = 1$ and $\tan 0 = 0$ and $\sec(\pi/4) = \sqrt{2}$ and $\sec 0 = 1$, so we have

$$\int_0^{\pi/4} \sec^6 x \; dx = \tfrac{1}{15}(1)(12 + 8 + 8) - 0 = \tfrac{28}{15}.$$

Exercise 9.B.10

In these answers, I shall quote the results we have just found in this section. You might need to show these yourself for exam purposes.

(1)   This becomes $\displaystyle\int_0^1 \left(\dfrac{1}{2 + 3\left(\dfrac{2t}{1+t^2}\right) + 2\left(\dfrac{1-t^2}{1+t^2}\right)} \right) \dfrac{2dt}{1+t^2}$

(using $t = \tan \pi/4 = 1$ if $x = \pi/2$, and $t = 0$ if $x = 0$)

$$= \int_0^1 \frac{2dt}{2(1+t^2) + 6t + 2(1-t^2)} = \int_0^1 \frac{dt}{3t+2} = \left[\tfrac{1}{3} \ln |3t + 2| \right]_0^1$$

$$= \tfrac{1}{3}(\ln 5 - \ln 2) = 0.305 \text{ to 3 d.p.}$$

(2)   This becomes $\displaystyle\int_0^1 \left(\dfrac{1}{1 + \left(\dfrac{1-t^2}{1+t^2}\right)} \right) \dfrac{2 \, dt}{1+t^2} = \int_0^1 dt \text{ (nice!)} = \left[t \right]_0^1 = 1.$

(3)   This becomes $\displaystyle\int_0^{\sqrt{3}}\left(\dfrac{1}{2+\left(\dfrac{1-t^2}{1+t^2}\right)+2\left(\dfrac{2t}{1+t^2}\right)}\right)\dfrac{2\,dt}{1+t^2}$

$$= \int_0^{\sqrt{3}}\dfrac{2\,dt}{t^2+4t+3} = \int_0^{\sqrt{3}}\left(\dfrac{-1}{t+3}\right)+\left(\dfrac{1}{t+1}\right)dt$$

$$= \Big[-\ln|t+3|+\ln|t+1|\Big]_0^{\sqrt{3}} = \ln\,(\sqrt{3}+1)-\ln\,(\sqrt{3}+3)+\ln 3$$

$$= \ln\left(\dfrac{3(1+\sqrt{3})}{3+\sqrt{3}}\right) = \ln\left(\dfrac{3(1+\sqrt{3})\,(3-\sqrt{3})}{(3+\sqrt{3})\,(3-\sqrt{3})}\right) = \ln\,(\sqrt{3}).$$

Exercise 9.C.1

(1) $\displaystyle\int y\,dy = \int \cos x\,dx$ so $\tfrac{1}{2}y^2 = \sin x + C.$

Putting $y = 1$ when $x = \pi/6$ gives $C = 0$ so we have $y^2 = 2\sin x.$

(2) $\displaystyle\int 1/x\,dx = \int 3\,dt$ so $\ln|x| = 3t + C.$

If we let $C = \ln A$, we get $\ln|x/A| = 3t$ so $x/A = e^{3t}$ and $x = Ae^{3t}.$
Putting $x = 1$ when $t = 0$ gives $A = 1$ so $x = e^{3t}.$

(3) $\displaystyle\int 1/y^2\,dy = \int 1/x^2\,dx$ so $-1/y = -1/x + C$ or $1/y + C = 1/x.$

Putting $y = \tfrac{1}{2}$ and $x = 1$ gives $C = -1.$

(4) $\displaystyle\int y^2\,dy = \int x^2\,dx$ so $\tfrac{1}{3}y^3 = \tfrac{1}{3}x^3 + C$ or $y^3 = x^3 + K$ where $K = 3C.$

Putting $y = 2$ when $x = 1$ gives $K = 7.$

(5) $\displaystyle\int\left(\dfrac{2y}{y^2+1}\right)dy = \int\dfrac{dx}{x}$ so $\ln|y^2+1| = \ln|x| + \ln A = \ln|Ax|.$

From this, $y^2 + 1 = Ax.$ Putting $y = 0$ and $x = 1$ gives $A = 1$ so $y^2 = x - 1.$

(6) $\displaystyle\int 1/y\,dy = 2\int\dfrac{dx}{x-5}$ so $\ln|y| = 2\ln|x-5| + \ln A = \ln|A(x-5)^2|$ so $y = A(x-5)^2.$

Putting $y = 2$ when $x = 4$ gives $A = 2.$

(7) $\displaystyle\int \cot y\,dy = \int\dfrac{\cos y}{\sin y}\,dy = \int x\,dx$ so $\ln|\sin y| = \tfrac{1}{2}x^2 + C.$

Putting $y = \pi/2$ when $x = \sqrt{2}$ gives $C = -1.$

(8) $\displaystyle\int\dfrac{d\theta}{\cos^2\theta} = \int \sec^2\theta\,d\theta = \int\dfrac{1}{r}\,dr$ so $\tan\theta = \ln|r| + \ln A = \ln|Ar|.$

Putting $r = 2$ when $\theta = 0$ gives $\ln 2A = 0$ so $2A = 1$ and $A = \tfrac{1}{2}.$

(9)   You need to be a bit crafty here, and say

$$e^{x+y} = e^x\,e^y \quad\text{so}\quad \int e^y\,dy = \int xe^{-x}\,dx.$$

Using integration by parts gives

$$e^y = -xe^{-x} + \int e^{-x}\,dx \quad\text{so}\quad e^y = -xe^{-x} - e^{-x} + C.$$

Putting $y = 0$ when $x = 0$ gives $C = 2.$

(10) $\displaystyle\int\dfrac{dy}{1+y^2} = \int (1+x^2)\,dx$ so $\tan^{-1}y = x + \tfrac{1}{3}x^3 + C$

Putting $y = 0$ when $x = 0$ gives $C = 0.$

Exercise 9.C.2

(1) (a) $-\dfrac{dc}{dt} = kc^2$ so $\displaystyle\int -\dfrac{dc}{c^2} = \int k\,dt$ so $\dfrac{1}{c} = kt + A$

where A is a constant. $c = c_0$ when $t = t_0$ so $A = 1/c_0$.

(b) $\dfrac{dx}{dt} = k(a-x)(b-x)$ so $\displaystyle\int \dfrac{dx}{(a-x)(b-x)} = \int k\,dt.$

$$\dfrac{1}{(a-x)(b-x)} = \dfrac{P}{a-x} + \dfrac{Q}{b-x} \text{ to find partial fractions.}$$

so $1 = P(b-x) + Q(a-x).$

Putting $x = b$ gives $Q = \dfrac{1}{a-b}$ and putting $x = a$ gives $P = \dfrac{1}{b-a} = -Q$

so $\dfrac{1}{(a-x)(b-x)} = \dfrac{1}{a-b}\left(\dfrac{-1}{a-x} + \dfrac{1}{b-x}\right).$

Now we can say that

$$\dfrac{1}{a-b}\int\left(\dfrac{-1}{a-x} + \dfrac{1}{b-x}\right)dx = k\int dt$$

so $\left(\dfrac{1}{a-b}\right)(\ln(a-x) - \ln(b-x)) = kt + C.$

(We know physically that $a > x$ and $b > x$.)
Using the second law of logs gives

$$\left(\dfrac{1}{a-b}\right)\ln\left(\dfrac{a-x}{b-x}\right) = kt + C.$$

$x = 0$ when $t = 0$ so $C = \left(\dfrac{1}{a-b}\right)\ln(a/b)$, so we have

$$kt = \left(\dfrac{1}{a-b}\right)\left(\ln\left(\dfrac{a-x}{b-x}\right) - \ln\left(\dfrac{a}{b}\right)\right) = \left(\dfrac{1}{a-b}\right)\ln\left(\dfrac{b(a-x)}{a(b-x)}\right)$$

which takes some getting through but it's all right if you do it slowly and steadily!

(2) We start with $dv/dt = a$ so $\displaystyle\int dv = \int a\,dt$ so $v = at + C.$

We know that $v = u$ when $t = 0$ so $C = u$ and we now have (a) $v = u + at$.
(The integration here is so straightforward that you don't have to write in the step with the integral signs unless you want to.)

Now, $v = \dfrac{ds}{dt} = u + at$ so $\displaystyle\int ds = \int (u + at)\,dt$ so $s = ut + \tfrac{1}{2}at^2 + K.$

Since $s = 0$ when $t = 0$, we have $K = 0$ and this gives us (b) $s = ut + \tfrac{1}{2}at^2$.

From (a) we have $a = \dfrac{v-u}{t}.$ Substituting this in (b) gives

$$s = ut + \tfrac{1}{2}t^2\left(\dfrac{v-u}{t}\right) = ut + \tfrac{1}{2}vt - \tfrac{1}{2}ut = \left(\dfrac{u+v}{2}\right)t. \tag{c}$$

To get equation (d), you can substitute for t in (c) using (a). This gives

$$s = \left(\dfrac{u+v}{2}\right)\left(\dfrac{v-u}{a}\right) \text{ so } 2as = v^2 - u^2 \text{ or } v^2 = u^2 + 2as.$$

You can also get (d) by using the Chain Rule to say $dv/dt = dv/ds \; ds/dt = v \; dv/ds$. From this,

$$\frac{dv}{dt} = v\frac{dv}{ds} = a \quad \text{so} \quad \int v \; dv = \int a \; ds \quad \text{so} \quad \tfrac{1}{2}v^2 = as + C.$$

$v = u$ when $s = 0$ so $C = \frac{1}{2}u^2$ and we get (d) $v^2 = u^2 + 2as$.

It is quite likely that these four equations of motion will be very familiar to you. They describe the simplest case of what happens if a particle is accelerating, which is when this acceleration is constant.

They are only true if the acceleration *is* constant!
We've now done enough other examples for you to know that, if the amount of acceleration depends, for example, on the length of time you have been travelling, then you will get very different equations for the velocity and distance after a time of t.

One reason why the four equations above are physically useful is that they describe the motion of a body which is falling freely to the ground because of gravity, provided that we think that air resistance will have too slight an effect to matter. Whether this is true will depend on the type of object falling; a feather is designed to behave differently from a stone. (If they were falling in a vacuum, their behaviour would be identical; physics will tell you why this amazing thing is so.) I used the second of these equations of motion in the example of the ball thrown up in the air in Section 2.D.(g). In this particular case, $a = -g$ because we are taking the upwards direction as positive, so the acceleration due to gravity is negative since it is downwards.

If the acceleration is constant, we can see particularly nicely how the gradients and areas relate to each other when we compare the graphs of velocity against time and distance against time.

I show these two graphs below.

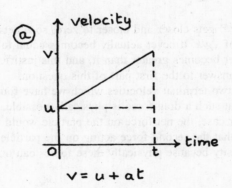
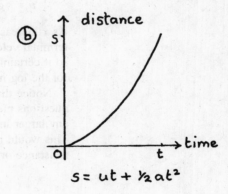

In (a), you can see that the acceleration (which is the rate of change of velocity with time) is given by the gradient of the line. This means that

$$a = \frac{v - u}{t} \quad \text{so} \quad v = u + at.$$

The area under the line in graph (a) is given by adding the rectangle and the triangle, so it is $\quad ut + \frac{1}{2}t(v - u) = ut + \frac{1}{2}at^2 \quad$ which is the same as the distance s.

The gradient of the curve of the distance–time graph shown in (b) gives us the rate of change of distance with time, that is, the velocity v at any time t. We have $s = ut + \frac{1}{2}at^2$ so $ds/dt = u + at = v$.

These two graphs have the same backwards and forwards links between them as the two graphs I used in Section 9.A.(c) to show you how differentiation and integration are the reverse processes of each other.

(3) $\dfrac{dv}{dt} = g - kv$ so $\displaystyle\int \dfrac{dv}{g-kv} = \int dt$ so $-\dfrac{1}{k}\ln|g-kv| = t + C.$

$v = 0$ when $t = 0$ so $C = -\dfrac{1}{k}\ln g$ and we have $\ln g - \ln|g-kv| = kt$

so $\ln\left|\dfrac{g}{g-kv}\right| = kt$ so $\dfrac{g}{g-kv} = e^{kt}$ and $ge^{-kt} = g - kv.$

This gives $v = g/k\,(1 - e^{-kt}).$

Since e^{-kt} gets closer and closer to zero as t increases, v gets closer and closer to what is called its terminal value of g/k.

(4) We have $v\dfrac{dv}{dx} = g - kv^2$ so $\displaystyle\int \dfrac{v\,dv}{g-kv^2} = \int dx$

so $-1/(2k)\ln|g-kv^2| = gx + C.$ Now, $v = 0$ when $x = 0$ so $C = -1/(2k)\ln g$

so $x = \dfrac{1}{(2k)}(\ln g - \ln|g-kv^2|) = \dfrac{1}{(2k)}\ln\left|\dfrac{g}{g-kv^2}\right|.$

We'll show in the second part of the question that g is always greater than kv^2, so we can use ordinary brackets for this log since the expression inside the brackets is always positive.

The second part of this question starts by asking us to write v in terms of x. We have just shown that

$$2kx = \ln\left|\dfrac{g}{g-kv^2}\right| \quad \text{so} \quad e^{2kx} = \dfrac{g}{g-kv^2}$$

so $g - kv^2 = ge^{-2kx}$ and $v^2 = g/k\,(1 - e^{-2kx}).$

This means that

$$v = \sqrt{g/k\,(1 - e^{-2kx})}.$$

As x increases, e^{-2kx} gets closer and closer to zero, so v gets closer and closer to the terminal velocity of $\sqrt{g/k}$. It never actually becomes equal to it (this is the limiting value), so it certainly never becomes greater than it, and this justifies the use of ordinary brackets for the log in the answer to the first part of this question.

Notice that the two terminal velocities which we have found for v in the last two questions mean that such a drag force is physically possible. If v had been able to become any larger in either case, the net force on the particle would be pushing it back upwards. This would mean that the second force acting on the particle couldn't be due to air resistance or viscosity because physically these forces can only slow down motion.

Chapter 10
Exercise 10.A.1

(1) $z^2 - 2z + 2 = 0$ so $z = \dfrac{2 \pm \sqrt{4-8}}{2} = \dfrac{2 \pm 2\sqrt{-1}}{2} = 1 \pm j$

(2) $z^2 - 4z + 13 = 0$ so $z = \dfrac{4 \pm \sqrt{16-52}}{2} = \dfrac{4 \pm 6\sqrt{-1}}{2} = 2 \pm 3j.$

(3) $z^2 + 4z + 5 = 0$ so $z = \dfrac{-4 \pm \sqrt{16-20}}{2} = \dfrac{-4 \pm 2\sqrt{-1}}{2} = -2 \pm j.$

(4) $z^2 + 2z + 6 = 0$ so $z = \dfrac{-2 \pm \sqrt{4-24}}{2} = \dfrac{-2 \pm 2\sqrt{-5}}{2} = -1 \pm \sqrt{5}j.$

Exercise 10.A.2

(1) (a) $|3| = 3$ and $\arg 3 = 0$ (b) $|-2| = 2$ and $\arg(-2) = \pi$

(c) $|2j| = 2$ and $\arg(2j) = \pi/2$ (d) $|-5j| = 5$ and $\arg(-5j) = -\pi/2$

(e) $|5 + 12j| = 13$ and $\arg(5 + 12j) = \tan^{-1}\left(\frac{12}{5}\right) = 1.18$ radians

(f) $|-5 - 12j| = 13$ and $\arg(-5 - 12j) = -\pi + \tan^{-1}\left(\frac{12}{5}\right) = -1.97$ radians

(g) $|7 - 24j| = 25$ and $\arg(7 - 24j) = \tan^{-1}\left(-\frac{24}{7}\right) = -1.29$ radians

(h) $|-7 + 24j| = 25$ and $\arg(-7 + 24j) = \pi - \tan^{-1}\left(\frac{24}{7}\right) = 1.85$ radians

(i) $|-\sqrt{3} + j| = 2$ and $\arg(-\sqrt{3} + j) = \pi - \tan^{-1}\left(\frac{1}{\sqrt{3}}\right) = 5\pi/6$ radians

(j) $|-1 - j| = \sqrt{2}$ and $\arg(-1 - j) = -\pi + \tan^{-1}(1) = -3\pi/4$ radians

(2) All four roots have a modulus of 1.

arg root (a) $= 0$, arg root (b) $= \pi/2$, arg root (c) $= \pi$, and arg root (d) $= -\pi/2$.

(3) For equation (1), $z_1 = 1 + j$ and $z_2 = 1 - j$, so $|z_1| = |z_2| = \sqrt{2}$.

$\arg z_1 = \tan^{-1}(1) = \pi/4$ radians and $\arg z_2 = -\pi/4$ radians, by symmetry.

For equation (2), $z_1 = 2 + 3j$ and $z_2 = 2 - 3j$ so $|z_1| = |z_2| = \sqrt{13}$.

$\arg z_1 = \tan^{-1}\left(\frac{3}{2}\right) = +0.98$ radians, so $\arg z_2 = -0.98$ radians by symmetry.

For equation (3), $z_1 = -2 + j$ and $z_2 = -2 - j$ so $|z_1| = |z_2| = \sqrt{5}$.

$\arg z_1 = \pi - \tan^{-1}\left(\frac{1}{2}\right) = 2.68$ radians, so $\arg z_2 = -2.68$ radians by symmetry.

For equation (4), $z_1 = -1 + \sqrt{5}j$ and $z_2 = -1 - \sqrt{5}j$ so $|z_1| = |z_2| = \sqrt{6}$.

$\arg z_1 = \pi - \tan^{-1}\sqrt{5} = 1.99$ radians, so $\arg z_2 = -1.99$ radians by symmetry.

Exercise 10.B.1

(1) $1 + 7j$ (2) $-6 + 5j$ (3) $5 + 4j$ (4) $-10 - 7j$

Exercise 10.B.2

(1) (a) $9 + 6j$ (b) $7j$ (c) 1 (d) $6 + 23j$

(2) (a) You should have (i) $w = 3 + 4j$ and $|w| = 5$, (ii) $w = 15 + 8j$ and $|w| = 17$ and (iii) $w = 7 + 24j$ and $|w| = 25$.

These gives the three right-angled triangles with sides of lengths 3, 4 and 5 units, 8, 15 and 17 units, and 7, 24 and 25 units respectively.

(b) $w = z^2 = (x + jy)^2 = x^2 - y^2 + 2xyj$ so $u = x^2 - y^2$ and $v = 2xy$.

If x and y are whole numbers, then $|w|$, u and v must also be whole numbers. (In case (iii), $|z|$ is also a whole number.)

(c) You should find that drawing the Argand diagram for w gives you a triangle with sides of 119, 120 and 169 units, which is almost isosceles.

Drawing the Argand diagram for W gives you a remarkably long thin triangle with sides of 239, 28 560 and 28 561 units.

Both triangles have sides given by whole numbers and, of course, fit Pythagoras' Theorem.

(3) You will know if you have the right answers for this question since you are just checking what you have already done.

I will go through one as an example in case you have any problems.

The roots of $z^2 + 2z + 6 = 0$ from question (4) of Exercise 10.A.1 are $-1 \pm \sqrt{5}j$.
Putting $z = -1 + \sqrt{5}j$ back into the equation gives

$$(-1 + \sqrt{5}j)^2 + 2(-1 + \sqrt{5}j) + 6 = 1 - 2\sqrt{5}j + 5j^2 - 2 + 2\sqrt{5}j + 6 = 1 - 5 - 2 + 6 = 0.$$

The other root checks similarly.
For this equation, $a = 1$, $b = 2$ and $c = 6$.
The sum of the roots is

$$(-1 + \sqrt{5}j) + (-1 - \sqrt{5}j) = -2 = -b/a.$$

The product of the roots is

$$(-1 + \sqrt{5}j)(-1 - \sqrt{5}j) = 1 - \sqrt{5}j + \sqrt{5}j - 5j^2 = 6 = c/a.$$

(4)   The three roots are symmetrically placed so they will look like the ones in my drawing below.

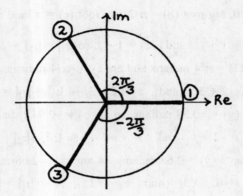

Each root has a modulus of one and the arguments are 0, $2\pi/3$ and $-2\pi/3$. From this,

root (1) = 1,

$$\text{root (2)} = 1\left(\cos\left(\frac{2\pi}{3}\right) + j\sin\left(\frac{2\pi}{3}\right)\right) = -\frac{1}{2} + \frac{\sqrt{3}}{2}j,$$

$$\text{root (3)} = 1\left(\cos\left(-\frac{2\pi}{3}\right) + j\sin\left(-\frac{2\pi}{3}\right)\right) = -\frac{1}{2} - \frac{\sqrt{3}}{2}j.$$

(These results use the sin and cos of $\pi/3$ from Section 4.A.(g) and the symmetry of the figure.)
If root (2) = ω, then $|\omega \times \omega| = 1$ since $|\omega| = 1$, and $\arg(\omega \times \omega) = 2\arg\omega = 4\pi/3$.
The principal value of this argument is $-2\pi/3$, so we see that $\omega \times \omega = \omega^2$ or multiplying root (2) by itself gives root (3).
Equally, we could say that

$$\omega \times \omega = \left(-\frac{1}{2} + \frac{\sqrt{3}}{2}j\right)\left(-\frac{1}{2} + \frac{\sqrt{3}}{2}j\right) = \frac{1}{4} - \frac{\sqrt{3}}{4}j - \frac{\sqrt{3}}{4}j + \frac{3}{4}j^2 = -\frac{1}{2} - \frac{\sqrt{3}}{2}j = \omega^2.$$

Similarly, we can show that $\omega^2 \times \omega^2 = \omega$, using either of the above methods.

The thinking point at the end of Exercise 10.B.2 is answered in Section 10.D.(a).

Exercise 10.B.3

The conjugates are all shown drawn in on this diagram. They are all reflections of each other in the real axis. Notice that this means that z_5 is its own conjugate.

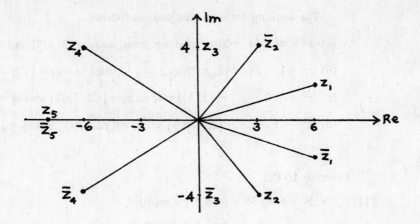

Exercise 10.B.4

The working for the first four questions is exactly the same as in the example given just before the exercise. For each one, you just multiply the top and the bottom by the conjugate of the bottom and then tidy up.

The answers you should have are as follows.

(1) $\frac{1}{5}(6 - 3j)$ (2) $\frac{1}{13}(12 + 8j)$ (3) j (4) $-\frac{1}{34}(1 + 13j)$

Notice that the answer to question (3) is the same as the result which we got by using the mod/arg form in Section 10.B.(c).

(5) Multiplying out the bottom gives $\dfrac{2 - j}{11 - 3j}$. Simplifying this gives $\frac{1}{26}(5 - j)$.

(6) The first stage gives $\dfrac{3(2 + j)}{5} + \dfrac{2(3 - 2j)}{13}$.

Adding these gives $\dfrac{39(2 + j) + 10\,(3 - 2j)}{65} = \dfrac{108 + 19j}{65}$.

(See Section 1.C.(c) if necessary.)

Exercise 10.C.1

(1) The working for each of these is the same as in the example which I have already given, so I shall just give you the three pairs of answers here.

    (a) $2 + j$ and $-2 - j$ (b) $4 + j$ and $-4 - j$ (c) $-3 + 2j$ and $3 - 2j$.

    Each pair of roots forms a straight line on its Argand diagram.

(2) The four Argand diagrams are shown below.

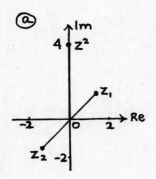
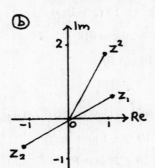
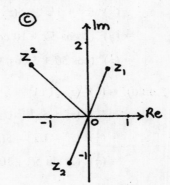
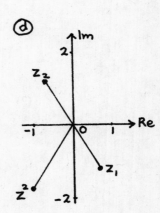

The working for each part goes as follows.

(a) $z^2 = 4j = [4, \pi/2]$ in mod/arg form, so $z_1 = [2, \pi/4]$ and $z_2 = [2, -3\pi/4]$.

(b) $z^2 = 1 + \sqrt{3}j = [2, \pi/3]$ so $z_1 = [\sqrt{2}, \pi/6]$ and $z_2 = [\sqrt{2}, -5\pi/6]$.

(c) $z^2 = -\sqrt{2} + \sqrt{2}j = [2, +3\pi/4]$ so $z_1 = [\sqrt{2}, 3\pi/8]$ and $z_2 = [\sqrt{2}, -5\pi/8]$.

(d) $z^2 = -1 - \sqrt{3}j = [2, -2\pi/3]$ so $z_1 = [\sqrt{2}, -\pi/3]$ and $z_2 = [\sqrt{2}, 2\pi/3]$.

Exercise 10.C.2

(1) $\cos 3\theta + j \sin 3\theta = (\cos\theta + j\sin\theta)^3$
$$= \cos^3\theta + 3\cos^2\theta(j\sin\theta) + 3\cos\theta(j\sin\theta)^2 + (j\sin\theta)^3$$

(a) Equating the real parts gives
$$\cos 3\theta = \cos^3\theta - 3\cos\theta\sin^2\theta = \cos^3\theta - 3\cos\theta(1-\cos^2\theta) = 4\cos^3\theta - 3\cos\theta.$$

(b) Equating the imaginary parts gives
$$\sin 3\theta = 3\cos^2\theta\sin\theta - \sin^3\theta = 3\sin\theta - 4\sin^3\theta.$$

(2) $\cos 7\theta + j\sin 7\theta = (\cos\theta + j\sin\theta)^7$
$$= \cos^7\theta + 7\cos^6\theta(j\sin\theta) + 21\cos^5\theta(j\sin\theta)^2 + 35\cos^4\theta(j\sin\theta)^3$$
$$+ 35\cos^3\theta(j\sin\theta)^4 + 21\cos^2\theta(j\sin\theta)^5 + 7\cos\theta(j\sin\theta)^6$$
$$+ (j\sin\theta)^7$$

Equating the real parts gives
$$\cos 7\theta = \cos^7\theta - 21\cos^5\theta\sin^2\theta + 35\cos^3\theta\sin^4\theta - 7\cos\theta\sin^6\theta$$
$$= \cos^7\theta - 21\cos^5\theta(1-\cos^2\theta) + 35\cos^3\theta(1-\cos^2\theta)^2 - 7\cos\theta(1-\cos^2\theta)^3$$
$$= \cos^7\theta - 21\cos^5\theta + 21\cos^7\theta + 35\cos^3\theta - 70\cos^5\theta + 35\cos^7\theta$$
$$- 7\cos\theta(1 - 3\cos^2\theta + 3\cos^4\theta - \cos^6\theta)$$
$$= 64\cos^7\theta - 112\cos^5\theta + 56\cos^3\theta - 7\cos\theta.$$

Exercise 10.C.3

(1) $\cos^5\theta = \left(\tfrac{1}{2}\right)^5 (z + 1/z)^5$
$$= \left(\tfrac{1}{2}\right)^5 (z^5 + 5z^4(1/z) + 10z^3(1/z)^2 + 10z^2(1/z)^3 + 5z(1/z)^4 + (1/z)^5)$$
$$= \left(\tfrac{1}{2}\right)^5 (z^5 + 1/z^5 + 5(z^3 + 1/z^3) + 10(z + 1/z))$$
$$= \left(\tfrac{1}{2}\right)^5 (2\cos 5\theta + 10\cos 3\theta + 20\cos\theta)$$
$$= \left(\tfrac{1}{2}\right)^4 (\cos 5\theta + 5\cos 3\theta + 10\cos\theta)$$

(2) $(j\sin\theta)^5 = \left(\tfrac{1}{2}\right)^5 (z - 1/z)^5$
$$= \left(\tfrac{1}{2}\right)^5 (z^5 + 5z^4(-1/z) + 10z^3(-1/z)^2 + 10z^2(-1/z)^3 + 5z(-1/z)^4 + (-1/z)^5)$$
$$= \left(\tfrac{1}{2}\right)^5 (z^5 - 1/z^5 - 5(z^3 - 1/z^3) + 10(z - 1/z))$$
$$= \left(\tfrac{1}{2}\right)^5 (2j\sin 5\theta - 10j\sin 3\theta + 20j\sin\theta)$$

so $\sin^5\theta = \left(\tfrac{1}{2}\right)^4 (\sin 5\theta - 5\sin 3\theta + 10\sin\theta)$ since $j^5 = j$.

Exercise 10.D.1

I have used the general rule for finding roots for these answers, but your answers will be equally good if you have used the geometry of your sketches. In either case, these sketches make a check on your arithmetic because you can quickly get a good idea of whether your calculated roots fit with what you have drawn.

(1) Let $w = z^3 = 27j$ so $|w| = 27$ and $\arg w = \pi/2 + 2k\pi$. Then

$$z = 27^{1/3} \exp\left[j\,\frac{(\pi/2 + 2k\pi)}{3}\right] = 3 \exp\left[j\left(\frac{\pi + 4k\pi}{6}\right)\right]$$

giving the three distinct solutions of $3e^{j(\pi/6)}$, $3e^{j(5\pi/6)}$ and $3e^{-j(\pi/2)}$.

(2) Let $w = z^4 = 1 + j$ so $|w| = \sqrt{2}$ and $\arg w = \pi/4$. This gives us

$$z = (\sqrt{2})^{1/4} \exp\left[j\,\frac{(\pi/4 + 2k\pi)}{4}\right] = 2^{1/8} \exp\left[j\left(\frac{\pi + 8k\pi}{16}\right)\right]$$

giving the four solutions $2^{1/8} \exp[j(\pi/16)]$, $2^{1/8} \exp[j(9\pi/16)]$, $2^{1/8} \exp[j(-7\pi/16)]$ and $2^{1/8} \exp[j(-15\pi/16)]$.

(3) Let $w = z^3 = \frac{1}{5}(3 - 4j)$ so $|w| = \frac{5}{5} = 1$ and $\arg w = \tan^{-1}\left(-\frac{4}{3}\right) = -0.927$ radians to 3 d.p.

So $|z| = 1$ and $z = \exp\left[j\,\frac{(-0.927 + 2k\pi)}{3}\right]$

giving $\exp(-0.309j)$, $\exp(1.785j)$ and $\exp(-2.403j)$.
    The three roots in $a + bj$ form to 2 d.p. are

$$\cos(-0.309) + j\sin(-0.309) = 0.95 - 0.30j,$$

$$\cos(1.785) + j\sin(1.785) = -0.21 + 0.98j,$$

$$\cos(-2.403) + j\sin(-2.403) = -0.74 - 0.67j.$$

(4) Let $w = z^5 = -4 + 4j$ so $|w| = \sqrt{32} = \sqrt{2^5}$ and $\arg w = 3\pi/4$, giving us $|z| = (\sqrt{2^5})^{1/5} = \sqrt{2}$, so that

$$z = \sqrt{2} \exp\left[j\,\frac{(3\pi/4 + 2k\pi)}{5}\right] = \sqrt{2} \exp\left[j\left(\frac{3\pi + 8k\pi}{20}\right)\right]$$

giving $\sqrt{2} \exp[j(3\pi/20)]$, $\sqrt{2} \exp[j(11\pi/20)]$, $\sqrt{2} \exp[j(19\pi/20)]$, $\sqrt{2} \exp[j(-5\pi/20)]$ and $\sqrt{2} \exp[j(-13\pi/20)]$.

(5) Let $w = z^6 = -4 - 4\sqrt{3}j$ so $|w| = \sqrt{16 + 48} = 8$ and $\arg w = -2\pi/3$ so $|z| = 8^{1/6} = 2^{1/2}$, giving

$$z = \sqrt{2} \exp\left[j\,\frac{(-2\pi/3 + 2k\pi)}{6}\right] = \sqrt{2} \exp\left[j\left(\frac{-\pi + 3k\pi}{9}\right)\right]$$

with $k = 0, -1, -2, 1, 2$ and 3 to give the six distinct answers.

(6) Let $\left(\dfrac{z - 1}{z + 1}\right)^4 = w.$

Now we know from the beginning of this section that $w^4 = -1$ has the four roots of $e^{j\pi/4}$, $e^{j(3\pi/4)}$, $e^{j(-\pi/4)}$ and $e^{j(-3\pi/4)}$.
    $w^4 = -4$ will have the same four roots except that each one will be multiplied by $\sqrt{2}$.
    Writing the first of these in $a + bj$ form gives $\sqrt{2}(\cos(\pi/4) + j\sin(\pi/4)) = 1 + j$.
    Similarly, the other three roots are $-1 + j$, $1 - j$ and $-1 - j$.
    Now, using the first root, we have

$$\frac{z - 1}{z + 1} = 1 + j \text{ so } z - 1 = (z + 1)(1 + j) = z + zj + 1 + j \text{ so } z = 1/j\,(-2 - j) = 2j - 1$$

multiplying top and bottom by $-j$.

The second root gives

$$z - 1 = (z + 1)(-1 + j) = -z + zj - 1 + j \quad \text{so} \quad z(2 - j) = j$$

giving

$$z = \frac{j}{2-j} = \left(\frac{j}{2-j}\right)\left(\frac{2+j}{2+j}\right) = \tfrac{1}{5}(2j - 1).$$

In a similar way, the other two roots come out as $-1 -2j$ and $\tfrac{1}{5}(-2j - 1)$. You could save the extra working here by noticing that, if you had started by multiplying out, you would have got an equation for z with real coefficients. This means that its roots must come in conjugate pairs, which you will see these roots do indeed do.

If you are at all doubtful about any of your sketches, use my solutions to draw yourself a new one so that you can check your own version.

Exercise 10.D.2

(1) $a = 1$, $b = -(3 - 2j)$ and $c = -(1 + 3j)$ so

$$z = \frac{(3 - 2j) \pm \sqrt{(3 - 2j)^2 + 4(1 + 3j)}}{2} = \frac{(3 - 2j) \pm \sqrt{5 - 12j + 4 + 12j}}{2}$$

$$= \tfrac{1}{2}(3 - 2j \pm 3) = 3 - j \quad \text{or} \quad -j.$$

(2) $a = 1$, $b = 3 - 4j$ and $c = -(22 + 6j)$ so

$$z = \frac{-3 + 4j \pm \sqrt{(3 - 4j)^2 + 4(22 + 6j)}}{2} = \frac{-3 + 4j \pm \sqrt{-7 - 24j + 88 + 24j}}{2}$$

$$= \tfrac{1}{2}(-3 + 4j \pm 9) = 3 + 2j \quad \text{or} \quad -6 + 2j.$$

(3) $a = 1$, $b = 2 + j$ and $c = -(3 + j)$ so

$$z = \frac{-(2 + j) \pm \sqrt{(2 + j)^2 + 4(3 + j)}}{2} = \frac{-(2 + j) \pm \sqrt{3 + 4j + 12 + 4j}}{2}.$$

From (1)(b) from Exercise 10.C.1 we know that the square roots of $15 + 8j$ are $4 + j$ and $-4 - j$ so we have

$$z = \frac{-(2 + j) \pm (4 + j)}{2} = 1 \quad \text{or} \quad -3 - j.$$

(4) $a = 1$, $b = 1$ and $c = -(\tfrac{1}{2} + j)$ so

$$z = \frac{-1 \pm \sqrt{1 + 4(\tfrac{1}{2} + j)}}{2} = \frac{-1 \pm \sqrt{3 + 4j}}{2}.$$

We know that the square roots of $3 + 4j$ are $\pm(2 + j)$ from (1)(a) from Exercise 10.C.1 so

$$z = \frac{-1 \pm (2 + j)}{2} = \tfrac{1}{2} + \tfrac{1}{2}j \text{ or } -\tfrac{3}{2} - \tfrac{1}{2}j.$$

(5) $a = 1$, $b = -1$ and $c = -1 + 3j$ so

$$z = \frac{1 \pm \sqrt{1 + 4 - 12j}}{2} = \frac{1 \pm (3 - 2j)}{2} = 2 - j \quad \text{or} \quad -1 + j$$

using (1)(c) from Exercise 10.C.1.

(6) $a = 1$, $b = -2(1 + j)$ and $c = 5(1 - 2j)$ so

$$z = \frac{2(1 + j) \pm \sqrt{4(1 + j)^2 - 20(1 - 2j)}}{2} = \frac{2(1 + j) \pm \sqrt{-20 + 48j}}{2}$$

$$= \frac{2(1 + j) \pm 2\sqrt{-5 + 12j}}{2} = (1 + j) \pm \sqrt{-5 + 12j}.$$

Using the methods of Section 10.C.(a), the square roots of $-5 + 12j$ are $\pm(2 + 3j)$, or you could also get these from multiplying the two roots of $5 - 12j$ from the previous answer by j. This gives

$$z = (1 + j) \pm (2 + 3j) = 3 + 4j \quad \text{or} \quad -1 - 2j.$$

(7)   If we put $w = z^2$, we have $w^2 - (1 + j)w + j = 0$. Using the formula gives

$$w = \frac{1 + j \pm \sqrt{(1 + j)^2 - 4j}}{2} = \frac{1 + j \pm \sqrt{-2j}}{2}.$$

The square roots of $-2j$ are $1 - j$ and $-1 + j$, either from using the methods of Section 10.C.(a) or drawing a little picture as I have below.

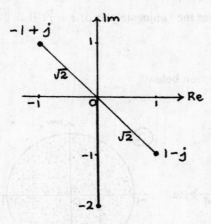

This gives $w = \frac{1}{2}(1 + j \pm (1 - j)) = 1$ or j so $z^2 = 1$ or j so $z = \pm 1$ or $\pm j$.

Exercise 10.D.3

(1) $z = 1 + 2j$ is also a root of $f(z) = z^4 - 3z^3 + az^2 + bz - 30 = 0$.
Therefore $(z - (1 - 2j))\, (z - (1 + 2j))$ is a factor of $f(z)$.
Multiplied together, this gives $(z^2 - 2z + 5)$ is a factor, so now we have

$$z^4 - 3z^3 + az^2 + bz - 30 = (z^2 - 2z + 5)\,(z^2 + pz - 6).$$

Matching the terms in z^3 gives us $-3z^3 = pz^3 - 2z^3$ so $p = -1$. Therefore

$$z^4 - 3z^3 + az^2 + bz - 30 = (z^2 - 2z + 5)(z^2 - z - 6).$$

Matching the terms in z^2 gives us $az^2 = 5z^2 - 6z^2 + 2z^2$ so $a = 1$.
Matching the terms in z gives us $bz = 12z - 5z$ so $b = 7$.
The other two roots come from $z^2 - z - 6 = (z - 3)(z + 2) = 0$ so they are $z = 3$ and $z = -2$.

(2) $-j$ is also a root so $(z - j)\, (z - (-j))$ is a factor of the equation.
Multiplying these gives us

$$z^4 + az^3 + bz^2 - 4z + 13 = (z^2 + 1)\,(z^2 + pz + 13).$$

Matching terms in z gives $-4z = pz$ so $p = -4$.
We now have

$$z^4 + az^3 + bz^2 - 4z + 13 = (z^2 + 1)\,(z^2 - 4z + 13).$$

Matching terms in z^3 gives $az^3 = -4z^3$ so $a = -4$.
Matching terms in z^2 gives $bz^2 = z^2 + 13z^2$ so $b = 14$.
The other two roots come from the solution of $z^2 - 4z + 13 = 0$.
Using the formula (see Section 10.D.(b) if necessary) gives us the conjugate pair of $z = 2 \pm 3j$.

(3) $z = 1 + j$ is also a root of the equation $f(z) = 0$ so $(z - (1 - j))\,(z - (1 + j)) = z^2 - 2z + 2$ is a factor of $f(z)$.

This gives us

$$z^4 + az^3 + bz^2 + 1 = (z^2 - 2z + 2)(z^2 + pz + \tfrac{1}{2}).$$

Matching terms in z gives $0 = -z + 2pz$ so $p = \tfrac{1}{2}$.
We now have

$$z^4 + az^3 + bz^2 + 1 = (z^2 - 2z + 2)\,(z^2 + \tfrac{1}{2}z + \tfrac{1}{2}).$$

Matching the terms in z^3 gives $az^3 = -2z^3 + \tfrac{1}{2}z^3$ so $a = -\tfrac{3}{2}$.

Matching the terms in z^2 gives $bz^2 = 2z^2 + \tfrac{1}{2}z^2 - z^2$ so $b = \tfrac{3}{2}$.

For the remaining two roots, we solve $z^2 + \tfrac{1}{2}z + \tfrac{1}{2} = 0$ or $2z^2 + z + 1 = 0$.

Using the formula gives the conjugate pair of $z = \tfrac{1}{4}\,(-1 \pm \sqrt{7}j)$.

Exercise 10.E.1

I show the sketch for each question below.

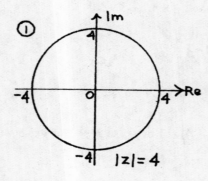

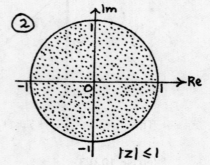

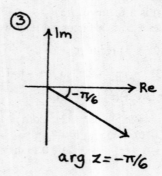

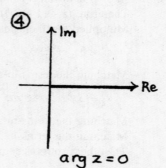

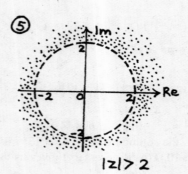

I show the sketches for each of the answers below.

(5) is the impossible question. You can't have a length of −2 units!

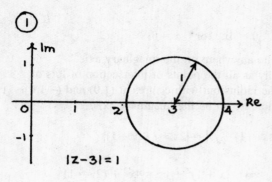

① $|z-3|=1$

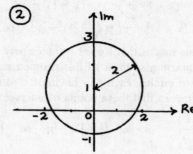

② $|z-j|=2$

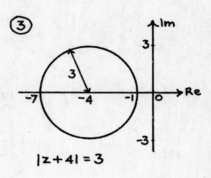

③ $|z+4|=3$

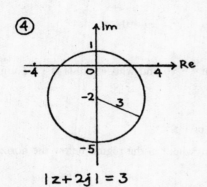

④ $|z+2j|=3$

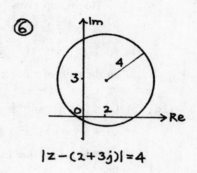

⑥ $|z-(2+3j)|=4$

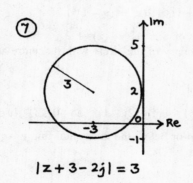

⑦ $|z+3-2j|=3$

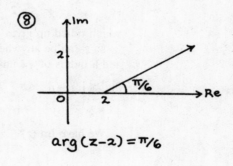

⑧ $\arg(z-2)=\pi/6$

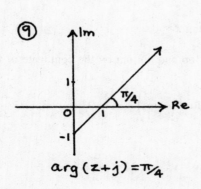

⑨ $\arg(z+j)=\pi/4$

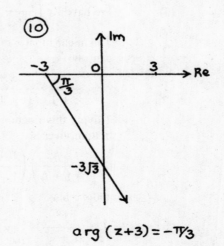

⑩ $\arg(z+3)=-\pi/3$

Exercise 10.E.3

(1) (a) Putting $z = x + jy$ gives

$$|(x - 1) + jy| = |(x + 1) + jy|$$

so $\quad \sqrt{(x-1)^2 + y^2} = \sqrt{(x+1)^2 + y^2}$

so $\quad (x - 1)^2 + y^2 = (x + 1)^2 + y^2 \quad$ so $\quad 0 = 4x \quad$ or $\quad x = 0.$

This means that Re $(z) = 0$ so z may lie anywhere on the imaginary axis.

You can also think of this geometrically as all the points of intersection of lots of pairs of circles, each pair having the same radius but with centres at $(1,0)$ and $(-1,0)$. This means that these points of intersection are all on the imaginary axis.

(b) $\left| \dfrac{z + 1}{2z - j} \right| = 1$ so $|z + 1| = |2z - j| \quad$ so $\quad |(x + 1) + jy| = |2x + j(2y - 1)|$

so $\quad \sqrt{(x + 1)^2 + y^2} = \sqrt{4x^2 + (2y - 1)^2} \quad$ so $\quad (x + 1)^2 + y^2 = 4x^2 + (2y - 1)^2.$

Tidying up gives

$$3x^2 - 2x + 3y^2 - 4y = 0 \quad \text{so} \quad x^2 - \tfrac{2}{3}x + y^2 - \tfrac{4}{3}y = 0$$

so $\quad (x - \tfrac{1}{3})^2 + (y - \tfrac{2}{3})^2 = \tfrac{5}{9}$

so the path of z is a circle whose centre is at $\tfrac{1}{3} + \tfrac{2}{3}j$ and whose radius is $\sqrt{5}/3$ units.

(2) (a) $|z| \leq |z - j| \quad$ so $\quad |x + jy| \leq |x + j(y - 1)|$

so $\quad x^2 + y^2 \leq x^2 + (y - 1)^2 \quad$ so $\quad 2y \leq 1$ or $y \leq \tfrac{1}{2}.$

This means that Im $(z) \leq \tfrac{1}{2}$ so z can lie anywhere in the region below the horizontal line $y = \tfrac{1}{2}.$

(b) $|z| \leq \sqrt{2}|z - j| \quad$ so $\quad |x + jy| \leq \sqrt{2}|x + j(y - 1)|$

so $\quad \sqrt{x^2 + y^2} \leq \sqrt{2}\sqrt{x^2 + (y - 1)^2} \quad$ so $\quad x^2 + y^2 \leq 2(x^2 + (y - 1)^2)$

which tidied up gives $\quad 0 \leq x^2 + y^2 - 4y + 2 \quad$ or $\quad x^2 + (y - 2)^2 \geq 2$

so z can be anywhere outside or on the boundary of the circle with its centre at $2j$ and a radius of $\sqrt{2}$ units.

(c) $2|z| \leq |2z - j| \quad$ so $\quad 2|x + jy| \leq |2x + j(2y - 1)|$

so $\quad 2\sqrt{x^2 + y^2} \leq \sqrt{4x^2 + (2y - 1)^2} \quad$ so $\quad 4(x^2 + y^2) \leq 4x^2 + (2y - 1)^2 \quad$ so $\quad 0 \leq 1 - 4y$ or $y \leq \tfrac{1}{4}.$

We have Im $(z) \leq \tfrac{1}{4}$ so z can lie anywhere on or below the horizontal line $y = \tfrac{1}{4}.$

(3) We have Re $\left(\dfrac{z}{z + 1} \right) = \tfrac{3}{2} + jz = \tfrac{3}{2} + jx - y.$

This means that x must be equal to zero since the LHS of this equation is real. So we can say

$$\frac{z}{z + 1} = \frac{x + jy}{(x + 1) + jy} = \frac{jy}{1 + jy} \quad \text{putting } x = 0.$$

Tidying this fraction up by multiplying the top and bottom by the conjugate of the bottom gives

$$\left(\frac{jy}{1 + jy} \right) \left(\frac{1 - jy}{1 - jy} \right) = \frac{jy + y^2}{1 + y^2} \quad \text{so} \quad \text{Re} \left(\frac{z}{z + 1} \right) = \frac{y^2}{1 + y^2}$$

and we have

$$\frac{y^2}{1 + y^2} = \tfrac{3}{2} - y \quad \text{so} \quad y^2 = \tfrac{3}{2} + \tfrac{3}{2}y^2 - y - y^3$$

so $2y^3 - y^2 + 2y - 3 = 0$.

$y = 1$ is a solution, so we get $(y - 1)(2y^2 + y + 3) = 0$.

Now $2y^2 + y + 3 = 0$ has no real roots. Therefore, since y is real, there is just the one solution to the given equation of $z = j$.

Exercise 10.E.4

(1)   We can't have $z = 1$ as this would give a zero underneath the fraction.

(2)   (a) If $z = 2$ then $w = 2 + j$.

(b) If $z = 1 - j$ then $w = \dfrac{1}{-j} = j$ multiplying top and bottom by j.

(3) $w = u + jv = \dfrac{z + j}{z - 1} = \dfrac{x + j(y + 1)}{(x - 1) + jy}$

Simplifying the fraction by multiplying it top and bottom by $(x - 1) - jy$ gives

$$\left(\frac{x + j(y + 1)}{(x - 1) + jy}\right)\left(\frac{(x - 1) - jy}{(x - 1) - jy}\right) = \frac{x^2 - x + y^2 + y + j(x - y - 1)}{(x - 1)^2 + y^2}$$

so $u = \dfrac{x^2 - x + y^2 + y}{(x - 1)^2 + y^2}$ and $v = \dfrac{x - y - 1}{(x - 1)^2 + y^2}$.

(4)   If $u = v$ it must be true that $x^2 - x + y^2 + y = x - y - 1$.

Therefore the path of z is given by the equation $x^2 - 2x + y^2 + 2y + 1 = 0$.

This is the same as $(x - 1)^2 + (y + 1)^2 = 1$ so, if $u = v$ then z must lie on the circle whose centre is at $1 - j$ and whose radius is 1.

(5)   I show the two Argand diagrams below. Again the little circle at $z = 1$ is to show that we have to exclude this point.

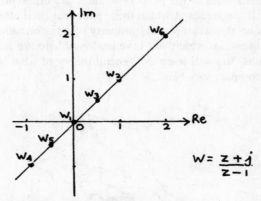

Argand diagram for $w = u + jv$
with $u = v$

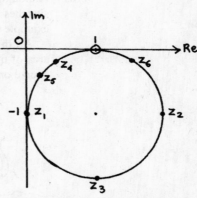

$w = \dfrac{z + j}{z - 1}$

Argand diagram showing
the corresponding path for z

(6)   (a) If $z_1 = -j$ then $w_1 = 0$.

(b) If $z_2 = 2 - j$ then $w_2 = \dfrac{2}{1 - j} = \dfrac{2(1 + j)}{(1 - j)(1 + j)} = 1 + j$.

(c) If $z_3 = 1 - 2j$ then $w_3 = \dfrac{1 - j}{-2j} = \dfrac{2j + 2}{4} = \tfrac{1}{2} + \tfrac{1}{2}j$.

(d) If $z_4 = \tfrac{2}{5} - \tfrac{1}{5}j$ then

$$w_4 = \frac{\tfrac{2}{5} + \tfrac{4}{5}j}{-\tfrac{3}{5} - \tfrac{1}{5}j} = \frac{2 + 4j}{-3 - j} = \left(\frac{2 + 4j}{-3 - j}\right)\left(\frac{-3 + j}{-3 + j}\right) = \frac{-10 - 10j}{10} = -1 - j.$$

(e) $w_5 = \dfrac{\frac{1}{5} + \frac{3}{5}j}{-\frac{4}{5} - \frac{2}{5}j} = -\frac{1}{2} - \frac{1}{2}j$ tidied up similarly.

(f) $w_6 = \dfrac{\frac{8}{5} + \frac{4}{5}j}{\frac{3}{5} - \frac{1}{5}j} = 2 + 2j$ tidied up similarly.

I have shown all these pairs of points marked on the two Argand diagrams above. It looks as though the part of the z-circle which turns anticlockwise or positively from $(0, -1)$ to the point which we aren't allowed of $(1,0)$ is giving the positive values on $u = v$, and the part which turns clockwise or negatively between these two points is giving the negative values on $u = v$.

It also looks as if the points are coming in the same order when they transfer from $u = v$ to the circle of z, but we can see that equal distances apart on $u = v$ don't mean that we get equal distances apart on the z-circle. The circle is made up of a kind of variegated elastic.

You might like to see for yourself where points like $w_7 = 10 + 10j$ or $w_8 = -10 - 10j$ will go. (If you do this by first writing z in terms of w, you will find that this particular relationship has a rather special property.)

(7) If $v = 0$ then from (3) we get $x - y - 1 = 0$ so z lies on the line $y = x - 1$.

We just have to exclude the point $(1,0)$ where $z = 1$.

(8) If $u = -v$ then from (3) we get

$$x^2 - x + y^2 + y = -x + y + 1 \text{ or } x^2 + y^2 = 1.$$

This means that z lies on the unit circle about the origin, again excluding the point $(1,0)$ where $z = 1$.

(9) You have sufficient information above to check your two sketches.

(10) Although the shapes of the paths of w and z are different, you will see from your two sketches that the angles at which these paths cut each other are unchanged. This is an example of an important general property of transformations like this.

All the questions which you have answered here are just on the edge of a whole new area of maths. You will learn the general theory of what is happening here if you do a course on complex variables.

Index

(**Note:** If you are searching for a broad topic, you may find it helpful to look at the very detailed contents list at the beginning of the book.)

using completing the square 391–2
constant of 373, 375
definite integral 373, 374
indefinite integral 374
using inverse functions 389–92
limits for 373, 377
using partial fractions 393–5
by parts 395–400
using reduction formulas 400–404
by substitution 382–7
using the $t = \tan(x/2)$ substitution 404–6
inv (calculator) 119, 134, 183, 322
inverse functions   *see under* functions
inverse proportion 88–9
irrational numbers 30, 64
isosceles 137

Kelvin temperature scale 88
kinetic energy 85
Kirchhoff's Junction Rule 447

l.c.m. 202
limits, limiting values 110, 169, 234, 272,
    289, 297, 548
    see also integration
linear factor 252
linear forms and logs 124–9, 412
ln 121, 304, 378–80
locus 457
logs
    antilog 408
    base 25, 116–17, 305
    changing bases 305–6, 311
    and differentiation 306–7, 311
    and integration 378–80, 386, 398, 402
    inverse of 116–17, 119, 304
    the (three) laws of 118–20
    and linear forms 124–8, 412
    ln 121, 304
    natural 121, 304, 311, 359
    and series 365–7
long division in algebra 77–8, 255
lotteries 272–3

Maclaurin series 361–7, 438
mapping 105, 465–7
maximum and minimum 336–8, 376
    applications of 341–6
midpoint of a line 45–6
modulus   *see* complex numbers

natural logs   *see* logs
natural numbers 278

Newton's Law of Cooling 413–14
Newton's Law of Gravitation 90
Newton–Raphson Rule 346–51
Normal distribution 100–2
numerator 14

odd functions 113, 175, 182, 326
one-to-one functions 94, 104
ordered pairs 45
origin 45
Osborn's Rule 320–1, 441

parabola 59, 63
parallelogram 139
parameter 329
partial fractions 247–56, 276
    in integration 393–5
partial sum 235, 274
Pascal's Triangle 260–1, 267–9, 442–4
pendulum 41, 86, 198, 296
perfect squares 34, 64
perimeter 165
period, periodic 175, 182, 195, 196, 198,
    202
permutations (arrangements) 261–3
phase difference 208, 447–8
phasor 448
polar coordinates 427
polygon 140
powers 5–6, 24–7, 118
prime numbers 33–4
principal value
    of argument 425
    of the inverse of cosh 325
    of trig functions 215, 216, 218, 220
Product Rule for differentiation 311–13
proof by induction 277–83
Pythagoras' Theorem 135–6, 143
Pythagorean triples 136, 432
Pythagoreans and rational numbers 30, 136

quadrant 152, 177, 426
quadratic equations 58–70, 159–60
    completing the square 61
    with complex roots 423–4, 426, 431,
        452–3
    factorising 60
    the formula for solving 63–4
    the sum and product of the roots of 66,
        73, 431, 453
quadrilateral 156, 171–2
quotient (long division) 77
Quotient Rule for differentiation 313–15

radians 163–6, 364
    properties of small angles measured in
        167–9, 295
radioactive decay 123, 409–11
range 94, 112
rate of change 289, 316, 334, 337, 351,
    371, 411, 414
rational numbers 28
rationalising the denominator 35, 435
real numbers 29–30, 421
reciprocal functions   *see under* functions
recurrence relation 228, 242
recurring decimals 240–1
reduction formulas 400–404
remainder from long division 77
Remainder Theorem 78–9
rhombus 139
roots 26–7
    of equations 62–3, 65–7, 76, 92, 160,
        420, 437. 448–56
    sign change showing presence of 349
    see also cubic equations, quadratic
        equations, trig functions *and*
        complex numbers

sectors and segments 153, 165–7
selections (combinations) 263–4
self-inverse functions 105
separation of variables (differential
    equations) 407–9
sequence 224
series 229
    comparison of 257–8
    convergence and divergence of 234–5,
        237–8, 248, 249, 256–8, 301, 364,
        367, 380
    giving functions 361–7, 438, 441
    partial sums of 235
    summing 229, 245–9, 256–8
    see also arithmetic series, geometric
        series, binomial expansions *and*
        terms
shift (calculator) 119, 134, 183, 322
SHM (simple harmonic motion) 181,
    196–8, 295–6, 417–18, 445–6
Σ (sigma) notation 244–5
similar shapes 138
simultaneous equations 54–8, 159
Sine Rule for triangles 144–6, 178
    ambiguous case 488
$\sinh x$ 317
    see also hyperbolic functions
sketching curves   *see* graph sketching
slope   *see* gradient
sphere 43, 86, 342